Soil Carbon

Science, Management and Policy for Multiple Benefits
SCOPE Series Volume 71

Soil Carbon

Science, Management and Policy for Multiple Benefits
SCOPE Series Volume 71

Edited by

Steven A. Banwart

The University of Sheffield, UK

Elke Noellemeyer

The National University of La Pampa, Argentina

Eleanor Milne

Colorado State University, USA and University of Leicester, UK

CABI is a trading name of CAB International

CABI	CABI
Nosworthy Way	38 Chauncy Street
Wallingford	Suite 1002
Oxfordshire OX10 8DE	Boston, MA 02111
UK	USA

Tel: +44 (0)1491 832111 　　　　　　　　Tel: +1 800 552 3083 (toll free)
Fax: +44 (0)1491 833508 　　　　　　　　E-mail: cabi-nao@cabi.org
E-mail: info@cabi.org
Website: www.cabi.org

© CAB International 2015. All rights reserved. No part of this publication may be reproduced in any form or by any means, electronically, mechanically, by photocopying, recording or otherwise, without the prior permission of the copyright owners.

A catalogue record for this book is available from the British Library, London, UK.

Library of Congress Cataloging-in-Publication Data

Soil carbon : science, management, and policy for multiple benefits / edited by Steven A. Banwart, The University of Sheffield, UK, Elke Noellemeyer, The National University of La Pampa, Argentina, Eleanor Milne, University of Leicester, UK, and Colorado State University, USA.
　　pages cm. -- (SCOPE series ; 71)
　Includes bibliographical references.
　ISBN 978-1-78064-532-2 (alk. paper)
　1. Soils--Carbon content. 2. Carbon sequestration. 3. Soil fertility. 4. Climate change mitigation. I. Noellemeyer, Elke. II. Milne, Eleanor, 1970-

　S592.6.C36S643 2015
　578.75'7--dc23

　　　　　　　　　　　　　　　　　　2014021246
ISBN-13: 978 1 78064 532 2

Commissioning editor: Victoria Bonham
Assistant editor: Alexandra Lainsbury
Production editor: Tracy Head

Typeset by SPi, Pondicherry, India.
Printed and bound by CPI Group (UK) Ltd, Croydon, CR0 4YY.

Contents

Contributors	ix
Foreword	xiii
Acknowledgements	xv
Executive Summary	xxi

*Steven A. Banwart, Helaina Black, Zucong Cai, Patrick T. Gicheru,
Hans Joosten, Reynaldo Luiz Victoria, Eleanor Milne,
Elke Noellemeyer and Unai Pascual*

PART I   INTRODUCTION, OVERVIEW AND INTEGRATION

1 The Global Challenge for Soil Carbon — 1
*Steven A. Banwart, Helaina Black, Zucong Cai, Patrick T. Gicheru,
Hans Joosten, Reynaldo Luiz Victoria, Eleanor Milne,
Elke Noellemeyer and Unai Pascual*

**2 Soil Carbon: a Critical Natural Resource – Wide-scale
Goals, Urgent Actions** — 10
*Generose Nziguheba, Rodrigo Vargas, Andre Bationo, Helaina Black,
Daniel E. Buschiazzo, Delphine de Brogniez, Hans Joosten, Jerry Melillo,
Dan Richter and Mette Termansen*

**3 Soil Carbon Transition Curves: Reversal of Land Degradation
through Management of Soil Organic Matter for Multiple Benefits** — 26
*Meine van Noordwijk, Tessa Goverse, Cristiano Ballabio,
Steven A. Banwart, Tapas Bhattacharyya, Marty Goldhaber,
Nikolaos Nikolaidis, Elke Noellemeyer and Yongcun Zhao*

**4 From Potential to Implementation: An Innovation Framework
to Realize the Benefits of Soil Carbon** — 47
*Roger Funk, Unai Pascual, Hans Joosten, Christopher Duffy,
Genxing Pan, Newton la Scala Jr, Pia Gottschalk, Steven A. Banwart, Niels H.
Batjes, Zucong Cai, Johan Six and Elke Noellemeyer*

5   A Strategy for Taking Soil Carbon into the Policy Arena   60
    Bas van Wesemael, Michael Stocking, Francesca Bampa,
    Martial Bernoux, Christian Feller, Patrick T. Gicheru,
    Philippe Lemanceau, Eleanor Milne and Luca Montanarella

PART II   SOIL CARBON IN EARTH'S LIFE SUPPORT SYSTEM

6   Soil Formation   82
    Marty Goldhaber and Steven A. Banwart

7   Soil Carbon Dynamics and Nutrient Cycling   98
    David Powlson, Zucong Cai and Philippe Lemanceau

8   Soil Hydrology and Reactive Transport of Carbon and Nitrogen
    in a Multi-scale Landscape   108
    Christopher Duffy and Nikolaos Nikolaidis

PART III   THE MULTIPLE BENEFITS OF SOIL CARBON

9   Climate Change Mitigation   119
    Martial Bernoux and Keith Paustian

10 Soil Carbon and Agricultural Productivity: Perspectives from Sub-Saharan Africa   132
    Andre Bationo, Boaz S. Waswa and Job Kihara

11 Soil as a Support of Biodiversity and Functions   141
    Pierre-Alain Maron and Philippe Lemanceau

12 Water Supply and Quality   154
    David Werner and Peter Grathwohl

13 Wind Erosion of Agricultural Soils and the Carbon Cycle   161
    Daniel E. Buschiazzo and Roger Funk

14 Historical and Sociocultural Aspects of Soil Organic Matter
    and Soil Organic Carbon Benefits   169
    Christian Feller, Claude Compagnone, Frédéric Goulet and Annie Sigwalt

15 The Economic Value of Soil Carbon   179
    Unai Pascual, Mette Termansen and David J. Abson

PART IV   QUANTIFICATION AND REPORTING OF SOIL CARBON

16 Measuring and Monitoring Soil Carbon   188
    Niels H. Batjes and Bas van Wesemael

17 Modelling Soil Carbon   202
    Eleanor Milne and Jo Smith

18 Valuation Approaches for Soil Carbon   214
    David J. Abson, Unai Pascual and Mette Termansen

PART V   INFLUENCE OF HUMAN ACTIVITY ON SOIL CARBON

19 Current Soil Carbon Loss and Land Degradation Globally:
    Where are the Hotspots and Why There?   224
    Hans Joosten

20	Climate Change and Soil Carbon Impacts *Pete Smith, Pia Gottschalk and Jo Smith*	235
21	Impacts of Land-use Change on Carbon Stocks and Dynamics in Central-southern South American Biomes: Cerrado, Atlantic Forest and Southern Grasslands *Heitor L.C. Coutinho, Elke Noellemeyer, Fabiano de Carvalho Balieiro, Gervasio Piñeiro, Elaine C.C. Fidalgo, Christopher Martius and Cristiane Figueira da Silva*	243

PART VI MANAGING SOIL CARBON FOR MULTIPLE BENEFITS

22	Basic Principles of Soil Carbon Management for Multiple Ecosystem Benefits *Elke Noellemeyer and Johan Six*	265
23	Managing Soil Carbon for Multiple Ecosystem Benefits – Positive Exemplars: Latin America (Brazil and Argentina) *Carlos Eduardo P. Cerri, Newton La Scala Jr, Reynaldo Luiz Victoria, Alberto Quiroga and Elke Noellemeyer*	277
24	Managing Soil Carbon for Multiple Benefits – Positive Exemplars: North America *Rich Conant*	287
25	Managing Soil Carbon in Europe: Paludicultures as a New Perspective for Peatlands *Hans Joosten, Greta Gaudig, René Krawczynski, Franziska Tanneberger, Sabine Wichmann and Wendelin Wichtmann*	297
26	Managing Soil Organic Carbon for Multiple Benefits: The Case of Africa *Peter T. Kamoni and Patrick T. Gicheru*	307
27	Benefits of SOM in Agroecosystems: The Case of China *Genxing Pan, Lianqing Li, Jufeng Zheng, Kun Cheng, Xuhui Zhang, Jinwei Zheng and Zichuan Li*	314
28	Assessment of Organic Carbon Status in Indian Soils *Tapas Bhattacharyya*	328

PART VII GOVERNANCE OF SOIL CARBON

29	Policy Frameworks *Luca Montanarella, Francesca Bampa and Delphine de Brogniez*	343
30	National Implementation Case Study: China *Yongcun Zhao*	353
31	Avoided Land Degradation and Enhanced Soil Carbon Storage: Is There a Role for Carbon Markets? *Meine van Noordwijk*	360

Index	381

Contributors

Abson, David J., FuturES Research Center, Leuphana Universität, Germany. E-mail: abson@leuphana.de

Ballabio, Cristiano, European Commission Directorate General Joint Research Centre, Italy. E-mail: cristiano.ballabio@jrc.ec.europa.eu

Bampa, Francesca, European Commission Directorate General Joint Research Centre, Italy. E-mail: francesca.bampa@gmail.com

Banwart, Steven A., Kroto Research Institute, The University of Sheffield, UK. E-mail: s.a.banwart@sheffield.ac.uk

Bationo, Andre, Alliance for a Green Revolution in Africa (AGRA), Kenya. Present address: Action for Integrated Rural Development, Accra, Ghana. E-mail: abationo@outlook.com

Batjes, Niels H., ISRIC – World Soil Information, the Netherlands. E-mail: niels.batjes@wur.nl

Bernoux, Martial, French Research Institute for Development (IRD), France. E-mail: martial.bernoux@ird.fr

Bhattacharyya, Tapas, National Bureau of Soil Survey and Land Use Planning (Indian Council of Agricultural Research), India. E-mail: tapas11156@yahoo.com

Black, Helaina, The James Hutton Institute, UK. E-mail: helaina.black@hutton.ac.uk

Buschiazzo, Daniel E., National Institute for Agronomic Research of Argentina (INTA), National University of La Pampa, Argentina. E-mail: buschiazzo@agro.unlpam.edu.ar/debuschiazzo@yahoo.com

Cai, Zucong, School of Geography Science, Nanjing Normal University, China. E-mail: zccai@njnu.edu.cn

Cerri, Carlos Eduardo P., Department of Soil Science, University of São Paulo (USP), Brazil. E-mail: cepcerri@usp.br

Cheng, Kun, Institute of Resource, Ecosystem and Environment of Agriculture, Nanjing Agricultural University, China. E-mail: kuncheng@aliyun.com

Compagnone, Claude, AgroSup, Dijon – INRA, France. E-mail: c.compagnone@agrosupdijon.fr

Conant, Rich, Natural Resource Ecology Laboratory and Department of Ecosystem Science and Sustainability, Colorado State University, USA. E-mail: conant@nrel.colostate.edu

Coutinho, Heitor L.C., National Centre for Soil Research (Embrapa Soils), Brazil. E-mail: heitor.coutinho@embrapa.br

de Brogniez, Delphine, European Commission Directorate General Joint Research Centre, Italy. E-mail: delphine.de-brogniez@jrc.ec.europa.eu

de Carvalho Balieiro, Fabiano, National Centre for Soil Research (Embrapa Soils), Brazil. E-mail: fabiano.balieiro@embrapa.br

Duffy, Christopher, National Science Foundation (NSF) Critical Zone Observatory, Pennsylvania State University, USA. E-mail: cxd11@psu.edu

Feller, Christian, French Research Institute for Development (IRD), France. E-mail: christian.feller@ird.fr

Fidalgo, Elaine C.C., National Centre for Soil Research (Embrapa Soils), Brazil. E-mail: elaine.fidalgo@embrapa.br

Figueira da Silva, Cristiane, Rural Federal University of Rio de Janeiro (UFRRJ), Brazil. E-mail: cfigueirasilva@yahoo.com.br

Funk, Roger, Institute for Soil Landscape Research, Leibniz Centre for Agricultural Landscape Research (ZALF), Germany. E-mail: rfunk@zalf.de

Gardi, Ciro, European Commission Directorate General Joint Research Centre, Italy. E-mail: ciro.gardi@jrc.ec.europa.eu

Gaudig, Greta, Institute of Botany and Landscape Ecology, Ernst Moritz Arndt University of Greifswald, Germany. E-mail: gaudig@uni-greifswald.de

Gicheru, Patrick T., Kenya Agricultural and Livestock Research Organization (KALRO), Kenya. E-mail: patrick.gicheru@kalro.org

Goldhaber, Marty, US Geological Survey, USA. E-mail: mgold@usgs.gov

Gottschalk, Pia, Potsdam Institute for Climate Impact Research, Germany. E-mail: pia.gottschalk@pik-potsdam.de

Goulet, Frederic, CIRAD Agricultural Research for Development, France. E-mail: frederic.goulet@cirad.fr

Goverse, Tessa, Division of Early Warning and Assessment, United Nations Environment Programme (UNEP), Kenya. E-mail: tessa.goverse@unep.org

Grathwohl, Peter, Center for Applied Geosciences, Tuebingen University, Germany. E-mail: grathwohl@uni-tuebingen.de

Hiederer, Roland, European Commission Directorate General Joint Research Centre, Italy. E-mail: roland.hiederer@jrc.ec.europa.eu

Joosten, Hans, Institute of Botany and Landscape Ecology, Ernst Moritz Arndt University of Greifswald, Germany. E-mail: joosten@uni-greifswald.de

Kamoni, Peter T., Kenya Agricultural and Livestock Research Organization (KALRO), Kenya. E-mail: pkamoni@gmail.com

Kihara, Job, International Center for Tropical Agriculture (CIAT), Kenya. E-mail: j.kihara@cgiar.org

Krawczynski, Rene, Chair of General Ecology, Brandenburg University of Technology, Germany. E-mail: rene.krawczynski@tu-cottbus.de

La Scala Jr, Newton, Universidade Estadual Paulista (Unesp), Brazil. E-mail: lascala@fcav.unesp.br

Lemanceau, Philippe, Plant Health and the Environment Division, INRA-University of Burgundy Joint Research Unit for Soil Microbiology and the Environment, France. E-mail: philippe.lemanceau@dijon.inra.fr

Li, Lianqing, Institute of Resource, Ecosystem and Environment of Agriculture, Nanjing Agricultural University, China. E-mail: lqli@njau.edu.cn

Li, Zichuan, Institute of Resource, Ecosystem and Environment of Agriculture, Nanjing Agricultural University, China. E-mail: lizichuan2004@163.com

Lugato, Emanuele, European Commission Directorate General Joint Research Centre, Italy. E-mail: emanuele.lugato@jrc.ec.europa.eu

Maron, Pierre-Alain, Environment and Agronomy Division, INRA-University of Burgundy Joint Research Unit for Soil Microbiology and the Environment, France. E-mail: Pierre-alain.maron@dijon.inra.fr

Martius, Christopher, Center for International Forestry Research (CIFOR), Indonesia. E-mail: c.martius@cgiar.org

Melillo, Jerry, The Ecosystems Center, Marine Biological Laboratory, USA. E-mail: jmelillo@mbl.edu

Milne, Eleanor, Colorado State University, USA/University of Leicester, UK. E-mail: eleanor.milne@colostate.edu

Montanarella, Luca, European Commission Directorate General Joint Research Centre, Italy. E-mail: luca.montanarella@jrc.ec.europa.eu

Nikolaidis, Nikolaos, Department of Environmental Engineering, Technical University of Crete, Greece. E-mail: nikolaos.nikolaidis@enveng.tuc.gr

Noellemeyer, Elke, Facultad de Agronomía, Universidad Nacional de La Pampa, Argentina. E-mail: noellemeyer@agro.unlpam.edu.ar/enoellemeyer@gmail.com

Nziguheba, Generose, International Institute of Tropical Agriculture, Nairobi, Kenya. E-mail: g.nziguheba@cgiar.org

Pan, Genxing, Institute of Resource, Ecosystem and Environment of Agriculture, Nanjing Agricultural University, China. E-mail: pangenxing@aliyun.com

Pascual, Unai, Basque Centre for Climate Change, Spain. E-mail: unai.pascual@bc3research.org

Paustian, Keith, Department of Soil and Crop Sciences, Natural Resource Ecology Laboratory, Colorado State University, USA. E-mail: keith.paustian@colostate.edu

Piñeiro, Gervasio, Department of Natural Resources, University of Buenos Aires, Argentina. E-mail: pineiro@ifeva.edu.ar

Powlson, David, Department of Sustainable Soils and Grassland Systems, Rothamsted Research, UK. E-mail: david.powlson@rothamsted.ac.uk

Quiroga, Alberto, Facultad de Agronomía, Universidad Nacional de La Pampa, Argentina. E-mail: aquiroga@anguil.inta.gov.ar

Richter, Dan, Division of Environmental Sciences and Policy, Duke University, USA. E-mail: drichter@duke.edu

Sigwalt, Annie, Département Economie et Sciences Sociales, Groupe ESA, France. E-mail: a.sigwalt@groupe-esa.com

Six, Johan, Sustainable Agroecosystem Group, Swiss Federal Institute of Technology Zurich (ETH), Switzerland. E-mail: jsix@ethz.ch

Smith, Jo, Institute of Biological and Environmental Sciences, School of Biological Sciences, University of Aberdeen, UK. E-mail: jo.smith@abdn.ac.uk

Smith, Pete, Institute of Biological and Environmental Sciences, University of Aberdeen, UK. E-mail: pete.smith@abdn.ac.uk

Stocking, Michael, School of International Development, University of East Anglia, UK. E-mail: m.stocking@uea.ac.uk

Tanneberger, Franziska, Institute of Botany and Landscape Ecology, Ernst Moritz Arndt University of Greifswald, Germany. E-mail: tanne@uni-greifswald.de

Termansen, Mette, Department of Environmental Science, Aarhus University, Denmark. E-mail: mter@dmu.dk

van Noordwijk, Meine, ICRAF (World Agroforestry Centre), Indonesia. E-mail: m.vannoordwijk@cgiar.org

van Wesemael, Bas, Georges Lemaître Centre for Earth and Climate Research, Earth and Life Institute, Université catholique de Louvain, Belgium. E-mail: bas.vanwesemael@uclouvain.be

Vargas, Rodrigo, Plant and Soil Sciences, University of Delaware, USA. E-mail: rvargas@udel.edu

Victoria, Reynaldo Luiz, Universidade de São Paulo, Brazil. E-mail: reynaldo.victoria@gmail.com

Waswa, Boaz S., International Center for Tropical Agriculture (CIAT), Kenya. E-mail: b.waswa@cgiar.com

Werner, David, School of Civil Engineering and Geosciences, Newcastle University, UK. E-mail: david.werner@ncl.ac.uk

Wichmann, Sabine, Institute of Botany and Landscape Ecology, Ernst Moritz Arndt University of Greifswald, Germany. E-mail: wichmann@uni-greifswald.de

Wichtmann, Wendelin, Michael Succow Foundation for the Protection of Nature, Ernst Moritz Arndt University of Greifswald, Germany. E-mail: wendelin.wichtmann@succow-stiftung.de

Zhang, Xuhui, College of Resources and Environmental Sciences, Nanjing Agricultural University, China. E-mail: xuhuizhang@njau.edu.cn

Zhao, Yongcun, Institute of Soil Science, Chinese Academy of Sciences, China. E-mail: yczhao@issas.ac.cn

Zheng, Jinwei, Institute of Resource, Ecosystem and Environment of Agriculture, Nanjing Agricultural University, China. E-mail: zhengjw@njau.edu.cn

Zheng, Jufeng, Institute of Resource, Ecosystem and Environment of Agriculture, Nanjing Agricultural University, China. E-mail: zhengjufeng@njau.edu.cn

Foreword

The resource demands of growth in human population and wealth are creating enormous pressure on soils worldwide through ever-increasing intensification of land use. Global soil threats include widespread desertification caused by the loss of soil carbon, significant physical erosion from mechanical disturbance of the land surface through agriculture practices, habitat degradation and loss of biodiversity including impacts on organisms that transform and supply nutrients to plants. A major threat to the global ambitions of food and water security is the further loss of soil fertility through salinization from irrigation with groundwater, sealing by urban development, compaction, which also reduces water storage and transmission, and industrial pollution.

Climate change adds to the challenge. Current predictions of climate change impacts indicate crop yields will be affected adversely in many areas due to declining water availability. Recent greenhouse gas measurements show the highest-ever levels of atmospheric methane and the fastest rate of increase in the levels of atmospheric CO_2 in the modern record. This trajectory demands urgency, and soil management will play an enormous role in risk mitigation and adaptation to global change.

The *UNEP Year Book 2012* identified the benefits of soil carbon as an emerging opportunity for positive action. Many of the benefits that soils deliver correlate positively with the content of soil carbon as organic matter, and many of the soil threats can be mitigated through optimal management of soil carbon. The year book identified an urgent need to draw together the complex, fragmented and multidisciplinary science base and to draw on this knowledge to identify and help implement solutions.

This Scientific Committee on Problems of the Environment (SCOPE) Volume 71 represents an intensive effort to mobilize the international scientific community and seize the opportunity for positive action. The volume compiles scientific evidence to support innovations in policy and practice for soil management that will reverse land degradation and enhance the functioning of soils worldwide. The volume represents the principal output from an international Rapid Assessment Process (RAP) project carried out under the aegis of SCOPE. The RAP was initiated in response to the *UNEP Year Book 2012* chapter on the benefits of soil carbon, and was reported in the *UNEP Year Book 2014* as an update since 2012 on international action for the improvement of soil management.

The project began in June 2012 with the convening of a Scientific Advisory Committee to agree the key subject areas of the RAP, recruit expert authors worldwide from the necessarily broad range of disciplines and geographic regions, and to produce a timetable for project completion. Authors were approached in late 2012, and 27 background chapters of diverse scientific evidence, prepared by 75 scientific experts from 17 countries, formed the knowledge base for a RAP workshop convened in March 2013. Forty participating experts engaged in debate, analysis and the drafting of four cross-cutting chapters of new policy evidence and recommendations for action.

The opening section of the volume includes an introduction to the global challenge and the four cross-cutting chapters with recommendations for action; the subsequent sections present the background chapters. A key aim of SCOPE Volume 71 is to help bridge the gap between science and policy on some of the key challenges and opportunities for beneficial soil and land management. This volume provides a link between the complexity of the scientific knowledge on soil carbon, and how this knowledge can be applied for multiple benefits, and the complexity of the policy and practice arenas where soil and land management impact many sectors: environment, farming, energy, water, economic development and urban planning.

Above all, this volume presents a positive message that improved management of soil carbon for multiple benefits is achievable and within reach. The science evidence supports the view that soil is a strategic asset. Improved soil and land management offers a vital pathway to tackle simultaneously the major challenges of environmental degradation, food security, clean water supply and climate change, and is an essential component of sustained economic development.

Steven A. Banwart
Elke Noellemeyer
Eleanor Milne

Acknowledgements

We wish to thank all authors and contributors to this global scientific collaboration, which would not have been possible without their generous effort and tireless commitment to this Rapid Assessment Process (RAP) project on the benefits of soil carbon.

We are extremely grateful to those who provided funding and support in kind, without which we could not have held the workshop, 18–22 March 2013, gathering experts from around the world to integrate the complex body of scientific evidence from a broad range of disciplines. We owe a debt of gratitude to Delphine de Brogniez for outstanding local organization of the workshop and also to Luca Montanarella and the Soil Action team at the European Commission Directorate General Joint Research Centre (JRC), Ispra (Italy), whose hospitality and workshop support made the week extremely enjoyable and productive. The JRC provided to the RAP project the meeting facilities and local organization for the entire week.

We gratefully acknowledge the considerable time and expertise of the external reviewers, Sally Bunning and Winfried Blum, who provided highly constructive and critical feedback, which greatly improved and helped shape this volume.

We acknowledge the leading role of Debbie Hill, The University of Sheffield, UK, as Managing Editor, planning and overseeing the schedule of work for all sections of the volume, liaising with the 75 authors throughout the RAP, managing the peer-review process, developing and maintaining the project website, handling funding applications and project finances, handling workshop organization, closely with JRC staff, on behalf of SCOPE and the Scientific Advisory Committee (SAC), overseeing the production style of the chapters, ensuring completion to specification of all chapters, handling business and production liaison with the publisher, CABI, and providing executive support to the scientific editors and to the SAC. We acknowledge the essential support from the Vice-Chancellor's Office, The University of Sheffield, for funding this core role on the project.

Special thanks go to the members of the SAC for their foresight, ambition, leadership and tireless efforts to develop and deliver the RAP project, together with the SCOPE executive and, in particular, the SCOPE secretariat, Susan Greenwood-Etienne, who has shared her RAP expertise and provided essential guidance and hands-on support throughout. We particularly thank Jerry Melillo, Harry Kroto, Michael Stocking and Tessa Goverse for their strong support and advice to the RAP.

The editors acknowledge the support of the following individuals and organizations:

Scientific Committee on Problems of the Environment (SCOPE)
President: Jon Samseth, Oslo and Akershus University College (HiOA) and Norwegian University of Science and Technology
Secretariat: Susan Greenwood-Etienne, SCOPE, France

RAP Co-Chairs
Steven A. Banwart, Kroto Research Institute, The University of Sheffield, UK
Elke Noellemeyer, Facultad de Agronomía, Universidad Nacional de La Pampa, Argentina

Scientific Advisory Committee
Steven A. Banwart, Kroto Research Institute, The University of Sheffield, UK
Helaina Black, The James Hutton Institute, UK
Zucong Cai, School of Geography Science, Nanjing Normal University, China
Patrick T. Gicheru, Kenya Agricultural and Livestock Research Organization (KALRO), Kenya
Hans Joosten, Institute of Botany and Landscape Ecology, Ernst Moritz Arndt University of Greifswald, Germany
Elke Noellemeyer, Facultad de Agronomía, Universidad Nacional de La Pampa, Argentina
Eleanor Milne, Colorado State University, USA/University of Leicester, UK
Unai Pascual, Basque Centre for Climate Change, Spain
Reynaldo Luiz Victoria, Universidade de São Paulo, Brazil

Ex-Officio
Tessa Goverse, Division of Early Warning and Assessment, United Nations Environment Programme (UNEP), Kenya
Sir Harry Kroto, Global Education Outreach, Florida State University, USA
Jerry Melillo, The Ecosystems Center, Marine Biological Laboratory, USA
Michael Stocking, School of International Development, University of East Anglia, UK

Editing Team
Scientific Editors
Steven A. Banwart, Kroto Research Institute, The University of Sheffield, UK
Eleanor Milne, Colorado State University, USA/University of Leicester, UK
Elke Noellemeyer, Facultad de Agronomía, Universidad Nacional de La Pampa, Argentina

Managing Editor
Debbie Hill, Kroto Research Institute, The University of Sheffield, UK

Lead Authors and Co-authors
David Abson, FuturES Research Center, Leuphana Universität, Germany
Cristiano Ballabio, European Commission Directorate General Joint Research Centre, Italy
Francesca Bampa, European Commission Directorate General Joint Research Centre, Italy
Steven A. Banwart, Kroto Research Institute, The University of Sheffield, UK
Andre Bationo, Alliance for a Green Revolution in Africa (AGRA), Kenya; present address: Action for Integrated Rural Development, Accra, Ghana
Niels H. Batjes, ISRIC – World Soil Information, the Netherlands
Martial Bernoux, French Research Institute for Development (IRD), France
Tapas Bhattacharyya, National Bureau of Soil Survey and Land Use Planning (Indian Council of Agricultural Research), India
Helaina Black, The James Hutton Institute, UK
Daniel E. Buschiazzo, National Institute for Agronomic Research of Argentina (INTA), National University of La Pampa, Argentina
Zucong Cai, School of Geography Science, Nanjing Normal University, China
Carlos Eduardo Cerri, Department of Soil Science, University of São Paulo (USP), Brazil

Kun Cheng, Institute of Resource, Ecosystem and Environment of Agriculture, Nanjing Agricultural University, China
Claude Compagnone, AgroSup, Dijon – INRA, France
Rich Conant, Natural Resource Ecology Laboratory and Department of Ecosystem Science and Sustainability, Colorado State University, USA
Heitor L.C. Coutinho, National Centre for Soil Research (Embrapa Soils), Brazil
Delphine de Brogniez, European Commission Directorate General Joint Research Centre, Italy
Fabiano de Carvalho Balieiro, National Centre for Soil Research (Embrapa Soils), Brazil
Christopher Duffy, National Science Foundation (NSF) Critical Zone Observatory, Pennsylvania State University, USA
Christian Feller, French Research Institute for Development (IRD), France
Elaine C.C. Fidalgo, National Centre for Soil Research (Embrapa Soils), Brazil
Cristiane Figueira da Silva, Rural Federal University of Rio de Janeiro (UFRRJ), Brazil
Roger Funk, Institute for Soil Landscape Research, Leibniz Centre for Agricultural Landscape Research (ZALF), Germany
Greta Gaudig, Institute of Botany and Landscape Ecology, Ernst Moritz Arndt University of Greifswald, Germany
Patrick T. Gicheru, Kenya Agricultural and Livestock Research Organization (KALRO), Kenya
Marty Goldhaber, US Geological Survey, USA
Pia Gottschalk, Potsdam Institute for Climate Impact Research, Germany
Frederic Goulet, CIRAD Agricultural Research for Development, France
Tessa Goverse, Division of Early Warning and Assessment, United Nations Environment Programme (UNEP), Kenya
Peter Grathwohl, Center for Applied Geosciences, Tuebingen University, Germany
Hans Joosten, Institute of Botany and Landscape Ecology, Ernst Moritz Arndt University of Greifswald, Germany
Peter T. Kamoni, Kenya Agricultural and Livestock Research Organization (KALRO), Kenya
Job Kihara, International Center for Tropical Agriculture (CIAT), Kenya
Rene Krawczynski, Chair of General Ecology, Brandenburg University of Technology, Germany
Newton La Scala Jr, Universidade Estadual Paulista (Unesp), Brazil
Philippe Lemanceau, Plant Health and the Environment Division, INRA-University of Burgundy Joint Research Unit for Soil Microbiology and the Environment, France
Lianqing Li, Institute of Resource, Ecosystem and Environment of Agriculture, Nanjing Agricultural University, China
Zichuan Li, Institute of Resource, Ecosystem and Environment of Agriculture, Nanjing Agricultural University, China
Pierre-Alain Maron, Environment and Agronomy Division, INRA-University of Burgundy Joint Research Unit for Soil Microbiology and the Environment, France
Christopher Martius, Center for International Forestry Research (CIFOR), Indonesia
Jerry Melillo, The Ecosystems Center, Marine Biological Laboratory, USA
Eleanor Milne, Colorado State University, USA/University of Leicester, UK
Luca Montanarella, European Commission Directorate General Joint Research Centre, Italy
Nikolaos Nikolaidis, Department of Environmental Engineering, Technical University of Crete, Greece
Elke Noellemeyer, Facultad de Agronomía, Universidad Nacional de La Pampa, Argentina
Generose Nziguheba, International Institute of Tropical Agriculture, Nairobi, Kenya
Genxing Pan, Institute of Resource, Ecosystem and Environment of Agriculture, Nanjing Agricultural University, China
Unai Pascual, Basque Centre for Climate Change, Spain

Keith Paustian, Department of Soil and Crop Sciences, Natural Resource Ecology Laboratory, Colorado State University, USA
Gervasio Piñeiro, Department of Natural Resources, University of Buenos Aires, Argentina
David Powlson, Department of Sustainable Soils and Grassland Systems, Rothamsted Research, UK
Alberto Quiroga, Facultad de Agronomía, Universidad Nacional de La Pampa, Argentina
Dan Richter, Division of Environmental Sciences and Policy, Duke University, USA
Annie Sigwalt, Département Economie et Sciences Sociales, Groupe ESA, France
Johan Six, Sustainable Agroecosystem Group, Swiss Federal Institute of Technology Zurich (ETH), Switzerland
Jo Smith, Institute of Biological and Environmental Sciences, School of Biological Sciences, University of Aberdeen, UK
Pete Smith, Institute of Biological and Environmental Sciences, School of Biological Sciences, University of Aberdeen, UK
Michael Stocking, School of International Development, University of East Anglia, UK
Franziska Tanneberger, Institute of Botany and Landscape Ecology, Ernst Moritz Arndt University of Greifswald, Germany
Mette Termansen, Department of Environmental Science, Aarhus University, Denmark
Meine van Noordwijk, ICRAF (World Agroforestry Centre), Indonesia
Bas van Wesemael, Georges Lemaître Centre for Earth and Climate Research, Earth and Life Institute, Université catholique de Louvain, Belgium
Rodrigo Vargas, Plant and Soil Sciences, University of Delaware, USA
Reynaldo Luiz Victoria, Universidade de São Paulo, Brazil
Boaz Waswa, International Center for Tropical Agriculture (CIAT), Kenya
David Werner, School of Civil Engineering and Geosciences, Newcastle University, UK
Sabine Wichmann, Institute of Botany and Landscape Ecology, Ernst Moritz Arndt University of Greifswald, Germany
Wendelin Wichtmann, Michael Succow Foundation for the Protection of Nature, Ernst Moritz Arndt University of Greifswald, Germany
Xuhui Zhang, College of Resources and Environmental Sciences, Nanjing Agricultural University, China
Yongcun Zhao, Institute of Soil Science, Chinese Academy of Sciences, China
Jinwei Zheng, Institute of Resource, Ecosystem and Environment of Agriculture, Nanjing Agricultural University, China
Jufeng Zheng, Institute of Resource, Ecosystem and Environment of Agriculture, Nanjing Agricultural University, China

Workshop
Host: European Commission, Directorate General Joint Research Centre (JRC), Ispra, Italy
Host coordinators and workshop participants: Delphine de Brogniez, Ciro Gardi, Roland Hiederer, Emanuele Lugato and Luca Montanarella (JRC)
Keynote speaker: Professor Jerry Melillo
Opening address: Professor Sir Harry Kroto
Discussion leaders: Meine van Noordwijk, Mette Termansen, Genxing Pan, Michael Stocking
Rapporteurs: Tessa Goverse, Generose Nziguheba, Rodrigo Vargas, Roger Funk, Bas van Wesemael

The United Nations Educational, Scientific and Cultural Organization (UNESCO) is gratefully acknowledged for hosting the Scientific Advisory Committee inaugural meeting in Paris, 11–12 June 2012.

Funders
Aarhus University
Basque Centre for Climate Change
Earth and Life Institute, Université catholique de Louvain
EcoFINDERS, European Commission FP7
Ernst-Moritz-Arndt-University Greifswald
French Research Institute for Development (IRD)
ICRAF (World Agroforestry Centre)
ISRIC – World Soil Information
National Science Foundation (NSF)
National Science Foundation China (NSFC)
Pennsylvania State University
São Paulo Research Foundation – FAPESP
Scientific and Technical Advisory Panel (STAP), Global Environment Facility (GEF)
SoilTrEC, European Commission FP7
Technical University of Crete
The James Hutton Institute
The University of Sheffield

External reviewers
Sally Bunning, Senior Land/Soils Officer, Land and Water Division, Natural Resources Management and Environment Department, Food and Agriculture Organization of the United Nations (FAO), Rome, Italy
Winfried Blum, Emeritus Professor of Soil Science, Department of Forest and Soil Sciences, University of Natural Resources and Applied Life Sciences (BOKU), Vienna, Austria

Executive Summary

Steven A. Banwart, Helaina Black, Zucong Cai, Patrick T. Gicheru,
Hans Joosten, Reynaldo Luiz Victoria, Eleanor Milne,
Elke Noellemeyer and Unai Pascual

Introduction

The global challenge of soil and land degradation is both daunting and immediate. It also provides an outstanding opportunity to take positive action rapidly to improve Earth's soils and enhance the multiple, essential benefits that they provide to humanity. There are both short- and long-term improvements that can be achieved through the application and development of new knowledge and the implementation of informed policy and management practice.

The challenge is clear. By 2050, the human population is expected to reach 9.6 billion, with an accompanying quadrupling in the global economy, a doubling in demand for food and fuel and a 50% increase in demand for clean water; all while dealing with climate change and declining biodiversity. Resource demand is set against the reality of limited land availability and the need for the world's soils to deliver productivity while avoiding the environmental costs of intensifying land use.

The 2014 United Nations Environment Programme (UNEP, 2014) report, *Assessing Global Land Use*, estimates an increase in demand for productive land of between 320 and *c*.850 Mha by 2050. This scenario overshoots the estimated environmentally sustainable capacity of Earth's land resources by 10–45%. Unprecedented intensification in the productivity of currently used land in order to achieve economic growth while avoiding exceptional environmental degradation will be required, and this must be achieved without further environmental costs. The beneficial management of soil carbon, along with water management, is perhaps the most important means of human intervention that is currently technically available against this daunting challenge.

Urgent Short-term Actions

Soil carbon, specifically in the form of soil organic matter, plays a central role in the functioning of soils to produce a wide range of vital environmental goods and services. Figure 2.1 illustrates five essential services delivered by soil which are directly dependent on the role of soil carbon. Loss of soil carbon degrades these services, resulting in decreased market value, such as reduced crop yields, and environmental and social costs that may be external to markets, such as increased soil greenhouse gas emissions and the pollution of drinking water. Targeted soil and land management methods can enhance soil carbon, and thereby offer substantial opportunities to increase the market and non-market value of land use, both locally and worldwide. Proactive intervention now could increase economic productivity, reverse global land degradation and provide an improved local and global environment.

Short-term actions to achieve these long-term goals include the following (detailed in Nziguheba et al., Chapter 2, this volume):

1. *Enhance agricultural productivity through increased plant carbon inputs to soil*, including optimal crop fertilization that does not oversupply mineral N, P and K. Measures include improved crop rotation, reduced tillage, organic matter addition, fallow cover crops, agroforestry and improved livestock management practices that recover nutrients and reduce compaction and excessive vegetation removal.

2. *Improve soil water infiltration and storage* by improving soil porosity and structure. This can be done through cover vegetation, which increases soil carbon stocks, helping to bind soil particles into larger aggregates, yielding greater soil porosity and permeability. These changes in structure buffer soil moisture levels, increase groundwater recharge, provide flood storage on land and reduce runoff, erosion and transport of soil nutrients and agrichemicals directly to surface waters.

3. *Improve biodiversity through the variety of organic carbon inputs to soil* including plant litter, the biomass and exudates of roots and of their symbiotic fungi, and the direct input from surface photosynthetic microorganisms. All of these sources contribute to the carbon and energy supply that supports the soil food web, including its base of decomposer microorganisms. Other functional populations in the food include litter fragmenters, fungi and other decomposers of complex organic constituents, nitrogen fixers and processors, other nutrient-transforming microbes, methane producers and consumers, and soil fauna as ecosystem engineers.

4. *Mitigate climate change and manage the impacts by storing carbon as soil organic matter*. This carbon storage reduces net carbon greenhouse gas emissions while utilizing the role of soil organic matter to form larger aggregates that protect the organic carbon. The intraparticle pores of aggregates hold plant available water and the larger interparticle pores drain excess bulk water and speed O_2 diffusion for root respiration, thus enhancing the resilience of soils to both flood and drought conditions that limit plant growth.

5. *Develop life-cycle analysis and full carbon accounting of biofuel crops* in order to quantify fully the carbon and environmental costs of marginal land transition from native vegetation to biofuel crops. Additional factors to quantify are the likely land competition for food production and the potential for crop management that increases soil carbon stocks. Increased stock provides a net reduction of soil carbon greenhouse gas emissions, while producing bioenergy sources that reduce the demand for hydrocarbon fuel and improving the soil to sustain crop production.

6. *Establish a global research programme to reduce uncertainties* associated with those soil processes that produce beneficial effects and those that degrade environmental services due to soil carbon decline. The research programme should reduce uncertainties at both local and global scales.

Creating Synergies for Multiple Benefits

The urgent short-term actions listed above must be implemented locally at the scale of the landowner and land manager. The multiple benefits of the proactive management of soil carbon accrue through the flows of carbon and other materials, energy and biodiversity through the landscape. These flows create an impact chain that propagates the cumulative effects of local action by many individuals to the larger scales of positive regional and global impact.

The impact chain begins by understanding that soil functioning is also related intimately to soil structure; i.e. the binding together of the primary soil components of mineral fragments, decaying biomass and living organisms into larger aggregates that are stable under the influence of wetting and water flow. Soil organic matter establishes cohesion between the particles; it supplies carbon and energy to support the microorganisms that decompose organic matter, transform the nutrient contents for plant growth and provide the base of the soil food web and its biodiversity. Aggregate formation provides physical protection for organic matter that is bound within the particles; it stores plant available water within the aggregate pores and it provides improved soil drainage through the larger pores between aggregates. Improved drainage allows greater ingress of O_2 from the atmosphere to support root respiration.

This beneficial impact chain, through the development of soil structure, begins with carbon addition from plants to soil. An example of beneficial downstream impacts is the reduction in physical erosion that results from the formation of water-stable aggregates during soil carbon addition and the ensuing development of soil structure (Fig. 3.1).

Most of the benefits that are linked to soil carbon content are aligned positively with the soil organic matter content. One notable exception is the productivity of agricultural land that supplies nutrients to crops through the mineralization of soil organic matter. However, the supply of soil carbon is limited and will become depleted along with a decline in soil fertility if soil carbon stocks are not replenished. If soil carbon content becomes too low, the soil structure collapses and soils become susceptible to erosion, compaction and flooding. This sequence of land-use development is described by a generalized soil carbon transition curve that describes three stages of soil fertility linked to carbon content: Stage I, the initial decline; Stage II, the collapse; and, if soil carbon is restored, Stage III, the rise (Fig. 3.2).

Soil carbon content under native vegetation varies massively worldwide depending on climate and the clay content of the soil. In spite of the enormous variability, the generalized temporal pattern represented by the soil carbon transition curve appears to hold across a wide variety of environments.

Examples (detailed in van Noordwijk *et al.*, Chapter 3, this volume) include the:

- decline in the productivity of the soils of the Indo-Gangetic Plains and the black soil region of India following the removal of forests for agricultural cultivation;
- decreased soil carbon content of the fertile black soil region of north-eastern China following agricultural intensification;
- increased soil erosion in the loess soils in semi-arid regions of South America;
- dust bowl conditions of the central plains of the USA following tillage of the native prairie soils – and the subsequent recovery;
- recent increase of soil carbon content in agricultural lands on the island of Java, Indonesia;
- improved methods of cultivation and increased soil carbon content since 1980 in the Indo-Gangetic Plains; and
- improvement of soil carbon stocks in paddy soils of China between 1980 and 1994 following the increased use of mineral fertilizers to increase crop productivity.

If the soil transition curve proves to hold generally, then the capacity to achieve the Stage III conditions of soil recovery compels efforts to improve soil carbon stocks wherever they are threatened or degraded through agricultural practices. Following agricultural development, soil degradation is not inevitable and can be reversed through improved soil carbon management. The vulnerability of the land to soil carbon loss during the transition is a critical challenge as Earth's marginal lands come under increasing pressure to be brought into production. This vulnerability must be addressed, and requires methods to eliminate or strongly mitigate soil carbon losses if marginal lands are brought into production.

Key recommendations to avoid the Stage II collapse and to achieve the Stage III recovery where necessary are listed here (detailed in van Noordwijk et al., Chapter 3, this volume).

- Identify the local thresholds and transition points for the three stages of the transition curve due to substantial variation in soil type and climatic conditions around the world.
- Use improved agricultural agroecological practices that replenish soil carbon.
- Maintain soil functions in mineral soils through necessary changes in tillage and crop residue management and/or return trees to agricultural landscapes and maintain adequate nutrient levels to support vegetation.
- Control grazing pressure in grasslands in order to recover and maintain vegetation, and possibly the addition of organic matter stocks derived elsewhere.
- Utilize external carbon inputs such as from agroindustrial waste where available, in order to trigger the threshold levels of soil carbon where on-site photosynthesis may not generate sufficient organic inputs to the soil following marginal land transition.
- Enable the necessary societal change to collect urban organic waste for land application in order to recycle sufficient organic matter and close the nutrient loop between urban and rural areas.

Achieving Multiple Benefits from Local to Global Scales

Previous conditions for significant progress in meeting global environmental goals and targets have been characterized by clear problem definition, agreed goals, specific targets and indicators to track progress. An essential step towards protecting and enhancing soil functions worldwide will be founded on a science-based contribution to help develop a target for soil carbon stocks and mechanisms to track progress toward sustainable land use. Severe bottlenecks to implanting innovation for soil carbon management exist and need to be addressed. The key factors and approaches to enable innovation are detailed by Funk et al. (Chapter 4, this volume) and are summarized here.

There is a profound mismatch in the scale of engagement at local and global scales. The beneficial impacts of improved soil carbon management to restore degraded land and enhance soil productivity hold the potential to benefit an enormous number of local farmers and landowners worldwide. The environmental challenges of managing land and producing food are starkly manifest in their daily lives and in the pressures to maintain seasonal profit margins.

At the global scale, however, decision makers are remote from these individual struggles, and only a few multilateral agencies and multinational companies are dealing with these challenges. An even greater bottleneck exists at the national level, where often there are only a few individuals or small teams with policy missions that deal with soil. These activities are generally fragmented, with efforts dispersed between agencies dealing with farming, environment, water and energy. Social innovations are needed to bridge the gap between short-term, local impacts and the long-term, global impacts that should provide feedback to incentivize best practice nationally and locally.

These innovations will require new governance structures in both the public and private sectors, so that policies at higher spatial scales, i.e. in governments and large companies, have impact at the local scale. The aim must be to influence the behaviour of landowners, farmers and consumers positively, in order to bring about improved soil carbon levels. This innovation will require an economic trigger such as unacceptable increases in food prices or unacceptable environmental or reputational costs of land degradation. In the absence of policy interventions, it is likely that local soil carbon management will remain suboptimal.

Improved soil carbon management at the local level will require access to information and technology, financial incentives and capable personnel to implement solutions. Schemes to offer financial and technical support to landowners will create immediate direct advantages that are compatible with both local and global long-term interests.

The key innovations that are required include (Funk *et al.*, Chapter 4, this volume):

- low-cost sensors for soil carbon levels;
- local demonstration sites for best practices;
- online web-based technical training;
- payments for soil carbon services;
- monetization of soil carbon benefits in national policy;
- carbon offset markets for landowners;
- legal frameworks to regulate soil carbon;
- remote sensing methods to monitor soil carbon;
- international trade agreements and certification that reward good soil management practice; and
- increased integration of national policy teams relating to soil.

A Policy Strategy for Soil Carbon

A strategy to realize the multiple benefits of soil carbon worldwide requires that the narrative of soil decline and the global threat of land degradation become embedded in the policy discourse of environmental sustainability, economic prosperity and international development. Key science evidence (detailed in van Wesemael *et al.*, Chapter 5, this volume) with the potential to raise the profile of soil carbon in the policy discourse includes quantifying soil carbon content as:

- a smart, integrating indicator of soil functions and benefits;
- hotspots and thresholds for soil degradation and soil and water conservation;
- a measure to diversify agricultural systems that are adapted to climate change;
- a mechanism to monitor and verify climate change mitigation;
- a core component of plant nutrition management and decisions on agrichemical inputs;
- contributing directly to food and water security and international development; and
- an indicator of belowground biodiversity reflecting and supporting aboveground biodiversity.

Several issues must be addressed urgently (van Wesemael *et al.*, Chapter 5, this volume) to integrate soil carbon in the policy discourse. First, there is a missing communications link between scientists and policy makers. The scientific language must be clearer and more accessible and effective in order to convince policy leaders and influence public opinion and political decisions. This SCOPE volume and the additional policy briefs that flow from it aim to help address this gap.

Second, it is essential to connect clearly the cross-cutting role of soil organic matter and soil carbon with the high-profile global challenges of food security, water supply, international development and climate change.

Third, there is a pressing need to move from discussing the protection of soil as an environmental material and instead draw attention to the protection of soil functions and the many benefits they provide.

Fourth, the valuation of soil must move far beyond the current approaches to market trading in CO_2 equivalents for soil carbon. Although carbon storage in soil clearly reduces greenhouse gas emissions, this market value is only traded voluntarily and it currently reflects only a small fraction of the value of soil as natural capital and as strategic national assets worldwide.

Finally, recommendations that connect the scientific evidence and the policy discourse include addressing the policy imperative and profile, the policy discourse and rationale, the necessary advocacy and institutional support and governance (Table 1).

Table 1. Recommendations for a policy strategy for soil carbon. (From van Wesemael et al., Chapter 5, this volume.)

Priority message	Local scale	National scale	International scale
Policy imperative	• Resource capacity • Socio-economic conditions • Field production conditions	• Land planning • Economic incentives • Legal frameworks • Regulatory mechanisms	Cross-cutting role of soil carbon in: • Food security • International development • Climate change
Policy profile and discourse	• Holistic perception of soil fertility • Economic access to resources	• Training programmes • Public awareness of soil as heritage • Importance of organic matter in soil management	• Join fragmented efforts together under, for example, Global Soil Partnership • Broaden media communications to address cultural, services, society and educational concerns • Establish an international agenda for soil with clear goals and steps
Policy support	• Collate information on cropping systems and practices • Extend this to developing regions • Build local capacity to use this information	• Harmonize local Soil Monitoring Networks for cross-comparison • Include parameters for estimation of soil organic carbon content	• Facilitate data exchange between regions and nations • Access to remote sensing data
Policy rationale	• Develop a methodology to value soil and soil organic carbon as natural capital in order to communicate the future value of this resource		
Advocates and institutions	• Scientists must communicate results suitably for policy makers • Extend soil organic carbon as an indicator beyond the UN Climate Change Convention to address conventions on desertification and biodiversity and food and water security		
Governance	• Develop flexible narratives and discourses that create a strong argument in compelling language for multiple policy arenas • Embed the relevance of soil carbon in all levels of decision making and action, via high-profile global accords, national action plans and institutions • Principles of good governance must underpin all initiatives to prioritize soil carbon, particularly those involving incentives, subsidies and penalties		

Reference

UNEP (2014) Assessing global land use: balancing consumption with sustainable supply. In: Bringezu, S., Schütz, H., Pengue, W., O'Brien, M., Garcia, F., Sims, R., Howarth, R., Kauppi, L., Swilling, M. and Herrick, J. (eds) *A Report of the Working Group on Land and Soils of the International Resource Panel*. United Nations Environment Programme, Nairobi.

1 The Global Challenge for Soil Carbon

Steven A. Banwart*, Helaina Black, Zucong Cai, Patrick T. Gicheru,
Hans Joosten, Reynaldo Luiz Victoria, Eleanor Milne,
Elke Noellemeyer and Unai Pascual

Abstract

Soil carbon in the form of organic matter is a key component of the soil ecosystem structure. The soil carbon content is an important contributing factor in the many flows and transformations of matter, energy and biodiversity – the essential soil functions that provide ecosystem services and life-sustaining benefits from soil. These goods and services include food production, water storage and filtration, carbon storage, nutrient supply to plants, habitat and biodiversity. Soil functions provide natural capital as a means of production for the ongoing supply of the essential goods and services. Soil carbon content and soil functions are under threat worldwide due to resource demands and the increasing intensification of land use. Land degradation is characterized by soil carbon losses, loss of soil structure and associated loss of fertility, and the physical loss of bulk soil by erosion. Soil carbon accumulation is associated with plant productivity, wet conditions that ensure water supply to vegetation and lack of physical disturbance to the soil. Carbon accumulation is also associated with decreased organic matter decomposition in the soil, created by cool conditions that reduce the rate of microbial activity and wet conditions that create an O_2 diffusion barrier from the atmosphere and reduced aerobic microbial respiration during organic matter decomposition. The environmental conditions for the accumulation of soil carbon also provide important clues to management approaches to reverse soil carbon losses and to increase soil carbon content under widely different environmental conditions around the world. Soil management strategies can be developed from the natural cycling of soil carbon, by reducing physical disturbances to soil, enhancing vegetation cover and productivity and through improved water management. These approaches are essential in order to prevent and reverse the loss of soil functions where land is degraded and to enhance soil functions where actively managed land is undergoing intensification of use. Improved soil carbon management provides an important opportunity in land management worldwide, to meet increasing resource demands and to create resilience in soil functions that arise from the intense pressures of land use and climate change.

Introduction

By 2050, the world's population is expected to reach 9.6 billion (United Nations, 2013). This enormous demographic pressure creates four major global challenges for Earth's soils over the coming four decades.

*E-mail: s.a.banwart@sheffield.ac.uk

This '4 × 40' challenge for global soils is to meet the anticipated demands of humanity (Godfray et al., 2010) to:

1. Double the food supply worldwide;
2. Double the fuel supply, including renewable biomass;
3. Increase by more than 50% the supply of clean water, all while acting to
4. Mitigate and adapt to climate change and biodiversity decline regionally and worldwide.

The demographic drivers of environmental change and the demand for biomass production are already putting unprecedented pressure on Earth's soils (Banwart, 2011). Dramatic intensification of agricultural production is central among proposed measures to potentially double the global food supply by 2050. An urgent priority for action is to ensure that soils will cope worldwide with these multiple and increasing demands (Victoria et al., 2012).

Soils have many different essential life-supporting functions, of which growing biomass for food, fuel and fibre is but one (Blum, 1993; European Commission, 2006; Victoria et al., 2012). Soils store carbon from the atmosphere as a way to mediate atmospheric greenhouse gas levels; they filter contaminants from infiltrating recharge to deliver clean drinking water to aquifers; they provide habitat and maintain a microbial community and gene pool that decomposes and recycles dead organic matter and transforms nutrients into available forms for plants; they release mineral nutrients from parent rock; and they store and transmit water in ways that help prevent floods. These functions underpin many of the goods and services that can lead to social, economic and environmental benefit to humankind. Specific land uses can create trade-offs by focusing on the delivery of one or a few of these functions at the expense of others. Under the pressures of increasingly intensive land use, when decisions are made on land and soil management, it is essential to protect and to enhance the full range of the essential life-sustaining benefits that soils provide.

The build-up of organic matter and carbon is one of the key factors in the development of ecosystem functions as soil forms and evolves. Thus, carbon loss is one of the most important contributions to soil degradation. Furthermore, this central role of carbon across the range of soil functions establishes a buffer function for soil organic matter whereby loss of soil carbon results in a decline in the soil functions, and maintaining or enhancing soil carbon confers resilience to these under pressure from environmental changes (van Noordwijk et al., Chapter 3, this volume). In the ensuing chapters of this volume, significant detail is provided to illustrate and quantify the uniquely central role of soil carbon in the delivery of ecosystem services and the opportunity that this presents in managing soil and land use positively to enhance the multiple benefits that soil carbon provides. Set against these opportunities to reverse, conserve and even enhance soil functions is the operational cost implied in the proactive management of soil carbon.

The global soil resource is already showing signs of serious degradation from human use and management. Soil degradation has escalated in the past 200 years with the expansion of cultivated land and urban dwelling, along with an increasing human population. Degradation continues, with soil and soil carbon being lost through water and wind erosion, land conversion that is associated with accelerated emissions of greenhouse gases and the burning of organic matter for fuel or other purposes. Significant degradation has taken place since the industrial revolution; recent and ongoing degradation is substantial; bulk soil loss from erosion remains severe in many locations, with the accompanying loss of soil functions; and the release of carbon and nitrogen from soil as the greenhouse gases CO_2, CH_4 and N_2O continues to contribute to global warming (Table 1.1).

The capacity of soils to deliver ecosystem goods and services which lead to human benefits, and the degree to which these benefits are lost due to soil degradation, varies significantly with geographical location (Plate 1). The global results in Plate 1 provide a first

Table 1.1. Global soil carbon fact sheet. (From Banwart et al., 2014.)

Amount of carbon in top 1 m of Earth's soil[b]	2200 Gt
2/3 as organic matter	
Organic C is around 2× greater C content than Earth's atmosphere	
Fraction of antecedent soil and vegetation carbon characteristically lost from agricultural land since 19th century[c]	60%
Fraction of global land area degraded in past 25 years due to soil carbon loss[d]	25%
Rate of soil loss due to conventional agriculture tillage[e]	~1 mm year^{-1}
Rate of soil formation[e]	~0.01 mm year^{-1}
[a]Global mean land denudation rate[f]	0.06 mm year^{-1}
Rate of peatlands loss due to drainage compared to peat accumulation rate[g]	20× faster
Equivalent fraction of anthropogenic greenhouse gas emissions from peatland loss[g]	6% annually
Soil greenhouse gas contributions to anthropogenic emissions, in CO_2 equivalents[h]	25%

[a]Rate of land lowering due to chemical and physical weathering losses; [b]Batjes (1996); [c]Houghton (1995); [d]Bai et al. (2008); [e]Montgomery (2007); [f]Wilkinson and McElroy (2007); [g]Joosten (2009); [h]2004 data not including CH_4, IPCC (2007).

indication of the regional and national pressures on soil and the associated trends in the gain or loss of soil functions. What is noteworthy is the broad geographical extent of areas associated with strong degradation.

Soil Carbon in Soil Functions and Ecosystem Services

The process of adding photosynthate carbon to rock parent material and the development of subsurface biodiversity and the formation of soil aggregates is the foundation of soil development and the establishment of soil functions.

Soil forms from parent rock material that is exposed at Earth's surface, receives infiltrating precipitation and is colonized by photosynthesizing organisms (Brantley, 2010): chiefly plants, but also symbiont algae in lichens and photosynthetic cyanobacteria. Organic carbon that is fixed in biomass by photosynthesis is rooted, deposited and mixed and transported by soil fauna into the soil layer, providing carbon and energy for heterotrophic decomposer microorganisms. Other functional groups of microorganisms transform N, P, K and other nutrient elements of decomposing biomass into forms that are available to plants for further biological productivity. Symbiotic fungi that draw energy from plant photosynthate carbon that passes from roots create the pervasive growth of hyphal networks. These proliferate when they encounter nutrient resources such as P- and K-bearing minerals and organic N and P in decomposing plant debris. Grazing and predator organisms including protozoa and soil fauna are sustained by the active microbial biomass. Soil fauna such as worms, termites, ants and other invertebrates play an important role in the initial processing of biomass and for physical mixing and transport through bioturbation, particularly at the surface, but for some organisms throughout the full depth of the soil profile.

Advanced decomposition of biomass by soil organisms yields humic material, which chemically binds to the smallest soil particles with the greatest surface area per mass: clay minerals and Fe and Al oxides. This mineral-adsorbed carbon is chemically more stable and less bioavailable, and produces a hydrophobic coating on the mineral surfaces. The smallest particles aggregate into micron-sized fragments, and decomposition of fresh organic matter by active heterotrophic microorganisms produces microbial extracellular polymers that help bind these intermediate aggregates with rock fragments, decomposing plant debris, biofilms of living microorganisms and fungal hyphae and root surfaces (Tisdall and Oades, 1982; Jarvis et al., 2012, and included references). This bound mixture of mineral, dead and living biomass and pore fluids, forms larger aggregates, resulting in a system called soil structure, which produces a pore volume distribution that allows both water storage in small throats and pores and free drainage of water and ingress of atmospheric O_2, to support root and microbial respiration, through connected networks of larger pores.

Soil organic matter and soil carbon are thus central to all of the underpinning physical, chemical and biological processes of soil functions. At the landscape scale, the resulting transformations and flows of material, energy and genetic information are delivered as ecosystem services that provide enormous benefits for humans (Fig. 1.1). This view of the environment is embodied in the concept of soil as natural capital that provides a means of production for the ongoing supply of beneficial goods and services (Robinson et al., 2013). Indeed, Robinson et al. (2013) noted soil carbon, along with soil organisms, embodied biogeochemical energy and the structural organization of soil as the key components of this natural capital. This view is broadly held through the following chapters of this volume and inherently introduces an anthropocentric view of natural processes that describes these in terms of economic services with an instrumental value for human well-being. The latter is not analysed explicitly, but this departs from a broader biocentric perspective of the intrinsic value of Earth's environment and the ongoing processes that it supports.

Drawing on the concepts of ecosystem services within the Millennium Ecosystem Assessment defines the following services arising from soil functions (MEA, 2005; Black et al., 2008; Robinson et al., 2013).

1. Supporting services are the cycling of nutrients, the retention and release of water, the formation of soil, provision of habitat for biodiversity, the exchange of gases with the atmosphere and the degradation of plant and other complex materials.

2. Regulating services for climate, stream and groundwater flow, water and air quality and environmental hazards are: the sequestration of carbon from the atmosphere, emission of greenhouse gases, the filtration and purification of water, attenuation of

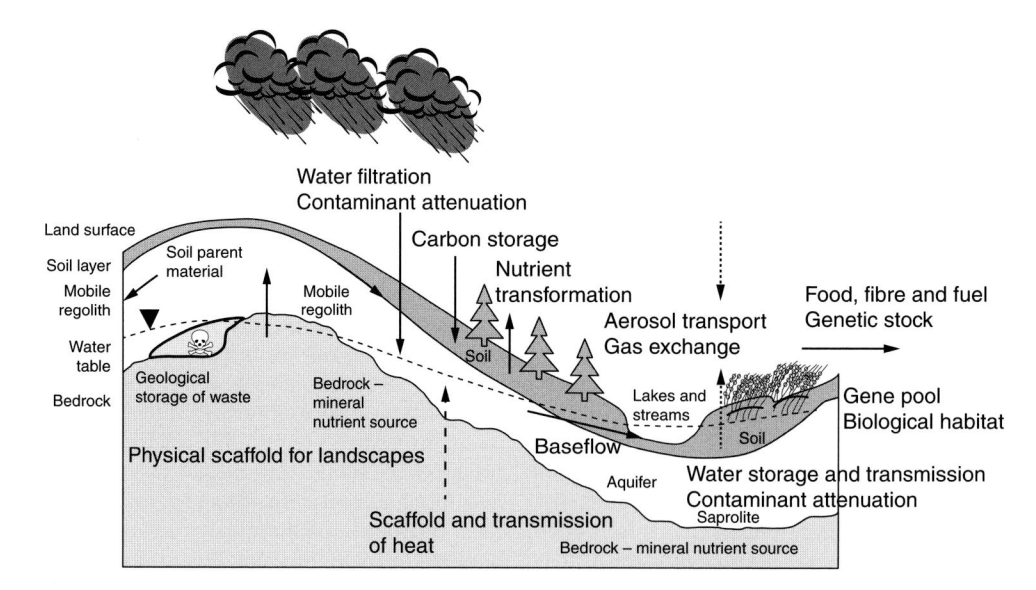

Fig. 1.1. Soil functions and ecosystem services are at the heart of Earth's critical zone; the thin outer layer of the planet that supports almost all human activity. Within this hill slope diagram, the arrows illustrate important flows of material, energy and genetic information that support ecosystem services and provide essential benefits. These flows create a chain of impact that propagates changes in the aboveground environment (e.g. changing climate, land use), via the soil layer, throughout the critical zone. Thus, considering decisions that affect soil requires understanding consequences along the entire chain of impact; and the full consideration of all costs and benefits whether intended or not. (From Banwart et al., 2012.)

pollutants from atmospheric deposition and land contamination, gas and aerosol emissions, slope and other physical stability, and storage and transmission of infiltrating water.
3. **Provisioning services** are food, fuel and fibre production, water availability, non-renewable mineral resources and as a platform for construction.
4. **Cultural services** are the preservation of archaeological remains, outdoor recreational pursuits, ethical, spiritual and religious interests, the identity of landscapes and supporting habitat.

Soil carbon plays a key role in all four classes of soil ecosystem services. The flows arising from environmental processes (Fig. 1.1) depend on ecosystem structure, where soil carbon is a key component, along with the environmental conditions and human interventions that can influence the produced services, goods and benefits strongly (Fig. 1.2; Fisher et al., 2009; Bateman et al., 2011; Robinson et al., 2013).

Subsequent chapters delve into the underlying processes, the impacts of environmental change, the chains of impact and the consequences that arise, and methods to intervene in order to influence these impacts beneficially. Decisions on land use and soil management that affect the stocks and composition of soil carbon therefore incur costs and benefits through changes to these ecosystem services. In many cases, the value of these changes is not reflected in markets,

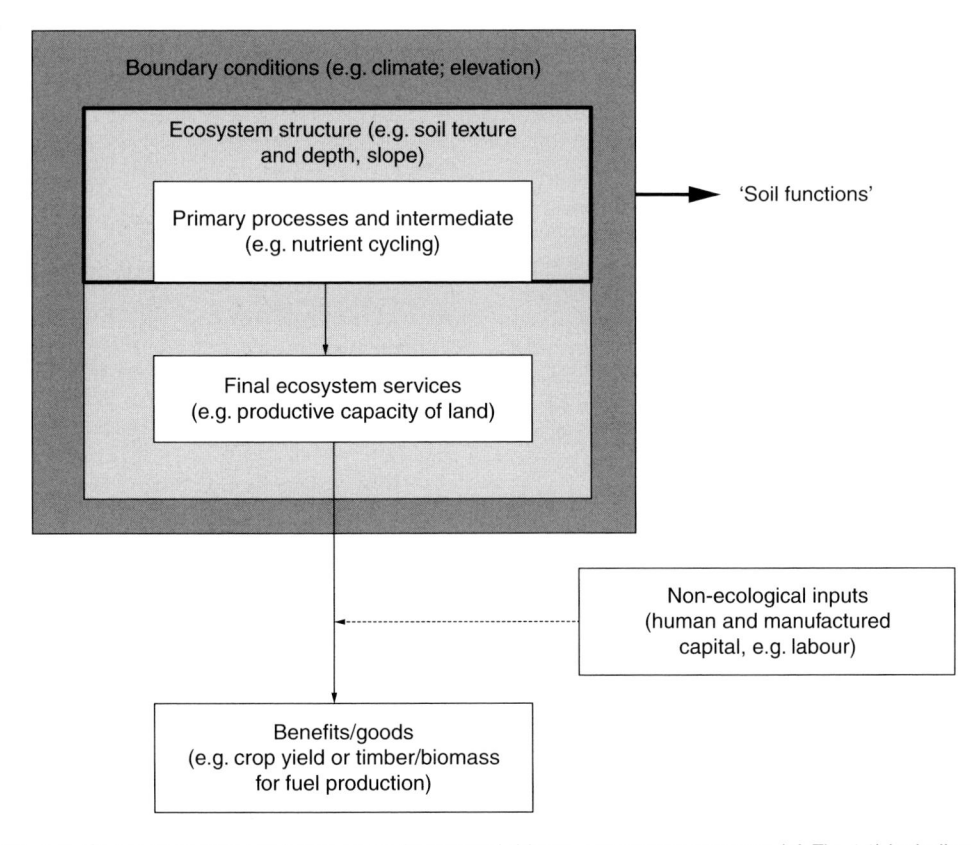

Fig. 1.2. Conceptual linkages between environmental drivers, ecosystem structure (cf. Fig. 1.1) including soil conditions, the soil processes that produce the environmental flows that characterize soil functions and their potential to provide ecosystem services, goods and benefits. (Adapted from Fisher et al., 2009 and Bateman et al., 2011.)

such that management decisions are made without full information on the consequences of change. Such market and informational failure necessarily leads to the suboptimal allocation of effort to conserve soil from a social perspective. The failure of markets and policy to prevent soil carbon loss and land degradation is therefore a key component of the global challenge to provide sufficient life-sustaining resources.

Threats to Soil Carbon

The global stocks of soil carbon are under threat (Table 1.1 and Plate 1), with consequences for the widespread loss of soil functions and an increase in greenhouse gas emissions from land and acceleration of global warming (Lal, 2010a,b). In many locations, soil functions are already compromised. Some of the consequences include increased erosion, increased pollution of water bodies from the N and P loads that arise from erosion, desertification, declining fertility and loss of habitat and biodiversity. The primary control on the global distribution of soil carbon is rainfall, with greater accumulation of soil organic matter in more humid regions. A secondary control is temperature, with greater organic matter accumulation in colder regions when otherwise sufficiently humid conditions persist regardless of temperature. Under similar climatic conditions, wetter soils help to accumulate soil carbon by limiting rates of microbial respiration (Batjes, 2011), since O_2 ingress is restricted by the gas diffusion barrier created by greater water content. Relatively drier conditions favour O_2 ingress and aeration of soil, thus accelerating soil carbon decomposition. Furthermore, physical disturbance such as tillage breaks up larger soil aggregates and exposes occluded carbon within aggregates to O_2 and biodegradation, thus creating conditions that allow greater soil carbon loss.

With sufficient water, nutrients and O_2 supply, biological processes are relatively faster at higher temperature; hence, greater rates of productivity and decomposition. Thus, warm, humid conditions favour soil carbon accumulation due to high productivity, while cool, humid conditions favour soil carbon accumulation due to low decomposition rates. Soil carbon varies substantially geographically with land cover (Plate 2). For example, savannah has relatively low soil carbon content but covers a large area globally. On the other hand, peatlands have extremely high carbon content but cover less than 0.3% of the global land surface. By inspection of Plates 1 and 2, it is clear that degraded land coincides in large part with Earth's drylands, due to low productivity from low water availability and relatively high decomposition due to dry, well-aerated and warm soils.

From these controls on soil carbon content, it is clear that predicted changes to regional as well as global climate in the coming decades will create important impacts on soil carbon (Schils et al., 2008; Conant et al., 2011). Drier, warmer conditions are expected to coincide with greater potential for loss of soil carbon and the associated loss of soil functions. Loss of permafrost will expose accumulated carbon in cold regions to much greater rates of microbial decomposition (Schuur and Abbot, 2011). Furthermore, the demographic drivers of more intensive land use raise the prospect of greater physical disturbance of soils, e.g. tilling of grasslands. More intense tillage and greater areas of mechanical tillage are expected to coincide with higher loss of soil carbon due to greater exposure of soil carbon to O_2 (Powlson et al., 2011).

Managing Soil Carbon for Multiple Benefits

Maintaining and increasing soil carbon content yields substantial multiple benefits. Greater soil carbon helps to maintain soil structure by forming stable larger aggregates and larger inter-aggregate pores that create greater soil permeability and drainage for root growth. Smaller interior pores within aggregates, on the other hand, provide water-holding capacity to sustain biological processes. Increasing soil carbon provides carbon and energy to support microbial activity, provides a reservoir of organic N, P and other nutrients for plant productivity and creates more physically cohesive soil to resist soil losses by physical erosion and by protecting occluded organic matter within the larger aggregates. Carbon that enters soil is removed from the atmosphere; any gains in soil carbon mitigate greenhouse gas emissions, with caveats about impacts on the N cycle and N_2O production and the production of CH_4 from the anaerobic decomposition of organic matter in waterlogged soils.

The factors that control soil carbon levels offer clues to strategies that can maintain and increase soil carbon content. Increasing carbon levels may be achieved by reducing soil carbon losses by measures to reduce physical erosion by wind or overland water flow, measures to prevent the mechanical disturbance of aggregates and measures to increase the water content of organic soils. Increasing input of soil carbon can be achieved by measures that increase the aboveground production of vegetation, the increased allocation of carbon below ground through greater root density and associated carbon input and microbial biomass, increased plant residue return to soil and the addition of imported organic matter such as compost.

Soil carbon is lost rapidly when soils are disturbed through land-use conversion from grassland and forest to arable, and when land is drained. However, building up soil carbon is slow. The risks of losing soil carbon are great because of the potential consequences of:

- the loss of soil fertility and agricultural production;
- increased greenhouse gas emissions and accelerated climate change; and
- diminished soil functions across the full range of the ecosystem services described above.

There is considerable knowledge and data on the role of soil organic matter in specific soil functions, particularly related to biomass production, water and contaminant filtration, and CO_2 emissions. There is considerably less known about the interactions between soil organic matter, biodiversity, transformations of nutrients and soil structure, and the physical stability of soil structure and aggregates. The knowledge of the role of soil carbon, and the existing methods and innovation potential to manage it effectively for this wide range of benefits, is collectively substantial but is fragmented between many different disciplines. The subsequent chapters of this volume seek to summarize this wide knowledge base and showcase regional examples of beneficial management of soil carbon with the potential to expand such practices greatly worldwide.

Beneficial management of soil carbon offers the opportunity not only to avoid the negative consequences but also to enhance the wide range of available soil functions and ecosystem services. For these reasons, policies are essential that encourage protecting, maintaining and enhancing soil carbon levels.

A new focus on soil carbon at all levels of governance for soil management would better enable the full potential of soil ecosystem services to be realized. This advance is urgent and essential. To meet successfully the '4 × 40' challenge laid out in the introduction (Godfray et al., 2010), there is significant opportunity through soil carbon management to help meet the demand for food, fuel and clean water worldwide. It is also an essential step towards soil management that establishes enhanced soil functions that last – in order to meet the needs of future generations; not only to meet the demands anticipated in the coming four decades.

References

Bai, Z.G., Dent, D.L., Olsson, L. and Schaepman, M.E. (2008) Proxy global assessment of land degradation. *Soil Use and Management* 24, 223–234.

Banwart, S.A. (2011) Save our soils. *Nature* 474, 151–152.

Banwart, S., Menon, M., Bernasconi, S.M., Bloem, J., Blum, W.E.H., de Souza, D., Davidsdotir, B., Duffy, C., Lair, G.J., Kram, P. et al. (2012) Soil processes and functions across an international network of Critical

Zone Observatories: introduction to experimental methods and initial results. *Comptes Rendus Geoscience* 344, 758–772.

Bateman, I.J., Mace, G.M., Fezzi, C., Atkinson, G. and Turner, K. (2011) Economic analysis for ecosystem service assessment. *Environmental and Resource Economics* 48, 177–218.

Batjes, N.H. (1996) Total carbon and nitrogen in the soils of the world. *European Journal of Soil Science* 47, 151–163.

Batjes, N.H. (2011) Soil organic carbon stocks under native vegetation: revised estimates for use with the simple assessment option of the Carbon Benefits Project system. *Agriculture, Ecosystems and Environment* 142, 365–373.

Black, H.I.J., Glenk, K., Towers, W., Moran, D. and Hussain, S. (2008) Valuing our soil resource for sustainable ecosystem services. Scottish Government Environment, Land Use and Rural Stewardship Research Programme 2006–2010. http://www.programme3.ac.uk/soil/p3-soilsposter-2.pdf (accessed 22 March 2013).

Blum, W.E.H. (1993) Soil protection concept of the Council of Europe and Integrated Soil Research. In: Eijsackers, H.J.P. and Hamers, T. (eds) *Soil and Environment, Volume I. Integrated Soil and Sediment Research: A Basis for Proper Protection*. Kluwer Academic Publisher, Dordrecht, the Netherlands, pp. 37–47.

Brantley, S. (2010) Weathering rock to regolith. *Nature Geoscience* 3, 305–306.

Conant, R.T., Ryan, M.G., Agren, G.I., Birge, H.E., Davidson, E.A., Eliasson, P.E., Evans, S.E., Frey, S.D., Giardina, C.P., Hopkins, F.M. et al. (2011) Temperature and soil organic matter decomposition rates – synthesis of current knowledge and a way forward. *Global Change Biology* 17, 3392–3404.

European Commission (2006) *Thematic Strategy for Soil Protection*. Commission of the European Communities, Brussels.

Fisher, B., Turner, R.K. and Morling, P. (2009) Defining and classifying ecosystem services for decision making. *Ecological Economics* 68(3), 643–653.

Godfray, H.C.J., Beddington, J.R., Crute, I.R., Haddad, L., Lawrence, D., Muir, J.F., Pretty, J., Robinson, S., Thomas, S.M. and Toulmin, C. (2010) Food security: the challenge of feeding 9 billion people. *Science* 327, 812–818.

Houghton, R.A. (1995) Changes in the storage of terrestrial carbon since 1850. In: Lal, R., Kimble, J., Levine, E. and Stewart, B.A. (eds) *Soils and Global Change*. Lewis Publishers, Boca Raton, Florida.

IPCC (2007) Climate Change 2007: Synthesis Report. Intergovernmental Panel on Climate Change. http://www.ipcc.ch/pdf/assessment-report/ar4/syr/ar4_syr.pdf (accessed 22 March 2013).

Jarvis, S., Tisdall, J., Oades, M., Six, J., Gregorich, E. and Kögel-Knabner, I. (2012) Landmark papers. *European Journal of Soil Science* 63, 1–21.

Joosten, H. (2009) *The Global Peatland CO_2 Picture. Peatland Status and Drainage Associated Emissions in All Countries of the World*. Wetlands International, Ede, the Netherlands.

Lal, R. (2010a) Managing soils and ecosystems for mitigating anthropogenic carbon emissions and advancing global food security. *BioScience* 60(9), 708–721.

Lal, R. (2010b) Managing soils for a warming earth in a food-insecure and energy-starved world. *Journal of Plant Nutrition and Soil Science* 1

MEA (Millennium Ecosystem Assessment) (2005) *Ecosystems and Human Well-Being: Synthesis*. Island Press, Washington, DC.

Montgomery, D.R. (2007) Soil erosion and agricultural sustainability. *Proceedings of the National Academy of Sciences USA* 104, 13268–13272.

Nachtergaele, F.O., Petri, M., Biancalani, R., van Lynden, G., van Velthuizen, H. and Bloise, M. (2011) Global Land Degradation Information System (GLADIS) Version 1.0. An Information Database for Land Degradation Assessment at Global Level, LADA Technical Report No 17. FAO, Rome.

Noellemeyer, E., Banwart, S., Black, H., Cai, Z., Gicheru, P., Joosten, H., Victoria, R., Milne, E., Pascual, U., Nziguheba, G. et al. (2014) Benefits of soil carbon: report on the outcomes of an international Scientific Committee on Problems of the Environment Rapid Assessment Workshop. *Carbon Management* 5(2), 185–192.

Powlson, D.S., Whitmore, A.P. and Goulding, K.W.T. (2011) Soil carbon sequestration to mitigate climate change: a critical re-examination to identify the true and the false. *European Journal of Soil Science* 62, 42–55.

Robinson, D.A., Hockley, N., Cooper, D.M., Emmett, B.A., Keith, A.M., Lebron, I., Reynolds, B., Tipping, E., Tye, A.M., Watts, C.W. et al. (2013) Natural capital and ecosystem services, developing an appropriate soils framework as a basis for valuation. *Soil Biology and Biochemistry* 57, 1023–1033.

Schils, R., Kuikman, P., Liski, J., Van Oijen, M., Smith, P., Webb, J., Alm, J., Somogyi, Z., Van den Akker, J., Billett, M., *et al*. (2008) Review of existing information on the interrelations between soil and climate change (ClimSoil). Final Report. European Commission, Brussels.

Schuur, E.A.G. and Abbot, B. (2011) High risk of permafrost thaw. *Nature* 480, 32–33.

Tisdall, J.M. and Oades, J.M. (1982) Organic matter and water-stable aggregates in soils. *European Journal of Soil Science* 33, 141–163.

UNEP-WCMC (2009) Updated global carbon map. Poster presented at the UNFCCC COP, Copenhagen. UNEP-WCMC, Cambridge, UK.

United Nations (2013) *World Population Prospects: The 2012 Revision, Key Findings and Advance Tables*. Working Paper No ESA/P/WP.227. United Nations, Department of Economic and Social Affairs, Population Division, New York.

Victoria, R., Banwart, S.A., Black, H., Ingram, H., Joosten, H., Milne, E. and Noellemeyer, E. (2012) The benefits of soil carbon: managing soils for multiple economic, societal and environmental benefits. In: *UNEP Year Book 2012: Emerging Issues in Our Global Environment*. UNEP, Nairobi, pp. 19–33. http://www.unep.org/yearbook/2012/pdfs/UYB_2012_CH_2.pdf (accessed 22 March 2013).

Wilkinson, B.H. and McElroy, B.J. (2007) The impact of humans on continental erosion and sedimentation. *Geological Society of America Bulletin* 119(1–2), 140–156.

2 Soil Carbon: a Critical Natural Resource – Wide-scale Goals, Urgent Actions

Generose Nziguheba*, Rodrigo Vargas, Andre Bationo, Helaina Black, Daniel Buschiazzo, Delphine de Brogniez, Hans Joosten, Jerry Melillo, Dan Richter and Mette Termansen

Abstract

Across the world, soil organic carbon (SOC) is decreasing due to changes in land use such as the conversion of natural systems to food or bioenergy production systems. The losses of SOC have impacted crop productivity and other ecosystem services adversely. One of the grand challenges for society is to manage soil carbon stocks to optimize the mix of five essential services – provisioning of food, water and energy; maintaining biodiversity; and regulating climate. Scientific research has helped develop an understanding of the general SOC dynamics and characteristics; the influence of soil management on SOC; and management practices that can restore SOC and reduce or stop carbon losses from terrestrial ecosystems. As the uptake of these practices has been very limited, it is necessary to identify and overcome barriers to the adoption of practices that enhance SOC. Actions should focus on multiple ecosystem services to optimize efforts and the benefits of SOC. Given that depleting SOC degrades most soil services, we suggest that in the coming decades increases in SOC will concurrently benefit all five of the essential services.

The aim of this chapter is to identify and evaluate wide-scale goals for maximizing the benefits of SOC on the five essential services, and to define the short-term steps towards achieving these goals. Stopping the losses of SOC in terrestrial ecosystems is identified as the overall priority. In moving towards the realization of multiple SOC benefits, we need to understand better the relationships between SOC and individual services. Interactions between services occur at multiple spatial scales, from farm through landscape to subnational, national and global scales. Coordinated national and international responses to SOC losses and degradation of the five essential services are needed to empower SOC actions at local levels that have benefits on the larger scales. We propose the creation of a global research programme to expand the scientific understanding of SOC and its contribution to the five essential services. This should address the challenges and uncertainties associated with the management of SOC for multiple benefits. This research programme must include a strong education and outreach component to address concerns to different communities outside academia.

Introduction

Soil organic matter (SOM) is an essential component of Earth's life support system. Soil organic carbon (SOC), which makes up half of the SOM by weight, plays a crucial role in the regulation of the global carbon cycle and its feedbacks within the Earth system

*E-mail: g.nziguheba@cgiar.org

(Trumbore, 1997; Lal, 2003). Humans rely on SOC stocks to help meet their needs for food, water, climate and biodiversity on our planet (Hooper *et al.*, 2000). Land degradation resulting in carbon losses is of great concern because it threatens our capacity to meet the demands of the world population, which is estimated to grow to over 9 billion by 2050. The resulting increased demand for food, water and energy will put an increasingly heavy pressure on land resources and the global climate.

Scientific research has given us clear and compelling evidence that SOC stocks have been reduced in many regions of the world, with these reductions often associated with agriculture and land degradation (Amundson, 2001; Sanderman and Baldock, 2010). One of the grand challenges for society is to manage soil carbon stocks to optimize the mix of five essential services – provisioning of food, water and energy; maintaining biodiversity and regulating climate (Fig. 2.1). These essential services and their interaction with SOC could be seen in an Anthropocene perspective (Richter, 2007). The global changes in SOC provide evidence that human activities are indeed having a global impact on the Earth system and on these five essential services underpinned by SOC.

For this chapter, SOM reflects the range of all organic materials found in the soil profile that influence the physical (e.g. soil bulk density, water infiltration rates), chemical (e.g. pH, nutrients) and biological (e.g. biomass, exogenous substrates) properties of soils. In this context, SOC can be increased by the addition of organic materials into the soil profile by means of different management for different purposes (Ingram and Fernandes, 2001; Swift, 2001).

Scientific research has helped develop an understanding of both the general SOC dynamics and characteristics and the influence of soil management on SOC at different temporal scales. This combined information can be used to motivate new research efforts to identify and promote best SOC management practices at local management units and to facilitate improvements at regional to global scales. Moving forward, there is a need to identify and overcome barriers to the adoption of practices that enhance SOC. Here, we argue for the necessity of an ambitious global

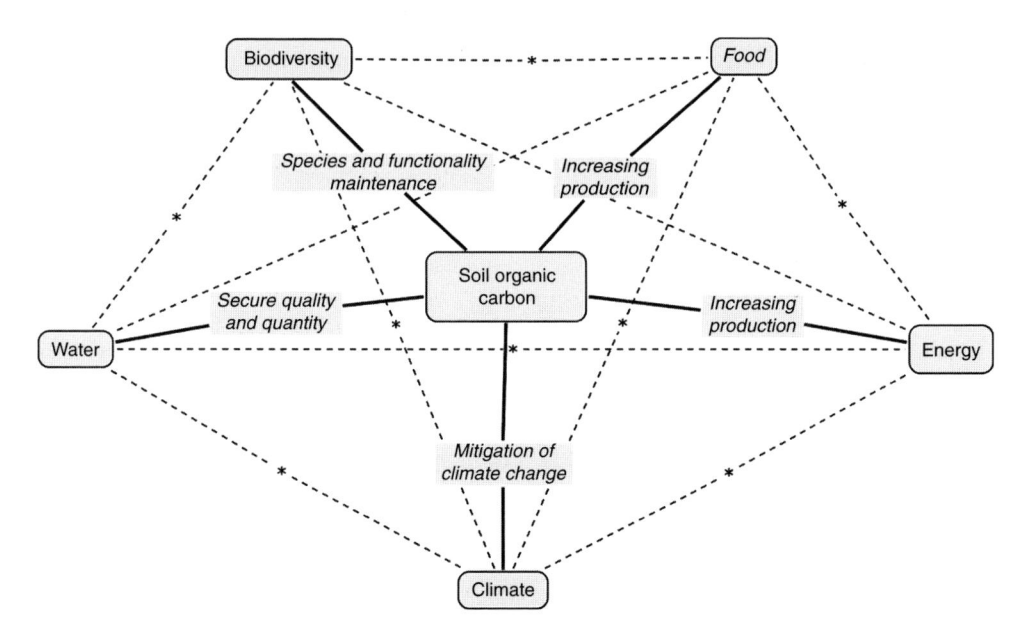

Fig. 2.1. Interactions between soil organic carbon (SOC) and the five essential services. Solid lines represent links discussed in this manuscript that refer directly to SOC. Dashed lines are interactions among essential services to show the interconnectivity.

research programme to expand the scientific understanding of SOC and its contribution to multiple environmental services, including management options towards the optimization of these services. These efforts should lead to coordinated national and international responses to SOC losses and degradation of the five essential services and empower SOC actions at local levels but be beneficial at larger scales. Thus, in moving towards the realization of multiple SOC benefits, we need to understand better the relationships between SOC and individual services to achieve long-term goals through new policy regulation and the research and development of economic incentive schemes.

The aim of this chapter is to identify wide-scale goals for maximizing the benefits of SOC on the five essential services and to define the short-term steps towards achieving these goals. First, we discuss the current knowledge on SOC and identify the feedbacks between increasing SOC and the five essential services. Second, we define the main long-term (next 25 years) challenges and uncertainties for managing SOC. We recognize that 25 years is not long term for soil carbon processes but is long term for policy and management actions towards maximizing the five essential services. Third, we outline a set of priorities and actions that will begin to move us towards optimizing the mix of benefits from these five essential services.

Wide-scale Goals and Urgent Actions

Food production

It is known that conventional agriculture reduces SOC in surface layers by up to 50% compared with natural vegetation (Jolivet et al., 1997; Mishra et al., 2010). In many parts of the world, degradation resulting from human activities has reduced the capacity of land to produce food. Underlying this degradation and declining agricultural productivity is the loss of SOC (Lefroy et al., 1993; Cheng et al., 2013). It is estimated that, on one-quarter of the global land area, soil carbon losses have caused a decline in productivity and in the ability to provide ecosystem services (Bai et al., 2008). In light of these facts, the goal is to increase and sustain food production to meet the demand of a growing population at both the local and global scale while increasing and sustaining SOC and the services it provides.

Soil organic C is imperative for food production because several SOC-related processes govern the availability of nutrients, water and toxins that control plant growth (Bationo et al., 2007). Soil carbon is the source of energy and substrate for soil microorganisms, which in turn regulate the decomposition and mineralization/immobilization processes responsible for nutrient availability (Insam, 1996; Bot and Benites, 2005). Soil organic C also improves the structure of soils by increasing the formation of soil aggregates, which enhances water infiltration and retention, thus reducing nutrient losses through leaching and runoff (Rawls et al., 2003; Blanco-Canqui and Lal, 2007).

It is important to acknowledge that the challenges faced in terms of increasing food production vary considerably across the globe. Increasing food production is particularly urgent in areas where current levels of food production are far below the potential levels (i.e. mainly in food-deficient regions such as sub-Sahara Africa). Food-deficient regions are characterized by low crop and livestock productivity, due mainly to soil degradation resulting from intensive land exploitation without adequate inputs of nutrients and from overgrazing (Drechsel et al., 2001). Low SOC affects vital soil functions such as nutrient cycling and microbial activity, both required for nutrient availability to crops. Current initiatives for fighting hunger in line with Millennium Development Goal 1, such as the African Green Revolution, need to take increasing SOC as a core component of interventions to ensure an efficient use of inputs and a sustainable increase of food production. Management practices that increase SOC and food production include fertilization, crop rotation, reduced tillage, organic matter addition, fallow, cover crops, agroforestry and improved livestock management.

Food-secure regions, predominant in developed countries, are often characterized by excess nutrient inputs in their farming systems, which can affect other ecosystem services negatively through pollution and greenhouse gas emissions (Csathó *et al.*, 2007; Vitousek *et al.*, 2009). Optimizing and sustaining current and future food production by maintaining the functionality of soils and minimizing the negative impact on other ecosystem services must be the major aim of a bold new programme of technical research and agricultural land management.

Water

Land use affects the quality and quantity of water strongly in many watersheds (Swallow *et al.*, 2009). One of the most important water pollution problems related to land use are the excess nutrients applied for agricultural production but which flow into surface and coastal waters (Ahrens *et al.*, 2008). Nitrate and phosphate contamination are well-known examples, but also pesticides enter both groundwater and surface-water bodies. Nutrients in surface waters can cause eutrophication, hypoxia, algal blooms and other infestations (such as of water hyacinth), which have been observed in coastal areas and many inland water bodies on all continents (Swallow *et al.*, 2009; Mateo-Sagasta and Burke, 2010). Water pollution has increased with the increased use of mineral fertilizers and higher concentrations of livestock (FAO, 2011). In light of these facts, the goal is to ensure the provision of sufficient quantity and quality of water needed for multiple uses by increasing SOC.

Soil organic carbon and protective vegetative cover are critical to maintaining the quality and quantity of water available for human consumption and plant production in the long term, because SOC determines soil properties that regulate in multiple ways the hydrological pathways within the soil. Soil organic carbon increases soil aggregates, which improves water infiltration and decreases the susceptibility of soil to water and wind erosion (Blanco-Canqui and Lal, 2007). The decrease in runoff and increase in infiltration contribute to recharging aquifers, and to preventing water pollution by decreasing the transport of nutrients and other contaminants to fresh waters. Soil organic carbon also improves water quality by acting as a filter of herbicide and pesticide residues and other pollutants that contaminate water reservoirs and streams (Lertpaitoonpan *et al.*, 2009; Rodriguez-Liébana *et al.*, 2013).

At the catchment scale, practices that increase SOC are required to improve water recharge (quantity) and purification (quality). In the short term, regulations at national or subnational levels, mainly in developing countries, must stimulate water erosion control measures in order to reduce the pollution of stream water and the effects of disasters such as hurricanes on the downstream population and infrastructure and to ensure the availability of potable water for human consumption (Bradshaw *et al.*, 2007; Brandimarte *et al.*, 2009). Adequate practices for increasing SOC at the catchment scale must be adopted by the farmers of the catchment area. Farmers could be grouped in farmer organizations, advised by experts from local, national and international institutions, including private organizations, and legally regulated and stimulated by the government. Practices to increase SOC that can be implemented immediately to reduce runoff and increase water infiltration include no till, cover crops, agroforestry, afforestation and others, complemented by specific technologies like terraces, contours and strip cropping (Mishra *et al.*, 2010; Powlson *et al.*, 2012). The cumulative effects of these and newer practices are hot topics for research. Land tenure policies that favour increases in SOC are needed to accompany these practices, particularly at the catchment level.

Once regulations are implemented, there is a need to monitor changes in SOC, in order to quantify its effects on the improvement of water quality and quantity. This should include the monitoring of the water table, hydrological regime and sediment loads in stream water. The results of this

Energy supply

Increasingly, plants are being grown to produce bioenergy, especially as the price of fossil fuels increases and efforts to mitigate climate change grow. The use of biomass for energy production is considered a promising way to reduce net carbon emissions and mitigate climate change (Don et al., 2012). The role of biomass in energy supply is expected to rise dramatically over the coming decades as cellulosic biofuel production becomes widespread. Reilly et al. (2012) project that an aggressive global biofuels programme could meet 40% of the world's primary energy needs by 2100. A large land area, perhaps as much as 21×10^6 km^2, would be required to produce biomass fuel crops at this large scale (Wise et al., 2009; Reilly et al., 2012). In light of these facts, the goal is to increase biomass fuel production to meet the demand for energy while increasing SOC.

As for food production, sustainable biomass fuel crop production will rely on an increase of SOC as a driver of processes regulating nutrient availability for use by these crops. However, land-use change to biomass fuel crops, particularly the conversion of native vegetation or peatlands, can result in carbon emissions from soil and vegetation in amounts that would take decades or centuries to compensate (Anderson-Teixeira et al., 2009; Gasparatos et al., 2011). The potential losses of soil carbon can counteract the benefits of fossil fuel displacement to the extent that biomass fuels from drained peatlands lead to emissions that, per unit of energy produced, exceed by far those from burning fossil fuels (Couwenberg, 2007; Couwenberg et al., 2010).

Maintaining or increasing biomass fuel production per unit area will require the careful management of soil carbon stocks over vast areas of the global landscape. Soil carbon management must be considered explicitly in carbon accounting efforts associated with biomass fuel production. This accounting should include both indirect effects on land use and fertilizer use and its consequences, including the release of nitrous oxide, a powerful heat-trapping gas, to the atmosphere (Melillo et al., 2009).

There is also evidence that some native vegetation (e.g. native grassland perennials) for biofuels could provide more usable bioenergy, larger reductions of greenhouse gas emissions and less agrichemical pollution than if the land were to be converted to producing annual bioenergy crops (Tilman et al., 2006; Don et al., 2012). Targeting degraded lands for biomass fuel production has been suggested as a potential way to reduce competition with food production and the negative effects of clearing natural vegetation and forest, particularly if perennial biomass fuel crops were grown (Kgathi et al., 2012). These perennial crops, if well identified, could contribute to increasing SOC on those degraded lands. There is therefore a need for full cycle analyses of biomass fuel production technologies and management regimes that take full account of the losses and gains of SOC (Davis et al., 2009; Gnansounou et al., 2009). Research should focus on monitoring the impact of land-use change for biomass fuel crop production on SOC losses and gains for proper guidelines on management for long-term benefits.

Biodiversity

Soil carbon is a primary ecosystem energy source that underpins the structure and function of terrestrial ecosystems, and thus the capacity of these ecosystems to maintain biodiversity. As illustrated in Fig. 2.2, decline of SOC comes as a second threat to soil diversity (Jeffery et al., 2010). Additionally, most of the other identified threats such as soil compaction and soil erosion are related to SOC losses and can be counteracted by an increase of SOC. Restoration projects around the world demonstrate that increasing SOC in degraded soils enhances not only biodiversity per se but also a range of ecosystem goods and services that can

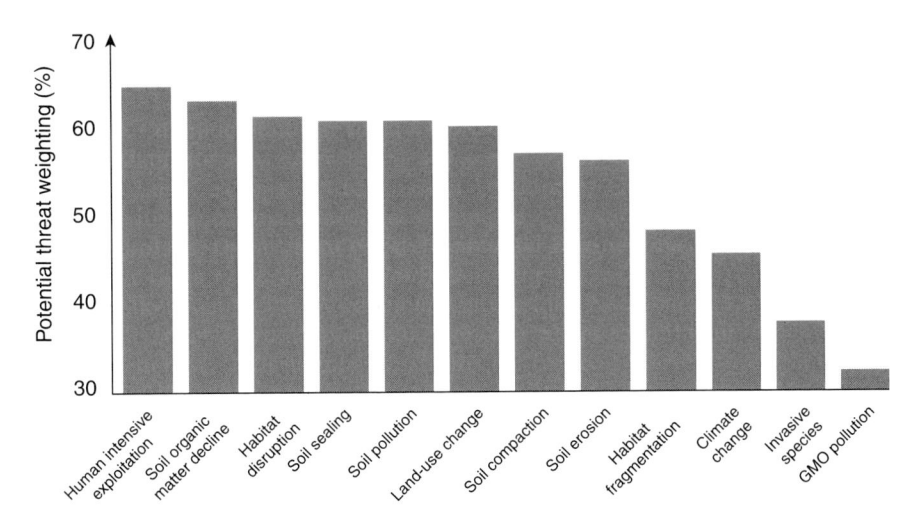

Fig. 2.2. Relative importance of possible threats to soil biodiversity in Europe as estimated by 20 soil biodiversity experts. (From Jeffery et al., 2010.)

benefit local people and wider communities (George et al., 2012). The goal here is to maintain or enhance the biodiversity of ecosystems by increasing SOC.

To date, conservation efforts to halt ongoing losses of global biodiversity have largely ignored critical interactions between the above- and belowground components of biodiversity. In part, this reflects a historical lack of information on the detailed composition and biogeography of soil communities. The application of molecular methods in large-scale surveys has begun to address this knowledge gap (Coleman and Whitman, 2005). The soil is estimated to be the largest terrestrial reserve of biodiversity (Fitter et al., 2005), with over one-quarter of the species on Earth living in the soil (Jeffery et al., 2010). The soil biota make up a complex food web consisting of microorganisms (e.g. bacteria, fungi, archaea, protozoa) through invertebrates (from nematodes to earthworms and termites) to mammals and reptiles (e.g. moles, snakes).

Soil biodiversity is important to soil quality since it has critical functional roles in the cycling of nutrients, organic matter and water, and in regulating soil structure, greenhouse gas fluxes, pest control and the degradation of pollutants. It is the presence of functional groups rather than taxonomic richness that appears to be important in soil C dynamics (Nielsen et al., 2011). Some of the main functional groups include litter fragmenters, decomposers of complex organic compounds, nitrifiers/denitrifiers, methanogens/methanotrophs and ecosystem engineers. Although we know these groups exist and we are rapidly gaining understanding about their roles in above- and belowground processes (Cornelissen et al., 2001; van der Heijden et al., 2008; Strickland et al., 2009), we still lack the ability to predict how, when and where these functional groups determine the capacity of soils to capture and store carbon and exchange greenhouse gases (Hunt and Wall, 2002).

This soil system derives its primary energy from carbon substrates obtained from root exudates, direct photosynthesis and the decomposition of organic matter from litter and plant roots. Thus, the quantity and quality of soil carbon is a key factor in determining the structure and activity of the soil community, and vice versa (Schulze, 2006). Changes in agricultural practices for food, livestock or bioenergy production affect SOC and disrupt both the below- and aboveground biodiversity. Practices to increase or maintain biodiversity include the protection of natural resources, halting land-use changes that affect natural vegetation and the restoration of degraded lands, all of which result in maintaining or increasing SOC.

Climate

Soils play a major role in the global carbon cycle, the dynamics of which have a large effect on Earth's climate system. Today, the top 1 m of soil worldwide contains about twice as much carbon in organic forms as does the atmosphere, and three times as much as does the vegetation (Batjes, 1996). Over the past three centuries, land clearing and land management for agriculture have resulted in the acceleration of soil organic matter decay and the transfer of more than 100 Pg carbon from the soil to the atmosphere as carbon dioxide (CO_2) (Sabine et al., 2004). In light of these facts, the goal is to mitigate climate change by practices towards ecosystem-level carbon sequestration including increasing SOC.

The extraction of peat and its use as fuel, litter or a soil improver has also resulted in substantial transfers of CO_2 (>20 Pg C) to the atmosphere over the same period (Gasparatos et al., 2011; Leifeld et al., 2011). Once in the atmosphere, CO_2 has a long half-life and it functions as a powerful heat-trapping gas that is the primary cause of the global temperature increases (IPCC, 2007). These temperature increases, in turn, accelerate SOC decay and create a self-reinforcing feedback, with warming begetting further warming (Heimann and Reichstein, 2008).

Practices that increase SOC, such as mulching and reduced tillage, increase and retain soil moisture, providing resilience to in-season rain shortages (dry spells), which are expected to occur more often in some regions as a consequence of climate change. The management of global soil carbon stocks with best practices has the potential to increase the magnitude of the SOC pool over decadal timescales to help mitigate climate change and climate variability. Two major soil science and management challenges are to: (i) minimize further losses of SOC to the atmosphere; and (ii) increase the soil carbon stocks. These two goals apply to the problem at local (catchment) and global scales, and in the short term as well as the longer term.

Interactions and Trade-offs Between Services

As illustrated above, there are many wide-scale goals and short- and long-term actions that must be implemented to meet growing human demands for food, water, energy, climate change mitigation and biodiversity in the coming decades at local and global scales. Soil organic carbon is central to these essential services and could be an important determinant of maintenance, buffering and enhancement of the supply of many ecosystem goods and other services under changing socio-economic and environmental conditions, as implied by the interactions in Fig. 2.1. Soil organic carbon, as a key component in ecosystem functioning, provides a useful mechanism to address jointly the threats to various ecosystem services. A focus on SOC enables us to set out the interactions between individual services and to assess appropriate synergies associated with actions to enhance SOC from local to global scales.

Actions affecting SOC long-term goals will inevitably have interactions and feedbacks. For example, as previously discussed, one interaction is between SOC and climate. In this case, management that induces SOC losses contributes to increasing greenhouse gas concentrations in the atmosphere, which in turn will increase air temperature and create a feedback by accelerating SOC decomposition and further losses (Heimann and Reichstein, 2008). Actions focusing on increasing the provision of one ecosystem service individually often impact various other ecosystems services negatively. We must learn from the past, where a focus on single services has led to significant reductions in the supply of other services (Tilman et al., 2006; Don et al., 2012). Typical examples are the focus on agriculture intensification for food production, which has led to water pollution and losses of biodiversity due to excess nutrients and pesticides

(Chappell and LaValle, 2011), and the clearance of native vegetation or drainage of peatlands for biomass fuel production, which also led to losses of biodiversity, water quality and quantity and contributed to climate changes through significant release of CO_2 to the atmosphere (Bessou et al., 2011). Focusing land management towards a range of benefits rather than one single benefit (as is often done) is a way forward in minimizing trade-offs and maximizing synergies. It is also proposed that losses in SOC have increased the vulnerability of these services to climate change (Reilly and Willenbockel, 2010; Don et al., 2012). Thus, restoring, increasing or protecting SOC could play a major role in buffering ecosystem goods and services in the future.

One view of interactions is that each essential service has an optimal operational range of SOC (Fig. 2.3). For example, while food production can, and continues to, operate at relatively low levels of SOC, there is a general hierarchy with other services requiring higher levels of SOC to be maintained effectively and for people to reap the benefits. The window for sustainable livelihoods is defined as the optimum range of C stocks that are adequate to supply all essential services. Currently, we are operating at SOC levels far below these windows, as demonstrated by global losses of biodiversity and problems with water quality and quantity (Powlson et al., 2011). The boundaries to these operational limits will vary at the local scale but ultimately are tied by the global potential to store SOC. As the current stock of SOC is below the optimal stock from a societal perspective (Fig. 2.3), managing soils for multiple services implies working towards levels of SOC that will allow all services to be delivered adequately.

Interactions between services occur at multiple spatial scales, from local (e.g. farm) through landscape (e.g. catchment) to subnational, national and global scales. The inducement of most interactions takes place at farm and catchment scales, where

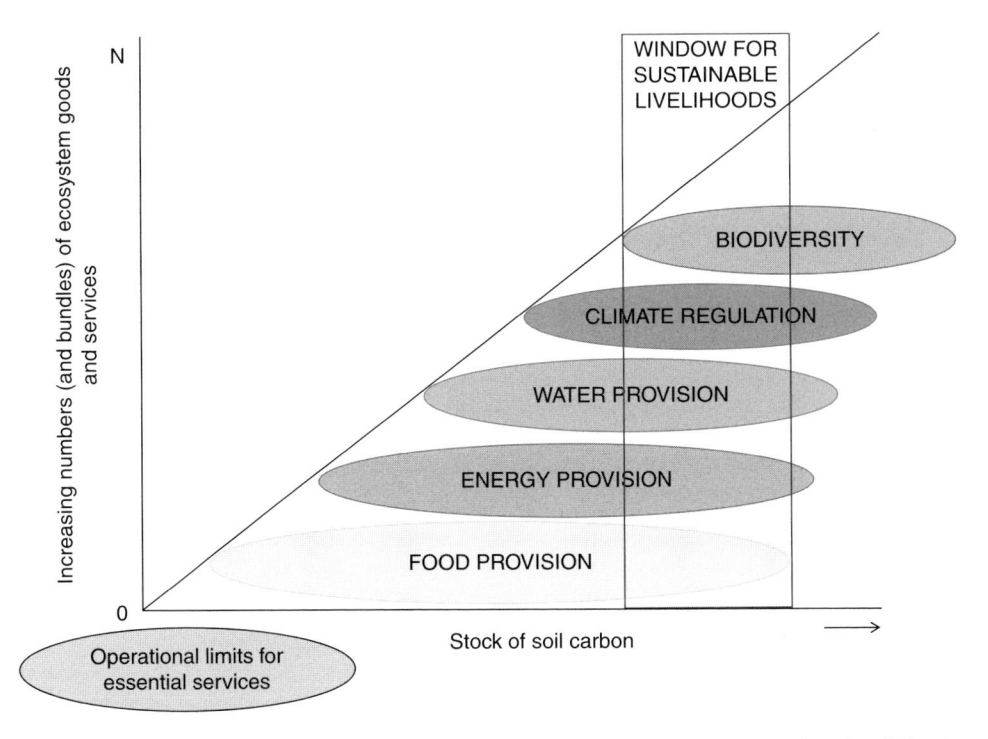

Fig. 2.3. Conceptual representation of operational ranges of essential ecosystem services in relation to SOC stocks.

people can implement management. The implications of the local management of SOC and its interactions with environmental services can have broader significance. Nowadays, given the degraded status of SOC in most managed soils and the ongoing threats to soils rich in carbon (e.g. peatlands, tropical forests), there are clear and immediate synergies between services in terms of SOC management. For example, at the farm level in low-carbon agricultural soils, there could be far-reaching co-benefits such as increased crop productivity, reduced runoff for water protection, enhanced soil biological functions and carbon sequestration. Therefore, increasing SOC could include landscape-derived benefits from water quality and quantity improvements and benefits from maintaining biodiversity by restoring soils and habitats. At the global level, improved farm- and catchment-level management to increase or maintain SOC could translate into a mitigation action for climate. However, none of the positive roles of increasing SOC for environmental services would be understood without scientific research. Therefore, a synergy must exist between academic institutions, research programmes and local communities to create public awareness and to communicate relevant findings quickly.

Uncertainties and Challenges

Across the world, there is evidence that managed soils have decreased their SOC due to changes in land use such as the conversion of natural systems to food or biofuel production systems (Leifeld et al., 2011; Powlson et al., 2011). The losses of SOC have adversely impacted crop productivity and other ecosystem services such as water resources, biodiversity, bioenergy and climate regulation (Bai et al., 2008). Much is known about management practices that can restore the organic matter contents of soils and can reduce or stop carbon losses from terrestrial ecosystems. In many regions and cropping systems, relatively small changes in land management practices can have relatively large impacts on SOC and its derived benefits. However, the adoption of these management practices has been very limited. There is an urgent need for identifying and overcoming the barriers to the adoption of practices that enhance SOC through appropriate policies, investment and land-use planning at various scales. Furthermore, tools are needed to enhance the measurement and analysis of the costs and benefits/valuation of various practices and farming systems on the range of ecosystem services at various temporal and spatial scales, including the economic, social and environmental benefits of increasing SOC.

Given that most soils and services can benefit from reversing their depleted state of SOC, we suggest that in the coming few decades increases in SOC will concurrently improve the five essential services (Fig. 2.1). However, the potential of soils to increase SOC is dependent on time and is constrained by different factors (Fig. 2.4). It is known that under given climatic, substrate, relief and hydrologic conditions there will

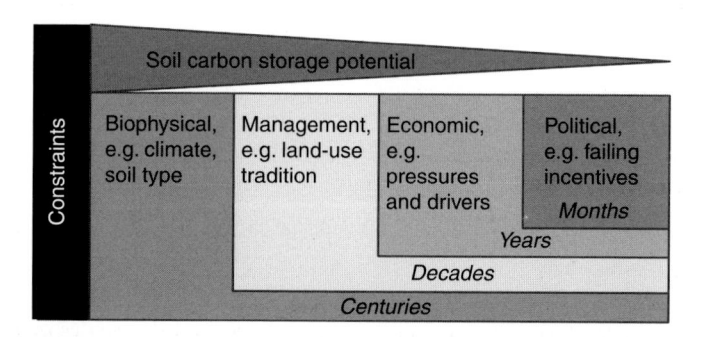

Fig. 2.4. Main constraints to soil carbon accumulation and the time frames over which they may be addressed.

be biophysical limits to how much carbon a soil can store naturally. However, the knowledge on a soil's inherent capacity to sequester carbon is absent, as natural reference soils are missing as a result of intensive land use. The biophysical limits are further constrained by land-use routines, which often have a strong historical/traditional bearing and are slow to change.

Economic drivers, in contrast, may change the cultivated crop or the land-use type (e.g. forest to grassland) rapidly, with possibly grave consequences for the soil carbon balance. Examples of the latter are changing market demands for food, fodder and energy crops. Changes in policies with implication for land use from one cultivation period to the next occur quickly and can lead to rapid and severe losses of soil carbon, as illustrated by governmental biofuel and bioenergy subsidies that stimulate the ploughing of grassland for maize cultivation or the drainage of peatlands. In view of the various constraints, a research management plan must be implemented along with management actions to monitor and adapt practices and goals according to site-specific conditions at different spatial and temporal scales. We propose to create a global research programme that focuses on developing robust SOC management and policies for multiple benefits across terrestrial ecosystems.

Despite current knowledge on SOC processes, there are still multiple uncertainties and challenges for the management of SOC that call for a research action programme. Uncertainties include, but are not limited to: the quantification of synergies between the different benefits of SOC, defining critical thresholds for achieving gains by individual and multiple benefits, and establishing the time frame needed to reach the level required for significant impact on an environmental service. In addition, the significance of change in SOC towards a social benefit is not well understood. Research is needed to measure and assess better the supplies and benefits of SOC for agricultural productivity, water, biodiversity, bioenergy and climate regulation. Other uncertainties of importance include the precise rates of change in SOC, especially across the full rooting zone of the soil system, and the quantification of the impact of future land conversions to agriculture, the abandonment of degraded land and deforestation on SOC. Finally, the lack of methodologies for quantifying the effects of land management and SOC on multiple benefits is a handicap for promoting initiatives towards enhancing SOC stocks. However, these uncertainties should not stand in the way of the critical need to increase SOC and of research that runs across terrestrial ecosystems (Seastedt et al., 2008).

The research community is exploring a wide range of technologies to reduce uncertainty on the benefits of SOC. A variety of geographic information system (GIS) tools and ecosystem models are being used to explore the spatial interactions between services from fields and across landscapes (Hayes et al., 2012; Aide et al., 2013). These tools and models can be used to identify where one service negates the ability to have other services (in the past, agriculture and biodiversity conservation). Such tools could be expanded to include SOC. The key limitations here are effective representation of soil carbon–services relationships, sufficient data to represent these services over space and the capacity to predict changes to interactions over time. While there is evidence of the positive impact of management practices for enhancing SOC on some services such as food production and water quality at local (plot) and catchment level, other services such as climate regulation occur at a larger scale (subnational, regional or global) and are even more difficult to quantify. Despite these uncertainties, failing to act towards increasing SOC on the basis of limited current scientific evidence is much more dangerous than the risks associated with continuous decline in SOC stocks.

Finally, an overriding challenge is the communication between scientists, policy makers and the public. Educating the public about the critical importance of SOC to food, water, bioenergy and climate requires a revolution in communication, specifically about the multiple benefits of SOC for daily life. Translating knowledge of the management and benefits of SOC into communications that inform and engage with societal debates and values can be a key part of the network of scientific and education centres. Therefore, a global research programme to reduce the uncertainty associated with SOC management across terrestrial ecosystems must include a strong educational and outreach component to address practical concerns to different communities outside academia.

Priorities and Actions

We argue that the overall priority is to stop losses of SOC in terrestrial ecosystems. To achieve this goal, we insist that there is a need to create a global research programme to address the challenges and uncertainties associated with increasing SOC for multiple benefits. The fundamental science questions should focus on reducing the uncertainties associated with large-scale assessments and the monitoring of SOC change and benefits at local and global scales. Therefore, urgent actions and new approaches are needed to answer key multi-purpose and multi-scale relationships, thresholds and trade-offs between soil carbon and the essential services (Fig. 2.1). First, we need to understand the recovery rates of SOC better as they are usually non-linear (i.e. have hysteresis effects), making it difficult to forecast the effects of a decision/management made today. Second, research efforts should focus on how to optimize the benefits of soil carbon across various spatial scales where management strategies will vary at the farm/plot, catchment and global level. Third, there is a need to identify the critical ranges/thresholds of SOC losses and recoveries for management purposes and to include the ability to estimate the economic value of investments in soil carbon. All these fundamental research priorities must inform public and economic interests and provide information for policy and actions towards reducing soil carbon losses. Finally, any of these priorities will not be possible without committed long-term funding support and missions by national research agencies and international organizations.

We propose that these research efforts should be linked with specific goals and priorities and actions tailored towards each one of the five essential services (Table 2.1). Here, we discuss specific goals for each essential service.

In order to meet the increasing *food* demand, at both the local and global level, there is a need to increase and sustain food production through better management of soils while improving environment quality. For this, current SOC losses must be stopped and practices to increase SOC must be adopted, including dormant-season cover crops, agroforestry systems, fallows, reduced tillage and applications of mulch, compost and safe biosolids, and in the case of organic soils, paludicultures (Lal *et al.*, 2007; Smith *et al.*, 2013). Examples include the proposed climate-smart agriculture approach, which

Table 2.1. Summary of wide-scale goals and urgent actions for the five essential services related to soil organic carbon.

Environmental service	Long-term goal	Priority/action
Food	Increase food production	Reduce soil organic carbon losses substantially
Water	Secure sufficient water quantity and quality	Restore hydrological pathways
		Improve water infiltration
		Prevent water pollution
Energy	Increase biofuel production	Increase biomass production considering full carbon cycle
Biodiversity	Maintain or enhance below- and aboveground biodiversity	Protect ecological hotspots
		Restore habitats
Climate	Mitigate climate change	Stop losses of soil organic carbon
		Increase soil organic carbon

aims at enhancing food productivity while reducing greenhouse gas emissions and enhancing SOC sequestration (FAO, 2013).

In order to enhance *water* quality and quantity, increases in SOC must be targeted to restore hydrological pathways, improve water infiltration management and prevent water pollution (Ahrens *et al.*, 2008; Thomas *et al.*, 2009). Soil and water conservation measures are required to accompany SOC management practices, particularly on sloping lands.

In order to increase immediate *energy production* to meet local demands, we have to focus simultaneously on maximizing the yield of bioenergy crops while preserving or restoring natural ecosystems and soil carbon stocks. Policies on biofuels and the installation of instrumentation for harvesting alternative energy (e.g. wind and solar power) need to be evaluated in light of their effects on soil carbon. For example, the initial conversion of land for biofuel production can result in immediate carbon loss, and the establishment of large deployments of solar and wind power could affect soil carbon storage (Anderson-Teixeira *et al.*, 2012).

In order to enhance *biodiversity*, new management practices that minimize damage and stimulate soil biological activity (e.g. reduced tillage, incorporation of plant residues, cover crops, careful pesticide use) must be applied. In the longer term, we must have sufficient understanding about the global distribution and role of soil biodiversity in ecosystem function, in particular carbon dynamics, to develop and implement sound guidance and policy. Efforts should be targeted towards the protection of ecological hotspots, habitat restoration and maintaining genetic and functional soil biodiversity (Carney and Matson, 2005; Pickles *et al.*, 2012).

To address *climate* change and propose climate mitigation strategies, SOC losses must be minimized through appropriate land-use practices. These include slowing and eventually eliminating the conversion of natural ecosystems such as forest to agricultural uses, slowing and eventually eliminating the use of drained peat soils and slowing and eventually eliminating the use of peat as an energy source and a raw material for horticultural substrates. Increases in soil carbon stocks can be achieved through the careful management of agricultural soils, including the use of reduced tillage, through the implementation of paludicultures on organic soils and through afforestation (Smith *et al.*, 2008; Tschakert *et al.*, 2008; Joosten, 2012).

Furthermore, efforts should be directed to communicate better in new ways to the general public and policy makers the value of increasing SOC. Thus, there is a high priority to increase the communication and education of SOC to permeate into the policy realm and the action plans of local managers/farmers. These actions could lead to public and transparent reports that communicate the state (gains or losses) of SOC and address needs accurately at the local or national scale. In fact, this simple reporting mechanism could be seen as an analogue of the gross domestic product (GDP) used as an economic development indicator. Such mechanism will require new monitoring, verification and reporting schemes for the regulatory, research and economical purposes of soil carbon. This chapter highlights that one of the most significant underlying reasons for lack of investment in SOC is the mismatch between short- and long-term objectives in land management (see also Chapter 4, this volume). It follows that irrespective of the favourable long-term economic case for investment in soil carbon, such investments are unlikely to come about without policy intervention. Soil carbon could be promoted through the payment of ecosystem schemes to reduce the intertemporal trade-offs between short- and long-term objectives. Ultimately, we emphasize that any of these priorities cannot be attained without extensive education efforts on the benefits of SOC to increase public understanding of the need to protect soils around the world.

Conclusions

This chapter has highlighted the need for managing SOC to optimize the mix of five essential services – the provisioning of food, energy and water, regulating climate and maintaining biodiversity (Fig. 2.1). The interaction of SOC with these services shows that they are interconnected and that actions focusing on single services, without considering SOC, impact other services negatively. This calls for a systems approach in order to maximize the benefits on all relevant spatial and temporal scales (Figs 2.3 and 2.4).

We highlight the wide-scale goals and urgent actions towards maximizing the benefits of SOC (Table 2.1) and conclude that the critical priorities are centred on stopping the current losses of SOC. This requires the involvement of various players at local, national and global levels. We propose that in order to quantify better the benefits of SOC, there is a need for complete analyses of the potential actions towards each of the services, including economic, political and environmental implications. Such analyses aim at assessing both the impacts of the actions on each individual service and the co-benefits or adverse impacts on other services. This is needed to maximize SOC gains and to optimize essential services.

We recognize the key uncertainties in managing SOC towards the essential services. However, we conclude that these uncertainties should not stand in the way of the critical need to increase SOC. We propose to take advantage of current scientific knowledge on SOC characteristics, its dynamics and complexity, and managements that affect it, to direct research efforts towards key missing areas and to improve knowledge and practices towards the long-term goals of increasing SOC.

A new vision of soil carbon science that enhances the understanding of the policy and economics of soil services is needed urgently. Such vision will help create a better public understanding of SOC and its societal benefits, which is needed to develop policies that protect soils around the world. Therefore, we call for a global research and education programme focused on the multiple benefits of SOC and with a strong outreach component to share the findings and communicate practical concerns with different communities outside academia. Finally, we recognize that the proposed research and education programme will not be possible without committed long-term funding support.

References

Ahrens, T.D., Beman, J.M., Harrison, J.A., Jewett, P.K. and Matson, P.A. (2008) A synthesis of nitrogen transformations and transfers from land to the sea in the Yaqui Valley agricultural region of northwest Mexico. *Water Resources Research* 44, W005, 1–13.

Aide, T.M., Clark, M.R., Grau, H.R., Lopez-Carr, D., Levy, M.A., Redo, D., Bonilla-Moheno, M., Riner, G., Andrade-Nuñez, M.J. and Muñoz, M. (2013) Deforestation and reforestation of Latin America and the Caribbean (2001–2010). *Biotropica* 45, 262–271.

Amundson, R. (2001) The carbon budget in soils. *Annual Review of Earth and Planetary Sciences* 29, 535–562.

Anderson-Teixeira, K.J., Davis, S.C., Masters, M.D. and Delucia, E.H. (2009) Changes in soil organic carbon under biofuel crops. *GCB Bioenergy* 1, 75–96.

Anderson-Teixeira, K.J., Duval, B.D., Long, S.P. and DeLucia, E.H. (2012) Biofuels on the landscape: is 'land sharing' preferable to 'land sparing'? *Ecological Applications* 22, 2035–2048.

Bai, Z.G., Dent, D.L., Olsson, L. and Schaepman, M.E. (2008) Proxy global assessment of land degradation. *Soil Use and Management* 24, 223–234.

Bationo, A., Kihara, J., Vanlauwe, B., Waswa, B. and Kimetu, J. (2007) Soil organic carbon dynamics, functions and management in West African agro-ecosystems. *Agricultural Systems* 94, 13–25.

Batjes, N.H. (1996) Total carbon and nitrogen in the soils of the world. *European Journal of Soil Science* 47, 151–163.

Bessou, C., Ferchaud, F., Gabrielle, B. and Mary, B. (2011) Biofuels, greenhouse gases and climate change: a review. *Agronomy, Sustainability and Development* 31, 1–79.

Blanco-Canqui, H. and Lal, R. (2007) Soil structure and organic carbon relationships following 10 years of wheat straw management in no-till. *Soil and Tillage Research* 95, 240–254.

Bot, A. and Benites, J. (2005) The importance of soil organic matter: key to drought-resistant soil and sustained food and production. *Food and Agriculture Organization Soils Bulletin* 80, pp. 78.

Bradshaw, C.J.A., Sodhi, N.S., Peh, K.S.H. and Brook, B.W. (2007) Global evidence that deforestation amplifies flood risk and severity in the development world. *Global Change Biology* 13, 1–17.

Brandimarte, L., Brath, A., Castellarin, A. and Baldassarre, G.D. (2009) Isla Hispaniola: a trans-boundary flood risk mitigation plan. *Physics and Chemistry of the Earth* 34, 209–218.

Carney, K.M. and Matson, P.A. (2005) Plant communities, soil microorganisms, and soil carbon cycling: does altering the world belowground matter to ecosystem functioning? *Ecosystems* 8, 928–940.

Chappell, M.J. and LaValle, L.A. (2011) Food security and biodiversity: can we have both? An agroecological analysis. *Agriculture and Human Values* 28, 3–26.

Cheng, X., Yang, Y., Li, M., Dou, X. and Zhang, Q. (2013) The impact of agricultural land use changes on soil organic carbon dynamics in the Danjiangkou reservoir area of China. *Plant and Soil* 366, 415–424.

Coleman, D.C. and Whitman, W.B. (2005) Linking species richness, biodiversity and ecosystem function in soil systems. *Pedobiologia* 49, 479–497.

Cornelissen, J., Aerts, R., Cerabolini, B., Werger, M. and Van Der Heijden, M. (2001) Carbon cycling traits of plant species are linked with mycorrhizal strategy. *Oecologia* 129, 611–619.

Couwenberg, J. (2007) Biomass energy crops on peatlands: on emissions and perversions. *IMCG Newsletter* 2007/3, 12–14.

Couwenberg, J., Dommain, R. and Joosten, H. (2010) Greenhouse gas fluxes from tropical peatlands in south-east Asia. *Global Change Biology* 16, 1715–1732.

Csathó, P., Sisák, I., Radimszky, L., Lushaj, S., Spiegel, H., Nikolova, M., Čermák, P., Klir, J., Astover, A., Karklins, A. et al. (2007) Agriculture as a source of phosphorus causing eutrophication in central and eastern Europe. *Soil Use and Management* 23, 36–56.

Davis, S.C., Anderson-Teixeira, K.J. and DeLucia, E.H. (2009) Life-cycle analysis and the ecology of biofuels. *Trends in Plant Science* 14, 140–146.

Don, A., Osborne, B., Hastings, A., Skiba, U., Carter, M.S., Drewer, J., Flessa, H., Freibauer, A., Hyvönen, N., Jones, M.B. et al. (2012) Land-use change to bioenergy production in Europe: implications for the greenhouse gas balance and soil carbon. *GCB Bioenergy* 4, 372–391.

Drechsel, P., Gyiele, L., Kunze, D. and Cofie, O. (2001) Population density, soil nutrient depletion, and economic growth in sub-Saharan Africa. *Ecological Economics* 38, 251–258.

FAO (2011) *Climate Change, Water and Food Security*. FAO Land and Water, Rome.

FAO (2013) *Climate-smart Agriculture Sourcebook*. FAO, Rome.

Fitter, A., Gilligan, C., Hollingworth, K., Kleczkowski, A., Twyman, R. and Pitchford, J. (2005) Biodiversity and ecosystem function in soil. *Functional Ecology* 19, 369–377.

Gasparatos, A., Stromberg, P. and Takeuchi, K. (2011) Biofuels, ecosystem services and human wellbeing: putting biofuels in the ecosystem services narrative. *Agriculture, Ecosystems and Environment* 142, 111–128.

George, S.J., Harper, R.J., Hobbs, R.J. and Tibbett, M. (2012) A sustainable agricultural landscape for Australia: a review of interlacing carbon sequestration, biodiversity and salinity management in agroforestry systems. *Agriculture, Ecosystems and Environment* 163, 28–36.

Gnansounou, E., Dauriat, A., Villegas, J. and Panichelli, L. (2009) Life cycle assessment of biofuels: energy and greenhouse gas balances. *Bioresource Technology* 100, 4919–4930.

Hayes, D.J., Turner, D.P., Stinson, G., McGuire, A.D., Wei, Y., West, T.O., Heath, L.S., Jong, B., McConkey, B.G., Birdsey, R.A. et al. (2012) Reconciling estimates of the contemporary North American carbon balance among terrestrial biosphere models, atmospheric inversions, and a new approach for estimating net ecosystem exchange from inventory-based data. *Global Change Biology* 18, 1282–1299.

Heimann, M. and Reichstein, M. (2008) Terrestrial ecosystem carbon dynamics and climate feedbacks. *Nature* 451, 289–292.

Hooper, D.U., Bignell, D.E., Brown, V.K., Broussard, L., Dangerfield, J.M., Wall, D.H., Wardle, D.A., Coleman, D.C., Giller, K.N., Lavelle, P. et al. (2000) Interactions between aboveground and belowground biodiversity in terrestrial ecosystems: patterns, mechanisms, and feedbacks. *BioScience* 50, 1049–1061.

Hunt, H. and Wall, D. (2002) Modelling the effects of loss of soil biodiversity on ecosystem function. *Global Change Biology* 8, 33–50.

Ingram, J. and Fernandes, E. (2001) Managing carbon sequestration in soils: concepts and terminology. *Agriculture, Ecosystems and Environment* 87, 111–117.

Insam, H. (1996) Microorganisms and humus in soils. In: Piccolo, A. (ed.) *Humic Substances in Terrestrial Ecosystems*. Elsevier, Amsterdam, pp. 265–292.

IPCC (Intergovernmental Panel on Climate Change) (2007) *Climate Change 2007: Impacts, Adaptation and Vulnerability*. Contribution of Working Group II to the Fourth Assessment Report of the IPCC. Cambridge University Press, Cambridge, UK, 976 pp.

Jeffery, S., Gardi, C., Jones, A., Montanarella, L., Marmo, L., Miko, L., Ritz, K., Peres, G., Rombke, J.R. and van der Putten, W. (2010) *European Atlas of Soil Biodiversity*. Publication Office of the European Commission, Luxembourg.

Jolivet, C., Arrouays, D., Andreux, F. and Lévèque, J. (1997) Soil organic carbon dynamics in cleared temperate forest spodosols converted to maize cropping. *Plant and Soil* 191, 225–231.

Joosten, H. (2012) Zustand und Perspektiven der Moore weltweit (Status and prospects of global peatlands). *Natur und Landschaft* 87, 50–55.

Kgathi, D.L., Ngwenya, B.N. and Sekhwela, M.B.M. (2012) Potential impacts of biofuel development on biodiversity in Chobe district, Bostwana. In: Janssen, R. and Rutz, D. (eds) *Bioenergy for Sustainable Development in Africa*. Springer, New York, pp. 247–260.

Lal, R. (2003) Global potential of soil carbon sequestration to mitigate the greenhouse effect. *Critical Reviews in Plant Sciences* 22, 151–184.

Lal, R., Follett, R.F., Stewart, B.A. and Kimble, J.M. (2007) Soil carbon sequestration to mitigate climate change and advance food security. *Soil Science* 172(12), 943–956.

Lefroy, R.D., Blair, G.J. and Strong, W.M. (1993) Changes in soil organic matter with cropping as measured by organic carbon fractions and 13C natural isotope abundance. *Plant and Soil* 155, 399–402.

Leifeld, J., Müller, M. and Fuhrer, J. (2011) Peatland subsidence and carbon loss from drained temperate fens. *Soil Use and Management* 27, 170–176.

Lertpaitoonpan, W., Ong, S.K. and Moorman, T.B. (2009) Effect of organic carbon and pH on soil sorption of sulfamethazine. *Chemosphere* 76, 558–564.

Mateo-Sagasta, J. and Burke, J. (2010) *Agriculture and Water Quality Interactions: A Global Overview. SOLAW Background Thematic Report-TR08*. FAO, Rome.

Melillo, J.M., Reilly, J.M., Kicklighter, D.W., Gurgel, A.C., Cronin, T.W., Paltsev, S., Felzer, B.S., Wang, X., Sokolov, A.P. and Schlosser, C.A. (2009) Indirect emissions from biofuels: how important? *Science* 326, 1397–1399.

Mishra, U., Ussiri, D.A.N. and Lal, R. (2010) Tillage effects on soil organic carbon storage and dynamics in corn belt of Ohio, USA. *Soil and Tillage Research* 107, 88–96.

Nielsen, U.N., Ayres, E., Wall, D.H. and Bardgett, R.D. (2011) Soil biodiversity and carbon cycling: a review and synthesis of studies examining diversity–function relationships. *European Journal of Soil Science* 62, 105–116.

Pickles, B.J., Egger, K.N., Massicotte, H.B. and Green, D.S. (2012) Ectomycorrhizas and climate change. *Fungal Ecology* 5, 73–84.

Powlson, D.S., Whitmore, A.P. and Goulding, K.W.T. (2011) Soil carbon sequestration to mitigate climate change: a critical re-examination to identify the true and the false. *European Journal of Soil Science* 62, 42–55.

Powlson, D.S., Bhogal, A., Chambers, B.J., Coleman, K., Macdonald, A.J., Goulding, K.W.T. and Whitmore, A.P. (2012) The potential to increase soil carbon stocks through reduced tillage or organic material additions in England and Wales: a case study. *Agriculture, Ecosystems and Environment* 146, 23–33.

Rawls, W.J., Pachepsky, Y.A., Ritchie, J.C., Sobecki, T.M. and Bloodworth, H. (2003) Effect of soil organic carbon on soil water retention. *Geoderma* 116, 61–76.

Reilly, J., Melillo, J., Cai, Y., Kicklighter, D., Gurgel, A., Paltsev, S., Cronin, T., Sokolov, A. and Schlosser, A. (2012) Using land to mitigate climate change: hitting the target, recognizing the trade-offs. *Environmental Science and Technology* 46, 5672–5679.

Reilly, M. and Willenbockel, D. (2010) Managing uncertainty: a review of food system scenario analysis and modeling. *Philosophical Transactions of the Royal Society B* 365, 3049–3063.

Richter, D.D. (2007) Humanity's transformation of earth's soil: pedology's new frontier. *Soil Science* 172, 957–967.

Rodriguez-Liébana, J.A., Mingorance, M.D. and Peña, A. (2013) Pesticide sorption on two contrasting mining soils by addition of organic wastes: effect of organic matter composition and soil solution properties. *Colloids and Surfaces A: Physicochemical and Engineering Aspects* 435, 71–77.

Sabine, C.L., Heimann, M., Artaxo, P., Bakker, D.C., Chen, C.-T.A., Field, C.B., Gruber, N., Le Quere, C., Prinn, R.G., Richey, J. *et al.* (2004) Current status and past trends of the global carbon cycle. In: Field, C.B. (ed.) *Scope 62, The Global Carbon Cycle: Integrating Humans, Climate, and the Natural World*. Scope 62, Island Press, Washington, DC, pp. 17–44.

Sanderman, J. and Baldock, J.A. (2010) Accounting for soil carbon sequestration in national inventories: a soil scientist's perspective. *Environmental Research Letters* 5, 034003.

Schulze, E.D. (2006) Biological control of the terrestrial carbon sink. *Biogeosciences* 3, 147–166.

Seastedt, T.R., Hobbs, R.J. and Suding, K.N. (2008) Management of novel ecosystems: are novel approaches required? *Frontiers in Ecology and the Environment* 6, 547–553.

Smith, J., Pearce, B.D. and Wolfe, M.S. (2013) Reconciling productivity with protection of the environment: is temperate agroforestry the answer? *Renewable Agriculture and Food Systems* 28, 80–92.

Smith, P., Fang, C.M., Dawson, J.J.C. and Moncrieff, J.B. (2008) Impact of global warming on soil organic carbon. *Advances in Agronomy* 97, 1–43.

Strickland, M.S., Lauber, C., Fierer, N. and Bradford, M.A. (2009) Testing the functional significance of microbial community composition. *Ecology* 90, 441–451.

Swallow, B.M., Sang, J.K., Nyabenge, M., Bundotich, D.K., Duraiappah, A.K. and Yatich, T.B. (2009) Tradeoffs, synergies and traps among ecosystem services in the lake Victoria basin of east Africa. *Environmental Science and Policy* 12, 504–519.

Swift, R.S. (2001) Sequestration of carbon by soil. *Soil Science* 166, 858–871.

Thomas, M.A., Asce, S.M., Angel, B.A. and Chaubey, I. (2009) Water quality impact of corn production to meet biofuel demands. *Journal of Environmental Engineering* 135, 1123–1135.

Tilman, D., Hill, J. and Lehman, C. (2006) Carbon-negative biofuels from low-input high-diversity grassland biomass. *Science* 314, 1598–1600.

Trumbore, S.E. (1997) Potential responses of soil organic carbon to global environmental change. *Proceedings of the National Academy of Sciences USA* 94, 8284–8291.

Tschakert, P., Huber-Sannwald, E., Ojima, D.S., Raupach, M.R. and Schienke, E. (2008) Holistic, adaptive management of the terrestrial carbon cycle at local and regional scales. *Global Environmental Change* 18, 128–141.

van der Heijden, M.G., Bardgett, R.D. van Straalen, N.M. (2008) The unseen majority: soil microbes as drivers of plant diversity and productivity in terrestrial ecosystems. *Ecology Letters* 11, 296–310.

Vitousek, P.M., Naylor, R., Crews, T., David, M.B., Drinkwater, L.E., Holland, E., Johnes, P.J., Katzenberger, J., Martinelli, L.A., Matson, P.A. *et al.* (2009) Nutrient imbalances in agricultural development. *Science* 324, 1519–1520.

Wise, M., Calvin, K., Thomson, A., Clarke, L., Bond-Lamberty, B., Sands, R., Smith, S.J., Janetos, A. and Edmonds, J. (2009) Implications of limiting CO_2 concentrations for land use and energy. *Science* 324, 1183–1186.

3 Soil Carbon Transition Curves: Reversal of Land Degradation through Management of Soil Organic Matter for Multiple Benefits

Meine van Noordwijk*, Tessa Goverse, Cristiano Ballabio, Steven A. Banwart, Tapas Bhattacharyya, Marty Goldhaber, Nikolaos Nikolaidis, Elke Noellemeyer and Yongcun Zhao

Abstract

Soils provide important ecosystem services at the local, landscape and global level. They provide the basis for crop, livestock and forestry production and help mitigate climate change by storing carbon. With expectations of a growing bioenergy supply to meet global energy demand added to the imperative to feed a global population of 9 billion people by mid-century and beyond, coupled with higher per person food demands than currently provided, the challenges to keep agricultural and rangeland soils healthy and productive are daunting. In this paper, we explore the existence of a common pattern in the use of soils under increasing demand for productivity – here described as a *soil carbon transition curve*: a rapid decline of soil carbon due to human clearing of natural vegetation for agricultural land use and management practices, followed by a 'crisis' phase of diminished soil fertility and finally by recovery once agricultural practices improve. We test this pattern in its ability to convey the impact of major land-use changes on soils, with examples from arable, grazing and forest land in different parts of the world. The initial stage of the curve represents a trade-off between extractive productivity (growing crops, extracting biomass, excessive burning and grazing) on the one hand and the farm-level, landscape *plus* global benefits of soil carbon storage on the other hand, based on the loss of an initial endowment of soil organic matter. The second, turnaround stage of the curve often following a crisis tends to be driven by locally relevant loss of ecosystem services, manifest as flooding, pests and crop diseases, erosion and nutrient deficiencies. In the final, recovery stage, local, landscape and global benefits coincide, and synergy between local and global stakeholders' interests dominates the trajectory. In mineral soils used for crop production, recovery generally requires a change in tillage and residue management, crop–livestock integration and/or a return of (agroforestry) trees to agricultural landscapes, plus maintaining adequate nutrient levels. In grasslands, a control of grazing pressure is generally needed in order to recover the vegetation. While global soil carbon storage is linked positively to other ecosystem functions in the recovery stage, the magnitude of the services involved differs substantially between soil types. Incentives for carbon emission reduction and prevention may be drawn towards a small fraction of soils with very high emission potential, especially peatlands, while incentives to maintain or enhance soil carbon storage will be most effective for soil in Stages I and II of the transition curve. Increased carbon storage to mitigate climate change, however, should not be considered the main purpose for improved soil organic carbon management but could be seen as a *co-benefit* of actions that seek local and watershed-level benefits from a full set of improved ecosystem services provided by soil organic carbon.

---

*E-mail: m.vannoordwijk@cgiar.org

Introduction

Soils are – literally and figuratively – the foundation of civilization (Janzen et al., 2011). We derive benefits from soils that are essential for humankind and human development (Chapter 14, this volume). One of these benefits is the production of food of the type, quality and quantity to feed a global population that will exceed 9 billion by 2050 (UN, 2011). A key question facing humanity is whether the quality and health of our planet's soils can be preserved while providing adequate food for this growing population (Lal and Stewart, 2012). Food production through crops and livestock can (but does not have to) lead to soil degradation, such as compaction, erosion, loss of nutrients and organic matter – depending on soil type and management (Chapters 10, 17 and 21, this volume). In addition, there are growing demands for land for various non-agricultural uses, such as urban expansion and soil sealing associated with infrastructure development. Moreover, biofuel demands will compete with food production for the most productive lands, especially with potential incentives for substituting fossil fuels by biofuels as a policy measure to mitigate climate change and ease energy security risks (Rajagopal et al., 2007; Reilly et al., 2013).

Soil formation involves the accumulation, over long periods, of organic carbon, which can be depleted rapidly by various processes (Victoria et al., 2012). It can be released in the atmosphere as gaseous carbon dioxide (CO_2), leaching of soluble organic compounds and lost by wind or water erosion (Amundson, 2001; Chapters 6 and 13, this volume). These mass transformation and transport processes occur at field and watershed scales, but are also key components of the global cycling of carbon and nutrients. Benefits, such as food production and food security, that are linked both locally and globally to soil processes, include supporting biodiversity, maintaining water quality and contributing to (bio)energy security (Chapter 23, this volume). They illustrate not only the connection between local processes and global mass fluxes but also that the interactions that may occur can result in synergies or trade-offs. There are many examples of good carbon management at the local scale for food, fibre and energy production (Chapters 22 and 23, this volume). However, in order to improve land management for greater carbon storage, we need to know better how these processes leading to benefits are aligned.

Our hypothesis is that there is a general set of local–global feedback processes that influence and depend on soil carbon stocks and that occur along a chain of impact between local and global scales. We argue that there is an evolutionary pathway of soil utilization by humanity at local scales that – although differing in detail from place to place – has had a commonality over much of our history since the dawn of agriculture. This pathway, driven by increasing biomass production and intensity of land use, has resulted in a loss of soil organic matter, nutrient and water retention and decreases in productivity. This pathway of soil degradation has global implications for food security, climate change, biodiversity, water quality and biofuel production for energy security for the entire planet, and is a common thread that links soil processes at all scales. The variable that is a dominant control and integrates the impacts of environmental change across all of the soil functions is the organic carbon content in the soil. We now know how to manipulate the delivery of soil functions through improved agroecological practices. Manipulation occurs through methods that will enhance soil carbon and the resulting benefits it accrues. These benefits include greater food production, due to increased nutrient availability in soil organic matter, and buffering of water and nutrients, which generally supports terrestrial ecosystem productivity, increased organic carbon sequestration and greenhouse gas mitigation (Minasny et al., 2011; Chapter 1, this volume).

We explain the evolutionary pathways of local land management as a series of soil carbon transition curves showing local and global implications for soil functions and ecosystem services. The local and global decisions and consequences are subsequently linked through a conceptual model of land (soil) degradation and its recovery. Finally, we integrate these concepts to derive policy implications for global environmental sustainability, while defining and seeking synergies between soil functions that develop through increasing soil carbon stocks.

Land-use Change Pathways and Soil Carbon Transition Stages

Local and global benefits and critical soil services

Land-use decisions are driven by the expected benefits and costs from the perspective of the primary land owner, but they can be shaped by rules and incentives that reflect external stakes in how land is managed. The benefits that humans derive from agroecosystems can be analysed under four groups of ecosystem services: provisioning (allowing the harvesting of 'goods'), regulating, cultural and supporting services (MA, 2005). Many of these are related to soil organic matter content and are often described collectively as a state of 'soil health' (Chapter 14, this volume) or as soil functions (COM(2006) 231, 2006). Table 3.1 gives examples of ecosystem services, distinguishing benefits at three scales: local (the level at which the primary land users operate), landscape or watershed, and global scale (cf. Chapter 15, this volume).

In search of ways to mitigate climate change, the potential of soil to sequester CO_2 at the global level has put a spotlight on the multiple and substantial benefits that can be achieved through increasing soil carbon stocks worldwide (UNEP, 2012a). Additional benefits of soil carbon, beyond carbon storage for mitigation of greenhouse gas emissions, arise principally from the central role of soil organic matter in a range of soil functions that include food production, filtering water, transforming nutrients and maintaining habitat and biodiversity (Chapters 1 and 14, this volume). The role of soil carbon in delivering benefits at each scale can be assessed by considering the impact chain; the linkages between sets of biological, chemical and physical processes that propagate the impact of local changes in soil carbon stocks through the environment. This impact chain begins with the understanding that soil function is related intimately to soil structure – the formation of water-stable soil aggregates that can be influenced positively by increasing soil carbon stocks (Chapter 6, this volume) (Fig. 3.1).

Water-stable aggregates provide physical protection and regulate habitat for all sizes of biota and the associated transformation of carbon and nutrients within the soil, increasing nutrient availability due to improved soil drainage and aeration, as well as energy for microbial heterotrophic processes and belowground biodiversity (Chapters 1, 8 and 11, this volume). The synergy of the above physical, chemical and biological characteristics leads to higher land productivity, improved water infiltration and retention, and nutrient transformations, among other processes. Good soil structure (i.e. high amounts of water-stable aggregates) improves the capacity of soils to resist changes, offering a buffering function (or resilience), resulting in improved regulation of the processes that sustain soil functions and services (Box 3.1).

Fundamental to the above is the recognition that an impact chain exists for each of the ecosystem services listed in the rows of Table 3.1. These impact chains link changes in soil organic matter to the delivery of benefits at different scales. Narratives of a number of these impact chains have been outlined in the literature, demonstrating that gains and losses in soil organic matter at the local level generally are expected to lead to an associated gain or loss of benefits for a range of services at the local, landscape and global levels (Victoria *et al.*, 2012). If this correlation generally holds, it implies a wide-ranging (water and nutrient) buffering role for soil organic matter (Chapter 8, this volume).

Soil carbon transition stages

Most of the ecosystem services that are linked to soil carbon are aligned *positively* with soil organic matter content, and thus

Table 3.1. Ecosystem services for human well-being at various scales as related to soil organic carbon. (From authors, using various chapters in this volume.)

Ecosystem services	Local scale	Landscape/watershed scale	Global scale
Provisioning	Harvested yield based on: (i) decay of soil organic matter that supplies nutrients that feed into crop, timber, animal yield; or (ii) sustainable soil management	Local/regional value chains[a]: conversion, transport, consumption	Global value chains[a] for conversion, production and consumption of food, fibre and bioenergy
	Soil biodiversity (e.g. mushrooms, termites), peat as fuel source	Value chains[a] for specific products	Value chains[a] for specific products
	Soil drainage, surface water, groundwater supplies for local use	Aquifers, drainage network, hydraulic conductivity, lateral flow of water	Blue and rainbow water relations (van Noordwijk et al., 2014)
Regulatory	Microclimatic effects through water balance, albedo, wind	Mesoclimatic effects through wind, albedo and water balance, and based on vegetation, surface properties of soil and soil moisture storage	Terrestrial carbon storage, which plays a role in the global carbon cycle as a part of the global climate system
	Infiltration of water, buffering of soil water	Regulating stream and river flow	
	Reduced erosion (water and wind)	Reduced sedimentation	Reduced impacts of river sediment flux on, for example, coral reefs, aquatic life/fish stocks
	Nutrient buffering	Regulated nutrient exports	Limited lacustrine coastal and marine zone eutrophication
	Water buffering (pH, sediment, chemical load)	Water quality	
	Biological buffering: pest and disease control	Reduced pest and control agent dispersal and migration	Protection of global biodiversity, reduced damage/losses
	Ecological plant protection measures	Reduced agrochemical exposure	Reduced trophic chemical accumulation
Cultural	Sense of place (linked to vegetation, productivity, metaphor of 'being rooted'); stewardship; ecological knowledge	Cultural diversity (including indigenous) linked to integrity of ecosystems; religious expression; archaeological and landscape value connecting to history	Cultural value of food and water; wealth creation, global heritage
Supporting	Soil formation: soil structure and stability	Sedimentation control, soil formation and soil diversity; peatland subsidence and associated saltwater infiltration	River basin and coastal sediment budgets, including mangrove wetlands

[a]Value chains are the way producers and consumers are linked by processing, transport and other forms of value addition, generally with financial flows in return for material goods or value addition through information, quality control or risk insurance.

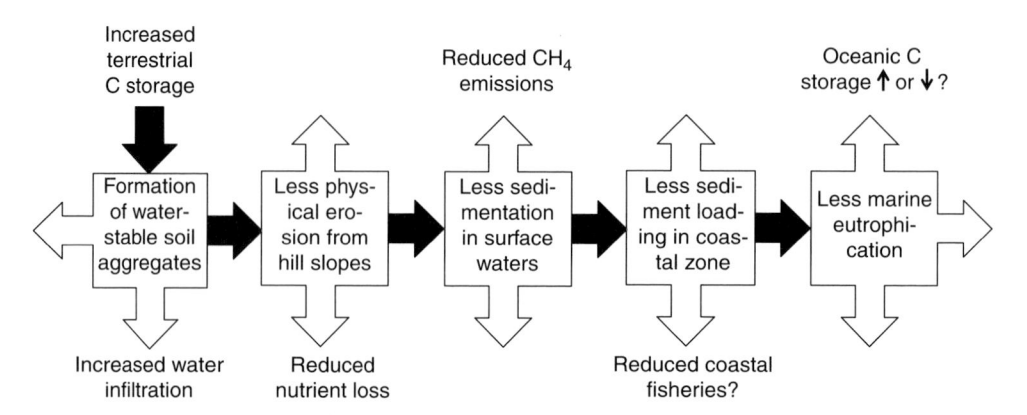

Fig. 3.1. Example of an impact chain that relates changes at the small scale of soil aggregate dynamics at the soil surface to long-term carbon storage at the global scale, with many further relationships unspecified.

Box 3.1. Quantifying the buffering function of soil organic matter

Under the heading of regulating functions, Table 3.1 includes a number of references to 'buffering'. Soil organic matter increases the soil's water-holding capacity in the plant-available range, and hence increases the buffering of vegetation against irregular rainfall or irrigation. It also binds nutrients and buffers plants against irregular nutrient inputs or mineralization. Organic matter may also increase root development and rooting depth, making vegetation more buffered. This buffer is no substitute for long-term shortfalls in supply, however. In advanced horticultural systems with fully regulated water and nutrient supply, there is no need for soil as such, as the buffer functions have been substituted. In a generic way, buffering is defined on the basis of its effect on a response variable of interest. Technically, the reduction of variance (1 − variance with/variance without) is a measure of buffering, independently of how it is achieved (van Noordwijk and Cadisch, 2002). The conceptual definition of buffer can be expressed quantitatively in mathematical models of coupled soil processes that use measured soil properties and where state variables correspond to flows of material and energy at the different scales of impact. This type of calculation quantifies the resilience of ecosystem services that are provided by these flows (biomass, nutrients, clean water, etc.) against changes in environmental conditions.

---

with each other. There is one notable exception, however. Many traditional agricultural systems derive their productivity from the nutrients mineralized during the decomposition of soil organic matter. Many food systems and regional economies have been based on such resource use, without a concurrent replenishment of the soil organic matter. In this first stage of human land use, a negative trend in soil carbon stocks is linked functionally to a positive result, namely agricultural production or harvested yield. However, the depletion of nutrients for crop production through enhanced soil carbon decomposition has limits. When soil organic carbon stocks become very low, the soil structure collapses, and soils become very susceptible to erosion, compaction and flooding due to poor drainage. The US Midwest dust bowl of the 1930s marked the end stage of grain production living off the prairie soil organic matter stocks and started a search for more sustainable soil management that increased soil carbon stocks (see discussion below). This pattern of decline followed by either a crash or a timely switch to more sustainable land management practices appears to be a common story across continents and ecological zones. The curve describing this story of transitions in soil carbon can be divided into three stages – Stage I: 'the fall', Stage II: 'the dip' (or collapse) and Stage III:

'the rise' (Fig. 3.2). The latter is a stage of rebuilding soil carbon stocks while maintaining or increasing agricultural production. The general pattern of 'degradation – crisis – recovery' in the soil carbon transition curve is similar to what has been referred to as the 'forest transition curve' (Meyfroidt and Lambin, 2011) or 'tree-cover transition curve' (van Noordwijk et al., 2011).

Diversity of soils and land use

Figure 3.2 represents a conceptual framework to guide thinking and help develop site-specific conceptual models to assess risks and explore solutions. Any claim of a generic, repeatable pattern such as depicted in Fig. 3.2 is challenged by the vast diversity of soil types and forms of land use. The guidelines of the Intergovernmental Panel for Climate Change (IPCC) for national greenhouse gas inventories provide a grouping of soils with respect to their soil organic matter content and dynamics that may represent the base minimum level of classification. It distinguishes three main texture classes (high-activity clays, low-activity clays, sandy soils) of the main upland soil types covering, respectively, 58.4, 14.08 and 10.05% of total land area, and four classes that have considerably higher carbon content but represent a much smaller area: spodosols, recent volcanic soils, wetlands and peatlands, with 1.63, 0.50, 5.27 and 1.47% of land area, respectively (Batjes, 2011).

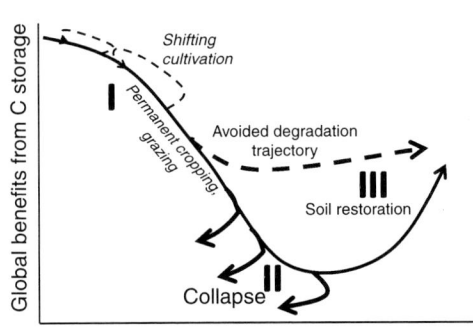

The relative share of these soil groups differs between climatic zones, with high-activity clays dominating in the temperate zone and dry tropics and low-activity clays (oxisols and ultisols) dominating the moist and wet tropics, where leaching and weathering rates are high (Table 3.2 and Plate 3). The mean soil carbon content in the top 30 cm of the profile under natural vegetation ranges from 9 to 143 Mg ha^{-1}, with tropical dry zones having the lowest soil carbon content and cooler regions the highest (Batjes, 2011). In terms of carbon emissions in response to agricultural land use, the tropical peatlands stand out (Chapter 19, this volume). The approach for reducing anthropogenic soil carbon emissions may thus differ among soil types and climatic zones when the focus is on food production and the local benefits of maintaining or restoring soil organic matter. We will explore the diverse experience with the three stages of the soil carbon transition curve in major parts of the world before returning to the issue of generic pattern versus site-specific responses.

The Soil Carbon Transition Curve in Various Parts of the World

The curve shown in Fig. 3.2 is typical for managed soils across all climate zones and arable soil types. In general, land-use alterations lead to organic carbon loss and a decline in soil fertility over periods ranging from a few years to decades or centuries (Stage I). Depending on the local conditions and intensity of land use, this initial decline is followed by a Stage II, which consists of either a collapse or – at best – a stable situation at low soil productivity. The initiation of agricultural practices to increase soil fertility often marks the beginning of a third stage (Stage III), which involves modifications of the original soil to bring more nutrients to the crops and improve other aspects of soil fertility. In the following sections, we give a few examples illustrative of the stages of impacts on both agricultural and rangelands from around the world.

Fig. 3.2. Schematic history of the decline of soil organic matter and loss of its associated ecosystem services as a result of agricultural production that relies on the mineralization of soil organic nutrient capital before alternate and more sustainable management practices arise.

Table 3.2. Soil carbon content (Mg C ha^{-1}) in the top 30 cm of the profile under undisturbed natural vegetation for major soil groups and climatic zones, based on means of global data sets with standardized methods; peatland data are not included. (From Batjes, 2011.)

	Tropical montane	Tropical wet	Tropical moist	Tropical dry	Warm temperate moist	Warm temperate dry	Cool temperate moist	Cool temperate dry	Boreal (undifferentiated)	Polar (undifferentiated)
High-activity clay	51	60	40	21	64	24	81	43	63	59
Low-activity clay	44	52	38	19	55	19	76	–	–	–
Sandy	52	46	27	9	36	10	51	13	–	27
Spodic	–	–	–	–	143	–	128	–	–	–
Volcanic	96	77	–	–	138	84	136	–	–	–
Wetland	82	49	68	22	135	74	128	–	116	–

Stage I: Decline in soil productivity and carbon

A decline in agricultural productivity has been documented historically in different parts of the world. One striking example comes from the Mediterranean region, where the philosopher Plato (427–347 BC) recognized loss of soil fertility in Greece as the effect of human action: 'The rich, soft soil has all gone away leaving the land nothing but skin and bone' (Plato, Critias 3.III). Agricultural decline in ancient Greece and Rome is characterized extensively by Huges (1994), who documented loss of soil fertility through concurrent accounts of the problem. Geoarchaeological surveys confirm the severity of the soil degradation Plato lamented about (Fuchs et al., 2004).

More contemporary accounts of regional Stage I soil productivity decline has been documented in, for example, India. Both the Indo-Gangetic Plains and the black soil region of India were brought to cultivation after removing forests. Between 1860–1920 and 1920–1978, nearly 49.1 and 59.6 million ha (Mha) of land was brought under cultivation (Richards et al., 1993). Conversion of land was completed by the time the Green Revolution began in 1965–1966 (Abrol et al., 2002). The Indo-Gangetic Plains, covering about 13% of the country, produce today nearly half of the food grains for 40% of the total population of India. However, reports of the land use and soils of the Indo-Gangetic Plains indicate a general decline in soil fertility, and long-term soil fertility studies have shown a reduction in soil organic matter and essential nutrients (Abrol and Gupta, 1998; Bhandari et al., 2002). The levelling off of crop yield in the Indo-Gangetic Plains following the Green Revolution started in the 1980s. This trend was linked to soil quality, and specifically to a decline in soil organic carbon, with levels dropping as low as 0.02% (Abrol and Gupta, 1998; Abrol et al., 2002) (for more detail see Chapter 28, this volume).

Conversion from native ecosystems to intensive agricultural cultivation has caused documented carbon losses in China as well. The changes in soil organic carbon content over time in the fertile black soil region in north-eastern China provide direct evidence for Stage I (Wang et al., 2002) (Fig. 3.3). The soil organic carbon content of the non-cultivated black soils (native grass-covered black soils) in the area is as high as 87.3 g kg^{-1}. However, within the first 20 years of agriculture, values dropped dramatically to levels in the order of 45 g kg^{-1} for areas under agriculture. Ultimately, after 100 years of cultivation, the soil carbon content had been reduced systematically to 30 g kg^{-1}.

Rangelands have also seen the effects of human intervention. Cattle grazing, population

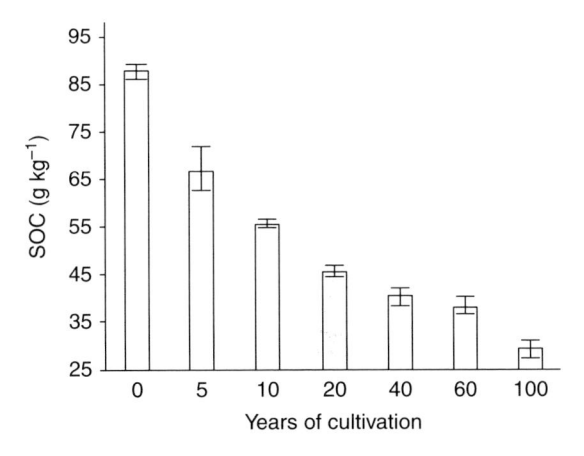

Fig. 3.3. Changes in soil organic carbon (SOC) content with the years of agricultural cultivation in the black soil region in the Beian area, north-eastern China. (From Wang et al., 2002.)

growth, inadequate (self-)regulation and increasing demand for food has led to overgrazing in many African savannah ecosystems (Lal, 2002; Oztas, 2003; Smet and Ward, 2006). The intensive use of these grasslands has led to the loss of valuable grass species (Glasscock et al., 2005), degradation of land cover, decreased carbon stocks and the associated decline of soil biological, chemical and physical quality (Teague, 2004; Noellemeyer et al., 2006; Savadogo et al., 2007; Steffens et al., 2008; Gili et al., 2010). Grazing reduced the total organic carbon content in the soils of the semi-arid rangelands of central Argentina, and specifically in more sandy textures, the carbon loss was noticeable (Fig. 3.4).

Semi-arid grasslands have also been converted extensively to arable agriculture (Priess et al., 2001; Nosetto et al., 2005; Coutinho et al., 2008). Especially in warmer regions, these conversions have triggered rapid and massive losses of soil organic carbon (Zach et al., 2006; Noellemeyer et al., 2008).

Stage II: Low or collapsing soil productivity

There are a number of examples of very rapid soil fertility collapse (Stage II). The dust bowl era in the great plains of the central USA is a case of extreme degradation of cropland resulting from a combination of environmental factors and improper farming practices. The area was covered by dense native grasses with deep rooting systems that had stabilized the loess (glacial windblown) soils and maintained soil organic matter. Deep ploughing to produce wheat crops disrupted this cover and reduced the soil organic matter contents and water-stable aggregates, and exposed tiny loose particles at the bare soil surface to wind erosion. There were initially good yields associated with a wet period in the 1920s, but when this area experienced a major drought in the 1930s, the dry soil was blown away as dust. An estimated 12 cm of topsoil was lost from 4 Mha within the dust bowl region (Montgomery, 2007). The resulting collapse of the farming sector led to the largest population migration in the history of the USA, involving hundreds of thousands of refugees (Worster, 2004).

Elsewhere, soils formed on loess have also exhibited widespread erosion, such as in the semi-arid regions of South America and South-east Asia (Chapter 13, this volume). Although erosion is a natural phenomenon in a semi-arid climate, it has been enhanced significantly by removal of the natural vegetation and cultivation (Shi and Shao, 2000).

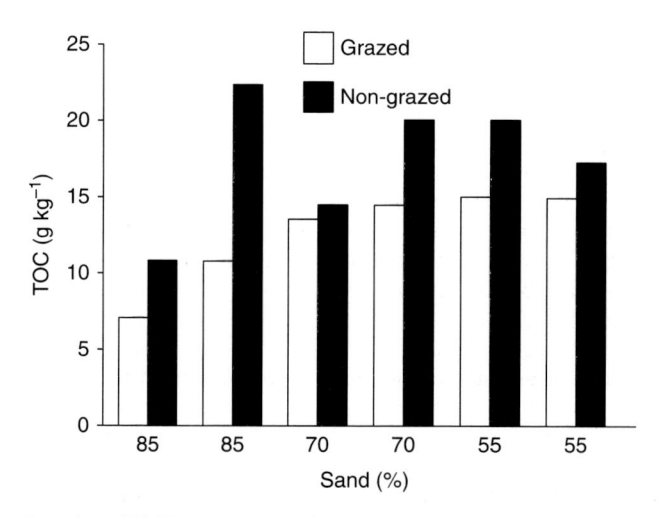

Fig. 3.4. Total organic carbon (TOC) content of surface 20 cm soil in different textures of the central Argentinean semi-arid rangelands under grazing and ungrazed. (From Gili et al., 2010.)

Arable and rangeland soils in semi-arid environments are vulnerable to wind erosion (Hevia *et al.*, 2007; Li *et al.*, 2008) and can exhibit compaction and other physical problems (Hamza and Anderson, 2005; Castellano and Valone, 2007), leading to diminished productivity and higher risk of crop failure (Stage II). Within this context, intrinsic soil properties, such as texture and mineralogy, and environmental conditions (specifically climate) affect the resilience or buffer capacity of soils to mitigate the degradation processes triggered by Stage I land use. Soils with high clay contents in semi-arid tropics sequester more soil organic carbon (Bhattacharyya *et al.*, 2000, 2006), and therefore might be better buffered against the impacts of declining soil organic carbon, whereas sandy soils, even in temperate climates, will degrade more rapidly and are more prone to collapse (Noellemeyer *et al.*, 2006).

Stage III: On the road to recovery

After decades of decline, Minasny *et al.* (2011) documented an increase in soil carbon in agricultural lands across the island of Java, Indonesia, in the past two decades. They attributed this to the effects of soil conservation plus increased intensity of cropping, which resulted in higher root inputs to the soil.

Table 3.3 shows the changes in carbon stock of the selected benchmark locations in the Indo-Gangetic Plains of India in 1980 and 2005, reflecting a positive response to improved methods of cultivation. In general, the increase in soil organic carbon stock was higher in semi-arid and subhumid dry portions of the Indo-Gangetic Plains. Carbon stock changes in the soils of selected benchmark spots in the black soil region due to intensive agriculture are also evident, but the relative increase of soil organic carbon is larger in the Indo-Gangetic Plains than in the black soil region (Bhattacharyya *et al.*, 2008). The application of fertilizer and a combination of fertilizer and farmyard manure has increased the soil organic carbon stock. This observation, found in many experimental sites, was also evaluated by a modelling exercise (Bhattacharyya *et al.*, 2007a).

Changes in soil organic carbon stocks of paddy soils in China indicate that the total stock of carbon stored in the upper 30 cm of soils increased from 2.51 Pg in 1980 to 2.65 Pg in 2008, with an average sequestration rate of 5.0 Tg carbon per year (Xu *et al.*, 2012). This notable increase of soil organic carbon stocks occurred mainly between 1980 and 1994, and was associated with an increase in chemical fertilizer applications. After 1994, the level of soil organic carbon stocked in the soil became relatively stable, with only a slight upward trend observed, although the use of chemical fertilizer still increased (Xu *et al.*, 2012).

In central Argentina, the introduction of no-till agriculture facilitated agricultural management for carbon accrual (Alvarez and Steinbach, 2009). Even when plant residues were partially removed by grazing, no-till soil maintained and recovered soil organic matter (Quiroga *et al.*, 2009).

Adequate grazing management can also reverse the degradation process and lead to a stabilization or even accumulation of soil organic carbon. This is done mainly through rotational grazing schemes that allow plant debris to be accumulated on the soil (Savory and Butterfield, 1998; Hooker and Stark, 2008) or through revegetation strategies (Jiao *et al.*, 2011). Higher net primary productivity achieved through fertilization and improved arable and pasture crops contributes to higher residue returns and a more positive carbon budget (Holeplass *et al.*, 2004).

In vast regions of agricultural land around the world, no-till agriculture was introduced which facilitated the accumulation of soil organic carbon in arable soils (Alvarez and Steinbach, 2009; Chapter 23, this volume). Nevertheless different land-use types condition the amount of soil organic carbon they can possibly attain, and considerably lower soil organic carbon contents are typical for cash crop production in intensive arable agriculture as compared to, for example, animal husbandry systems in extensive grasslands (Berhongaray *et al.*, 2013).

Table 3.3. Changes in carbon stock with time, comparing 2005 with 1980 in selected benchmark spots in the Indo-Gangetic Plains and the black soil region, India (0–150 cm). (From Bhattacharyya et al., 2007b.)

Bioclimatic systems	Mean annual rainfall (mm)	Benchmark spots	Production system (further details in Bhattacharyya et al., 2007b)	SOC stock (Mg ha^{-1}) 1980	SOC stock (Mg ha^{-1}) 2005	SOC change over 1980 (%)
Indo-Gangetic Plains						
Semi-arid	550–850	Phaguwala	Irrigated rice–wheat/mustard/potato/fodder	33.6	54.8	63
		Ghabdan	Irrigated rice–wheat/mustard/wheat–mustard intercropping	26.3	70.4	167
		ZarifaViran	Rice–wheat/mustard	41.3	53.8	30
		Fatehpur	Rice–wheat for four decades	11.1	55.0	395
		Sakit	Rice–wheat for about 12 years	40.5	85.5	111
		Dhadde	Rice–wheat/mustard, sugarcane	44.7	58.4	31
Subhumid	850–1200	Bhanra	Rice–wheat/mustard, etc	18.1	53.4	197
		Jagjitpur		25.2	87.6	248
		Haldi	Rice/maize/soybean–wheat	85.5	62.8	−26
Humid	>1200	Hanrgram	Rice–rice	69.3	110.2	59
		Madhpur	Rice–mustard/potato–rice	39.9	49.7	25
		Sasanga	Rice–mustard/potato–rice	52.5	84.2	61
Black soil region						
Arid	<550	Sokhda	Cotton–pearl millet/sesame	111.9	92.0	−18
Semi-arid	550–850	Asra	Rain-fed mustard/intercropping system	62.9	135.9	116
		Teligi	Monocropping of rice; lowland rice; 7–8 months fallow	74.1	152.0	105
		Semla	Cotton–groundnut rain-fed system	157.8	132.8	−16
		Vijaypura	Rain-fed groundnut–finger millet (3-year rotation period)	77.0	77.0	0
		Kaukantla	Rain-fed castor + pigeonpea strip cropping	47.1	102.5	118
		Patancheru	Fallow land under continuous native grassland	83.9	167.2	101
Subhumid	850–1200	Kheri	Irrigated soybean–wheat production double-cropping system	56.2	105.1	87
		Linga		96.6	129.2	34

SOC, soil organic carbon.

The restoration potential for soil organic carbon is dependent on soil and climate factors. In general terms, high-clay soils will recover more carbon (Riestra et al., 2012), while the capacity of sandy soils to recover is far more limited and in many cases will not be able to attain the original levels of soil organic carbon. Zach et al. (2006) showed that in the semi-arid Pampas, sandy soils levelled off rapidly in their carbon sequestration rate: within just a few years (Fig. 3.5).

In the cases described, the general trend for soil organic carbon to decrease has come to a halt due to the introduction of improved crop management, which includes better use of mineral and organic fertilizer and more timely and less aggressive tillage since herbicides are available to control weeds that before had to be eliminated by tillage.

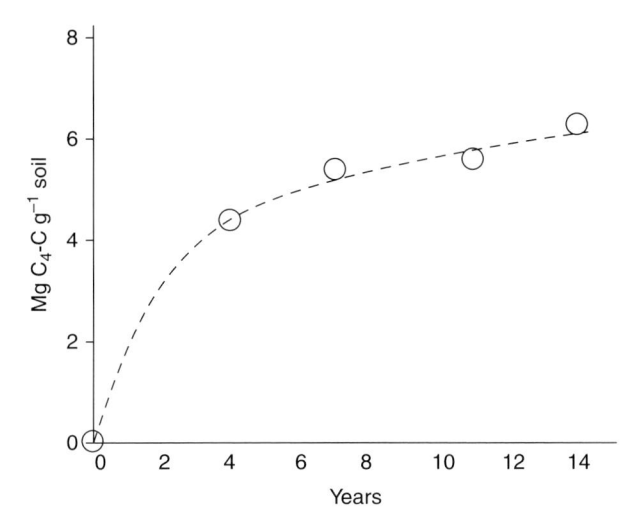

Fig. 3.5. Sequestration of pasture-derived C_4-C (mg C g^{-1} bulk soil) with time of pasture introduction on original C3 vegetation. (From Zach et al., 2006.)

Local versus global effects of land use

Land degradation, and specifically a decline in soil organic carbon, is a typical – if not universal – accompaniment of, or at least the initial stages of, human land occupation. The curve seems a reasonable approximation of reality at all locations supporting human agriculture and grazing; however, in many locations, the degradation has been stepwise, interspersed with partial recovery. Given the scale of human modification of the planet, it is probable that in aggregate these local and regional impacts can have global consequences. Over the past three centuries, 30–50% of global land surface and more than half of the extent of fresh water has been used by humans (Crutzen, 2002; Zalasiewicz et al., 2010). Croplands and pastures now rival forest cover as the major biome on Earth (Foley et al., 2005).

Given the extent of our manipulation, it is reasonable to conclude that the total sum of local and regional changes on the planetary landscape will add up to global effects. For example, the sum of soil organic carbon declines associated with Stage I of the curve of Fig. 3.2 is thought to have contributed to the overall increase of CO_2 in the atmosphere (Smith et al., 2013). Specifically, soils contain vast reserves (~1500 Pg) of organic carbon, about twice the carbon dioxide in the atmosphere and nearly three times as large as the carbon stocks of vegetation. Historically, soils in managed ecosystems have lost a significant portion of this carbon (40–90 Pg) through land-use change – from which some carbon has remained in the atmosphere. On the other hand, the capacity for increases in soil organic carbon during Stage III has the potential to sequester significant amounts of carbon to help mitigate the post-industrial rise in atmospheric CO_2. These competing possibilities present policy makers and scientists alike with important challenges and trade-offs, which will be discussed below.

Turning Trade-offs into Synergies at Local and Global Scales

Connecting the stages of the soil carbon transition curve

The previous section concluded that the soil carbon transition curve provided a conceptual framework for the dynamics of the decline and recovery of soil carbon in many agricultural land-use trajectories around the world. Agricultural productivity is achieved during the mineralization of organic carbon, and specific practices exist that can subsequently drive a turnaround that leads to increased carbon content. In this final section, we will discuss to what degree the many local soil carbon transition scenarios add up to a global pattern and what it implies for the local, national and global policy making that may be wanted or needed in order to manage the trade-offs stage of the curves and the choices to be made in order to enable recovery in Stage III, balanced with avoided damage in Stage I.

Qualitatively, the soil carbon transition curve paradigm aligns with agroecological experience from around the world, but the quantities involved in soil under the natural vegetation at the start of the curve (Fig. 3.2), the relative and absolute decline in Stage I (10–80%, depending on cropping practice and length of time), the thresholds of land abandonment in Stage II and the rates of recovery in Stage III can vary considerably. New questions arise: over what period can Stage III recover half of the Stage I losses? Can Stage III exceed the starting point at the farming system level? It certainly can locally, as shown by anthrosols (i.e. a soil that has been formed or heavily modified due to long-term human activity) and home garden sites (Chapter 10, this volume) with substantial organic inputs derived elsewhere in the landscape. Organic waste from the local processing of agricultural products can also be a source of local enrichment beyond that achieved by the native vegetation's interaction with climate and inherent soil properties.

The decline of soil carbon in Stage I is a consequence of land management decisions that are driven mainly by the opportunity to use productive soils, newly converted from natural vegetation, where the buffering functions provided by high soil organic matter content appear to be in excess of what is needed or appreciated by the farmer, where there has not yet been a build-up of soilborne crop diseases and where declining soil organic matter content provides a nutrient source that makes the use of expensive fertilizers unnecessary. While other types of mining of natural resources are taxed and regulated, this soil resource utilization (mining of soil carbon) tends to be free, or even stimulated and subsidized by governments that see an expansion of agricultural production as key to their development strategies. The subsequent decline of productivity, loss of soil structure, reduced infiltration, initiation of erosion problems and emergence of pest and disease problems tends to be left to the farmer to deal with – with land abandonment and migration to new land as an historical option for which the Earth is now too densely populated.

The examples given in the previous section speak of the potential for a rapid decline in soil organic carbon, whereas recovery may be slow. Preventive action on the decline is strategic and ultimately economical from a macro perspective, even if it involves short-term costs. There is a rationale for the stakeholders at the watershed or landscape scale to influence individual farmer decisions to increase infiltration effectively, if that can reduce the risks of erosion and spillover sedimentation effects, leading to loss of watershed functions. These influences normally come as a combination of three primary policy instruments: regulation, persuasion and incentives (Bemelmans-Videc et al., 2003; van Noordwijk et al., 2012). Community control over individual land access has been the historical base of such feedbacks; land-use planning and zoning decisions are the current ones – contested by a sense that farmers should be allowed to do what they want on land that they own.

Multiple positive processes occur once trajectories change towards enhancing soil organic carbon levels during Stage III. As the case studies illustrate, there are various ways to increase soil carbon levels and deal with trade-offs (Klapwijk et al., 2014). A wide range of practices/technologies and approaches (such as participatory processes in a watershed approach) are available to realize the potential (Chapter 4, this volume). An enabling environment is critical to allow their adaptation and adoption in Stage III. How can policy encourage the actions needed to move from Stage II to Stage III?

Are the direct benefits to the farmer at the local level sufficient; do farmers have long-term security of tenure or are regulatory actions and/or economic incentives needed in order to attain multiple benefits at the local, landscape and global levels? A strategy may be needed to move the importance of healthy, productive soils more to the core of mainstream decision making (Chapter 5, this volume).

Economic incentives to assist in the co-production of public goods often start as additional ways to initiate decisions that lead into the synergy stage of the soil carbon transition curve (Stage III), but they may become seen as entitlements (Swallow et al., 2009). The question is whether the impacts of local soil carbon transition warrant action on the global costs/benefits of reduced/enhanced soil carbon storage, above and beyond what makes sense at the landscape scale. The concept of a global carbon market is built on the idea that it might lead to efficient responses – but a more segmented partial market approach or a compensation/co-investment approach is an alternative (Namirembe et al., 2014; Chapter 31, this volume).

One way of turning trade-offs into synergies is the implementation of good agricultural practice, and increasingly this is becoming a precondition for acceptance in global commodity trade. Good agricultural practice is enforceable in World Trade Organization rules (as non-compliance is seen as an undue subsidy of farming) and as such has a much larger reach than the Fair Trade model, as exists, for example, in the tea, cocoa and coffee markets. The global benefits of increased soil carbon storage may be achieved on the back of such market-based responses – in ways that are unexpected by those who seek measurable increases in soil carbon as quid pro quo in targeted soil carbon payments. Good agricultural practice concepts are location specific. There is, therefore, a need to link practices and policies across scales. One challenge is that local decisions are often more focused on short-term gain rather than long-term benefits (Chapter 2, this volume). Yet a long-term focus is what may be required to move from a situation whereby humanity has lived off the land for decades without sufficient returns towards sustainable synergistic approaches to land use, leading to multiple ecosystem benefits across scales in both the short as well as the long term.

Good agricultural practices for strategic areas of high carbon sequestration potential

Worldwide, there are three situations that have the highest potential for sequestrating carbon in the soil. One case is the vast areas of degraded agricultural lands in semi-arid climates that originally were grasslands, savannahs or tropical dry forests. The second case is tropical wetlands or peats that have been drained and cultivated; beyond avoided losses, these also have a potential to gain carbon when good close-to-natural agricultural practices are carried out. As a third issue, temperate draining of wetlands and management of subarctic permafrosts are of global importance in emission terms.

Taking as an example the semi-arid agricultural lands of North America, the soil organic carbon contents have diminished by, on average, 50% during the arable agriculture period. Similar consideration is valid for arid land in China, Mongolia, Russia, Africa and South America. Despite the inherently low carbon contents of their soils, drylands represent 41% of the global land area, with their use characterized as 45% rangeland and 25% cultivated land (MA, 2005). These regions therefore collectively represent the highest potential for carbon sequestration. The potential to sequester soil carbon is estimated at 0.4–0.6 billion t year^{-1} if drylands degradation worldwide is completely reversed and arrested (Lal, 2001). This rationale is based on the magnitude of previous carbon loss and the intrinsic capacity of these soils to accumulate organic matter – as an indicator of potential future gains that are very high in these soils.

Good agricultural practices are at the core of realizing the potential to sequester carbon in semi-arid arable lands. Bhattacharyya et al. (2007b) have given an overview of some of the techniques and management practices that have been applied strategically in India in arable lands with high carbon potential.

'Subsidizing carbon' and closing the resource loop

Organic matter plays a catalysing role for soil structure, nutrient turnover and other soil functions. The fundamental strategy of restoring soil functions is based on agroecological land-care practices that entail carbon additions. These additions can be either through aboveground vegetation (crop–pasture rotation, cover crops, intercropping, etc.) or through inputs from elsewhere, such as urban and industrial organic waste. Typically, soil ecosystems are restored naturally unless they have passed a degradation tipping point. However, this process can take decades (Zach et al., 2006; Quiroga et al., 2009) or centuries (Nikolaidis, 2011). Reversing the degradation trend and enhancing soil ecosystem services requires significantly more organic matter additions in the short term compared to natural to long-term restoration conditions, and in most of the cases this process of restoration does not reach the original level of carbon stocks (Zach et al., 2006).

A significant increase in the demand for quantities of organic matter as a soil supplement makes the need for organic matter/nutrient recycling an urgent priority. It can also help to reduce resource demand for inorganic fertilizer sources for nitrogen and phosphorus (De Schutter and Vanloqueren, 2011), as the cost of nitrogen fertilizers – which is tied to energy prices for their production and security – is rising and global phosphate rock reserves are diminishing (Cordell et al., 2009; UNEP, 2011a). The global demand for organic matter additions will require a concerted effort in recycling and closing of the nutrient resource loop; for example, between urban areas and peri-urban areas. It will also require organic matter separation at the waste sources, composting of municipal solid wastes and the use of biosolids, as well as the collection and use of agricultural residues and livestock manure. Agroecological practices in conjunction with active organic matter amendments could bring about a synergy and enhance significantly the provision of ecosystem services and increase in food production globally, as illustrated by the African experience, where millions of farmers have doubled their yield following such practices.

Sustainably closing the yield gap

The 'yield gap' is the difference between the theoretical plant physiological maxima for production in the absence of environmental limitations and that which is achieved with the help of currently used technology. Current, climate-adjusted yields are all similarly in the order of 60% of the theoretical maxima for rice in South-east Asia, rain-fed wheat in central Asia and rain-fed cereals in Argentina and Brazil (Godfray et al., 2010). The yield gap is a factor in meeting the world demand for food, which is projected to double by 2050. The actual projected demand for additional food is somewhat less than double, because currently some 30–40% of food in both developed and developing economies is lost to waste. It is assumed that efforts to reduce waste will become an important factor contributing to meeting the projected demand for food (Godfray et al., 2010).

Improved crop and livestock management are needed to address food security issues and restore soil carbon. In doing so, there are multiple synergistic benefits: (i) the benefit of enhanced nutrient supply to crop plants; (ii) improved water use and water quality management; (iii) energy and carbon inputs to support soil biodiversity; (iv) the buffering role to help mitigate the negative impacts of fluctuations in environmental conditions such as extreme weather events and pest infestations; (v) reduced soil erosion; and (vi) the co-benefit of storing carbon, thereby sequestering CO_2 from the atmosphere (Bhattacharyya et al., 2004).

The potential for reduced benefits from soil carbon loss during (marginal) land transition (Stage I) means that methods to eliminate, or at least strongly mitigate, soil carbon loss is a critical challenge to be addressed as marginal lands are brought into production. The key synergies are: (i) gaining local benefits by increasing the income potential from agriculture while mitigating external costs to the environment (reduced water quality, loss of biodiversity, soil erosion); and (ii) contributing to the global benefits of meeting the growing demand for food and avoiding further losses of carbon to the atmosphere. In some cases, these synergies will include carbon sequestration to mitigate climate change (Box 3.2).

Critical thresholds?

At the local level, some measure of carbon stock over time is needed in order to track progress along the transition curve. A minimum level of carbon is critical for soils to be able to deliver multiple benefits (Table 3.1). Minimum carbon threshold values have been established at the local level; for example, for semi-arid Pampas arable lands (Box 3.3). The question is whether such a measure could also be identified for the global level.

Although it would certainly help to gain influence with relevant policy makers if quantitative targets for ecosystem functions could be formulated on the basis of thresholds (Chapter 29, this volume), at this stage there is probably not sufficient science evidence to define such thresholds beyond specific case studies and locations. Plate 4 sketches an outline of how thresholds relative to the carbon levels under native vegetation, supposedly aligned with climate and soil properties, might differ between functions (cf. Chapter 2, this volume). Local expertise and expert opinion will be needed to estimate threshold values, as formal science cannot yet be brought to this level of synthesis in the face of existing variation.

For a number of special soil types, the quantities of soil carbon that can be released to or sequestered from the atmosphere are of the magnitude (or beyond in the case of peat) of changes in aboveground carbon stock in forests. Global carbon mitigation efforts and their potential financial incentives are likely to focus on these areas where measurement and monitoring costs might be justified by the types of changes in carbon stock that are feasible. For the 80% of soils outside of these special categories, economic incentives

---

Box 3.2. Increasing biofuels and food

The potential for land management to harness carbon storage in the soil for climate change mitigation, while supporting increased biofuel production to partially replace fossil fuels, has been explored through scenario analysis using computational simulation (Reilly *et al.*, 2013). The results suggest that specific policy approaches can help to maintain global warming within a 2°C limit during the coming decades – a target that is considered to keep warming within safe planetary limits (Rockström *et al.*, 2009; UNEP, 2011b). The policy scenario allows biofuel production to compete with food production within a carbon market that provides incentives to store carbon in the soil and in vegetation and offset CO_2 emissions from oil and gas. One result projected by this specific policy scenario simulation is to reduce marginally the share of land supporting food production, with the consequence of increasing food prices. The scenario analysis assumed a 1% year^{-1} increase in land productivity.

The thought experiment presented in the text above indicates that even if the yield gap can be closed through improved agricultural technology and land management, there may well be additional requirements for the conversion of marginal lands into agricultural production, with special reference to more populous parts of the globe to feed more people. The potential competition of land within a highly valued biofuel market, as illustrated by the experiment of Reilly *et al.* (2013), presents a further constraint on food production. The co-benefits of maintaining and enhancing soil organic matter content during the conversion of marginal lands for either food or biofuel production are likely to be significant. In addition to mitigating soil carbon loss as greenhouse gas emissions during land conversion, the co-benefits can contribute to closing the yield gap and to reducing external environmental costs by protecting water quality and biodiversity and minimizing soil erosion.

> **Box 3.3.** Local indicators for soil organic carbon
>
> Since soil carbon contents within a region with similar climate, and in soils derived from the same parent material, depend mainly on the proportion of silt and clay size particles and human interventions, a good local indicator is the ratio of soil organic carbon content to silt-plus-clay contents multiplied by 100, using compatible units for carbon and silt-plus-clay contents (e.g. g kg^{-1}). Indicator values of 4.5 could be considered a threshold value below which soil functions would be severely affected, as, for example, shown by barley yield decreases below the average (Quiroga et al., 2006).

derived from global carbon storage issues will at most be a co-benefit of actions that primarily are motivated and financed on the basis of local and watershed benefits.

There is a range of threshold values where the buffer function of soil organic matter is affected negatively. With an increasing climatic variability, and constant or reduced tolerance to the variability of agricultural productivity, it is easy to see that both farmers and other stakeholders will need more buffering capacity to sustain soil benefits, while the actual trend in agricultural systems is towards less buffering. The decline of soil organic matter as the provider of buffering is part of such a broader trend, and the recovery of buffering needs to be part of the answer. The answer is not only of interest from a food production and climate change mitigation point of view, but also from the point of view of water quality, regulating stream flow and aquifer recharge and supporting below- and aboveground biodiversity.

Ecological–technical aspects of buffering soil water and nutrient availability, which influence plant growth and harvestable yield, interact with social and economic aspects of buffering at the level of livelihood strategies of farmers and other land managers. In a rapidly changing world, it is essential to retain the buffering function of soil carbon, to sustain benefits and to cushion these from global change or shock events.

Conclusions

The analysis presented in this chapter suggests the following:

- Soil ecosystem functions and services can be maintained only if sufficient carbon stocks exist in the soil in the form of soil organic matter.
- The 'threshold' or 'transition points' vary with different types of soils and climatic conditions around the world.
- To maintain soil functions, soils should be managed using improved agricultural or agroecological practices that replenish soil carbon.
- In mineral soils used for crop production, recovery generally requires a change in tillage and residue management and/or a return of trees to agricultural landscapes, plus maintaining adequate nutrient levels.
- In grasslands, a control of grazing pressure is generally needed in order to recover the vegetation, potentially followed by the addition of organic matter stocks derived elsewhere.
- Where low potential/marginal land has been converted to agriculture, on-site photosynthesis might not generate sufficient organic inputs to the soil to trigger a recovery, and active amendments, for example with agroindustrial organic waste, might be necessary.
- Societal changes to collect urban organic waste for land application are necessary to recycle sufficient organic matter to close the nutrient loop between urban and rural areas.

In order to integrate efforts and thinking to achieve the required management and policy changes, it may be desirable to work towards a common indicator for measuring minimum carbon threshold levels in the soil in relation to various functions and uses. Soil texture will probably need to be included to allow functional interpretation; alternatively, as in Plate 4, the carbon level under natural vegetation on a similar soil in the local climate and hydrology may serve as a reference. A recent analysis has shown that progress towards meeting environmental goals and targets over the past decades has been very limited (UNEP, 2012b). However, areas where significant progress has been achieved were characterized by a clear problem definition, agreed goals, specific targets and indicators to track progress. As countries and other stakeholders prepare themselves to follow up on the outcomes of the United Nations Conference on Sustainable Development (Rio+20), a science-based contribution to help develop a target for soil carbon stock and mechanisms to track progress towards meeting goals in land-use sustainability could be an important contribution. This effort would fill an important gap in the sustainability framework related to land-use changes and their impact and role in climate change mitigation, food security, biofuels and water supply.

References

Abrol, I.P. and Gupta, R.K. (1998) Indo-Gangetic Plains – issues of changing land use. *LUCC Newsletter* 1, 5–8.

Abrol, Y.P., Sangwan, S., Dadhwal, V.K. and Tiwari, M.K. (2002) Land use/land cover in Indo-Gangetic Plains – history of changes, present concerns and future approaches. In: Abrol, Y.P., Sangwan, S. and Tiwari, M.K. (eds) *Land Use – Historical Perspective Focus on Indo-Gangetic Plains*. Allied Publishers, New Delhi, pp. 1–28.

Alvarez, R. and Steinbach, H.S. (2009) A review of the effects of tillage systems on some soil physical properties, water content, nitrate availability and crops yield in the Argentine Pampas. *Soil and Tillage Research* 104, 1–15.

Amundson, R. (2001) The carbon budget in soils. *Annual Reviews of Earth and Planetary Science* 29, 535–562.

Batjes, N.H. (2011) Soil organic carbon stocks under native vegetation – revised estimates for use with the simple assessment option of the Carbon Benefits Project system. *Agriculture, Ecosystems and Environment* 142, 365–373.

Bemelmans-Videc, M.L., Rist, R.C. and Vedung, E. (2003) *Carrots, Sticks and Sermons: Policy Instruments and Their Evaluation*. Transaction Publishers, New Brunswick, New Jersey.

Berhongaray, G., Alvarez, R., De Paepe, J., Caride, C. and Cantet, R. (2013) Land use effects on soil carbon in the Argentine Pampas. *Geoderma* 192, 97–110.

Bhandari, A.L., Ladha, J.K., Pathak, H., Padre, A.T., Dawe, D. and Gupta, R.K. (2002) Yield and soil nutrient changes in a long-term rice-wheat rotation in India. *Soil Science Society of America Journal* 66, 162–170.

Bhattacharyya, T., Pal, D.K., Chandran, P. and Mandal, C. (2000) Total carbon stock in Indian soils: issues, priorities and management. In: *Land Resource Management for Food and Environment Security (ICLRM)*. Soil Conservation Society of India, New Delhi, pp. 1–46.

Bhattacharyya, T., Pal, D.K., Chandran, P., Mandal, C., Ray, S.K., Gupta, R.K. and Gajbhiye, K.S. (2004) Managing Soil Carbon Stocks in the Indo-Gangetic Plains, India. Rice–Wheat Consortium for the Indo-Gangetic Plains, New Delhi, p. 44.

Bhattacharyya, T., Pal, D.K., Lal, S., Chandran, P. and Ray, S.K. (2006) Formation and persistence of Mollisols on zeolitic Deccan basalt of humid tropical India. *Geoderma* 136, 609–620.

Bhattacharyya, T., Chandran, P., Ray, S.K., Pal, D.K., Venugopalan, M.V., Mandal, C. and Wani, S.P. (2007a) Changes in levels of carbon in soils over years of two important food production zones of India. *Current Science* 93, 1854–1863.

Bhattacharyya, T., Pal, D.K., Easter, M., Williams, S., Paustian, K., Milne, E., Chandran, P., Ray, S.K., Mandal, C., Coleman, K. *et al.* (2007b) Evaluating the Century C model using long-term fertilizer trials in the Indo-Gangetic Plains, India. *Agriculture Ecosystems and Environment* 122, 73–83.

Bhattacharyya, T., Pal, D.K., Chandran, P., Ray, S.K., Mandal, C. and Telpande, B. (2008) Soil carbon storage capacity as a tool to prioritise areas for carbon sequestration. *Current Science* 95, 482–494.

Castellano, M. and Valone, T. (2007) Livestock, soil compaction and water infiltration rate: evaluating a potential desertification recovery mechanism. *Journal of Arid Environments* 71, 97–108.

Cordell, D., Drangert, J.O. and White, S. (2009) The story of phosphorus: global food security and food for thought. *Global Environmental Change-Human and Policy Dimensions* 19, 292–305.

Coutinho, H.L.C., Noellemeyer, E., Jobbagy, E. and Jonathan, M. (2008) Impacts of land use change on ecosystems and society in the Rio de la Plata basin. In: Tiessen, H. and Stewart, J.W.B. (eds) *Applying Ecological Knowledge to Land Use Decisions*. Inter-American Institutue for Global Change Research and SCOPE, São Paulo and Paris, pp. 56–64 (http://www.iai.int/files/communications/publications/institutional/Applying_Ecological_Knowledge_to_Landuse_Decisions.pdf, accessed 15 July 2014).

Crutzen, P.J. (2002) Geology of mankind. *Nature* 415(3), 23.

De Schutter, O. and Vanloqueren, G. (2011) The new green revolution. How twenty-first-century science can feed the world. *Solutions* 2, 33–44.

Foley, J.A., DeFries, R., Asner, G.P., Barford, C., Bonan, G., Carpenter, S.R., Chapin, F.S., Coe, M.T., Daily, G.C., Gibbs, H.K. *et al.* (2005) Global consequences of land use. *Science* 309, 570–574.

Fuchs, M., Lang, A. and Wagner, G.A. (2004) The history of Holocene soil erosion in the Philous Basin, NE Peloponnese, Greece, based on optical dating. *Holocene* 14, 334–345.

Gili, A., Trucco, R., Niveyro, S., Balzarini, M., Estelrich, D., Quiroga, A. and Noellemeyer, E. (2010) Soil texture and carbon dynamics in Savannah vegetation patches of central Argentina. *Soil Science Society of America Journal* 74(2), 647–657.

Glasscock, S.N., Grant, W.E. and Drawe, D.L. (2005) Simulation of vegetation dynamics and management strategies on south Texas, semi-arid rangeland. *Journal of Environmental Management* 75, 379–397.

Godfray, H.C.J., Beddington, J.R., Crute, I.R., Haddad, L., Lawrence, D., Muir, J.F., Pretty, J., Robinson, S., Thomas, S.M. and Toulmin, C. (2010) Food security: the challenge of feeding 9 billion people. *Science* 327, 812–818.

Hamza, M. and Anderson, W. (2005) Soil compaction in cropping systems – a review of the nature, causes and possible solutions. *Soil and Tillage Research* 82, 121–145.

Hevia, G.G., Mendez, M. and Buschiazzo, D.E. (2007) Tillage affects soil aggregation parameters linked with wind erosion. *Geoderma* 140, 90–96.

Holeplass, H., Singh, B.R. and Lal, R. (2004) Carbon sequestration in soil aggregates under different crop rotations and nitrogen fertilization in an inceptisol in south-eastern Norway. *Nutrient Cycling in Agroecosystems* 70(2), 167–177.

Hooker, T.D. and Stark, J.M. (2008) Soil C and N cycling in three semi-arid vegetation types: response to an *in situ* pulse of plant detritus. *Soil Biology and Biochemistry* 40, 2678–2685.

Huges, J.D. (1994) *Pan's Travail, Environmental Problems of the Ancient Greeks and Romans*. The Johns Hopkins University Press, Baltimore, Maryland, 276 pp.

Janzen, H.H., Fixen, P.E., Franzluebbers, A.J., Hattey, J., Izaurralde, R.C., Ketterings, Q.M., Lobb, D.A. and Schlesinger, W.H. (2011) Global prospects rooted in soil science. *Soil Science Society of America Journal* 75(1), 1–8.

Jiao, F., Wen, Z.-M. and An, S.-S. (2011) Changes in soil properties across a chronosequence of vegetation restoration on the Loess Plateau of China. *Catena* 86, 110–116.

Klapwijk, C.J., van Wijk, M.T., Rosenstock, T.S., van Asten, P.J.A., Thornton, P.K. and Giller, K.E. (2014) Analysis of trade-offs in agricultural systems: current status and way forward. *Current Opinion in Environmental Sustainability* 6, 110–115.

Lal, R. (2001) Potential of desertification control to sequester carbon and mitigate the greenhouse effect. *Climatic Change* 51, 35–72.

Lal, R. (2002) Soil carbon dynamics in cropland and rangeland. *Environmental Pollution* 116, 353–362.

Lal, R. and Stewart, B.A. (2012) Sustainable management of soil resources and food security. In: Lal, R. and Stewart, B.A. (eds) *World Soil Resources and Food Security*. Advances in Soil Science, Vol 18. CRC Press, Boca Raton, Florida, pp. 1–10.

Li, J., Okin, G.S., Alvarez, L. and Epstein, H. (2008) Effects of wind erosion on the spatial heterogeneity of soil nutrients in two desert grassland communities. *Biogeochemistry* 88, 73–88.

MA (Millennium Ecosystem Assessment) (2005) *Ecosystems and Human Well-Being – A Framework for Assessment*. Island Press, Washington, DC.

Meyfroidt, P. and Lambin, E.F. (2011) Global forest transition: prospects for an end to deforestation. *Annual Review of Environment and Resources* 36, 343–371.

Minasny, B., Sulaeman, Y. and McBratney, A.B. (2011) Is soil carbon disappearing? The dynamics of soil organic carbon in Java. *Global Change Biology* 17, 1917–1924.

Montgomery, D.R. (2007) Soil erosion and agricultural sustainability. *Proceedings of the National Academy of Sciences USA* 104, 13268–13272.

Namirembe, S., Leimona, B., van Noordwijk, M., Bernard, F. and Bacwayo, K.E. (2014) Co-investment paradigms as alternatives to payments for tree-based ecosystem services. *Current Opinion in Environmental Sustainability* 6, 89–97.

Nikolaidis, N.P. (2011) Human impacts on soil: tipping points and knowledge gaps. *Applied Geochemistry* 26, 230–233.

Noellemeyer, E., Quiroga, A. and Estelrich, D. (2006) Soil quality in three range soils of the semi-arid Pampa of Argentina. *Journal of Arid Environments* 65, 142–155.

Noellemeyer, E., Frank, F., Alvarez, C., Morazzo, G. and Quiroga, A. (2008) Carbon contents and aggregation related to soil physical and biological properties under a land-use sequence in the semi-arid region of central Argentina. *Soil and Tillage Research* 99, 179–190.

Nosetto, M.D., Jobbagy, E.G. and Paruelo, J.M. (2005) Land-use change and water losses: the case of grassland afforestation across a soil textural gradient in central Argentina. *Global Change Biology* 11, 1101–1117.

Oztas, T. (2003) Changes in vegetation and soil properties along a slope on overgrazed and eroded rangelands. *Journal of Arid Environments* 55, 93–100.

Plato, *Critias* 3III. 2014 (427–347 BC)

Priess, J.A., de Koning, G.H.J. and Veldkamp, A. (2001) Assessment of interactions between land use change and carbon and nutrient fluxes in Ecuador. *Soil Science* 85, 269–279.

Quiroga, A., Funaro, D., Noellemeyer, E. and Peinemann, N. (2006) Barley yield response to soil organic matter and texture in the Pampas of Argentina. *Soil and Tillage Research* 90, 63–68.

Quiroga, A., Fernández, R. and Noellemeyer, E. (2009) Grazing effect on soil properties in conventional and no-till systems. *Soil and Tillage Research* 105, 164–170.

Rajagopal, D., Sexton, E.S., Roland-Holst, D. and Zilberman, D. (2007) Challenge of biofuel: filling the tank without emptying the stomach? *Environmental Research Letters* 2, 044004, doi:10.1088/1748-9326/2/4/044004.

Reilly, J., Paltsev, S., Strzepek, K., Selin, N.E., Cai, Y., Nam, K.-M., Monier, E., Dutkiewicz, S., Scott, J., Webster, M. and Sokolov, A. (2013) Valuing climate impacts in integrated assessment models: the MIT IGSM. *Climatic Change* 117, 561–573.

Richards, J.F., Olson, J. and Rotti, R.M. (1993) Development of a database for carbon dioxide release resulting from conversion of land to agricultural use, Oak Ridge (quoted in Abrol *et al.*, 2002).

Riestra, D., Noellemeyer, E. and Quiroga, A. (2012) Soil texture and forest species condition the effect of afforrestation on soil quality parameters. *Soil Science* 177, 279–287.

Rockström, J., Steffen, W., Noone, K., Persson, Å., Chapin, F.S. III, Lambin, E., Lenton, T.M., Scheffer, M., Folke, C., Schellnhuber, H.J. *et al.* (2009) Planetary boundaries: exploring the safe operating space for humanity. *Ecology and Society* 14(2), 32.

Savadogo, P., Sawadogo, L. and Tiveau, D. (2007) Effects of grazing intensity and prescribed fire on soil physical and hydrological properties and pasture yield in the savannah woodlands of Burkina Faso. *Agriculture, Ecosystems and Environment* 118, 80–92.

Savory, A. and Butterfield, J. (1998) *Holistic Management: A New Framework for Decision Making*. Island Press, Washington, DC.

Shi, H. and Shao, M. (2000) Soil and water loss from the Loess Plateau in China. *Journal of Arid Environments* 45(1), 9–20.

Smet, M. and Ward, D. (2006) Soil quality gradients around water-points under different management systems in a semi-arid savannah, South Africa. *Journal of Arid Environments* 64, 251–269.

Smith, P., Haberl, H., Popp, A., Erb, K., Lauk, C., Harper, R., Tubiello, F.N., de Siqueira Pinto, A., Jafari, M., Sohi, S. *et al.* (2013) How much land-based greenhouse gas mitigation can be achieved without compromising food security and environmental goals? *Global Change Biology* 19(8), 2285–2302.

Steffens, M., Kolbl, A., Totsche, K. and KögelKnabner, I. (2008) Grazing effects on soil chemical and physical properties in a semi-arid steppe of Inner Mongolia (P.R. China). *Geoderma* 143, 63–72.

Swallow, B.M., Kallesoe, M.F., Iftikhar, U.A., Van Noordwijk, M., Bracer, C., Scherr, S.J., Raju, K.V., Poats, S.V., Kumar Duraiappah, A., Ochieng, B.O., Mallee, H. and Rumley, R. (2009) Compensation and rewards for environmental services in the developing world: framing pan-tropical analysis and comparison. *Ecology and Society* 14(2), 26 (http://www.ecologyandsociety.org/vol14/iss2/art26/, accessed 15 July 2014).

Teague, W. (2004) Drought and grazing patch dynamics under different grazing management. *Journal of Arid Environments* 58, 97–117.

UN (2011) *World Population Prospects: The 2010 Revision, Highlights and Advance Tables*. Department of Economic and Social Affairs, Population Division (2011), United Nations, New York.

UNEP (2011a) *United Nations Environment Programme (UNEP) Year Book 2011*. Chapter 3: Phosphorus and Food Production. UNEP, Nairobi.

UNEP (2011b) *Bridging the Emission Gap*. United Nations Environment Programme (UNEP), Nairobi.

UNEP (2012a) *United Nations Environment Programme (UNEP) Year Book 2012*. UNEP, Nairobi.

UNEP (2012b) *Global Environment Outlook-5: Environment for the Future We Want*. United Nations Environment Programme, Nairobi.

van Noordwijk, M. and Cadisch, G. (2002) Access and excess problems in plant nutrition. *Plant and Soil* 247, 25–39.

van Noordwijk, M., Hoang, M.H., Neufeldt, H., Öborn, I. and Yatich, T. (eds) (2011) *How Trees and People Can Co-adapt to Climate Change: Reducing Vulnerability Through Multifunctional Agroforestry Landscapes*. World Agroforestry Centre (ICRAF), Nairobi, 134 pp.

van Noordwijk, M., Leimona, B., Jindal, R., Villamor, G.B., Vardhan, M., Namirembe, S., Catacutan, D., Kerr, J., Minang, P.A. and Tomich, T.P. (2012) Payments for environmental services: evolution towards efficient and fair incentives for multifunctional landscapes. *Annals of Review Environment and Resources* 37, 389–420.

van Noordwijk, M., Namirembe, S., Catacutan, D.C., Williamson, D. and Gebrekirstos, A. (2014) Pricing rainbow, green, blue and grey water: tree cover and geopolitics of climatic teleconnections. *Current Opinion in Environmental Sustainability* 6, 41–47.

Victoria, R., Banwart, S.A., Black, H., Ingram, H., Joosten, H., Milne, E. and Noellemeyer, E. (2012) The benefits of soil carbon: managing soils for multiple economic, societal benefits. In: *UNEP Year Book 2012: Emerging Issues in Our Global Environment*. UNEP, Nairobi, pp. 19–33 (http://www.unep.org/yearbook/2012/pdfs/UYB_2012_CH_2.pdf, accessed 15 July 2014).

Wang, J.K., Wang, T.Y., Zhang, X.D., Guan, L.Z., Wang, Q.B., Hu, H.X. and Zhao, Y.C. (2002) Effects of cultivation duration on the changes of black soil fertility qualities [in Chinese]. *Journal of Shengyang Agricultural University* 33(1), 43–47.

Worster, D. (2004) *Dust Bowl: The Southern Plains in the 1930s*. Oxford University Press, New York, 295 pp.

Xu, S.X., Shi, X.Z., Zhao, Y.C., Yu, D.S., Wang, S.H., Tan, M.Z., Sun, W.X. and Li, C.S. (2012) Spatially explicit simulation of soil organic carbon dynamics in China's paddy soils. *Catena* 92, 113–121.

Zach, A., Tiessen, H. and Noellemeyer, E. (2006) Carbon turnover and Carbon-13 natural abundance under land use change in semi-arid Savannah soils of La Pampa, Argentina. *Soil Science Society of America Journal* 70(5), 1541–1546.

Zalasiewicz, J., Williams, M., Steffen, W. and Crutzen, P. (2010) The new world of the anthropocene. *Environmental Science and Technology* 44, 2228–2231.

4 From Potential to Implementation: An Innovation Framework to Realize the Benefits of Soil Carbon

Roger Funk*, Unai Pascual, Hans Joosten, Christopher Duffy, Genxing Pan, Newton la Scala, Pia Gottschalk, Steven A. Banwart, Niels Batjes, Zucong Cai, Johan Six and Elke Noellemeyer

Abstract

This chapter addresses the mismatch between existing knowledge, techniques and management methods for improved soil carbon management and deficits in its implementation. The paper gives a short overview of the evolution of the concept of soil carbon, which illustrates the interactions between scientific, industrial, technical, societal and economic change. It then goes on to show that sufficient techniques are available for the large-scale implementation of soil organic carbon (SOC) sequestration. A subsequent analysis of the bottlenecks that prevent implementation identifies where issues need to be addressed in order to enable robust, integrated and sustainable SOC management strategies.

Introduction

In this chapter, we address the need for the wide-scale implementation of a strategy for improved soil organic carbon (SOC) management. Such a strategy can be denoted generally as innovation, but would also include new methods of governance. In this paper, we define innovation as the improved use of novel but readily available methods or technologies. In addition, it includes a change of the current behaviour by managing SOC in a different way. A broad range of SOC management methods and techniques has been proposed and tested. Many have already demonstrated their applicability and advantages, including implementation strategies (Reicosky, 2003; WOCAT, 2010; Batjes, 2011; FAO, 2011). However, there is still a gap between recognition of the positive aspects of these innovations and the extent of their implementation. We might ask: what are the reasons for delays in implementation? Obvious impediments are temporal and spatial mismatches across scales; that is, long time delays and the large physical distance between the places of implementation and the beneficial effects that result. Discrepancies between private and social benefits and the costs of SOC management are also crucial. To be successful, innovations must pass certain thresholds in the rate of adoption in order to be self-sustaining. The widespread adoption of an innovation is influenced strongly by cultural, social and economic factors such as issues surrounding communication and ways of managing the time delay associated with the achievement of benefits (Rogers, 1962).

*E-mail: rfunk@zalf.de

After a short sketch of the evolution of the soil carbon concept, illustrating the interactions between scientific, industrial/technical and societal/economic changes, we show that sufficient techniques are available for the implementation of large-scale SOC sequestration. Furthermore, we discuss the discrepancies between knowledge of available techniques and implementation. An analysis of the bottlenecks to implementation then identifies where decisions have to be made in order to enable robust, integrated, sustainable and overarching SOC management strategies. One example is provided by the potential to restore SOC stocks where they have been depleted on large geographical scales, such as the cropland soils of the USA (Ogle et al., 2003, 2006, 2010). The observed increase of SOC in these soils can be attributed mainly to the conversion of annual cropland to grassland, a strategy that has been supported by the US federally funded Conservation Reserve Program (Ogle et al., 2010).

A Short History of the Soil Carbon Concept

A comprehensive and detailed review on the history and evolution of soil organic matter (SOM) concepts is given by Manlay et al. (2007). The concept of SOC (see Chapter 14, this volume, for the distinction between SOC and SOM) and its importance to soil fertility and sustainability is now well established in the scientific and the agronomic community. However, this has not always been the case. The recognition of SOM as a concept is linked closely to the understanding of its functions, ecologically and in terms of sustainable and productive cropping.

The term 'humus' was regarded as a precursor for SOM. The word 'humus' was used descriptively for different constituents of the soil. Its meaning changed over time and varied in different parts of Europe. In the 19th century, a more structured approach to defining the meaning of 'humus' was taken, with efforts coming mainly from forestry research. Different 'humus horizons' and their development and functions for forest growth were analysed and described. At that time, understanding of the biological origins of humus began to be based on observations of the different humification stages of organic remains (Manlay et al., 2007). The 'humus principle' gave an even wider, more philosophical and holistic perception of humus, defining humus as a principle that signified the relationship between 'the living earth and other organisms' (Rusch, 1968).

Initial understanding of the role of SOM in plant growth did not differentiate between plant nutrients and carbon sources. It was believed that plants relied on soil as a source of carbon (carbon heterotrophy) and nutrients. This led to the assumption that soils should be tilled as frequently as possible to grind the soil particles as small as possible to enhance plant root uptake. At the end of the 18th century, some experimentalists demonstrated the gaseous origin of carbon and discovered the role of light in photosynthesis. However, their findings were highly disputed (Manlay et al., 2007). During that time, and independently of the ongoing dispute, Thaer developed the first comprehensive fertilization scheme for sustainable agriculture which was based on the whole plant–soil system. Thaer's systematic approach to assess soil fertility and to derive a rational fertilizer scheme included factors such as soil texture, lime and humus content, crop species, organic fertilization management and yield (Feller et al., 2003).

The concept of the cycling of organic and mineral substances was central in Liebig's humus theory of soils. However, the integrative role of SOM moved into the background through the recognition of the role of 'salts in soils' as plant nutrients. Liebig was the first to synthesize this new knowledge into his *Theory of Mineral Nutrition of Plants* (Liebig, 1840). Liebig's concepts can be regarded as the basis of modern agricultural science, accelerating the production and use of mineral fertilizers. However, the role of SOM as vital for plant growth was not abandoned completely, and many scientists at that time held an intermediate position regarding the role of mineral and organic fertilizers (Manlay et al., 2007).

Knowledge of the biogeochemical cycles of carbon and nitrogen advanced rapidly during the beginning of the 19th century when the role of microorganisms in the turnover of carbon and nitrogen was discovered. The appreciation of the integrative function of SOM was initiated with further insights into the role of clay in forming organomineral complexes and aggregate formation. This cumulated in the works of Waksman, who characterized humus as a 'source of Human wealth on this planet' (Waksman, 1938).

The widespread industrial production of chemical fertilizers improved crop yields tremendously, but also led to serious environmental degradation. The demand for food from the explosively growing cities created new carbon pathways from the rural production areas to the urban centres of consumption, breaking the traditional approaches of recycling organic matter. The dust bowl of the 1930s painfully showed the unintended consequences of new agricultural practices that were unsustainable. Severe drought combined with the conversion of grasslands to cropland, simplified crop rotation and deep ploughing of virgin soil contributed to the destruction of soil aggregates, loss of organic matter and wind erosion of topsoil, illustrating the significant relationships between soil, land surface and climate. In the 1940s, the environmental degradation legacy of the industrial era led to renewed interest in SOM research for improving understanding of nutrient cycling and ecosystem functioning. The emergence of ecosystem science, founded on the concept of energy and nutrient cycling, gave momentum to a more holistic approach to material and life sciences.

It was during the Industrial Revolution that the 'greenhouse gas effect' was proposed as a mechanism for controlling the temperature of the planet (Arrhenius, 1896; Callendar, 1938). The idea was controversial until Keeling et al. (1976) observed a rapid rise of CO_2 concentration in the atmosphere. The recognition of the linkage between atmospheric and soil carbon came even later when the exchange between these carbon stocks was identified.

With the Millennium Ecosystem Assessment (MA, 2005), ecosystem services became one of the core issues of modern ecology. Ecosystem services comprise the wide range of benefits that humans derive from ecosystems and provide a conceptual framework for sustainable management (MA, 2005). The MA distinguishes between provisioning, regulating, supporting and cultural services. Soils fulfil a wide variety of environmental services as a foundation for biomass production, a filter and buffer for water, an archive of natural and human history and an important store of carbon. In 1997, Daily et al. suggested six major services associated with soil organic matter (SOM), including physical support, water buffering, nutrient retention, waste/OM recycling, soil fertility renewal and regulation of element cycling. Schmidt et al. (2011) mentioned soil fertility, water quality, erosion resistance and climate mitigation as important ecosystem services related to SOM. Soil biodiversity is seen as an important prerequisite for providing high-quality soil ecosystem services (Brussaard et al., 2007). The concept of ecosystem services facilitates the assessment of the impacts of management (e.g. past innovations such as mineral fertilizer use) in an integrated, holistic way.

What Ought to be Done? A Summary of Best Practices

Figure 4.1 presents a conceptual model change in SOC over time based on the 'soil carbon transition curve' (Chapter 3, this volume). It shows the decline in SOC during conventional agricultural land use and its degree of restoration depending on the chosen carbon sequestration techniques. As a result of unsustainable land-use practices or disturbance (Fig. 4.1, arrow 1), the content of soil organic carbon will decrease substantially in conventional agricultural systems until a lower stable situation, where $C_{input} \sim C_{output}$, is reached (Fig. 4.1, arrow 2)

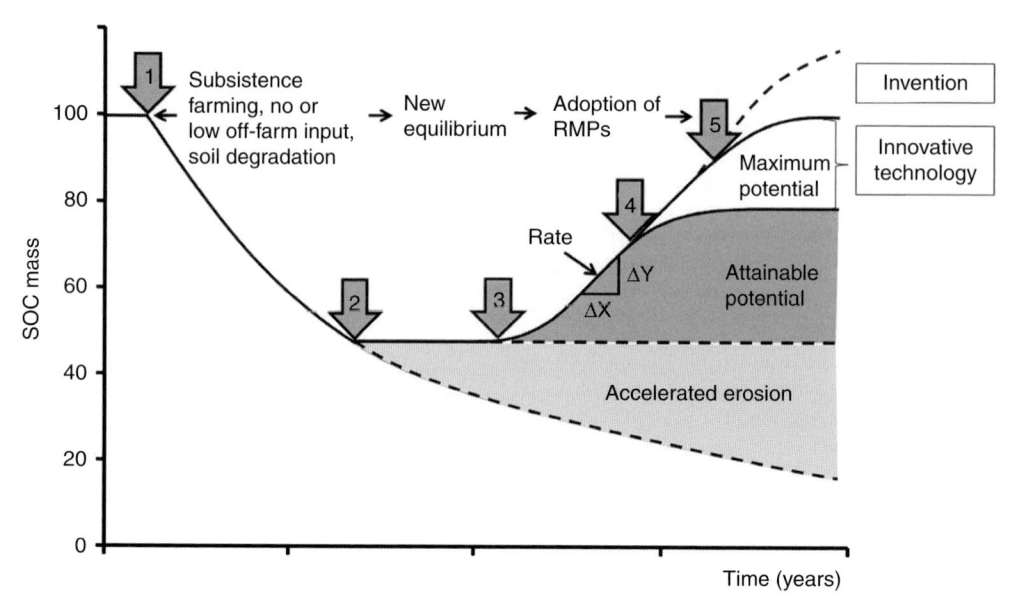

Fig. 4.1. The soil organic carbon transition curve: (1) naturally stable level; (2) reduced unstable level under conventional land use; (3) application of standard widely known and tested sequestration techniques; (4) application of existing innovative sequestration techniques; (5) application of hitherto unknown sequestration techniques. (From Lal, 2008.)

(Johnson, 1995). Alternatively, in a worst-case scenario where C_{input} remains smaller than C_{output}, the system may ultimately collapse as described by Lal (2008b) and others (Chapter 3, this volume). On adoption of recommended management practices (RMPs) (Fig. 4.1, arrow 3), whereby $C_{input} > C_{output}$, SOC levels may again increase until a new stable level is reached ($C_{input} \sim C_{output}$). This 'attainable potential' can be achieved with widely known and tested sequestration techniques (Table 4.1). Similar considerations apply when innovative sequestration techniques (Fig. 4.1, arrow 4) have to be implemented to reach the next level of restoration, termed maximum potential in Fig. 4.1. A rise in carbon content above the natural reference (Fig. 4.1, arrow 5) may require new unprecedented conditions.

The benefits of increasing SOC content in the soil are wide-ranging and well known (see Smith et al., 2008; Victoria et al., 2012; Bationo et al., 2014; Chapters 10 and 14, this volume). Suitable management practices to build up SOC are those that increase the input of organic matter to the soil and/or decrease the rate of SOM decomposition (e.g. Johnson, 1995; Paustian et al., 1998; FAO, 2011). The most appropriate practices are site specific, being adapted to soil type and land-use system (e.g. Batjes, 1998; Ingram and Fernandes, 2001). The magnitude and rate of SOC sequestration that may be achieved depends on several factors, including the reference SOC stock (for a given land unit), land-use history, soil type (depth of soil, clay content and mineralogy, internal drainage/aeration, soil nutrient status) and soil and water conservation practices. The total (project) area where intervention will take place should be divided into homogeneous units (e.g. climate, soils and land use) termed 'strata'. Defining these units, i.e. stratification, allows researchers to obtain precise estimations at a lower cost than treating the entire region as a homogeneous unit (see McKenzie et al., 2002; IPCC, 2006; Ravindranath and Ostwald, 2008).

Technological land management options for enhancing soil carbon sequestration generally include a judicious combination of:

Table 4.1. Examples of land-use and soil management strategies to sequester organic carbon in the soil.

System	Land-use practices
Land use	Afforestation; permanent crops; improved pastures with low stocking rate; multiple crop systems; land restorative measures (e.g. use of chemical fertilizers, planted fallows, erosion control); conversion of marginal arable land to grassland, forest or wetland; restricted or adapted agricultural use of organic soils; paludicultures; wetland restoration; intensification of prime agricultural land (e.g. erosion control, supplemental irrigation; soil fertility management; improved crop assortment; reduction of fallow)
Farming systems	Farming systems with high diversity (e.g. mixed farming, agroforestry, silvopastoral and agri-silvopastoral systems)
Tillage	Conservation tillage, mulch farming, reduction of plough-intensive systems
Fertility maintenance	Judicious use of fertilizers and organic amendments; improving fertilizer use efficiency; nutrient cycling through cover crops and planted fallows; enhanced biological nitrogen fixation
Pest management	Integrated pest management (IPM), selective use of chemicals

1. Tillage methods and residue management (e.g. conservation tillage, cover crops, mulch farming).
2. Soil fertility and nutrient management (e.g. for macronutrients and micronutrients, strengthening nutrient cycling mechanisms to minimize losses).
3. Water management (e.g. supplemental irrigation, surface and subsoil drainage, soil-water management, water harvesting).
4. Erosion control (e.g. runoff management with terraces, vegetative barriers, soil surface amendment and mulch farming).
5. Crop selection, plant breeding (improved varieties) and rotation.

Depending on the land system, and agroecological setting, various combinations of these strategies may need to be prioritized (Table 4.1).

Table 4.1 shows that numerous, mainly conventional, practices exist for improving soil and water management in order to increase the concentration of SOC. Any increase in SOC content, that is soil organic matter, is likely to have an overall beneficial impact on soil ecosystem services (Chapter 2, this volume). The question is therefore: if more SOC is associated with overall societal benefits, why is its sequestration not massively pursued? What limits the implementation of well-known techniques? What are the bottlenecks? And: what kind of innovative techniques and procedures are needed to change the situation?

What are the Bottlenecks to Implementation?

The question is why, despite the knowledge we have about how to enhance SOC technically in different land systems (see above), such knowledge is not being sufficiently put into practice? We posit that the key reason is the mismatch between private and social benefits and the costs of SOC management across temporal and spatial scales. While the longer-term interests largely run parallel among the larger spatial scales (from the global to the regional/national) and organizational scales (global to national political institutions), the shorter-term interests of localized individual land users diverge from sustainable SOC management.

This implies that it is necessary to consider SOC improvements across spatial and temporal scales (Chapter 3, this volume; Table 4.1):

- in space, i.e. on the farm, catchment, national and up to the global scale;
- in time, i.e. from seasonal management operations to intergenerational periods of decades that address sustaining soil ecosystem functions in the longer term.

The chain between decisions and impacts across spatial scales is presented in Chapter 3, this volume, and is summarized briefly here. Soil ecosystem functions operate and are managed directly at the scale of the farm and the catchment because local inhabitants (farmers, nature conservationists) undertake the soil management. Hence, the maintenance of soil fertility and soil health at the field/farm level is inherently linked to the maintenance of SOC at that level. However, different priorities for soil management prevail at the two different scales. While at farm scale the individual profit and subsistence assurance of farmers drive soil management actions, at catchment scale overarching soil functions such as filter and buffering functions move into the focus of societal demands, and issues such as large-scale erosion protection, provision of clean drinking water, protection from desertification and extreme events will drive decision making. Such soil functions are useful in a meaningful way only at this larger scale. At the catchment level, processes such as deforestation, desertification and extreme climate events have a direct impact on SOC through the reduction of C input and the increase of C output from the soil. Furthermore, erosion is increased by these processes and in turn affects the quantity, quality and distribution of SOC across the catchment (Lal, 2005; Doetterl et al., 2013).

Most actions which benefit SOC at the local scale provide benefits at the national and global scales and can simply be aggregated. If all single farms are prosperous, the catchment and the nation are also prosperous, and vice versa. However, some soil ecosystem functions only become meaningful at a larger scale; for example, climate change mitigation by avoiding SOC losses, reducing greenhouse gas (GHG) emissions and sequestration of SOC. Such ecosystem functions can only be realized when practices are implemented on many farms simultaneously.

All of the SOC management strategies mentioned above also need to be dealt with at the national level, as this is often the administrative unit at which political decisions are made. This can be done in an aggregated way if there are no properties emerging at the national level that are not already addressed at the catchment or farm level. At the global level, climate change is a major issue. At the local level, however, with respect to SOC management, climate change is considered much less important than land use and land-use change.

At the catchment level (or administrative 'region'), SOC stocks are influenced by land-use planning including the preservation of aesthetic values, urbanization linked to rural urban migration and the need for transportation corridors. Such regional topics are often addressed by local actors such as local-level public agencies (e.g. planning agencies), conservation and development NGOs and local business. A region is usually made up of a matrix of farms, forests, urban and infrastructure areas. The regional level is key, as it is at this spatial scale where farmers' behaviour and national policy interact.

The main issues at the national scale relate to improving trade balances: for example, through developing export-oriented agricultural incentives; national security, involving the need to produce sufficient cheap food to feed increasingly urban populations; and securing the provision of sufficient energy and water sources. All these objectives are linked to how SOC is managed at the lower scales. It is at the national scale where regulations are designed and implemented that affect directly the constraints and opportunities for alternative SOC management strategies at the farm and catchment level. Of course, such policy decisions are influenced by higher-order topics at the global level. For instance, migration and land acquisition processes can both constrain or enable national policy.

Crucially, the socio-economic issues cut across all scales in a nested way. What may be a prime concern for a farmer (e.g. livelihood vulnerability) may be influenced by rural urban migration processes at the catchment level or by a higher-order food security objective at the national level. The latter may, in turn, be affected by global issues, such as climate change policies and economic globalization. Likewise, farmers' land-use options are determined by global institutions feeding through national policies and regional land-use planning.

On the temporal scale, the main objectives of land use on different spatial scales operate less in parallel and depend more strongly on the cultural, social and economic context. In the short term, maintaining soil fertility may not be the primary concern and may even be jeopardized, as securing yields and maximizing income may have a higher direct priority (Table 4.2). Making decisions based on achieving long-term, stable revenues requires surplus/capital, as long-term benefits do not pay off immediately. If there is no buffer to protect short-term benefits, long-term actions are less likely to succeed. Payments for carbon can contribute to cash income and may enable smallholders to overcome some initial project investment costs (e.g. Palmer and Silber, 2012). However, with the current low prices for carbon on the carbon-offset market such incentives may prove inadequate (e.g. Smith *et al.*, 2007; Grace *et al.*, 2012).

In industrialized countries, farmers' key concern may be to secure the viability of their farming operations, while in developing countries or under pressure of poverty, short-term food and water security are higher priorities. Soil carbon may indeed be managed to enhance the adaptive capacity and to reduce the vulnerability at the farm level to exogenous changes (climatic, economic, demographic), but these objectives are often traded-off against managing the immediate short-term threat. It is thus important to note that biophysical and socio-economic concerns interact and sometimes counteract in complex ways. The farm system is a complex social-ecological system, and finding solutions that enhance aims across temporal scales requires integrated approaches.

Plate 5 illustrates that the socio-economic effects (in terms of actors affected) that arise from the problem of SOC loss (shaded red) are greatest at the bottom of the spatial scale; that is, where most individual land users are located. However, such socio-economic problems are also manifest at the global level, although fewer actors are dealing with such problems (e.g. multilateral development agencies). The biophysical issues show the opposite pattern as we move from the local to the global scale, i.e. the biophysical problems mostly become manifest at the local scale. The meso scale (national scale) is associated with fewer (national level) agents being affected by SOC loss (both in terms of socio-economic and biophysical considerations).

Most of the existing 'best practices' occur at the lower scales and are related mostly to biophysical/technological innovations. Fewer best practices are found as we move upwards in the spatial scale.

Plate 5 has the shape of an hourglass to illustrate the following points:

Table 4.2. Objectives of land use at various scales in space and time. Grey highlight indicates short-term objectives that may conflict with long-term objectives.

	Short-term objectives			Long-term objectives		
	Socio-economy	Soil system	Soil carbon	Socio-economy	Soil system	Soil carbon
Farm	Securing/ maximizing profit	Yield maximization	Utilization	Prosperity	Fertility resilience	Maintenance/ increase
Catchment	Securing/ maximizing profit	Resource optimization	Exploitation	Viability	Diversity/ stability	Maintenance/ increase
Nation	Economic growth/ employment	Balanced productivity	Maintenance	Prosperity, geopolitical influence	Sustainability	Maintenance/ increase
Global	Freedom from armed strife/peace	Balanced productivity	Maintenance	Security and stability	Sustainability	Maintenance/ increase

1. At the base (local scale), the majority of actors affected by the biophysical and socio-economic factors associated with soil carbon are the multitude of farmers that depend on SOC directly, and also consumers (albeit in an indirect way).

2. At the global level, actors are affected mostly by and perceive the problem of soil carbon loss as a socio-economic issue. At this level, the institutional (governance) context is dominated by a few actors (e.g. UN bodies, multilateral treaties, multinational companies, etc.), but their objectives and perception of the problem of SOC loss is fundamental, as they create the policy environment that constrains actors at the national scale. That is, the national level where the number of actors is reduced to few policy organizations effectively creates a bottleneck through which the supranational institutional constraints (international trade rules, international environmental agreements, etc.) are passed down to the lowest scales.

The shape of the hourglass reflects not only the number of actors that are associated with the potential to design and implement innovative solutions but also the degree to which such best practices are currently available (cf. the section above headed 'What ought to be done? A summary of best practices'). Hitherto, most innovation has been focused on the technological domain and designed to be applied at the farm and regional scale, although fewer and weaker, socio-economic practices exist at the national (land-use regulations) and fewer still at the global (e.g. international land-use conventions and corporate governance agreements) level.

As the biophysical limitations are concentrated at the farm and catchment level, innovations to address these limitations generally have to be applied at these scales, but in fact they often are not. Two main reasons constrain application: (i) a lack of information and knowledge about practices to maintain SOC in some parts of the world; and (ii) socio-economic constraints that prevent implementation and widespread adoption, even if information and knowledge is available and accessible. In both cases, innovations are required that address the social, cultural and economic barriers.

While current best practices are associated mainly with innovative technologies addressing biophysical aspects regarding SOC (green shading on Plate 5), these are unlikely to achieve their potential if robust social innovations (red shading on Plate 5) at higher scales are not put in place. The figure thus depicts a need to develop and implement these higher-scale social innovations, which will require the restructuring and adaptation of powerful institutional structures (e.g. the World Trade Organization, United Nations Framework Convention on Climate Change, multinational companies, etc.).

Innovations are particularly needed to bridge the current gaps between short- and long-term objectives. Such innovations must be both technological and social. Social innovations relate mainly to new types of governance structures in the public and private sectors so that policies at higher spatial levels, in governments or companies, filter down effectively towards the lower scales and ultimately reach the consumers and the farmers, who can bring about SOC sequestration efficiently.

Implementation of the necessary technical and institutional innovations will generally require a trigger, an economic feedback from a higher level than the local level where the carbon management activities are implemented (e.g. Koning et al., 2001). Changing prices, subsidies and taxes, laws and regulations and reacting on increasing awareness of environmental costs and unsustainability could be examples of these feedbacks (see Izac, 1997; Sanderman and Chappell, 2013; Chapter 3, this volume).

As noticed by Sanderman and Chappell (2013), the 'maximum feasible C sequestration potential at any given location will seldom be realized due to a series of biological, physical, social, and political constraints'. Often, in the absence of policy interventions, it will be rational for individual farmers to manage their SOC at levels that are suboptimal from a national and global natural capital perspective (Izac, 1997; Chapter 15, this volume).

Depending on the location on the 'degradation curve' (Fig. 4.2), innovative and more demanding types of interventions that combine financial, social and political actions will be needed to bring back SOC to its original levels or to avoid further degradation. The timescales required for these processes are yet unknown, but certainly are longer than the planning horizons of current environmental programmes.

Which Innovations are Needed?

In order to realize the potential of SOC benefits, innovations are required. Innovation, as defined here, differs from invention, as it includes the use of existing knowledge, approaches or technologies, assembling them in a new or more efficient way. Hence, innovation does not necessarily require the creation of new methods or technologies but can result from new ideas of how to integrate and assemble existing approaches in a novel, more effective way. In many degraded sites, SOC stocks could be reverted towards their natural levels once appropriate known technologies presented in the section above 'What ought to be done? A summary of best practices' are applied. Here, we discuss how to achieve enhanced soil carbon accumulation by the improved integration of existing management practices and policies.

Adoption of carbon sequestration technologies at the farm level (section above 'What ought to be done? A summary of best practices') needs a combination of access to information, technologies, financial incentives and capable personnel to link the benefits of individual farmers to those of the adjacent scale(s) (Fig. 4.3) and to make the farmer's direct advantages compatible with his own long-term interests. Such schemes would offer financial and technical support to the farms, while bringing environmental services from the farm scale to other higher scales (Fig. 4.3).

Such schemes could require farm-scale monitoring with low-cost sensors to provide a direct measure of the state of SOC, which could then be processed at a regional level in order to follow stock changes at regional, national and global scales. At the national scale, the evaluation of data from different regions could drive soil and water conservation practices to achieve agroecological goals at the regional level, whereas global financial support could be provided to those who have achieved SOC accumulation through adaptations. By means of trade agreements, international organizations would return to countries not only financial support but also knowledge and technology to stimulate national actions for SOC conservation. This support, received by nations, should go back to the farm level, as technical and financial aid.

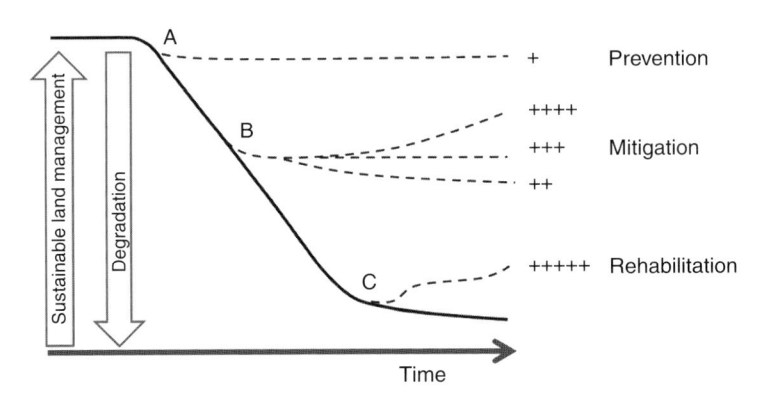

Fig. 4.2. Initial societal implementation investment cost (+) of sustainable land management interventions needed to restore SOC and reduce GHG emissions. (From WOCAT, 2007.)

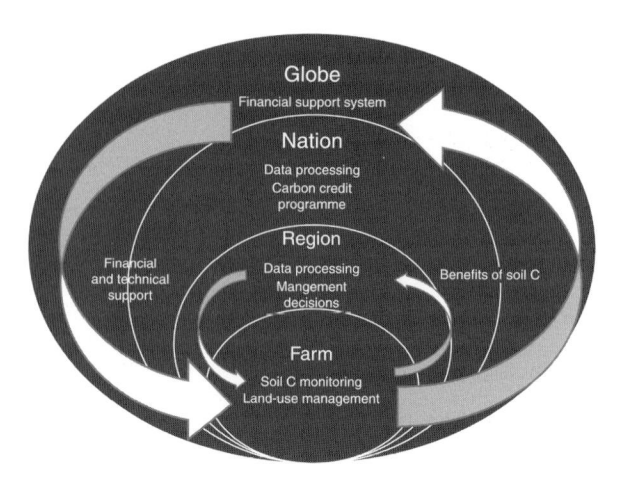

Fig. 4.3. Soil carbon accumulation mechanisms from farm to global level.

Table 4.3. Examples of innovations to be applied across spatial scales and the associated challenges.

Scale	Innovations	Challenges
Farm	Virtualized environmental services	Access to data
	Farmer-specific application	Data availability
	Site-specific management	Technical barriers
	Low-cost sensors	Financial access
	Local demonstration	Innovators/early adopters
Catchment region	Stakeholder accessible remote and proximal sensing monitoring	Access to data
		Mobilizing stakeholders
	Decision support system (DSS) for integrated watershed management	Financial incentives/support
		Capacity building
	Stakeholder networking	Online open access teaching facilities
	Virtual technical training	Network and facilities for recycling crop residues on soil
	New curricula for technicians	
	Soil carbon conservation programmes	Pressure on arable land resource
	Payments for soil carbon services (PSOCS)	
Nations	Economic incentives	Unfavourable macroeconomic conditions
	New legal frameworks (regulating soil protection)	High administrative costs
		Weak governance (lack of organization, corruption)
	Carbon offset markets	
	Supporting new research (agroecology, biotechnology)	Monitoring cost
		Tradition
	Monetarization of soil carbon benefits	
Supranational Global	International organizations	Power structures
	Regulatory framework	Geopolitical interests
	Trade agreements	
	Knowledge networks	

The socio-economic costs of restoring soil carbon require increased awareness of degradation processes, as well as increasing investment. Monitoring of SOC stocks is needed to track maintenance and enhancement, but also degradation (Table 4.3).

Hence, the innovation needed for improved soil carbon management implies bringing the farm to the global and the global to the farm level in a synergetic way, across scales. Examples of instruments that aim to operate in such a way include the

Kyoto Protocol Clean Development Mechanism. However, this type of approach could be more focused on soil carbon and the social, economic and ecological benefits derived from it.

Conclusions

The benefits of maintaining and increasing the SOC content of agricultural soils are clear and in demand worldwide. In particular, agriculture has the potential to contribute significantly to reduce carbon emissions from, and sequester carbon into, soils by land management practices. In addition, these practices increase soil fertility and the resistance to erosion and can improve water quality and flood protection, while enhancing biodiversity and habitats. Despite the clear benefits, the implementation of appropriate measures is far behind the need. Obvious impediments are temporal and spatial mismatches across scales, and discrepancies between the private and social benefits and costs of SOC management. To be successful, and self-sustaining, innovations have to overcome these impediments and address issues regarding communication, time delays from implementation to realizing benefits, and social systems.

This chapter points out the existence of a mismatch between the problems and impacts of the loss of SOC at different scales and the available best practices at those same scales. While most of the best practices are related to technological aspects applied at the local scale, there are few innovations at the higher levels, which are fundamental to address both the socio-economic and biophysical problems that arise with SOC loss at the lower scales.

The most critical innovation needed for improved SOC management requires bringing the farm to the global and the global to the farm level in a synergistic way, across scales. This challenge is daunting, but similar challenges are already being tackled in other policies with instruments such as the Kyoto Protocol Clean Development Mechanism. A similar approach could be applied to innovation in soil carbon management globally in order to gain the multiple social, economic and ecological benefits that are so clearly technically attainable.

References

Arrhenius, S. (1896) On the influence of carbonic acid in the air upon the temperature of the ground. *Philosophical Magazine* 41, 237–276.

Batjes, N.H. (1998) Mitigation of atmospheric CO_2 concentrations by increased carbon sequestration in the soil. *Biology and Fertility of Soils* 27, 230–235.

Batjes, N.H. (2011) Research needs for monitoring, reporting and verifying soil carbon benefits in sustainable land management and GHG mitigation projects. In: De Brogniez, D., Mayaux, P. and Montanarella, L. (eds) *Monitoring, Reporting and Verification Systems for Carbon in Soils and Vegetation in African, Caribbean and Pacific Countries*. European Commission, Joint Research Center, Brussels, pp. 27–39.

Brussaard, L., de Ruiter, P.C. and Brown, G.G. (2007) Soil biodiversity for agricultural sustainability. *Agriculture, Ecosystems and Environment* 121, 233–244.

Callendar, G.S. (1938) The artificial production of carbon dioxide and its influence on temperature. *Quarterly Journal of the Royal Meteorological Society* 64, 223–237.

Daily, G.C., Alexander, S.E., Ehrlich, P.R., Goulder, L.H., Lubchenco, J., Matson, P.A., Mooney, H.A., Postel, S., Schneider, S.H., Tilman, D. and Woodwell, G.M. (1997) Ecosystem services: benefits supplied to human societies by natural ecosystems. *Issues in Ecology* 2, 1–18.

Doetterl, S., Stevens, A., van Oost, K., Quine, T.A. and van Wesemael, B. (2013) Spatially-explicit regional-scale prediction of soil organic carbon stocks in cropland using environmental variables and mixed model approaches. *Geoderma* 204–205, 31–42.

FAO (2011) *The State of The World's Land and Water Resources for Food and Agriculture (SOLAW) – Managing Systems at Risk*. Food and Agriculture Organization of the United Nations and Earthscan, Rome.

Feller, C.L., Thurie's, L.J.-M., Manlay, R.J., Robin, P. and Frossard, E. (2003) "The principles of rational agriculture" by Albrecht Daniel Thaer (1752–1828). An approach to the sustainability of cropping systems at the beginning of the 19th century. *Journal of Plant Nutrition and Soil Science* 166, 687–698.

Grace, P.R., Antle, J., Aggarwal, P.K., Ogle, S., Paustian, K. and Basso, B. (2012) Soil carbon sequestration and associated economic costs for farming systems of the Indo-Gangetic Plain: a meta-analysis. *Agriculture, Ecosystems and Environment* 146(1), 137–146.

Ingram, J.S. and Fernandes, E.C.M. (2001) Managing carbon sequestration in soils: concepts and terminology. *Agriculture, Ecosystems and Environment* 87, 111–117.

IPCC (2006) *IPCC Guidelines for National Greenhouse Gas Inventories Volume 4: Agriculture, Forestry and other Land Use*. IPCC National Greenhouse Gas Inventories Programme, Hayama, Japan.

Izac, A.M.N. (1997) Developing policies for soil carbon management in tropical regions. *Geoderma* 79(1–4), 261–276.

Johnson, G. (1995) The role management in sequestering soil carbon. In: Lal, R., Kimble, J., Levine, E. and Stewart, B.A. (eds) *Soil Management and Greenhouse Effect*. Lewis Publishers, Boca Raton, Florida, pp. 351–363.

Keeling, C.D., Bacastow, R.B., Bainbridge, A.E., Ekdahl, C.A., Guenther, P.R. and Waterman, L.S. (1976) Atmospheric carbon dioxide variations at Mauna Loa Observatory, Hawaii. *Tellus* 28, 538–551.

Koning N., Heerink, N. and Kauffman, S. (2001) Food insecurity, soil degradation and agricultural markets in West Africa: why current policy approaches fail. *Oxford Development Studies* 29, 189–207.

Lal, R. (2005) Soil erosion and carbon dynamics. *Soil and Tillage Research* 81(2), 137–142.

Lal, R. (2008) Soils and sustainable agriculture. A review. *Agronomy for Sustainable Development* 28(1), 57–64.

Liebig, J.v. (1840) *Die organische Chemie in ihrer Anwendung auf Agricultur und Physiologie*. Vieweg, Braunschweig, Germany, 352 pp.

MA (Millennium Ecosystem Assessment) (2005) http://www.millenniumassessment.org/en/Global.html (accessed 2005).

McKenzie, N., Henderson, B. and McDonald, W. (2002) *Monitoring Soil Change: Principles and Practices for Australian Conditions*. CSIRO Land and Water, CSIRO Mathematical and Information Sciences, National Land and Water Resources Audit (http://nrmonline.nrm.gov.au/downloads/mql:3058/content, accessed 2002).

Manlay, R.J., Feller, C. and Swift, M.J. (2007) Historical evolution of soil organic matter concepts and their relationships with the fertility and sustainability of cropping systems. *Agriculture, Ecosystems and Environment* 119, 217–233.

Ogle, S.M., Breidt, F.J., Eve, M.D. and Paustian, K. (2003) Uncertainty in estimating land use and management impacts on soil organic carbon storage for US agricultural lands between 1982 and 1997. *Global Change Biology* 9, 1521–1542.

Ogle, S.M., Breidt, F.J. and Paustian, K. (2006) Bias and variance in model results associated with spatial scaling of measurements for parameterization in regional assessments. *Global Change Biology* 12, 516–523.

Ogle, S.M., Breidt, F.J., Easter, M., Williams, S., Killian, K. and Paustian, K. (2010) Scale and uncertainty in modelled soil organic carbon stock changes for US croplands using a process-based model. *Global Change Biology* 16, 810–822.

Palmer, C. and Silber, T. (2012) Trade-offs between carbon sequestration and rural incomes in the N'hambita Community Carbon Project, Mozambique. *Land Use Policy* 29(1), 83–93.

Paustian, K., Andrén, O., Janzen, H.H., Lal, R., Smith, P., Tain, G., Tiessen, H., van Noordwijk, M. and Woomer, P.L. (1998) Agricultural soils as a sink to mitigate CO_2 emissions. *Soil Use and Management* 13, 230–244.

Ravindranath, N.H. and Ostwald, M. (2008) *Carbon Inventory Methods – Handbook for Greenhouse Gas Inventory, Carbon Mitigation and Roundwood Production Projects, Advances in Global Change Research*, Vol 29, Springer, Heidelberg, Germany.

Reicosky, D.C. (2003) Conservation agriculture: global environmental benefits of soil carbon management. In: Garcia-Torres, L., Benites, J., Martinez-Viela, A. and Holgado-Cabrera, A. (eds) *Conservation Agriculture: Environment, Farmers Experience, Innovations, Socio-economy, Policy*. Springer, the Netherlands, pp. 3–12 (http://link.springer.com/chapter/10.1007/978-94-017-1143-2_1, accessed 1 July 2014).

Rogers, E.M. (1962) *Diffusion of Innovations*, 5th edn 2003. Free Press, New York.

Rusch, H.P. (1968) *Bodenfruchtbarkeit: eine Studie biologischen Denkens*. K.F. Haug Verlag, Heidelberg, Germany.

Sanderman, J. and Chappell, A. (2013) Uncertainty in soil carbon accounting due to unrecognized soil erosion. *Global Change Biology* 19(1), 264–272.

Schmidt, M.W.I., Torn, M.S., Abiven, S., Dittmar, T., Guggenberger, G., Janssens, I.A., Kleber, M., Kogel-Knabner, I., Lehmann, J., Manning, D.A.C. *et al.* (2011) Persistence of soil organic matter as an ecosystem property. *Nature* 478, 49–56.

Smith, P., Martino, D.L., Cai, Z., Gwary, D., Janzen, H., Kumar, P., McCarl, B., Ogle, S., O'Mara, F., Rice, C. *et al.* (2007) Policy and technological constraints to implementation of greenhouse gas mitigation options in agriculture. *Agriculture, Ecosystems and Environment* 118(1–4), 6–28.

Smith, P., Martino, D., Cai, Z., Gwary, D., Janzen, H., Kumar, P., McCarl, B., Ogle, S., O'Mara, F., Rice, C. *et al.* (2008) Greenhouse gas mitigation in agriculture. *Philosophical Transactions of the Royal Society B: Biological Sciences* 363(1492), 789–813.

Victoria, R., Banwart, S.A., Black, H., Ingram, H., Joosten, H., Milne, E. and Noellemeyer, E. (2012) The benefits of soil carbon. In: *UNEP Year Book 2012: Emerging Issues in Our Global Environment*. UNEP, Nairobi, pp. 19–33 (http://www.unep.org/yearbook/2012/pdfs/UYB_2012_CH_2.pdf, accessed 23 June 2014).

Waksman, S.A. (1938) *Humus. Origin, Chemical Composition and Importance in Nature*, 2nd edn. The Williams and Wilkins Company, Baltimore, London (revised).

WOCAT (2007) *Where the Land is Greener: Case Studies and Analysis of Soil and Water Conservation Initiatives Worldwide*. In: Liniger, H. and Critchley, W. (eds) CTA, UNEP, FAO and CDE, Berne (https://www.wocat.net/fileadmin/user_upload/documents/Books/WOOK_PART1.pdf, accessed 2007).

WOCAT (2010) World Overview of Conservation Approaches and Technologies (https://www.wocat.net/ accessed 2010).

5 A Strategy for Taking Soil Carbon into the Policy Arena

Bas van Wesemael*, Michael Stocking, Francesca Bampa,
Martial Bernoux, Christian Feller, Patrick T. Gicheru,
Philippe Lemanceau, Eleanor Milne and Luca Montanarella

Abstract

Soil organic carbon (SOC) has a relatively low profile in the policy arena. Here, we discuss the different steps of the policy-making process as well as the actors involved at the local, national and international scale. The first part analyses the policy-making process. The **policy imperative** consists of building up and maintaining SOC. The **policy profile and discourse** focuses on raising awareness. The **policy rationale** includes the economic and social benefits as well as the soil as capital. The **policy support** concerns the tools and programmes available. The second part of the chapter deals with the **actors,** from the **advocates and institutions** to the **governance**. For more detailed information, the reader is guided in each of these sections towards the chapters of the background document. Finally, recommendations are given at each of these levels for increasing the profile of SOC in the policy arena.

Introduction

Policy making must be understood as a political process as much as it is an analytical or problem-solving process.

> The policy-making process is by no means the rational activity that it is often held up to be in much of the standard literature. Indeed, the metaphors that have guided policy research over recent years suggest that it is actually rather messy, with outcomes occurring as a result of complicated political, social and institutional processes which are best described as 'evolutionary'.
> (Juma and Clark, 1995)

As Clay and Schaffer (1984) point out, the generation of policy is a chaos of purposes and accidents. It is never a matter of the rational implementation of scientific evidence through selected strategies.

Policy narratives are the specific 'stories' that bring the importance of a particular issue to life and public prominence. The narrative will often employ hyperboles and selective imagery, such as a polar bear perched on a tiny iceberg and a sea level rise catastrophe for major world cities. The policy discourse is distinct from the narrative, being how the issue refers to a wider set of values and a way of thinking. It is the way in which the evidence and importance of the topic is framed, examining the language and the people who understand the language. The discourse must be linked to the advocates and the institutions for the issue, a subject discussed later in this chapter. A narrative

*E-mail: bas.vanwesemael@uclouvain.be

can be part of a discourse if it reinforces in the policy maker's mind that here is an issue that demands attention and challenges current values and ideals. The narrative is a presentational challenge; how to make the 'story' so appealing and demanding that it spontaneously generates a policy uptake.

A first step in developing a strategy to take a scientifically derived priority issue such as soil organic carbon (SOC) into the policy arena is to decide on the discourse. Scientific evidence on its own is rarely sufficient; it is the impact, who is affected, how they are impacted and the consequences of inaction that grab the attention. The discourse matters because the environment is a social construction – as well as a physical entity – and issues about the environment such as soil carbon are contested, not just in their importance but also in their relevance to society and human well-being (Feindt and Oels, 2005).

Because SOC has a low profile at all levels – local, national and international – the context for the narrative and discourse has to be chosen carefully. Some of the evidence that is important for different contexts is covered elsewhere in this volume; it includes SOC:

- as a smart indicator for soil quality (Google maps, smart phone);
- as hotspots and thresholds for soil degradation, soil and water conservation and belowground biodiversity;
- as a component of diversified agricultural systems, adapted to climate change;
- in monitoring and verifying mechanisms for climate change mitigation;
- as risk and opportunities for inputs and residue management (fertilizer, chemicals, etc.) applied in agriculture;
- as a contributor to food security and sustainable development;
- as a protector of biodiversity – as a substrate for soil biodiversity and as a support for aboveground biodiversity.

It is, therefore, imperative that issues involving SOC must achieve a higher and more prominent profile, with their champions and self-evident financial worth. A strategy for taking SOC into the policy arena must have this high-profile springboard before diving into the murky waters of the complex arrangements for governance at various levels, from local to global. This chapter summarizes briefly the data and scientific demands for an SOC strategy into policy; it then examines the components of this strategy, setting out a framework for mainstreaming SOC issues in environmental and sustainable development policy. The chapter is structured according to: (i) what needs to be achieved by policy, i.e. imperative, profile and discourse, rationale and support; and (ii) who are the main stakeholders, i.e. advocates and institutions and governance (Table 5.1). Each of these topics is discussed in a different section at three scale levels, i.e. local, national and international.

Policy

Policy imperative

Introduction

In the developing and also in developed countries, farmers are faced with serious problems of sustainable soil management such as market forces, limited access to inputs and expertise, poverty and lack of capital. In developed countries, for example, plant productivity is high due to large use and sometimes excess of inputs (fertilizers, pesticides and energy), but often with low organic restitutions to soil. This has negative consequences for the environments, e.g. for air and water pollution, soil erosion, etc. (Lal *et al.*, 2003).

For centuries, many farming systems have relied on the soil organic matter (SOM) to sustain production. These systems were maintaining a stable SOM pool in soils by applying a closed farming system with a regular return of organic matter to the soil through manure, crop residues, kitchen waste, etc. However, with the adoption of industrialized farming systems implying increasing intensification, land degradation and climate change, the quantity of SOM has declined rapidly, thereby threatening the capacity of the land to produce sustainably (Lal, 2004; Zdruli *et al.*, 2004; Lal *et al.*, 2006). In developing

Table 5.1. Components of a policy process to raise the status of soil organic carbon (SOC).

	Section	Scale level		
		Local	National	International
Policy (What?)	Policy imperative (build-up and maintenance of SOC)	Agro (urban)-ecological alternative Avoid degradation Sustainable land management Increasing productivity/fertilization	SOC into national soil protection legislation SOC into NAP SOC into NAMAs (Chapter 31, this volume)	Sustainable development (under construction, Chapter 29, this volume) Climate adaptation + mitigation (UNFCCC, LDNW UNCCD) Conservation + sustainable use of biodiversity (CBD)
	Policy profile and discourse (raising awareness)	Adapt to local socio/cultural context (Chapter 14, this volume) Education	Value of SOC Regional patterns (Chapters 15 and 18, this volume)	Include SOC in sustainable development (mainstreaming) Hyperbole
	Policy rationale (economic/social benefits, soil as a capital)	Develop strategy for sustainable livelihoods (Chapter 14, this volume) Reduced risk	Multiple benefits (Chapters 15 and 18, this volume) Reduced costs from erosion, etc.	Maintaining SOC for future generations Reduce vulnerability of populations
	Policy support (tools and programmes)	Best practices demonstrated at local level (e.g. WOCAT) Field-scale SOC models (e.g. Comet VR, Cool farm; Chapter 17, this volume) Smartphones	Soil monitoring networks (Chapter 16, this volume) Modelling tools GEFSOC (Chapter 17, this volume) Google maps National incentives/PES (water, biodiversity, carbon)	Harmonization SMNs Develop research to valuate soils/SOC (Chapters 18 and 22, this volume) Climate adaptation and environmental funds
Actors (Who?)	Advocates and institutions	Farmers' organizations, CBOs, NGOs	Cross-compliance Ministries, focal points, NARS	Global conventions and partnerships, international NGOs, UN, GEF, World Bank, FAO, IFAD
	Governance	Agricultural extension Conservation districts Local producers/landcare associations/watershed committees	Agricultural strategy and sectors NAMAs (Chapter 31, this volume) Labels and markets Carbon footprint Soil certification	Conference of Parties, Global Soil Partnership (GSP), IPCC, IPBES

CBO, community-based organizations; FAO, Food and Agriculture Organization; GEF, Global Environment Facility; GEFSOC, Global Environment Facility Soil Organic Carbon; IFAD, International Fund for Agricultural Development; IPBES, Intergovernmental Science Policy Platform on Biodiversity and Ecosystem Services; IPCC, Intergovernmental Panel on Climate Change; LDNW, land-degradation neutral world; NAMA, nationally appropriate mitigation action; NAP, national action plans; NARS, national agricultural research systems; NGOs, non-governmental organizations; PES, payment of ecosystem services; SMN, soil monitoring network; UN, United Nations; UNCCD, United Nations Convention to Combat Desertification; UNFCCC, United Nations Framework Convention on Climate Change; WOCAT, World Overview of Conservation Approaches and Technologies. Chapters refer to others in this volume where a more detailed discussion of the issues can be found.

countries (e.g. Africa), productivity is very low due to insufficient input application (fertilizer, both organic and mineral). At the same time, the traditional systems are not being practised as a result of an uneven distribution of land in relation to the high population density, lack of tenure security and lack of investments. The consequence is a low level of organic restitutions to the soil by an increasingly urbanized population. The low level of organic matter and associated nutrients has prompted the resource-poor local population to convert forests into monocultures and to overgraze common lands in other cases, with severe environmental consequences such as high erosion rates, mining agriculture and greenhouse gas (GHG) emissions. Whatever the level of development, these numerous environmental problems need to be addressed.

Sustainable production and increasing productivity/fertilization at the local level

Agricultural productivity and household income depend on natural resources management (soil, water, forests, air, and flora and fauna biodiversity) and their quality. In most of the farming systems, their lower income is due to declining terms of trade between farm inputs and agricultural production. This decline has diminished the use of organic matter and fertilizer inputs for land- quality improvement, especially among smallholder farmers, leading to low biomass production. This also results in a decline of the land quality through a loss of SOC, and this can be attributed to crop removals, soil erosion and leaching of nutrients. Therefore, in order to increase smallholder productivity, it is necessary to introduce viable incentives and market regulating instruments, encouraging farmers to sustain soil and water resources as well as biodiversity.

Since land is a fragile and a finite resource on this planet, its use is bound to have conflicts. Among the farmers, conflict takes the form of competition for arable and grazing land. This means that as the population grows, so too does the quest to acquire more land to accommodate the increasing population. Countries with high population densities and limited national fertile soil resources have to rely on imports of 'virtual soils' or invest in foreign land acquisitions ('land grabbing'). Locally, forests have been converted into agricultural land, endangering soil biodiversity and severely diminishing SOC stocks. Forest ecosystems are interrupted and finally collapse due to the clearing activities of farmers. The often unscrupulous dealers, leaning on loose legislation, find a boon by corrupting their ways to owning forestland and selling to other third parties or farmers in need.

Sustainable land management (SLM) combines technologies, policies and activities aimed at integrating socio-economic principles with environmental concerns, so as simultaneously to maintain and enhance production, protect natural resources and improve incomes at socially acceptable levels. It is the use of land to meet changing human needs (agriculture, forestry and conservation) while ensuring long-term socio-economic and ecological functions of the land. Land provides an environment for agricultural production, but it is also an essential condition for improved environmental management, including source–sink functions for GHGs, recycling of nutrients, reduced use and filtering of pollutants, and transmission and purification of water as part of the hydrological cycle. Whatever technology advances, the land will always be necessary for humans to grow most of the food, and soils are non-renewable resources at the human timescale. Therefore, optimizing the biological productivity of the land to ensure ecological sustainability and environmental protection will always be a necessity for production and the improvement of rural livelihoods. A pre-condition for SLM implementation is the existence of secure land tenure rights, allowing intergenerational transfer of land rights, and therefore incentivizing sustainable management practices, preserving land for future generations.

Farmers' modifications of natural ecosystems will always have an impact on land quality. Hence, advice on land management must consider farmers' options according to their goals, resource capacity, socio-economic circumstances and field production conditions.

Sustainable land management implies the wise use of land, which encompasses the principle that mined nutrients must also be restored. Crops require an instant flow of nutrients at specific growth stages. Mineral fertilizers are the quickest and surest way of supplying the nutrients in known amounts, proportions and available forms ready for uptake by plants, and hence the surest way of also building C stocks. However, the amount of inorganic fertilizers used at local level is low. The cost of fertilizers coupled with that of seeds, pesticides and other requirements are out of reach to resource-poor farmers. Most of the small-scale farmers use organic fertilizers such as farmyard manure and compost and recycle crop residues as a means of sustaining and improving land productivity. However, the limited amounts of organic sources of plant nutrients and their very low fertilizer grades mitigate reliance on them as stand-alone sources of nutrient supply. Since farmers' operations will always have an impact on land quality and SOC benefits at the local level, recommendations on land management must consider their goals, resource capacity, socio-economic circumstances and field production conditions.

Enabling policy environment to promote sustainable land-use management at the national level

At the national level, there exist shortcomings in some parts of the world in policy development and implementation in the area of land use and management (e.g. lack of National Action Plans, or malfunctioning of these). Appropriate policies should focus on national land planning, market-oriented tools and legal frameworks for natural resources management as well as for land tenure (e.g. access, control and ownership). The implementation of an unfavourable policy and legal environment in potential areas for good land management in conjunction with local market distortions has exacerbated the problem of low land productivity. There is a lack of coherence between market drivers, land-use policy and formal institutional arrangements, which creates a lack of coordination and communication in the implementation of good management and strategies at the farmer/stakeholders level.

The institutions that are supposed to address land issues are lacking frameworks, especially in relation to SOC. The importance of SOC is mostly hidden within other related disciplines in most national governments and is not given the prominence it deserves. The role of traditional institutions and land rights is also not explicit. Investment in land improvement programmes is limited in some parts of the world. International organizations should be able to provide resources through national governments so that the agenda of SOC conservation can be realized at the national level. Appropriate policies should focus primarily on land planning, legal frameworks and regulatory mechanisms linked with efficient economic incentives to assure the compliance of soil and SOC improvement at the national level.

Sustainable development at the international level

At the global level, there is an increasing interest in land as a resource that should be preserved. The Multilateral Environmental Agreements (MEA from the United Nations Framework Convention on Climate Change (UNFCCC), Convention on Biological Diversity (CBD) and United Nations Convention to Combat Desertification (UNCCD)), revisited last year during the Rio+20 Sustainable Development Conference, identified the importance of land. During the conference, it has been agreed to aim towards a *Land Degradation Neutral World* as a global effort to fulfil soil protection and soil restoration activities (Chapter 29, this volume). More recently, during the second meeting of the Intergovernmental Science Policy Platform on Biodiversity and Ecosystem Services (IPBES) Plenary (IPBES-2) held in Antalya, Turkey, in December 2013, the work programme for 2014–2018 was adopted and it included a thematic assessment aiming to enhance the knowledge base for policies to address land degradation, desertification and the restoration of degraded land.

Nowadays, there are many international bodies that dedicate part of their mandate to SOC (World Bank, Food and Agriculture Organization (FAO), United Nations Environment Programme (UNEP), European Union (EU), etc.). A coordination among diverse institutions, harmonized policies and system approaches for synergy (soil, water, crop, livestock, forest, etc.) would be welcome. The recent Global Soil Partnership (GSP) established by FAO aims towards providing the platform for achieving effective coordination among the various initiatives. However, the protection, maintenance and enhancement of SOC can only be achieved through actions driven by economic incentives linked to main high-profile goals such as: food security, sustainable development and climate change negotiations.

Another issue to be taken into account is the missing link between policy makers and scientists. Scientific language is not clear and effective enough to convince and move political decisions, or even public awareness. There is an urgent need to ease communication through the use of an innovative approach supported by clear numbers, figures and maps. Furthermore, common agreed monitoring and accounting protocols or methodologies have not yet been adopted by the scientific community as a whole. Hence, the marketing of SOC is not moving ahead. For many years, working groups and strategy sessions have been focused on creating a real market for SOC credits or offsets. However, today, these efforts are reflected in a limited number of C projects only. The recently established Intergovernmental Technical Panel on Soils (ITPS) has been designed by the GSP partner countries and organizations as the science–policy interface addressing soil-related issues in support to global policy-making processes by translating available scientific results in policy-relevant information and assessments. The UNCCD adopted also, during a Conference of Parties (COP11) in September 2013, a science–policy interface to enable its researchers to communicate scientific findings to policy makers concerned with land degradation and desertification.

The international rhetoric of the importance of SOC for soil quality has not been translated into international action. The disconnection between the scientific and the policy arenas at international level can only be solved with an innovative language through clear and simple messages relating societal priorities like growth, income, jobs and social welfare with soil quality and SOC. The starting point should be the key cross-cutting role of SOC towards high-profile topics such as food security, sustainable development and climate change scenarios.

Policy profile and discourse

Local scale: adapt to local socio/cultural context

The perception of soil varies according to different cultures in the world. Not only on a biophysical basis but also in the context of different religions or cults, both in developing and developed countries. For instance, in some cultures of developing or emerging countries, religious habits support soil tillage and other religions discourage soil operations with sharp tools (Lahmar and Ribaut, 2001; Chapter 14, this volume). In contrast, developed countries have to face different influences, driven mostly by economic interests and political constraints. In some developed countries, even in a similar socio-economic context, farmers do not have the same perception of soil and of SOC benefits. As a consequence, this perception will orient them in the adoption of one or another ecoagricultural management alternative as, for example, organic farming or no tillage with cover plant systems (Compagnone et al., 2013; Chapter 14, this volume).

No matter which country in the world, raising awareness towards SOC benefits will depend strongly on socio-economic drivers and the cultural habits of the beneficiaries.

The policy profile and the discourse for raising awareness of the SOC benefits of given alternatives of land management have to be adapted to the cultural and socio-economic habits of the beneficiaries.

For the discourse, the following points should be taken into account:

- the holistic perception of the soil fertility from farmers, often linked to the cult/religious relationships between the society with soil/earth;
- the economic access of farmers to resources (e.g. fertilizers, tools, machines, etc.);
- the possible competition between management strategies for the same land resource (e.g. competition for plant residues between fertilization, feeding the cattle or building material);
- the removal of residues for livestock or fuel is a big issue that needs investments in fodder production and efficient or alternative fuel sources; and
- the need to establish a partnership between the growing urban population and the rural farming communities for returning the organic waste (biowaste) generated in cities to the soil of the rural areas.

National scale: value of soil and SOC – regional patterns

'SOC benefits' is a purely scientific concept and is often perceived in the scientific community only. Even though the multiple functions of SOM are well known in the scientific arena, these concepts generally do not reflect any human cultural dimension. As a consequence, the public does not incorporate SOC/SOM benefits in its own culture. To make the population more aware of the benefits of SOM/SOC, the proposal is to develop actions at the national level, starting from the concept of 'soil as a patrimony and a natural capital' (Costanza and Daly, 1992; Pascual et al., 2010; Chapters 15 and 18, this volume) that the society needs to transmit to the next generations. This message needs to be incorporated in school and university programmes focusing on the following points:

- Soil is a non-renewable resource, to be protected against erosion, decline in organic matter and biodiversity, etc. (EC, 2009, 2010).

- Soil protection needs to take into account the availability of other natural resources, pedo-climatic conditions and land management (EC, 2009, 2012).

Training programmes at secondary schools and universities for both students and teachers should focus on soil as a patrimony and on the importance of SOM in adequate soil management. Events and activities to address the public (soil celebration day, campaigns of awareness through media, etc.) will help in raising awareness.

International scale: the inclusion of SOC in sustainable development (mainstreaming)

The perception of soil as a patrimony for society is not well developed, given that most soils are in private property and are not perceived as a public good, like air and water. Soil as a resource is essential to provide goods and services not only to the landowner but also to all of us. There is the need to move from soil protection to the protection of soil functions. While soils are mostly in private property, soil functions are delivering public goods and therefore need to be protected beyond private property rights. Only an adequate SOC/SOM management will permit the delivery of these services. At the international level, hotspots of awareness activities exist that address small and often specialized audiences.[1] Unfortunately, there is a lack of communication and coordination between these hotspots that reduces the potential of an international awareness on the importance of SOC.

In order to facilitate the acceptance of SOC as a mainstream environmental concern, the level of SOC/SOM to be reached in a specific region could be used as one of the proxies. A teaming up of all failed efforts to raise awareness in a network or platform would no doubt increase its efficiency. Moreover, all types of media should be approached with an innovative approach: cultural services (exhibitions, movies), societal concerns (television programmes, journals) and educational targets (workshops, conferences).[2] Finally, an international agenda for a celebration of a soil day³/year could accelerate all previous recommendations. The FAO has submitted two resolutions to the United Nations General Assembly for the celebration of World Soil Day on 5 December and an International Year of Soil in 2015. These will be two key awareness-raising platforms and events. Concerning drylands, the French Scientific Committee on Desertification published at the end of 2013 a thematic study entitled 'Carbon in dryland soils – multiple essential functions' showing the multiple functions benefiting societies and the environment in drylands (Bernoux and Chevallier, 2013).

Policy rationale

On the one hand, the land containing the soil is in most cases private property and can thus be traded by its owners. On the other hand, soil functions deliver the economic and social benefits of the SOC capital to society as a whole and therefore need to be considered as public goods. As SOC is part of a natural resource, it cannot be traded directly on international markets; nor are the social effects of the SOC capital immediately visible (Chapter 15, this volume), although changes in the SOC capital can be traded on the emissions markets, as an increase in SOC can be considered to represent a sink of atmospheric CO_2. Currently, only voluntary markets for CO_2 trading exist, and prices are well below US$5 Mg^{-1} C (Lipper et al., 2010). Clearly, such trading considers only a part of the services that flow from SOC and by no means considers the value of soil as a natural capital.

In addition to the long-term effect of a decrease in SOC and its consequences for future generations, the economic and social benefits and costs flowing from the use of the SOC capital are discussed. First, the opportunity costs of increasing/maintaining SOC levels refer to examples of the impact of returning biomass to the soil and how this affects food sources for cattle (Antle and Stoorvogel, 2003). In other words, farmers, who tend to maximize income are often reluctant to compromise a service yielding direct income, such as using the residues for energy consumption for the long-term effect of the same residues that, when left on the field, may slowly increase SOC. The latter may produce a benefit for society, for instance in terms of reducing C emissions (Chapter 15, this volume). The second economic parameter concerns the externalities that can be either positive (e.g. no-till designed for C sequestration also protects the soil against erosion) or negative (e.g. no-till induces an increase in pesticide use). Finally, the willingness to pay for goods and services that flow from SOC are discussed in Chapter 15 of this volume.

Local scale

The main concern of farmers and land managers is to ensure a secure and sufficient income. If they have secure land tenure rights, they will also be concerned about maintaining a value of the soil capital for future generations. Unfortunately, in the current economic environment, the value of the soil capital is secondary to the value of the services that the soil, and more specifically the SOC, can provide (i.e. the externalities). In order to benefit from these services, some choices have to be made incurring costs (e.g. increase in SOC will require biomass input that cannot be fed to cattle). These choices lead to opportunity costs. The management of SOC has an (long-term) impact on soil quality through improving soil aggregate stability, water-holding capacity and reducing vulnerability to erosion (Govers et al., 2013). Furthermore, SOC has an influence on soil fertility through the nutrients (mainly N and P) that it contains. In most intensive agricultural systems, these nutrients are supplied in chemical fertilizers, and hence evidence of direct relationships between SOC additions and yield increases are rare. Several authors mention that the variability in yields decreases by a two-way interaction between SOC stocks and crop productivity (overview in Govers et al., 2013). van Noordwijk et al. (1997) argue that nutrient supply, nutrient buffering, water regulation, maintenance of soil structure and other functions such as pest control can, in principle, be replaced by technical solutions, leading to fertigation in hydroponic horticultures in the most extreme case. However, these technical solutions require investments that are only profitable in the case of high-value crops in greenhouses, such as salad vegetables, flowers and pot plants. The long-term sustainability of these high-input agricultural systems may be questioned. The majority of agricultural production still relies on most of the functions of SOM mentioned above that guarantee sustainable soil management for future generations.

The supply of nutrients such as reactive N and P contained within the SOM can be estimated using the C:N and C:P ratios. In the literature, the average rates of SOC increase by modified agricultural practices are around 0.4 Mg C ha^{-1} year^{-1} (Freibauer et al., 2004). At US fertilizer prices in 2012, this would result in a gain of US$40 for N and US$8 for P per hectare. In the case of nutrient mining of soils rich in C, these would be regarded as positive externalities. In poor soils, the increase in nutrients will be an investment in the soil capital, and will therefore require opportunity costs. Diels et al. (2002) concluded that for a case study in West Africa, SOC increases resulted in an enhanced buffer capacity that would lead to higher efficiency of mineral fertilizers. However, care should be taken for negative externalities. There are many examples that such an increase in SOC and fertility has been achieved by concentrating organic residues from forests/heathlands on fields close to the farm. For instance, Bationo et al. (2007) state that up to 30 ha of dry-season grazing may be required to produce the manure necessary to maintain the fertility of 1 ha of cropland in West Africa: promoting the use of manure may therefore increase grazing pressure and the degradation of SOC stocks on the grazing land, and may further reduce the amount of crop residues that might be applied directly to the cropland. This will lead to the fertility gradients mentioned in Chapter 10, this volume. Diels et al. (2002) have demonstrated that increasing the SOC content from 0.8 to 1.3% results in an increase in water-holding capacity of 1 mm for a study area in Kenya. Even though the increase in SOC is substantial, the effect is limited at 50 mm.

National scale

The multiple benefits derived from SOC interact at scales beyond the individual farm, and therefore should be addressed and remunerated through public incentives at scales ranging from the catchment to the nation. Natural capital is thus the stock of natural ecosystems that yields a flow of valuable ecosystem goods or services into the future (Chapter 22, this volume). Pascual et al. argue that:

> Valuing soil carbon from an economic point of view, however, is not straightforward, since most of the benefits derived from its conservation and sustainable use are not reflected in the markets. This is mostly because current markets only reveal sufficient information about the scarcity of a small subset of goods and services from nature. Most natural resources, ecosystem processes and functional components are not incorporated in transactions as commodities or services, and their economic value is not reflected in any market prices.
> (Chapter 15, this volume)

A more immediate way of considering the benefits of SOC is its value that is attributed by people through their willingness to pay for the goods and services that flow from it (Pascual et al., 2010). Costanza and Daly (1992) were the first to define the concept of ecosystem services. Soil, and in particular SOC, are considered to be an intermediate for final goods and services such as climate-regulation and water-regulation services. Noellemeyer and Six (Chapter 22, this volume) discuss the range of services that SOC can provide and detail the role of SOC in soil structure maintenance and improvement, erosion control, climate regulation, pollutant attenuation and degradation, pest and disease control and biodiversity conservation. Abson et al. (Chapter 18, this volume) argue that these ecosystem services should be estimated jointly for case studies, as in many cases the positive effects on one service can have negative impacts on another.

International scale

At this scale, costs and benefits from climate-regulation services are discussed as examples of local actions that will make a contribution to a global problem. Given the low prices for CO_2 on the voluntary trade markets in Chicago (US$5 Mg^{-1} C), the sequestration of 0.18–0.4 Mg C ha^{-1} that is most likely to be achieved will yield only US$0.9–2 ha^{-1}. Smith (2004) has calculated the costs of monitoring and verification, showing that these costs do not outweigh the benefits at these low CO_2 prices. A positive experience of trading SOC on the international markets comes from the first agricultural SOC project in Kenya, where smallholders use the Sustainable Agriculture Land Management (SALM) methodology from the Verified Carbon Standard (VCS) to certify C credits – which are currently purchased through the World Bank Biocarbon Fund. Smallholders have adopted mixed cropping systems, based on residue management, composting and agroforestry. The RothC model was parameterized in several farming systems and used to define the best management practices to be adopted. In another case study, the standard elaborated by Plan Vivo is used to support smallholders in applying sustainable management practices (Reducing Emissions from Deforestation and Forest Degradation (REDD) and agroforestry) and to generate payments for the ecosystem services provided (C credits). Additional C services with respect to a baseline are quantified by an independent methodology. The project coordinator enters into payments of ecosystem services in agreement with multiple participants. Staged payments are based on performances (http://www.planvivo.org).

Policy support

Tools and programmes

In order to implement strategies to enhance or maintain SOC stocks, information is needed on land management practices, where and how they should be implemented and their likely impacts on SOC. Some programmes already exist at the local, national and international level to provide this information, but substantial gaps exist. In a similar way, tools are needed to estimate the impacts of land management practices on SOC over a range of spatial and temporal scales. In this volume, the current status of programmes is discussed in Chapters 1 and 16, among others, and SOC models and how they are in use today is discussed in Chapter 17. Below, we consider the gaps in current programmes and tools at the local, national and international level.

Local

The World Overview of Conservation Approaches and Technologies (WOCAT) provides a global database for the storage, searching and exchange of land management practices for soil and water conservation and SLM (WOCAT, 2013). Although it is a resource with global coverage, information is aimed at the local scale, with users able to search for land management practices relevant to their local situation. Land Degradation Assessment in Drylands (LADA)-WOCAT mapping of land-use systems, land degradation and SLM at local, subnational and national scales provides information on land management practices used by more than 20 countries, and further decision support tools are to be developed (http://www.wocat.net). Many of the practices presented include data on soil loss prevention. However, currently, no inclusion is made of data on, or estimations of, how each land management practice will impact SOC in given soil and climate conditions. Inclusion of such information would allow local land managers to consider the potential impacts on SOC before implementing specific practices. Currently, there are plans to apply the project-scale tools from the Carbon Benefits Project (CBP) (Milne et al., 2010) to WOCAT practices to provide such estimations.

The FAO's MICCA (Mitigation of Climate Change in Agriculture) programme includes activities and resources relevant to SOC at the local level. MICCA has pilot projects which implement and quantify climate-smart agricultural practices and a database of agriculture, forestry and other land-use mitigation projects (AFOLU MPs). It includes few agricultural SOC projects, but these are expected to increase as the capacity to use new methods and tools increases (McCarthy et al., 2011). Other programmes include the Intergovernmental Panel on Climate Change (IPCC) Emission Factors database, which is a repository for the site-specific stock change and emission factors needed to make estimates of changes in C stocks in both biomass and soils, and the FAO's Harmonized World Soils database, which includes local-level information on SOC stocks. One of the main gaps in these programmes, which hinders the estimation of SOC stocks and changes, is information on different land management practices at the local level, particularly in developing country areas dominated by smallholders.

There are many examples of tools and models that can be used to make estimates of SOC stocks and changes at the local level (Chapter 17, this volume). Some are based on the IPCC's computational method (CBP Simple and Detailed Assessment, EX-ACT, Cool Farm), others use dynamic ecosystem models (COMET Farm, COMET VR). Constraints include the fact that most models have originally been developed using data from temperate areas and are therefore less applicable to tropical conditions. Two recent reviews considered tools that could be used to make landscape-scale estimates of C stock changes in soils and biomass. Colomb et al. (2013) consider 18 tools covering different geographic areas, and the authors are currently developing an online application that will allow users to choose the most appropriate tool. The second review considers tools that can be used in developing countries in areas dominated by smallholder agriculture, an important area for the future consideration of global SOC stock change (Milne et al., 2010, 2013). There is a need for capacity building for the use of local-level tools and methods and to develop local-level tools based on dynamic models that are applicable to conditions in tropical areas and other areas of rapid land-use change.

National

Making national-scale estimates of SOC stocks and changes is dependent on available soil survey information, which can be incomplete and of varying quality in many countries (Batjes et al., 2007). Parameters such as bulk density, which are needed to estimate SOC stocks, are often missing. Substantial work has been carried out by the International Soil Reference and Information Centre (ISRIC) to complete national-scale information on SOC in the soil and terrain (SOTER) database using collations of existing data sets, pedo-transfer functions and expert opinion (Batjes et al., 2007). In terms of the ongoing collation of SOC information at the national scale, many countries have programmes for national-scale survey and monitoring. van Wesemael et al. (2011) summarized existing Soil Monitoring Networks (SMNs). Networks were found to have different criteria in terms of land types sampled, sampling depth and frequency of sampling. Harmonization of SMNs would allow the comparison of different national-scale SOC stocks, and would also facilitate the use of national-scale models. A problem among almost all networks is continuity of the funding needed for ongoing monitoring.

In terms of the tools available for the estimation of SOC stocks at the national scale, empirical methods that use knowledge of soil type, climate and land use have been used frequently to estimate stocks (Lal et al., 2006; Batjes et al., 2007). There are also tools for estimating stock change at the national scale, including those based on computational methods where a stock at one point in time is subtracted from another. Tools also exist that employ dynamic models which attempt to capture the dynamic nature of the decomposition of organic matter over time. The World Bank Agriculture and Rural Development programme has an online tool that employs the RothC model to make national-scale estimates of SOC change over a 25-year period. Users can input information on different land management practices to alter estimates (ARD, 2013). A more rigorous but less user-friendly tool is the Global Environment Facilities GEFSOC tool

(Easter et al., 2007; Milne et al., 2007). This tool links two models (RothC and Century) to a geographical information system (GIS), which contains layers of soils, climate and land-use information. The tool has been used to estimate SOC stocks and changes in several countries (Bhattacharyya et al., 2007; Kamoni et al., 2007) but does require expertise in modelling and GIS. Development is needed of more user-friendly, national-scale tools with applicability to tropical areas and areas of rapid land-use change.

International

Globally, there are many regional and continental programmes considering SOC. Examples include the European Soil Portal of the EC Joint Research Centre (http://eusoils.jrc.ec.europa.eu/), which provides (among many things) online maps of organic carbon content in the surface horizon of soils in Europe (Jones et al., 2005a,b) and links to information and networks for other areas of the globe, including the GSP (see http://www.fao.org/globalsoilpartnership/en/ and action plan for Pillar 4 that was endorsed by the Plenary). Within the GSP, the ITPS was established and is fully operational (Montanarella and Vargas, 2012). The GSP is a major international initiative that has recently produced an analysis of the state of the art of soil information, including information on SOC (Omuto et al., 2012; Chapter 29, this volume). Another example is a new network for francophone Africa, 'Carbone des Sols', which aims to exchange information on the estimation of SOC storage and methods to achieve this (http://www.reseau-carbone-sol-afrique.org). A global consortium has been formed that aims to make a new digital soil map of the world using state-of-the-art and emerging technologies for soil mapping and predicting soil properties at fine resolution: http://www.globalsoilmap.net. Finally, the Global Soil Biodiversity Initiative aims at developing a coherent platform for promoting the translation of expert knowledge on soil biodiversity into environmental policy and sustainable land management for the protection and enhancement of ecosystem services including C storage in soils (http://globalsoilbiodiversity.org/).

Issues of data sharing and governmental restrictions on exchanging soils information remain a problem in many regions of the globe, and actions to facilitate data exchange are needed. This issue is currently seen as important within the International Council for Science (ICSU) World Data System and also in the data policy of ISRIC World Soils Information (Batjes et al., 2013). ISRIC have developed and made available many resources related to global information on SOC. This includes contributions to the Harmonized World Soil Database (HWSD) (FAO, 2013). ISRIC has provided a set of derived soil properties including SOC content and bulk density. For areas of the world covered by SOTER, the soil properties are 'best estimates' based on information from ISRIC's database of soil profile information (World Inventory of Soil Emission Potentials, WISE). Profile information is under-represented for many areas of the globe, notably the global south, and is needed urgently. Online facilities for collecting such profiles, through crowd sourcing, have recently been launched (see: http://worldsoilprofiles.org/). In conjunction with this, a suite of tools for analysing and mapping the data, at various resolutions, are being developed to help address a range of global issues, including SOC management (see Batjes et al., 2013). Other programmes such as the United Nations Environment Programme's World Conservation Monitoring Centre (UNEP–WCMC) have used the HWSD to generate an improved global map of SOC values to a 1 m depth at 1 km spatial resolution (Scharlaman et al., 2104).

In terms of tools to make global assessments of SOC stock changes, there are examples of General Circulation Models (GCMs) such as HadCM3LC that represent SOC using a single pool (Cox et al., 2000). The benefits of coupling such models to more complex multicompartmental models such as RothC for global assessments have been demonstrated; however, the main barrier to this approach is the availability of suitable data to initialize the different pools represented by the model. Global efforts are needed to collate such data for those areas of the globe where data are currently missing.

Actors

Advocates and institutions

Local scale

Local entry points to identify innovative practices, to advocate for their dissemination or simply to raise awareness, range from individual farmers (including those sometimes referred to as champion farmers) to unions of farmers at different administrative and geographical scales. For instance, in Africa, there has been in the past two decades an important development of farmers' organizations. These organizations have established five regional larger networks in Africa (Plate 6): the Union Maghrébine des Agriculteurs (UMAGRI); the Réseau des Organisations Paysannes et de Producteurs de l'Afrique de l'Ouest (ROPPA); the Eastern Africa Farmers Federation (EAFF); the Plateforme Sous-Régionale des Organisations Paysannes d'Afrique Centrale (PROPAC); and the Southern African Confederation of Agricultural Unions (SACAU). As it might be difficult to reach scattered farmers, a key issue at the local level is therefore to find a suitable entry point. That is where organizations, cooperatives and value chain federations constitute key institutions as farmers' aggregators at the local level. Those organizations can also permit linkage with national and international levels; thus, such organizations often represent thousands of farmers or regional networks of federations of farmers, which may represent over several million farmers. Until now, however, local farmers' organizations have rarely included soil-quality concepts in their communications to farmers, despite a growing awareness of some of them. A major distortion of this awareness (in Europe at least) is that farmers often receive 'technical' support from high-value income sellers promoting improved seeds or pesticides often linked to specific agricultural practices requiring high levels of inputs; like, for example, zero tillage in combination with specific herbicides.

National scale

At the national level, the best entry point should include ministries involved not only in agriculture and forestry but also in environmental management. Cross-sectorial ministry coordination is essential; thus, C sequestration can involve directly or indirectly different environmental issues, economic aspects and legislative necessities (food regulation, certification and labelling, and C footprinting). National focal points of the different international agreements should also be involved. Fortunately, in most countries, they are already linked with ministries.

At the national scale, most consumers in developed countries are already concerned by the notion of C footprinting. For instance, policies and legislation are being implemented in buildings, energy and transportation, but C footprinting of agriculture and food products is still in its infancy, with many debates concerning methodologies and communication to the public.

A major issue concerns land-use change (LUC) and whether LUC should be recognized as playing a major role in GHG emissions associated with products. LUC includes direct LUC (dLUC), occurring in the study area itself, and indirect LUC (iLUC), occurring outside the study area but resulting from changes within the study area. Including iLUC in environmental assessment involves many different assumptions about the socio-economic relationships between different areas. Several methodologies for quantifying the share of dLUC and iLUC are currently being proposed and discussed as part of the debate of bioenergy approaches that focuses on the impacts of changes and the relationships between production systems (Brander et al., 2009).

In addition, the national development or cooperation agencies could have a role to play (e.g. US Agency for International Development (USAID), Deutsche Gesellschaft für Internationale Zusammenarbeit (GIZ), the French Development Agency (AFD),

Japan International Cooperation Agency (JICA), Department for International Development (DFID)). Thus, priorities of actions set by those agencies will influence the implementation of technical options for mitigating climate change. National agricultural research systems (NARS) are key institutions in bringing the necessary sound science concerning the mitigation potential of different technical mitigation strategies. NARS should have in their prior references the promotion of mitigation strategies (Chapter 9, this volume) and/or improvement of the methodological aspects (Chapter 16, this volume).

The institutions at the level that works with 'mitigation' strategies are involved most of the time with 'adaptation'. Thus, the advocacy should consider the trade-offs between these two aspects. Hopefully, most mitigation strategies concerning SOC sequestration will also promote more resilient systems through the co-benefits associated with SOM management. Most OECD (Organisation for Economic Co-operation and Development) national agencies promote policy guidance for implementing adaptation strategies in their cooperative effort (OECD, 2009).

International scale

After the 2008 and 2009 food price crises and food riots, the international debates began to consider more strongly the agriculture sector and the SOC sequestration options (Chapter 9, this volume). SOC is now recognized as a global environmental issue, and policies should capitalize on UN institutions that promote SOC sequestration. At first, initiatives were promoted individually by different UN agencies, such as the GEF, FAO, UNEP, International Fund for Agricultural Development (IFAD) or the World Bank. Now, most UN agencies are promoting convergent strategies, e.g. the climate-smart agriculture initiative and the Global Donor Platform for Rural development (http://www.donorplatform.org). The GSP is an example of a convergent strategy, i.e. through the development of action plans for the five pillars.

It is important to maintain, or even give force to, SOC initiatives within such international institutions, as they are also responsible for the management of funds and can influence widely the implementation of activities (for a comprehensive list of funds linked with mitigation, adaptation and REDD, see http://www.climatefundsupdate.org/listing). They are either directly implementing programmes to mitigate, as the Bio-carbon Fund from the World Bank or the MICCA programme from FAO. Moreover, they are also searching to maximize synergies between rural development and mitigation (Branca *et al.*, 2013), promoting the use of simple *ex ante* tools. For instance, the *ex ante* C-balance tool (EX-ACT, Bernoux *et al.*, 2010), an FAO initiative, has appraised during the past 4 years more than 30 development programmes and national policies, representing the equivalent of c.US$3.5 billion of investment in rural development (see http://www.fao.org/tc/exact/ex-act-applications/en/).

More recently, the GSP is positioned as the perfect entry place to centralize the policy debates concerning soils, and thus SOC sequestration. The ITPS established within the GSP will support the science–policy interfaces of the relevant environmental conventions, e.g. the IPCC and the IPBES, with the necessary scientific knowledge and advice on SOC.

Other relevant international institutions that might be fully concerned include the EU as the single biggest donor organization in the world, which can promote policies influencing all EU members as well as other countries in the world. However, the entry point is delicate; it can be at the higher level or at the national level. Chapter 29, this volume, details the effort made for the implementation of the Soil Thematic Strategy for Soil Protection (COM (2006) 231 final (EC, 2006a); COM (2006) 232 (EC, 2006b); and COM (2012) 46 (EC, 2012)). The Common Agricultural Policy (CAP) is the agricultural and rural development policy of the EU concerned with ensuring sufficient food at reasonable and stable prices (EC, 2009, 2012). With respect to soil protection, the CAP contributes preventing and mitigating soil degradation in order to build up SOM, enhance soil biodiversity and reduce soil erosion, contamination and compaction (Chapter 29, this volume). For several years, debates have been engaged in at the EU level to mainstream cross-compliance within agriculture policies. The EU defines cross-compliance as 'a mechanism that links direct payments to compliance by farmers with basic standards concerning the environment, food safety, animal and plant health and animal welfare, as well as the requirement of maintaining land in good agricultural and environmental condition' (http://ec.europa.eu/agriculture/envir/cross-compliance/index_en.htm). This last point can be an efficient entry point to advocate for SOC sequestration, and for payment of associated ecosystems services (PES). Other international organizations, FAO for example, are also working to mainstream PES into agricultural development and poverty reduction (FAO, 2011).

Concerning the CAP policies, Henriksen et al. (2011) recognized that synergies between mitigation and agricultural production were currently not exploited fully at the EU policy level, and suggested that there was much scope to encourage soil management strategies in Europe that would mitigate GHG emissions (Chapter 29, this volume).

The equivalent of NARS at the international level, for example the Consultative Group on International Agricultural Research (CGIAR) centres and their associated programmes, such as the Research Programme on Climate Change, Agriculture and Food Security (CCAFS), have to be involved in the policy-making path. Recently, 33 national initiatives from different countries were merged into a common effort under the Global Research Alliance (GRA) on agricultural GHG. GRA members set three research groups, namely croplands, paddy rice and livestock, and two additional cross-cutting groups directly in link with SOC: 'carbon and nitrogen cycling' and 'inventories and measurement'. Research at the international level includes the aggregation of national initiatives within international coalitions (see GRA above) or initiatives. Different international research programmes are also under way (e.g. the EU Ecofinders project, http://ecofinders.dmu.dk/) on the relation between biodiversity and C storage/mineralization (Chapter 11, this volume).

Governance

Environmental governance may be defined as the rules, practices and institutions for the management of the environment and the standards, values and behavioural mechanisms used by citizens, organizations and interest groups for exercising their rights and defending their interests in using natural resources. Good environmental governance takes into account the role of all actors that impact on the environment – from governments, NGOs, private sector and civil society. As UNEP (2010) notes, 'cooperation between all actors is critical to achieving effective governance ... towards a more sustainable future'. However, because of its complex linkages, governance for SOC raises substantial logistical and practical problems.

Natural resources, including SOC, and the environment are global public goods. Nevertheless, they are goods that also exist in the private domain, such as on farms. Therefore, SOC is often managed privately, but has impacts on atmospheric C that is unambiguously global. The global nature of these goods stems from their relationship between those who look after the resources – in the case of SOC, mainly land users as the immediate custodians and 'guardians' – and those who benefit from the public goods – mainly nation states, some far away (such as Pacific islands at sea level) and the global community. This is well recognized in the whole climate change debate, where it is clear to most that incremental small changes towards climate change mitigation controlled by individuals or small groups in the release of GHGs potentially benefits everyone.

The equivalent case for SOC has not received the same attention, although it intersects closely with GHGs and climate change. This planetary dimension requires a collective management approach with governance arrangements that are targeted appropriately for different stakeholders at different levels. The governance challenge, then, is how to generate a conducive enabling and mainstreaming environment for SOC that will encourage and incentivize the 'guardians' and exact suitable controls and penalties on those who choose to ignore the policy imperative. In addition, governance has to work at multiple scales and in different political and social environments.

What are the barriers for governance of SOC? Some key principles for SOC governance need to be elaborated. Governance structures must:

- Embed SOC in all levels of relevant decision making and action, so that decisions on the allocation of resources or the development of new curricula, for example, must take account of the importance of SOC.
- Conceptualize rural and urban communities and economic and political life as collective interest groups for the environment, which means that SOC becomes a responsibility for all levels of society – and not just an interest of land users. This is particularly relevant where incentives and penalties are to operate as part of governance structures – as they will have to with SOC.
- Emphasize the connection of people to the ecosystems in which they live and their collective reliance on the SOC in the ecosystems which provide goods and services. Goods and services are derived privately but emanate from a complex of private and public processes. Governance – or more explicitly, the point of intervention to achieve sustainable control – must balance private and public responsibility and not just target the weakest in society, the poor rural land user.
- Promote a transition from relatively simple linear systems (such as the promotion of land-use practices that maximize SOC) to circular positive-feedback systems that generate co-benefits and complex reinforcing interactions with multiple entry points (such as organic agriculture, markets, incentives and nutrition). The concept of Zero Net Land Degradation, proposed by the UNCCD secretariat, is an example (Chapter 29, this volume).

Local

Because of its nature and management, SOC requires participative local environmental governance arrangements. The principal actors involved are land users as the immediate guardians of SOC, local professionals, local government and NGOs (Plate 6). Key considerations (based on Vidal, 2005) will include:

- Utilizing local social capital, including local knowledge on soil resources, local leaderships and local education and training. Building SOC must become part of the local soil management discourse, familiarly discussed by men and women, old and young.
- Participation and information access: decision making on soil management and land use must be based on adequate and legitimate information. Access to this information is critical.
- Government presence: at the local level, the promoters of SOC are primarily professionals such as agricultural extension workers. They have to be trained adequately, resourced and incentivized. In some countries, such local professionals act as gatekeepers to subsidize and incentivize as well as being the source of technical information.
- Local institutional framework: informal mechanisms need to be employed for decentralized environmental governance, including forums for social interaction and agreements acceptable to local stakeholders. The institutional framework needs to address both collective action, such as the management of organic residues that are normally available as community-wide free resources, and private action that needs to be regulated for the collective good.

National

According to Benjamin and Fulton (2011, p. 2): 'without effective environmental governance at the national level, none of our sustainability aspirations – international or domestic – can be realized.' This highlights the pivotal role of good governance by nation states, not only in filtering down to the local level but also in aggregating up to the global and international levels. Because of its complex nature and multiple stakeholders, the governance of SOC will be especially challenging.

The same paper elaborates seven precepts for effective environmental governance:

- Environmental laws should be clear, even-handed, implementable and enforceable.
- Environmental information should be shared with the public.
- Affected stakeholders should be afforded opportunities to participate in environmental decision making.
- Environmental decision makers, both public and private, should be accountable for their decisions.
- Roles and lines of authority for environmental protection should be clear, coordinated and designed to produce efficient and non-duplicative programme delivery.
- Affected stakeholders should have access to fair and responsive dispute resolution procedures.
- Graft and corruption in environmental programme delivery can obstruct environmental protection and mask results and must be actively prevented.

Governance for SOC at the national level must be subject to laws and regulations appropriate to national procedures and enforcement mechanisms relevant to other closely related environmental issues. This would often entail explicitly coupling SOC maintenance to regulations for controlling soil degradation and incentives for encouraging soil conservation. A 'carrot-and-stick' approach may be adopted, providing penalties for losing SOC along with subsidies for increasing SOC.

A forum for governance for SOC is in the various National Action Plans written to meet country obligations under several of the environmental and sustainable development conventions. Examples where SOC needs to be included and/or strengthened in its profile and importance are:

- National Action Plan on Climate Change (NAPCC); required by signatory countries to the UNFCCC.
- National Action Plans (NAPs) to combat desertification; required by the UNCCD.
- Nationally Appropriate Mitigation Actions (NAMA) of developing country partners; required by the UNFCCC.
- Agenda 21 national plans under the lead of the UN Commission on Environment and Development, and National Sustainable Development Strategies (NSDS).
- National and regional agricultural policies (e.g. EU Common Agricultural Policy).

There is an unfortunate tendency for these national action plans to have very limited circulation and be unknown outside the ministry tasked with delivering the text – often a relatively weak ministry such as environment. Governance at national level needs to be mainstreamed into ministries of agriculture, finance, planning and economic development. SOC suffers at two levels in national governance arrangements. First, there is no single obvious repository for policy on SOC to be included in national plans – therefore, it becomes nobody's clear responsibility. Finance ministries are unlikely to take up the challenge, while environment, agriculture, rural development, energy and other possible line agencies have many other priorities. Second, while SOC has a relatively inconspicuous profile at the national level, the accomplishment of better governance arrangements will remain unachievable as a single priority issue. In the meantime, national-level better governance for SOC will necessarily have to ride on the 'coat-tails' of more high-profile issues such as climate change, pollution and food security.

International

At the international level, global environmental governance is 'the sum of organizations, policy instruments, financing mechanisms, rules, procedures and norms that regulate the processes of global environmental protection'. Therefore, the international level is largely the sum of national governance and reporting arrangements, with global financing mechanisms such as the GEF (for land degradation, climate change and biodiversity) and the Global Mechanism (for desertification control). The GEF, for example, is currently negotiating a new strategy with its developed country parties – its donors – for the next replenishment (GEF-6; 2014–2018) that includes SOC. If soil carbon gets explicit mention as either or both of objectives and impact indicator, then some measure of compliance will inevitably follow. The financing incentive is probably the most influential mechanism in governance, alerting country parties to the global environmental conventions that money will only follow inclusion of SOC as something worth accumulating.

There is a large literature on international environmental governance – for example Aggarwal-Khan (2012) – and the difficulties of achieving compliance, when almost all agreements are non-binding. Compliance is primarily via the open release of national action plans on convention websites. Therefore, governance is essentially by national peer pressure, identifying the 'guilty' or the 'weak' and praising those countries that have comprehensive plans and strategies. For the same reasons as at the national level, SOC struggles to achieve a high profile and must necessarily be placed as a subset of more high-profile issues.

One of the possibilities for better governance arrangements internationally is to re-seek some way through international agreements to design carbon markets that include SOC accumulations by developing countries in return for payments by developed countries. REDD is an effort to create a financial value for the C stored in forests, offering incentives for developing countries to reduce emissions from forested lands and invest in low-carbon paths to sustainable development. 'REDD+' goes beyond deforestation and forest degradation and includes the role of conservation, the sustainable management of forests and the enhancement of forest C stocks. REDD+ may be a model for SOC also to be included in C accounting. Verification and compliance will need to be addressed and made more robust.

Recommendations

The disconnection between the scientific and the policy arenas at the international level can only be solved with an innovative language through clear and simple messages (see also the overview in Table 5.1).

Policy imperative

- At the **local level**, programmes aiming at restoring SOC should address the resource capacity, socio-economic circumstances and field production conditions of the farmers.
- At the **national level**, land planning, economic incentives, legal frameworks and regulatory mechanisms should stimulate soil protection and SOC improvement.
- The starting point for entering SOC into the policy arena should be to emphasize its key cross-cutting role in high-profile topics such as food security, sustainable development and climate change scenarios.

Policy profile and discourse

- At the **local level**, it is necessary to adopt an appropriate discourse that focuses on: (i) the holistic perception of the soil fertility from farmers; and (ii) the economical access of farmers to resources.
- At the **national level**, training programmes and events to address the public should be adopted, passing on the message of soil as a heritage and of the importance of SOM in adequate soil management.
- At the **international level**, the focus should be on coordination and communication: (i) all the jeopardized efforts to raise awareness by different hotspots should team up under a unique organization such as the GSP; (ii) all types of media should be explored and broadened with an innovative approach (cultural, services, society concerns, educational targets); (iii) an international agenda for soil should be established to promote and speed up the setting up of all the previous recommendations.

Policy rationale

- An economic valuation of SOC should be based on a methodology to quantify the soil capital, and in particular the contribution of SOC, in order to preserve this heritage for future generations.

Policy support

- At the **local level**, develop programmes that pull together information on cropping systems and practices in developing countries and build capacity at the local level to use tools and methods.
- Harmonize **national-level** SMNs so they can be compared with each other and include the parameters needed to make estimates of SOC (e.g. bulk density).
- Facilitate actions to enable data exchange where there are national or regional restrictions.

Advocates and institutions

- Strategy: move beyond UNFCCC using SOC as an indicator of other co-benefits, to link better with biodiversity, food security and sustainable development.

Governance

- SOC must find a set of flexible narratives and discourses that create a strong 'story' in compelling language suitable for multiple policy arenas.
- The relevance and importance of SOC must be embedded in all levels of decision making and action, via the high-profile global accords (such as UNCCD, United Nations Conference on Environment and Development (UNCED)), national action plans (NAPs, NAMAs) and institutions at all levels, such as NGOs and NARS.
- Good governance must underpin initiatives to prioritize SOC with arrangements that provide specifically targeted incentives/subsidies and penalties for non-compliance.

Acknowledgement

This work was partially supported by the European Union within the project EcoFINDERS (FP7-264465).

Notes

[1] As an example, the Commission 4.4 'Soil education and public awareness' of the International Union of Soil Science (IUSS) organize every 4 years within the International Congress of Soil Science some symposiums on this subject.
[2] First Global Soil Week 2012. Rapporteurs' Reports. Global Soil Forum. Institute for Advanced Sustainability Studies (IASS), Potsdam, Germany.
[3] 'Securing healthy soils for a food secure world', a day dedicated to soils (http://www.fao.org/globalsoilpartnership/world-soil-day/en/).

References

Aggarwal-Khan, S. (2012) *The Policy Process in International Environmental Governance*. Palgrave-Macmillan, London, 208 pp.
Antle, J.M. and Stoorvogel, J.J. (2003) Agricultural carbon sequestration, poverty, and sustainability. *Environment and Development Economics* 13(3), 327–352.

ARD (2013) *Soil Carbon Sequestration* (http://www-esd.worldbank.org/SoilCarbonSequestration, accessed 4 July 2014).

Bationo, A., Kihara, J., Vanlauwe, B., Waswa, B. and Kimetu, J. (2007) Soil organic carbon dynamics, functions and management in West African agroecosystems. *Agricultural Systems* 94, 13–25.

Batjes, N.H., Al-Adamat, R., Bhattacharyya, T., Bernoux, M., Cerri, C.E.P., Gicheru, P., Kamoni, P., Milne, E., Pal, D.K. and Rawajfih, Z. (2007) Preparation of consistent soil data sets for modelling purposes: secondary SOTER data for four case study areas. *Agriculture, Ecosystems and Environment* 122, 26–34.

Batjes, N.H., Reuter, H.I., Tempel, P., Hengl, T., Leenaars, J.G.B. and Bindraban, P.S. (2013) Development of global soil information facilities. *CODATA Data Science Journal* 12, 70–74. dx.doi.org/10.2481/dsj.WDS-009.

Bernoux, M. and Chevallier, T. (2013) Carbon in dryland soils – multiple essential functions. *CSFD Thematic Report no. 10*, CSFD/Agropolis International, Montpellier, France, 40 pp. (Available in French and English at http://www.csf-desertification.eu, accessed 4 July 2014).

Benjamin, A.H. and Fulton, S. (2011) *Effective national environmental governance – a key to sustainable development.* Paper to the First Preparatory Meeting of the World Congress on Justice, Governance and Law for Environmental Sustainability, Kuala Lumpur,.

Bernoux, M., Branca, G., Carro, A., Lipper, L., Smith, G. and Bockel, L. (2010) Ex-ante greenhouse gas balance of agriculture and forestry development programs. *Scientia Agricola* 67, 31–40.

Bhattacharyya, T., Pal, D.K., Easter, M., Batjes, N.H., Milne, E., Gajbhiye, K.S., Chandran, P., Ray, S.K., Mandal, C., Paustian, K. *et al.* (2007) Modelled soil organic carbon stocks and changes in the Indo-Gangetic Plains, India from 1980 to 2030. *Agriculture, Ecosystems and Environment* 122, 84–94.

Branca, G., Hissa, H., Benez, M.C., Medeiros, K., Lipper, L., Tinlot, M., Bockel, L. and Bernoux, M. (2013) Capturing synergies between rural development and agricultural mitigation in Brazil. *Land Use Policy* 30, 507–518.

Brander, M., Tipper, R., Hutchison, C. and Davis, G. (2009) *Consequential and Attributional Approaches to LCA: A Guide to Policy Makers with Specific Reference to Greenhouse Gas LCA of Biofuels.* Technical Paper TP-090403-A, Ecometrica Press, London.

Clay, E.J. and Schaffer, B.B. (eds) (1984) *Room for Manoeuvre. An Explanation of Public Policy in Agriculture and Rural Development.* Heinemann, London.

Colomb, V., Touchemoulin, O., Bockel, L., Chotte, J.-L., Martin, S., Tinlot, M. and Bernoux, M. (2013) Selection of appropriate calculators for landscape-scale greenhouse gas assessment for agriculture and forestry. *Environmental Research Letters* 8, 015029, doi:10.1088/1748-9326/8/1/015029.

Compagnone, C., Sigwalt, A. and Pribetich, J. (2013) Les sols dans la tête. Conceptions et modes de production des agriculteurs. *Etude et Gestion des sols* 20, 81–95.

Costanza, R. and Daly, H.E. (1992) Natural capital and sustainable development. *Conservation Biology* 6, 37–46.

Cox, P.M., Betts, R.A., Jones, C.D., Spall, S.A. and Totterdell, I.J. (2000) Acceleration of global warming due to carbon-cycle feedbacks in a coupled climate model. *Nature* 408, 184–187.

Diels, J., Aihou, K., Iwuafor, E.N.O., Merckx, R., Lyasse, O., Sanginga, N. and Deckers, J. (2002) Options for soil organic carbon maintenance under intensive cropping in the West African savannah. In: Vanlauwe, B., Diels, J., Sanginga, N. and Merckx, R. (eds) *Integrated Plant Nutrient Management in Sub-Saharan Africa.* CAB International, Wallingford, UK, pp. 299–312.

Easter, M., Paustian, K., Killian, K., Williams, S., Feng, T., Al-Adamat, R., Batjes, N.H., Bernoux, M., Bhattacharyya, T., Cerri, C.C. *et al.* (2007) The GEFSOC Soil Carbon modelling system: a tool for conducting regional scale soil carbon inventories and assessing the impacts of land use change on soil carbon. *Agriculture Ecosystems and Environment* 122(1), 13–25.

EC (2006a) (COM (2006) 231 final) Communication from the Commission to the Council, the European Parliament, the European Economic and Social Committee of the Regions, Thematic Strategy for Soil Protection, Brussels. European Commission, Brussels.

EC (2006b) (COM (2006) 232) Proposal for a Directive of the European Parliament and of the Council establishing a framework for the protection of soil and amending Directive 2004/35/EC. European Commission, Brussels.

EC (2009) (EC No 73\2009) Council Regulation establishing common rules for direct support schemes for farmers under the common agricultural policy and establishing certain support schemes for farmers, amending Regulations (EC) No 1290/2005, (EC) No 247/2006, (EC) No 378/2007 and repealing Regulation (EC) No 1782/2003. European Commission, Brussels.

EC (2010) (COM (2010) 672 final) Communication from the Commission to the European Parliament, the Council, the European Economic and Social Committee and the Committee of the Regions. The CAP towards 2020: Meeting the food, natural resources and territorial challenges of the future. European Commission, Brussels.

EC (2012) (COM (2012) 46 final) Report from the Commission to the European Parliament, the Council, the European Economic and Social Committee and the Committee of the Regions. The implementation of the Soil Thematic Strategy and ongoing activities. European Commission, Brussels.

FAO (2011) *Payments for Ecosystem Services and Food Security*. FAO, Rome, 300 pp. (http://www.fao.org/docrep/014/i2100e/i2100e.pdf, accessed 4 July 2014).

FAO (2013) Harmonized World Soil Database (http://www.fao.org/nr/lman/abst/lman_080701_en.htm, accessed 4 July 2014).

Feindt, P.H. and Oels, A. (2005) Does discourse matter? Discourse analysis in environmental policy making. *Journal of Environmental Policy and Planning* 7, 161–173.

Freibauer, A., Rounsevell, M., Smith, P. and Verhagen, J. (2004) Carbon sequestration in the agricultural soils of Europe. *Geoderma* 122, 1–23.

Govers, G., Merckx, R., Van Oost, K. and van Wesemael, B. (2013) *Managing Soil Organic Carbon for Global Benefits: A STAP Technical Report*. Global Environment Facility, Washington, DC.

Henriksen, C.B., Hussey, K. and Holm, P.E. (2011) Exploiting soil-management strategies for climate mitigation in the European Union: maximizing 'win–win' solutions across policy regimes. *Ecology and Society* 16(4), 22. (http://dx.doi.org/10.5751/ES-04176-160422, accessed 10 November 2014).

Jones, R.J.A., Hiederer, R., Rusco, E. and Montanarella, L. (2005a) Estimating organic carbon in the soils of Europe for policy support. *European Journal of Soil Science* 56, 655–671.

Jones, C., McConnell, C., Coleman, K., Cox, K., Falloon, P., Jenkinson, D. and Powlson, D. (2005b) Global climate change and soil carbon stocks; predictions from two contrasting models for the turnover of organic carbon in soil. *Global Change Biology* 11, 154–166.

Juma, C. and Clark, N. (1995) Policy research in sub-Saharan Africa: an emploration. *Public Administration and Development* 15, 121–137.

Kamoni, P.T., Gicheru, P.T., Wokabi, S.M., Easter, M., Milne, E., Coleman, K., Falloon, P. and Paustian, K. (2007) Predicted soil organic carbon stocks and changes in Kenya between 1990 and 2030. *Agriculture Ecosystems Environment* 122, 105–113.

Lahmar, R. and Ribaut, J.P. (2001) *Sols et sociétés*. Regards pluriculturels. Editions Charles Léopold Mayer, Paris, 218 pp.

Lal, R. (2004) Soil carbon sequestration impacts on global climate change and food security. *Science* 304(5677), 1623–1627.

Lal, R., Follet, R.F. and Kimble, J.M. (2003) Achieving soil carbon sequestration in the United States: a challenge for policy makers. *Soil Science* 168(12), 827–845.

Lal, R., Cerri, C.C., Bernoux, M., Etchevers, J. and Cerri, C.E.P. (2006) *Carbon Sequestration in Soils of Latin America*. Food Products Press, The Haworth Press Inc, Binghamton, New York, ISBN: 978-1-56022-136-4, 554 pp.

Lipper, L., Dutilly-Diane, C. and McCarthy, N. (2010) Supplying carbon sequestration from West African Rangelands: opportunities and barriers. *Rangeland Ecology and Management* 63, 155–166.

McCarthy, N., Lipper, L. and Branca, G. (2011) Climate-smart agriculture: smallholder adoption and implications for climate change adaptation and mitigation. *Mitigation of Climate Change in Agriculture* Series 4, Food and Agriculture Organization of the United Nations (FAO), Rome.

Milne, E., Al-Adamat, R., Batjes, N.H., Bernoux, M., Bhattacharyya, T., Cerri, C.C., Cerri, C.E.P., Colemen, K., Easter, M., Falloon, P. *et al*. (2007) National and sub-national assessments of soil organic carbon stocks and changes: the GEFSOC modelling system. *Agriculture Ecosystems and Environment* 122, 3–12.

Milne, E., Paustian, K., Easter, M., Batjes, N.H., Cerri, C.E.P., Kamoni, P., Gicheru, P., Oladipo, E.O., Stocking, M., Hartman, M. *et al*. (2010) Estimating the carbon benefits of sustainable land management projects: the carbon benefits project component A. In: Gilkes, R.J. and Prakongkep, N. (eds) *Proceedings of the 19th World Congress of Soil Science, Soil Solutions for a Changing World*. International Union of Soil Sciences, 1–6 August, Brisbane, Australia, pp. 73–75.

Milne, E., Neufeldt, H., Rosenstock, T., Smalligan, M., Cerri, C.E., Malin, D., Easter, M., Bernoux, B., Ogle, S., Casarim, F. *et al*. (2013) Methods for the quantification of GHG emissions at the landscape level for developing countries in smallholder contexts. *Environmental Research Letters* 8, doi:10.1088/1748-9326/8/1/015019.

Montanarella, L. and Vargas, R. (2012) Global governance of soil resources as a necessary condition for sustainable development. *Current Opinion in Environmental Sustainability* 4, 559–564.

OECD (Organisation for Economic Co-operation and Development) (2009) Integrating Climate Change Adaptation into Development Co-operation: Policy Guidance (http://www.oecd.org/dac/43652123.pdf, accessed 4 July 2014).

Omuto, C., Nachtergaele, F. and Vargas Rojas, R. (2012) State of the Art Report on Global and Regional Soil Information: Where Are We? Where to Go? FAO, Rome.

Pascual, U., Muradian, R., Brander, L., Gómez-Baggethun, E., Martín-López, M., Verman, M., Armsworth, P., Christie, M., Cornelissen, H., Eppink, F. et al. (2010) The economics of valuing ecosystem services and biodiversity. In: Kumar, P. (ed.) *The Economics of Ecosystems and Biodiversity Ecological and Economic Foundations,* Chapter 5. Earthscan, London, pp. 183–256.

Scharlaman, J.P.W., Hiederer, R. and Kapos, V. (2014) *Global Map of Terrestrial Soil Organic Carbon Stocks*, UNEP-WCMC & EU-JRC, Cambridge, UK.

Smith, P. (2004) How long before a change in soil organic carbon can be detected? *Global Change Biology* 10, 1878–1883.

UNEP (2010) Environmental governance. United Nations Environment Programme, Nairobi (http://www.unep.org/pdf/brochures/EnvironmentalGovernance.pdf, accessed 4 July 2014).

van Noordwijk, M., Cerri, C., Woomer, P.L., Nugroho, K. and Bernoux, M. (1997) Soil carbon dynamics in the humid tropical forest zone. *Geoderma* 79, 187–225.

van Wesemael, B., Paustian, K., Andrén, O., Cerri, C.E.P., Dodd, M., Etchevers, J., Goidts, E., Grace, P., Kätterer, P., McConkey, B.G. et al. (2011) How can soil monitoring networks be used to improve predictions of organic carbon pool dynamics and CO_2 fluxes in agricultural soils? *Plant and Soil* 338, 247–259.

Vidal, M.P. (2005) *Gobernanza Ambiental Decentralisada*. Fondo Mink'a de Chorlavi, Santiago do Chile, 6 pp.

WOCAT (2013) http://www.wocat.net (accessed 19 March 2013).

Zdruli, P., Jones, R.J.A. and Montanarella, L. (2004) *Organic Matter in the Soils of Southern Europe*. European Soil Bureau Technical Report, EUR 21083 EN. Office for Official Publications of the European Communities, Luxembourg, 16 pp.

6 Soil Formation

Marty Goldhaber* and Steven A. Banwart

> Essentially, all life depends upon the soil ... There can be no life without soil and no soil without life; they have evolved together.
>
> USDA Yearbook of Agriculture (1938) by Charles E. Kellogg

Abstract

Soil formation reflects the complex interaction of many factors, among the most important of which are (i) the nature of the soil parent material, (ii) regional climate, (iii) organisms, including humans, (iv) topography and (v) time. These processes operate in Earth's critical zone; the thin veneer of our planet where rock meets life. Understanding the operation of these soil-forming factors requires an interdisciplinary approach and is a necessary predicate to charactering soil processes and functions, mitigating soil degradation and adapting soil management to environmental change. In this chapter, we discuss how these soil-forming factors operate both singly and in concert in natural and human modified environments. We emphasize the role that soil organic matter plays in these processes to provide context for understanding the benefits that it bestows on humanity.

Introduction

In this paper, we examine the processes of soil formation, emphasizing factors that influence or are influenced by the organic matter. Soil organic matter is crucial to the functioning of natural and managed ecosystems. It is also a significant component of the global carbon (C) budget. Excluding carbonate rocks, it represents about 1500 Gt C in the top 1 m, and possibly another 900 Gt C at a depth of 1–2 m (e.g. Batjes, 1996). This mass is approximately twice that of the carbon held in the atmosphere and three times the amount held in terrestrial vegetation. For these reasons, the behaviour of soil organic matter during soil formation and erosion is an important topic at all scales, from local to global. In this paper, soil organic carbon (C_{org}) will be taken as a proxy for organic matter, unless specified otherwise.

We approach the subject of soil formation with a broad perspective rooted in geology. The reason is that soil formation is a reflection of processes occurring over a greater vertical interval of Earth's near surface, rather than the soil profile itself. In fact, soils are nested within Earth's critical zone (CZ), which represents

---

*E-mail: mgold@usgs.gov

a thin veneer of our planet where rock meets life. The critical zone is 'the heterogeneous, near-surface environment in which complex interactions involving rock, soil, water, air, and living organisms regulate the natural habitat and determine the availability of life-sustaining resources' (National Research Council, 2001; Figs 6.1 and 6.2).

In order to discuss soil formation in this CZ context, we adopt the definitions of the CZ system of Brantley et al. (2011), noting that other definitions may exist. The base of the CZ is the *parent material* – earth material that was present prior to soil formation, be it bedrock or unconsolidated sediment. Alteration of bedrock leads to *regolith*; the mantle of unconsolidated and altered material derived from the parent material. Regolith is the transportable material at Earth's surface. Unconsolidated parent material and regolith can be transported great distances and deposited, where soil formation subsequently proceeds, potentially far from the location of the initial exposure of the parent material to Earth's surface. Saprolite is the zone within the regolith where bedrock alters in place, generally retaining evidence of the parent material texture and fabric. Weathering refers to processes that turn parent material into regolith. In general, weathering is a consequence of the exposure of earth materials, formed at elevated temperature and pressure, to near-surface conditions at which they are thermodynamically unstable. The regolith–parent material interface is the *weathering front*.

The most important consequence of weathering is to form *soil*, which is the layer capping the regolith that has been altered extensively by chemical, physical and biological processes, often leading to the development of horizons (unique layers within the soil). There is substantial complexity in how these horizons develop, given the wide variety in the interaction of the state factors of soil formation (described below). Likewise, the terminology to describe soil layers is complex and differs around the globe. However, in simplified form, the layers from top to bottom commonly consist of: the O horizon, a layer of plant residues; the A horizon, a layer of mineral soil depleted in iron, clay, and aluminum and soluble constituents; the B horizon, which has accumulated iron, clay and aluminium; and the C horizon, the parent material. Typically, soil organic carbon is most elevated near the surface (O and A horizons) compared to deeper layers.

With this conceptual model in mind, we can summarize soil formation, by a mass balance approach, as primarily the product of weathering losses of mineral elements from

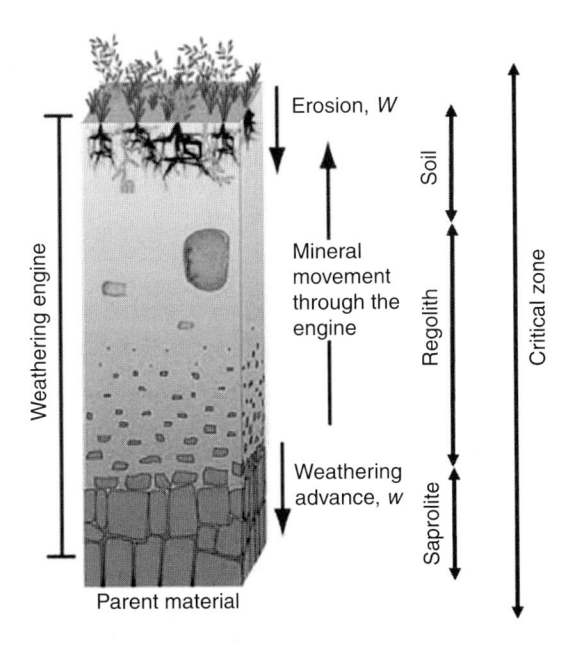

Fig. 6.1. Earth's weathering engine and the critical zone. Uplift of continents and exposure of rock provides a flux of reactive soil parent material at Earth's surface. This weathering engine is an enormous, planetary-scale, biogeochemical reactor. The other inputs to form soil are the fluxes of infiltrating precipitation and solutes, the addition of organic carbon that has been fixed by plant photosynthesis above the surface and the diffusion of atmospheric gases. As soil layers form, they are also removed by ongoing physical erosion and chemical denudation of elements from the soil profile – by aeolian and hydrological transport to depositional environments and to groundwater. (From Brantley et al., 2007.)

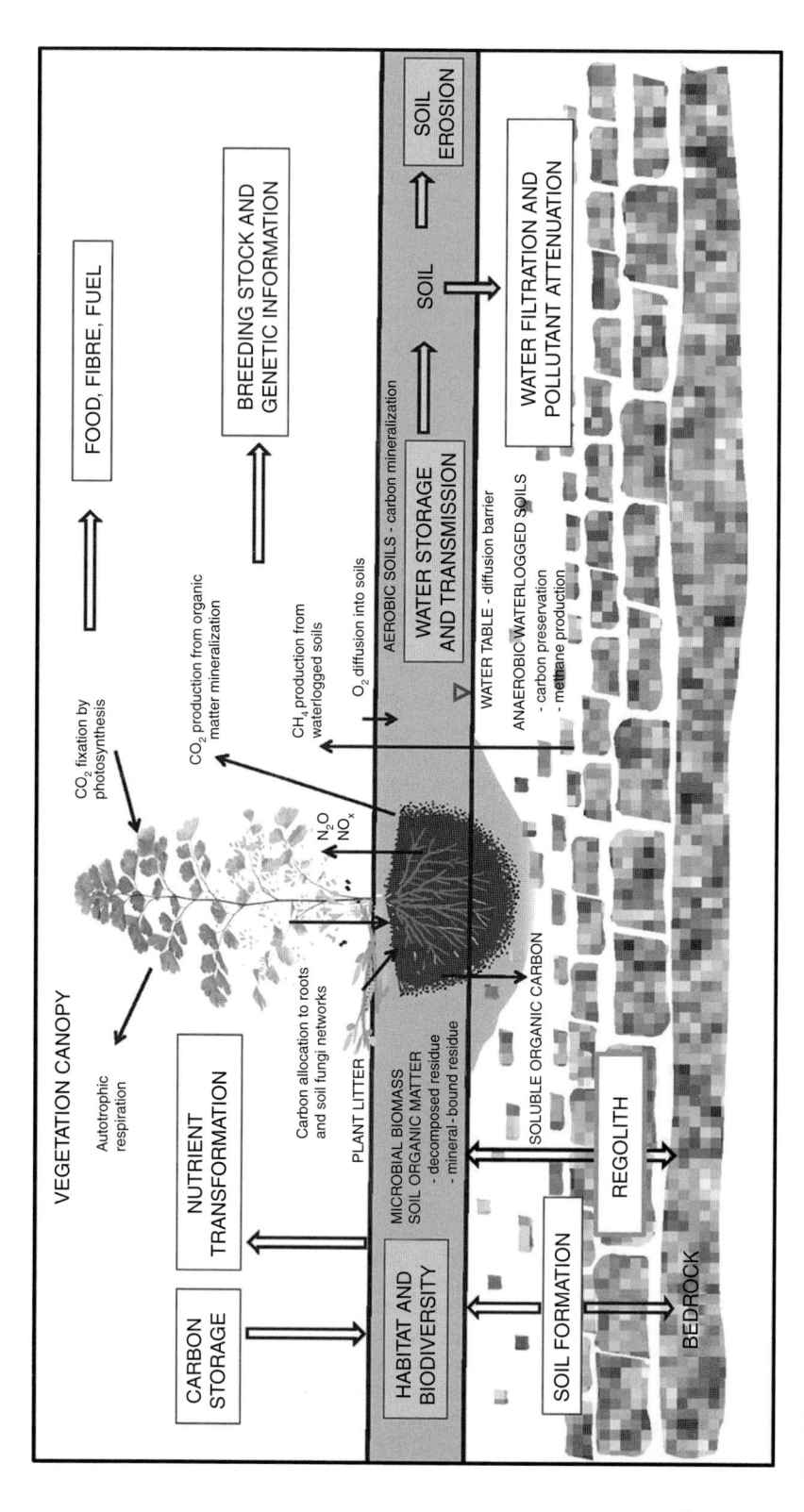

Fig. 6.2. As soil forms from the parent material dissolve, contributing solutes to the pore water solution that can be hydrologically transported from the soil profile. Within the soil profile, secondary mineral phases form by pseudomorphic mineral alteration or precipitation from solution, particularly micro- and nano-sized fragments of clay minerals and binary oxide minerals of Fe and Al. Organic carbon addition to the saprolite occurs by deposition of dead plant litter, which is vertically mixed by soil fauna, and through the allocation of photosynthate carbon belowground via root growth, allocation to microbial root symbionts such as mycorrhizal fungi and as organic exudates. All of these organic inputs provide a carbon and energy source to support heterotrophic organisms as organic matter decomposers and other functional communities of the microbial ecosystem of the soil. Living organisms bind with decaying biomass and mineral fragments into larger aggregates, resulting in the development of soil structure – the mass distribution of aggregates by size, also reflecting the pore size distribution. Thus, soil formation occurs at the reactive interface of Earth's surface where parent material meets plant inputs and infiltrating water. (From Victoria et al., 2012.)

the initial parent material and additions from plants (Amundson, 2005). To these we must also include alteration of parent material, as well as external inputs from sources other than plants. The soil production rate is the rate at which saprolite is converted to soil. To expand upon these two statements, it is first useful to recognize that weathering and soil formation are coupled through the 'state factors of soil formation', originally described by Dokuchaev (1880; cited by Fortescue, 1992). A description of the factors of soil formation that is highly relevant today was given by Jenny (1941). These are: (i) nature of soil parent material; (ii) regional climate; (iii) organisms including humans; (iv) topography; and (v) time. The combined influence of these soil-forming factors determines the properties of a soil. Understanding the complex interactions among these factors requires an interdisciplinary scientific approach (Brantley et al., 2007). Further, characterizing soil formation is complicated by the fact that the processes involved occur over distance scales from the submicroscopic to planetary (see Fig. 6.1) and timescales from microseconds to millions of years (Brantley et al., 2007; Brantley, 2008).

Separation of the State Variables

Although the processes leading to soil formation are interdependent, these can be separated under some circumstances, thus permitting clarification of the role of the individual variables.

The role of organisms

The recognition that vegetation plays an important role in weathering can be traced back to Belt (1874; cited by White, 2003). He wrote that 'the percolation through rocks of rain water charged with a little acid from decomposing vegetation' accelerated weathering. Plants are the main source of carbon to soils through tissue residues, root exudates and symbiotic fungi.

To appreciate the profound global role that plants have played in weathering and soil formation, it is useful to consider the broad sweep of Earth's history. As expressed by Amundson et al. (2007), 'The surface of our planet is the result of billions of years of feedback between biota and earth materials.' Paleosols, soils formed in Earth's past and preserved in the rock record, document the profound influence of biota on Earth surface processes. Weathering processes during Earth's early history differed from modern ones as the result of the planetary-scale evolution of Earth's biota and feedbacks of this evolution to the atmosphere. During the Archean (i.e. prior to 2.5 Ga), soil formed under a reducing atmosphere, resulting in loss of iron from the profile (Rye and Holland, 1998). None the less, these very early soils are thick, clay-bearing, exhibit loss of the base cations Ca^{2+}, Mg^{2+} and Na^+, and based on carbon isotopic evidence, show traces of life (see review by Retallack, 2005). The intensity of Archean weathering may reflect an elevated atmospheric CO_2 content compared to today. Later, colonization of land by mosses, fungi and liverworts by 700 Ma led to enhanced clay formation and burial of organic C at continental margins. Kennedy et al. (2006) proposed that the colonization of land surfaces by fungi and other organisms stabilized soil cover, created longer groundwater residence times, added organic acids and chelating agents and drastically enhanced secondary mineral formation, i.e. produced a 'clay factory'.

The advent of land plants was a major event in the Devonian that influenced weathering and soil formation dramatically. Land plants altered the balance of atmospheric CO_2 by the removal of C to plants and soil (Beerling and Berner, 2005). During the early Devonian period (about 420 Ma), small, leafless plants populated the landscape, but by the end of the Devonian (about 359 Ma), large trees had evolved. This late Devonian flora was associated with complex and deep rooting systems. Deep rooting likely led to the enhanced dissolution of bedrock (Berner, 1997; Retallack, 1997).

Recently, the roles of mycorrhizal fungi in weathering have received increased attention

(Taylor et al., 2009). These organisms are symbiotic to plants and play a key role in providing plants with inorganic and organic nutrients that they might not otherwise be able to obtain. Taylor et al. (2009) reviewed evidence that the rise of land plants during the Devonian and subsequent expansion of angiosperms from the Cretaceous onwards, and the concomitant drops in atmospheric CO_2, coincided with the co-evolution of land plants, respectively, with these fungi. Subsequent evidence was provided through mechanistic mathematical modelling of biological weathering processes (Taylor et al., 2011) and experimental studies under differing modern analogues to evolutionary tree physiology and mycorrhizal associations (Quirk et al., 2012). These studies demonstrated that these co-evolutionary trends could explain the enhanced biological weathering of the continents associated with soil formation as the mechanism for atmospheric CO_2 decline.

The role of time

The role of time in soil formation, with insights into biotic influences, can be illustrated through consideration of soil chronosequences. Chronosequences are suites of soils of different ages in topographically stable environments evolved under similar conditions of vegetation, climate and parent material (see Birkland, 1999, for an overview). Chronosequences permit the isolation of the variable time in weathering and soil formation and are, in effect, the result of a natural experiment. It also should be noted that the relative stability of the state factors of soil formation that define a chronosequence are not likely to be replicated for the majority of soil weathering environments. Nevertheless, the study of chronosequences has proved to be a powerful tool to understand critical zone processes and soil-forming processes.

A key observation based on chronosequence studies is that the early stages of soil formation are associated with volumetric increase due to rock break-up, whereas the later stages are characterized by volumetric decrease due to mass loss. These conclusions rest in significant part on the concepts introduced by Brimhall and co-workers (Brimhall and Dietrich, 1987; Brimhall et al., 1991). They defined soil strain ($\in$), which is the change in volume relative to the initial volume of parent material, by the equation:

$$\in = \left(\frac{\rho_p C_{i,p}}{\rho_s C_{i,s}} - 1 \right) \quad (6.1)$$

where ρ_p and ρ_s are the bulk density of the parent material and soil, respectively, and $C_{i,p}$ and $C_{i,s}$ are the concentration (wt %) of an immobile element i in the parent and soil, respectively. Normalization to an index element, for which immobility is assumed, is required because of variable gains and losses of both the major and minor elements during weathering. The elements zirconium (Zr) and titanium (Ti) are commonly employed as index elements, but niobium (Nb) and tantalum (Ta) might be better choices in strongly weathered soil (Kurtz et al., 2000).

As noted above, positive strain (expansion) has been shown to be characteristic of the early stages of weathering processes. For example, Egli et al. (2001) reported on a chronosequence of glacial moraines in the Swiss Alps ranging in age from 150 to 10,000 ybp. They found that expansion (positive strain) was a characteristic of young soils due to the increased porosity of soil compared to rock, as well as the incorporation of low-density organic matter near the surface. Likewise, initial expansion over times measured in thousands to tens of thousands of years characterizes both a 240 ka chronosequence formed on beach terraces in California (Merritts et al., 1992), and a ~4 My chronosequence formed on dated basalt flows in Hawaii (Vitousek et al., 1997). Thus, disaggregation of rock and increased porosity characterizes the initial processes of soil formation over a broad range of climatic conditions. Biological processes largely drive this expansion. The dominant biological processes are associated with the accumulation of roots and organic matter. Roots exert significant force that can break apart the parent rock, initially entering cracks as small as 100 μm and exerting pressures over 1 MPa (Gabet et al., 2003).

With time, volumetric collapse replaces expansion in many weathering sequences, as chemical weathering processes lead to elemental loss from the parent material. In topographically stable environments, low rates of physical denudation implies that volumetric loss is caused by chemical denudation – the removal of elements that are transported hydrologically from the soil profile as solutes (Brantley et al., 2011). This transition is in part facilitated by increased access to water and the increased surface area of mineral particles during the early expansion phase. In the California beach terrace and Hawaii chronosequence studies mentioned above, this transition from expansion to collapse required times exceeding 40 Ky. However, elemental gains may also occur, for example by the addition of dust at the soil surface (Derry and Chadwick, 2007). As is the case for volumetric change, characterizing element losses and gains requires normalization to an index element according to the equation (Brimhall and Dietrich, 1987):

$$\tau_j = \frac{\dfrac{C_{j,w}}{C_{j,p}}}{\dfrac{C_{i,w}}{C_{i,p}}} - 1 \quad (6.2)$$

where τ_j is the mass transfer coefficient of the jth element ($\tau_j = 0$ means no mobilization has occurred, -1 reflects complete removal and positive values reflect addition). White et al. (2002) gave an example illustrating the application of Eqn (6.2) for weathering of a kaolinitic soil overlying porous saprolite formed from granodioritic bedrock located in the state of Georgia, USA. They found that τ values for the major rock-forming elements clustered near zero in the deeper bedrock. However, calcium (Ca) and sodium (Na) were removed progressively within the shallow bedrock (with τ values approaching -1 within the bedrock itself). Potassium and magnesium (Mg) were mobilized within the overlying saprolite.

The depth dependency of these mass transfer coefficients reflects the weathering characteristics of specific minerals, particularly silicates that represent 90% of Earth's crust. These weathering reactions are a fundamental control on soil formation and reflect the influence of parent material. Brantley (2005) calculated that the mean lifetime of a 1 mm crystal at pH 5 might differ by over two orders of magnitude between fayalite (Fe_2SiO_4 – 1900 years) and K-feldspar ($KAlSi_3O_8$ – 740,000 years); removal of the mineral, quartz, might require 34,000,000 years under these conditions. In the case of granodiorite referenced above, Na and K removal in the early stages of regolith formation reflects removal of plagioclase ($NaAlSi_3O_8$) and K-feldspar, respectively. The rate and extent of element removal is governed by a range of factors, as summarized by Brantley (2005) and White (2003). These differences in the weathering rates of individual minerals are an important control that initial parent material can exert on soil formation.

Time, as well as the nature of parent material, also plays a significant role in the availability of the important soil nutrients, nitrogen (N) and phosphorus (P) (see review by Vitousek et al., 2010). Initially, soil P is supplied predominantly from the parent rock in the form of minerals such as apatite (($Ca_5(PO_4)_3(OH,F,Cl)$)). The P content of the parent rock is thus a variable that provides an initial condition on the P content of the formed soil. With time, the initial endowment of P is depleted by plant uptake, such that very old soils tend to be P deficient. In contrast, with the exception of some sedimentary rocks, N is nearly absent in the parent material. However, over time, symbiotic microbial N fixers develop in the soil and provide a mechanism to convert atmospheric N to nutrient forms for biomass production. As production occurs over time, the organic N content of the soil accumulates with C_{org} and provides a sustained reservoir of nutrient N. A sequence of soil N limitation followed by P limitation may thus develop with time (Lambers et al., 2008).

Data from chronosequences have also aided in the characterization of the evolution of organic carbon during soil formation. In the Egli et al. (2001) glacial moraine study, soil organic carbon content was correlated linearly with the degree of initial expansion of the parent rock. More generally, the organic carbon content of soil has been shown to increase with time in chronosequences. Typically, there is an initial rapid increase in organic carbon during the early stages of soil formation (e.g. Harden et al., 1992). In a recent study of soil formation processes in the forefield of the Damma Glacier in central Switzerland, the progressive stages of soil formation as the glacier has retreated over the past 150 years demonstrate increases in soil carbon to around 3% in the most developed soil profile studied at the site (Bernasconi et al., 2011). The C_{org} accumulation in soil may then continue for many millennia, but at a much lower rate. In support of this conclusion, Schlesinger (1990) derived long-term C_{org} accumulation rate data from (uncultivated) soil chronosequences spanning a wide range of ecosystem types. He documented that this slow accumulation accounted for <1% of terrestrial net primary production.

Understanding C_{org}-related processes in soil is confounded by the fact that there are multiple timescales of organic accumulation and decomposition. Characterizing the rates of these processes has benefited substantially from the application of radiocarbon (^{14}C) dating techniques (see Trumbore, 2009, for an overview). This is because different classes of organic compounds have vastly different turnover times. Soil litter has a short residence time measured in years. Longer-term stabilization of C_{org} is tied to inorganic soil evolution pathways that result in sorption to mineral phases with high surface area and surface charge. The resulting mineral-stabilized C_{org} may have residence times measured in millennia. Soil C_{org} may also be protected by the formation of aggregates (discussed in more detail below).

Role of topography

The low topographic gradient regolith-forming regime associated with chronosequences is modified significantly by the introduction of a topographic gradient, which changes the overall balance of weathering processes. Whereas chemical denudation removes material from throughout the weathering profile, the physical removal of material from the surface of the regolith (physical denudation) driven by topography removes only material from the uppermost layers. In addition to the export of material, erosion can also result in downslope deposition, thus thickening the downslope profile and burying material that was formerly at the surface. The thickness of the regolith along topographic gradients is determined by the competition between the kinetics of mineral weathering (chemical denudation rate) and the physical denudation rate. At one extreme, where the physical denudation rate is much more rapid than mineral weathering kinetics, the regolith will be thin, and relatively fresh material is then continuously exposed to chemical weathering. In effect, physical denudation has the effect of moving less weathered material upward in the profile (Fig. 6.1).

The balance between chemical and physical denudation has been the focus of considerable research (e.g. Yoo and Mudd, 2008; Gabet and Mudd, 2009; Hilley et al., 2010). A correlation between rates of physical denudation and chemical weathering has been recognized. This correlation may result from the fact that as the denudation rate increases, fresher, more readily weathered material is brought to the surface. One recent analysis (West, 2012) concluded that the effective thickness of the weathering zone varied relatively little across several orders of magnitude of denudation rate, but with increasing denudation rate, the contribution of bedrock weathering becomes increasingly significant.

A powerful advance in characterizing the mechanisms of erosion has been the analysis of cosmogenic nuclides such as ^{10}Be and ^{26}Al in river sediment, soil horizons, profiles and hill slopes (see review by von Blanckenburg, 2005). These data integrate denudation history on timescales of 10^3–10^5 years. Among the significant trends identified by von Blanckenburg, rates of weathering co-vary primarily with physical erosion rates and much less with temperature or precipitation.

Erosion and soil organic carbon content are linked. As summarized by Lal (2003), erosion can have a number of effects on soil carbon that may differ between the sites of erosion and deposition. Locally, soil organic carbon may be decreased by preferential removal of the organic-rich surface layer, by enhanced exposure of residual organic matter to near-surface oxidative conditions, by increased oxidation due to changes in soil temperature and moisture, and by breakdown of carbon protective aggregates (discussed further below). Conversely, revegetation may replenish organic matter at the erosional site. The organic matter transported to a depositional site may be oxidized during transport, or protected by burial sequestration and reformation of soil aggregates. As discussed below in the section on the role of humans, the importance of the soil carbon–erosion linkage has been increased significantly by greatly enhanced erosion due to anthropogenic activities. The overall complexity of these erosion-linked processes has led to a variety of estimates on their net effect on the global carbon budget that range from a global source to a global sink (see, for example, van Oost et al., 2007). The latter authors calculated that erosion was a net sink of carbon globally (see also Chapters 3 and 19, respectively, this volume).

A holistic view of near-surface processes has recently emerged that can be conceptualized as the operation of 'the critical zone reactor' (Anderson et al., 2007). In this interpretation, solid material enters through the lower regolith parent material boundary, the reactor is stirred at the top by biological and physical processes, sedimentary products are removed from the surface by erosion and dissolved products leave by fluid flow. Operating over Earth's entire land surface, this critical zone reactor constitutes our planet's weathering engine (Fig. 6.1).

The role of climate

Water availability and temperature are considered to be among the primary factors in regolith development and soil formation. The conceptual underpinning and a synthesis of data from weathering processes in granite terrain across a range of climatic conditions illustrate how these factors affect regolith development and soil formation (Rasmussen et al., 2011). The availability of water to penetrate the subsurface and to transport weathering products is reflected in the water balance at a site. Water availability for weathering is the difference between infiltration and evapotranspiration; with surplus infiltration providing groundwater recharge and subsurface discharge to surface waters. Greater subsurface flow volume removes dissolved weathering products from the locale of source minerals; preventing the back-reaction of kinetically inhibiting solutes and maintaining pore waters in a less saturated chemical state with respect to the dissolution and alteration of source minerals and the precipitation of secondary minerals. Temperature effects occur through the temperature dependence of the thermodynamic solubility equilibrium of mineral phases, and through the temperature dependence of the rates of mineral dissolution and precipitation reactions. In broad terms, greater flushing of the subsurface and warmer subsubsurface conditions should coincide with greater regolith development and soil formation rates.

This picture of rates of weathering and soil formation from regolith is confounded by the fact that physical erosion of the land surface occurs simultaneously with soil formation. Net rates of soil formation are reflected in this flux balance of soil formation and removal. The total flux of dissolved and solid material from the regolith is the denudation, expressed as the mass of material per area per time. This is often expressed as an equivalent rate of lowering of the land surface in units of depth per unit time. In areas of high relief, greater topography-driven water penetration and flushing of the regolith and physical erosion dominate. In this case, regolith alteration is limited by the rate of chemical alteration; i.e. weathering limited. In areas with low rates of physical erosion, soil accumulates and denudation is transport limited. Wilkinson and McElroy (2007) estimate globally averaged rates of 0.062 mm year^{-1} total land surface denudation. This compares with areas of rapid uplift that are estimated to be much higher; 10–20 mm year^{-1} (cf. Burbank, 2002; Montgomery and Brandon, 2002; Montgomery, 2007), compared with erosion rates for conventional tillage agriculture of ~1 mm year^{-1} (Montgomery, 2007) and with around 70% of Earth's land surface eroding at less than 0.05 mm year^{-1} (Hilley et al., 2010).

Rasmussen et al. (2011) were able to disentangle climatic and tectonic (i.e. relief) factors, utilizing as a weathering proxy the Na depletion relative to Zr (i.e. application of Eqn 6.2) throughout the full depth of regolith within weathering granite terrain. The data were constrained to moderately eroding terrain (0.005–0.072 mm year^{-1}) and the relative degree of water availability for the range of sites was estimated by defining the humidity index, (HI, dimensionless) as the mean annual precipitation (MAP, mm year^{-1}) divided by the annual potential evapotranspiration (PET, mm year^{-1}) (Rasmussen et al., 2011).

Sites with HI < 1 were defined as water limited, with the subsurface water availability determined primarily by the availability of precipitation. Sites with HI > 1 were defined as energy limited where precipitation exceeded the energy available for evapotranspiration. Separating the data sets for weathering rates into those for sites with HI greater and less than 1, respectively, yielded linear correlation of the natural logarithm of the weathering rates against temperature (Fig. 6.3; Rasmussen et al., 2011). The correlations allow apparent activation energies (Ea, kJ mol^{-1}) to be estimated, yielding a value of 69 kJ mol^{-1} and 136 kJ mol^{-1}, respectively, for conditions of HI > 1 (wet conditions) and HI < 1 (dry conditions) (Rasmussen et al., 2011). Furthermore, the sites with HI < 1 (dry conditions) exhibited weathering rates that were independent of total denudation rates, suggesting strongly weathering-limited conditions. The sites with HI > 1 (wetter conditions) exhibited a dependence of weathering rate with total denudation rate, but substantially less than a 1:1 relation, which would be indicative of strictly erosion-limited conditions (Fig. 6.4; Rasmussen et al., 2011).

Although other studies have demonstrated the effect of climate-related factors such as hydrological conditions and average temperature on soil pedogenesis, the above study is one of the first to demonstrate a methodology to separate the effects

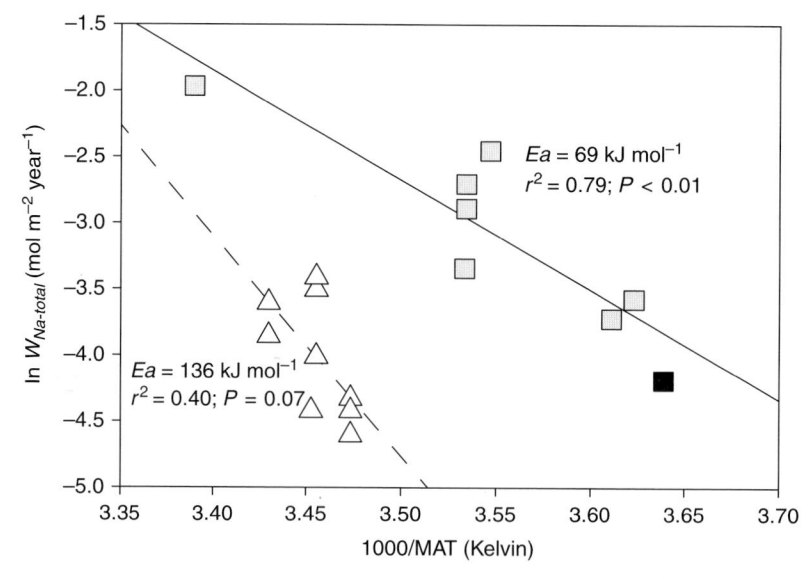

Fig. 6.3. Arrhenius plot of weathering rate (total Na weathering rate $W_{Na\text{-}total}$) as a function of mean annual temperature (MAT), illustrating the role of climate regime. Climate is represented by the humidity index (HI) for relatively wetter (HI > 1, filled square) and dryer (HI < 1, open triangles) locations (Rasmussen et al., 2011). The linear plot demonstrates that when hydrologic conditions are accounted for, the effect of temperature on chemical weathering and soil formation rates is discernible and significant.

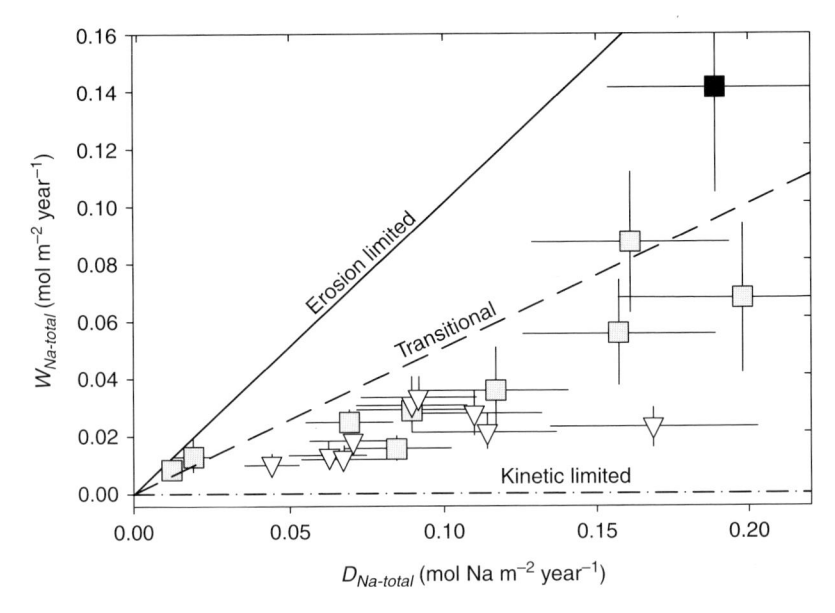

Fig. 6.4. Chemical weathering rate represented by the removal of Na from the regolith, plotted against the total Na denudation rate (from Rasmussen et al., 2011). Filled squares represent sites with HI > 1, while open triangles are sites with HI < 1. Sites plotting close to the 1:1 line corresponding to erosion-limited conditions for soil formation coincide with virtually complete depletion of Na from the regolith. Dryer sites (HI < 1) plot more closely to the line of kinetic-limited weathering, where erosion is minimal and the rate of soil formation is controlled by the rate of chemical alteration of the regolith.

of water availability and temperature on weathering as a fundamental set of soil-forming processes.

The Development of Soil Function

The operation of Earth's weathering engine (Fig. 6.1) as influenced by the interrelated factors of soil formation (Jenny, 1941) does not reveal fully the true nature of soil. As soil forms, the range of chemical, biological and physical processes that occur gives rise to important soil functions that support ecosystem services, with impacts far beyond creating soil as a geological material that accumulates at Earth's surface. These functions include the storage and transmission of water, filtration and transformation of pollutants to reduce contamination in infiltrating water, transformation and recycling of nutrients to enhance their bioavailability, storage of carbon and nitrogen and reduction of greenhouse gases, sustaining habitat and maintaining the gene pool of the terrestrial environment (Fig. 6.2).

Central to these functions is the development of soil structure, the building of soil aggregates. Soil carbon is a major factor in this process. As lichens and plants colonize the surface of parent material, the photosynthate organic carbon provides an energy and carbon source to support heterotrophic microorganisms with a vital role as decomposers. Living organisms and extracellular products of decomposition such as polysaccharide adhere to larger fragments of decomposing biomass, and also bind rock and mineral fragments. These accretions form the larger aggregates in the size fraction of 0.25–1 mm. These are associated with rapid turnover of carbon and associated nutrient elements from the decomposing organic material, thus rendering nutrients bioavailable for new plant productivity. The abundance of decomposers also supports grazing populations, presumably protozoa, but also soil fauna that further transform ingested biomass, produce faecal material and also physically transport this material through burrowing and turbation.

Fresh organic material is gradually transformed into humic material that binds to mineral surfaces and is rendered less bioavailable and less degradable, thus preserving the remaining organic carbon. The smallest mineral particles, clay minerals and nanometric-sized oxides provide the greatest specific surface area and associated adsorption capacity by mineral mass. Sorption of humic substances to these polar mineral surfaces renders them more hydrophobic and prone to aggregation due to the ionic composition of pore water solutions. These agglomerations, together with bound microorganisms and fragments of rock and decaying biomass, form particles of micron size and larger. These intermediate aggregates can also be incorporated into the larger accretions. The sorption surface area of rock fragments and minerals within aggregates also provides sorption and ion exchange surfaces that sequester ionic forms of mineral elements (PO_4^{3-}, K^+) as they are released during the decomposition of biomass; hence providing a reservoir of nutrients within the soil profile. The formation of larger aggregates also favours the drainage of pore waters and ingress of atmospheric O_2 to support root and microbial respiration in the subsurface.

During soil formation, aggregate formation and the development of soil structure are essential to establishing the full range of vital soil functions, and are also an indicator of the state of a soil and its capacity to deliver ecosystem services of value. Central to this are both the input of recently produced organic matter, to drive rapid decomposition and nutrient release associated with larger accretions of aggregates, and the role of decomposed organic matter, to be preserved as a surface-bound coating that also contributes to the formation of larger aggregates. All of these components and characteristics of aggregates and their formation contribute with the specific roles described here to deliver the full range of soil functions. Proactive management of soil carbon provides a key point of human intervention to enhance soil structure and the associated delivery of soil functions (Chapters 9–14, this volume).

The Role of Humans

The natural processes outlined above are being impacted and in some cases overwhelmed by anthropogenic ones. The relation between humans and the Earth has changed markedly over evolutionary history, from a minor influence on ecosystems to being major drivers of ecosystem change on a global scale. The geologist, A.P. Pavlov (1854–1929), was perhaps the first to recognize the magnitude of human influence on global ecosystems, referencing his contemporary time as the 'anthropogenic era' (see Vernadsky, 2005); many years later, the 'Anthropocene' has been proposed as a new geologic epoch (Zalasiewicz, 2008).

One example of the role of humans is the movement of sediment including soil on Earth's surface. Wilkinson and McElroy (2007) argue that geologic denudation of the continents occurs dominantly at headwaters (83% of global river flux is derived from the most elevated 10% of Earth's surface). In contrast, subarial erosion by humans (agriculture primarily) now dominantly occurs at lower elevations. The mean denudation over the past half-billion years of Earth history has lowered continental surfaces by a few tens of metres per million years. In comparison, construction and agricultural activities currently result in the transport of enough sediment and rock to lower all ice-free continental surfaces by a few hundred metres per million years. Mean cropland soil loss from the USA (amounting to about 11% of the global land area) is equivalent to a lowering of over 200 m per million years (Wilkinson and McElroy, 2007). Cropland erosion rates in Asia, Africa and South America are higher yet (Pimentel et al., 1995). As a result, humans are now an order of magnitude more important at moving sediment than the sum of all other natural processes operating on the surface of the planet (Wilkinson, 2005). Maps reconstructing the development of agriculture across the USA between 1790 and 1997 (Waisanen and Bliss, 2002) show that over a timespan that is miniscule compared to many geologic processes, agriculture has transformed the landscape. As a consequence, Amundson *et al.* (2003) estimate that in the USA, mollisols (which correlate with grassland vegetation) have lost 28% of their undisturbed area to land-use change – primarily agriculture. Pimentel (2006) and Montgomery (2007) estimate that erosion rates from agricultural fields average 1–2 orders of magnitude greater than rates of soil production, with potentially negative consequences for human food security and environmental quality. As described above, these elevated erosional rates have consequences for local soil function because soil erosion is associated with the destruction of soil aggregates, leading to organic C loss as previously protected C is exposed to weathering (Lal, 2003). Cumulatively, loss of organic C from agricultural and pasture lands may even have influenced the planetary carbon balance (e.g. Lal, 2003; Janzen, 2004; van Oost *et al.*, 2007; Chapter 20, this volume).

Humans have also had a significant role in altering soil organic matter at scales from local to planetary (Chapter 3, this volume). The authors of Chapter 3, this volume, propose the existence of a soil carbon transition curve. The curve consists of a rapid decline of the initial endowment of soil carbon due to human clearing of natural vegetation for agricultural land use and management practices such as conventional tillage, followed by a phase of greatly diminished soil fertility resulting from depletion of this initial stock, and finally by recovery of soil organic content once agricultural practices improve. Lal and Follett (2009) present generalized trends of relative soil carbon loss as a function of time since conversion from native to managed land status. This loss is rapid in the initial decade or so after conversion. Over periods of several decades, carbon loss (driven in part by enhanced erosion) may amount to up to 80%. The absolute amounts vary with the setting. For example, carbon loss may amount to 50–70 Mg ha^{-1} in tropical rainforests, 30–35 Mg ha^{-1} in prairies and as high as 200–220 Mg ha^{-1} in peatlands.

Conversely, recovery in soil organic carbon has taken place in many parts of the globe owing to practices such as no-till agriculture. Follett *et al.* (2009) reported that modern farming practices across a significant portion of the USA led to a mass of carbon in the top 100 cm that averaged 78% of that in paired native sites. This topic is explored much more comprehensively in Chapter 3, this volume.

The Importance of Critical Zone Observatories

The science of Earth's critical zone combined with environmental observatory research methodology provides a powerful framework to advance understanding of soil functions and soil threats. Critical Zone Observatories (CZOs) are field research facilities with intensely focused multidisciplinary research (cf. CZO Special Issue, *Vadose Zone Journal*, Volume 10, 2011). The experimental design and methodology is strongly hypothesis driven, with a focus on nested scales of observation from molecular-to-grain, profile-plot, catchment and basin scales. There is a strong emphasis on process understanding and integration of the multiple scales of observation with interpretation through mathematical modelling and computational simulation.

The vertical integration of process understanding, linking the aboveground vegetation with soil functions and deeper regolith and aquifer processes, is particularly powerful in assessing the chain of impact that results from environmental change. Aboveground changes in climate driven by global warming, or land-use change driven by demographic developments, set off a chain of impacts that are transmitted in many cases through the resulting changes in soil functions. Defining and quantifying this chain of impact allows assessment of the potential consequences of change, and the benefits of different intervention options. CZOs provide the essential data sets and process models to enable these assessments, and their link to important social and economic drivers of change.

One current experimental design tackles the links between soil functions and soil threats, with a network of CZOs that is organized conceptually along gradients of parent material, climate and land-use intensity (Fig. 6.5.; Banwart et al., 2011).

A meeting of international CZO teams in late 2011 identified 60 sites in 25 countries worldwide with either ongoing or planned research that aligned with the critical zone framework and CZO research approach described here. A particular point of discussion was the potential to tackle environmental change research by studying soil and other critical zone processes along planetary gradients of environmental change (outlined by Banwart et al., 2013). In broad terms, international networks of CZOs offer the potential to achieve several key advances related to soil functions and soil carbon management:

1. Field investigations to ground truth remote sensing data and proxies.
2. Nested measurements of soil process rates and their variation with environmental conditions.
3. Data sets to develop and test modelling and simulation methods to forecast soil functions.
4. Integrating multidisciplinary research efforts that link soil functions with ecosystem services, their social and monetary valuation and assessment of adaptation strategies for land, water and biomass resources during environmental change.

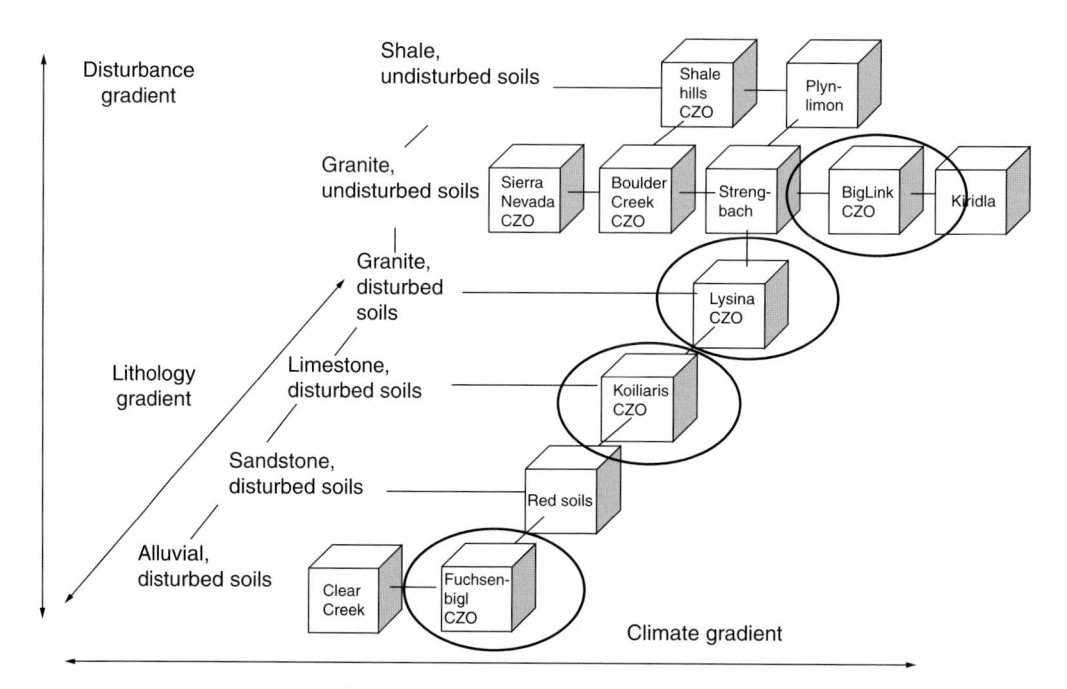

Fig. 6.5. Conceptual framework for experimental design using networks of Critical Zone Observatories (CZO) to study soil functions and soil threats along environmental gradients of climate, lithology and intensity of land use (disturbance) (see Banwart et al., 2011). Each block represents a CZO or associated field site. The circled sites represent different stages in the development of soil functions; BigLink CZO is at the forefront of the Damma Glacier in central Switzerland and represents the initial soil formation as the ice has retreated during the past two centuries. The Lysina (Czech Republic) and Fuchsenbigl (Austria) CZOs represent managed forest plantation and arable farming, and the Koiliaris (Crete, Greece) CZO represents mature land use during millennia of agriculture and imminent threat of desertification due to loss of soil organic carbon and future scenarios for warming of the Mediterranean Basin during this century. The additional sites represent a greater geographical spread of sites and environmental conditions that provide a wider envelope of data on soil processes.

Conclusions

Soil formation reflects the complex interaction of many factors, among the most important of which are: (i) the nature of soil parent material; (ii) regional climate; (iii) organisms, including humans; (iv) topography; and (v) time. These processes take place in Earth's critical zone, which is the thin veneer of our planet where rock meets life. Understanding the operation of these soil-forming factors requires an interdisciplinary approach and is a necessary predicate to charactering soil processes and functions, mitigating soil degradation and adapting soil management to environmental change.

References

Amundson, R. (2005) Soil formation. In: Drever, J.I. (ed.) *Surface and Groundwater, Weathering and Soils*. Series editors Holland H.D. and Turekian K.K. Treatise on Geochemistry, Vol 5. Elsevier-Pergamon, Oxford, UK, pp. 1–35.

Amundson, R., Guo, Y. and Gong, P. (2003) Soil diversity and land use in the United States. *Ecosystems* 6, 470–482.

Amundson, R., Richter, D. and Humphreys, G. (2007) Coupling between biota and earth materials in the critical zone. *Elements* 3, 327–332.

Anderson, S.P., von Blanckenburg, F. and White, A.F. (2007) Physical and chemical controls on the critical zone. *Elements* 3, 315–319.

Batjes, N.H. (1996) Total carbon and nitrogen in the soils of the world. *European Journal of Soil Science* 47, 151–163.

Banwart, S.A., Bernasconi, S., Bloem, J., Blum, W., Brandao, M., Brantley, S., Chabaux, F., Duffy, C., Kram, P., Lair, G. *et al.* (2011) Assessing soil processes and function across an international network of critical zone observatories: research hypotheses and experimental design. *Vadose Zone Journal* 10, 974–987.

Banwart, S.A., Chorover, J., Gaillardet, J., Sparks, D., White, T., Anderson, S., Aufdenkampe, A., Bernasconi, S., Brantley, S., Chadwick, O. *et al.* (2013) *Sustaining Earth's Critical Zone; Basic Science and Interdisciplinary Solutions for Global Challenges*. The University of Sheffield, UK, 45pp.

Beerling, D.J. and Berner, R.A. (2005) Feedbacks and the coevolution of plants and atmospheric CO_2. *Proceedings of the National Academy of Sciences USA* 102(5), 1302–1305.

Bernasconi, S.M., Bauder, A., Bourdon, B., Brunner, I., Bunemann, E., Christl, I., Derungs, N., Edwards, P., Farinotti, D., Frey, B. *et al.* (2011) Chemical and biological gradients along the Damma Glacier soil chronosequence (Switzerland). *Vadose Zone Journal* 10, 867–883.

Berner, R.A. (1997) Geochemistry and geophysics: the rise of plants and their effect on weathering and atmospheric CO_2. *Science* 276(5312), 544–546.

Birkland, P.W. (1999) *Soils and Geomorphology*. Oxford University Press, New York, 430 pp.

Brantley, S.L. (2005) Reaction kinetics of primary rock-forming minerals under ambient conditions. In: Drever, J.I. (ed.) *Treatise on Geochemistry*. Series editors H.D. Holland and K.K. Turekian. Elsevier-Pergamon, Oxford, UK, pp. 73–118.

Brantley, S.L. (2008) Understanding soil time. *Science* 321(9), 1454–1455.

Brantley, S.L., Goldhaber, M.B. and Ragnarsdottir, K.V. (2007) Crossing disciplines and scales to understand the critical zone. *Elements* 3, 307–314.

Brantley, S.L., Megonigal, J., Scatena F., Balogh-brunstad, Z., Barnes, R., Bruns, M., Van Cappellen, P., Dontsova, K., Hartnett, H., Hartshorn, A. *et al.* (2011) Twelve testable hypotheses on the geobiology of weathering. *Geobiology* 9(2), 140–165.

Brimhall, G.H. and Dietrich, W.E. (1987) Constitutive mass balance relations between chemical composition, volume, density, porosity, and strain in metasomatic hydrochemical systems: results on weathering and pedogenesis. *Geochmica et Cosmochimica Acta* 51(4), 567–587.

Brimhall, G.H., Lewis, C.J., Ford, C., Bratt, J., Taylor, G. and Warin, O. (1991) Quantitative geochemical approach to pedogenesis: importance of parent material reduction, volumetric expansion, and eolian influx in lateritization. *Geoderma* 51, 51–91.

Burbank, D.W. (2002) Rates of erosion and their implications for exhumation. *Mineralogical Magazine* 66(1), 25–52.

Derry, L. and Chadwick, O. (2007) Contributions from Earth's atmosphere to soil. *Elements* 3, 333–338.

Egli, M., Fitze, P. and Mirabella, A. (2001) Weathering and evolution of soils formed on granitic, glacial deposits: results from chronosequences of Swiss alpine environments. *Catena* 45(1), 19–47.

Follett, R.F., Kimble, J.M., Prussner, E.G., Samson-Liebig, S. and Waltman, S. (2009) Soil organic carbon stocks with depth and land use at various U.S. sites. In: Lal, R. and Follett, R.F. (eds) *Soil Carbon Sequestration and the Greenhouse Effect*. SSSA Special Publication 57, 2nd edn. Soil Science Society of America, Madison, Wisconsin, pp. 29–46.

Fortescue, J. (1992) Landscape geochemistry: retrospect and prospect—1990. *Applied Geochemistry* 7, 1–53.

Gabet, E.J. and Mudd, S.M. (2009) A theoretical model coupling chemical weathering rates with denudation rates. *Geology* 37(2), 95–103.

Gabet, E.J., Reichman, O.J. and Seabloom, E.W. (2003) The effects of bioturbation on soil processes and sediment transport. *Annual Review of Earth and Planetary Sciences* 31(1), 249–273.

Harden, J.W., Mark, R.K., Sundquist, E.T. and Stallard, R.F. (1992) Dynamics of soil carbon during deglaciation of the laurentide ice sheet. *Science* 258(5090), 1921–1924.

Hilley, G.E., Chamberlain, C.P., Moon, S., Porder, S. and Willett, S.D. (2010) Competition between erosion and reaction kinetics in controlling silicate-weathering rates. *Earth and Planetary Science Letters* 293(1–2), 191–199.

Janzen, H.H. (2004) Carbon cycling in earth systems – a soil science perspective. *Agriculture, Ecosystems and Environment* 104, 399–417.

Jenny, H. (1941) *Factors of Soil Formation: A System of Quantitative Pedology*. McGraw-Hill, New York/London, 281 pp.

Kennedy, M., Droser, M., Mayer, L.M., Pevear, D. and Mrofka, D. (2006) Late Precambrian oxygenation; inception of the clay mineral factory. *Science* 311(5766), 1446–1449.

Kurtz, A.C., Derry, L.A., Chadwick, O.A. and Alfano, M.J. (2000) Refractory element mobility in volcanic soils. *Geology* 28(8), 683–686.

Lal, R. (2003) Soil erosion and the global carbon budget. *Environment International* 29, 437–450.

Lal, R. and Follett, R.F. (2009) Soils and climate change. In: Lal, R. and Follett, R.F. (eds) *Soil Carbon Sequestration and the Greenhouse Effect*. SSSA Special Publication, Vol 57, 2nd edn. Soil Science Society of America, Madison, Wisconsin, pp. xxi–xxvii.

Lambers, H., Raven, J.A., Shaver, G.R. and Smith, S.E. (2008) Plant nutrient-acquisition strategies change with soil age. *Trends in Ecology and Evolution* 23, 95–103.

Merritts, D.M., Chadwick, O.A., Hendricks, D.M., Brimhall, G.H. and Lewis, C.J. (1992) The mass balance of soil evolution on late Quaternary marine terraces, northern California. *Geological Society of America Bulletin* 104, 1456–1470.

Montgomery, D.R. (2007) Soil erosion and agricultural sustainability. *Proceedings National Academy of Sciences USA* 104(33), 13268–13272.

Montgomery, D.R. and Brandon, M.T. (2002) Topographic controls on erosion rates in tectonically active mountain ranges. *Earth and Planetary Science Letters* 201(3–4), 481–489.

National Research Council (2001) *Basic Research Opportunities in the Earth Sciences*. National Academies Press, Washington, DC, 154 pp.

Pimentel, D. (2006) Soil erosion: a food and environmental threat. *Environment, Development and Sustainability* 8, 119–137.

Pimentel, D., Harvey, C., Resosudarmo, P., Sinclair, K., Kurz, D., McNair, M., Crist, S., Shpritz, L., Fitton, L., Saffouri, R. and Blair, R. (1995) Environmental and economic costs of soil erosion and conservation benefits. *Science* 267, 1117–1123.

Quirk, J., Beerling, D., Banwart, S.A., Kakonyi, G., Romero-Gonzalez, M. and Leake, J.R. (2012) Evolution of trees and mycorrhizal fungi intensifies silicate mineral weathering. *Biology Letters* 8(6), 1006–1011.

Rasmussen, C., Brantley, S., de B. Richter, D., Blum, A., Dixon, J. and White, A.F. (2011) Strong climate and tectonic control on plagioclase weathering in granitic terrain. *Earth and Planetary Science Letters* 301, 521–530.

Retallack, G.J. (1997) Early forest soils and their role in Devonian global change. *Science* 276(5312), 583–585.

Retallack, G.J. (2005) Soils and global change in the carbon cycle over geological time. In: Drever, J.I. (ed.) *Surface and Ground Water, Weathering, and Soils*. Treatise on Geochemistry, Vol 5. Elsevier-Pergamon, Oxford, UK, pp. 581–606.

Rye, R. and Holland, H. (1998) Paleosols and the evolution of atmospheric oxygen: a critical review. *American Journal of Science* 298(October), 621–672.

Schlesinger, W.H. (1990) Evidence from chronosequence studies for a low carbon-storage potential of soils. *Nature* 348, 232–234.

Taylor, L.L., Leake, J.R., Quirk, J., Hardy, K., Banwart, S.A. and Beerling, D.J. (2009) Biological weathering and the long-term carbon cycle: integrating mycorrhizal evolution and function into the current paradigm. *Geobiology* 7(2), 171–191.

Taylor, L.L., Banwart, S.A., Leake, J. and Beerling, D.J. (2011) Modelling the evolutionary rise of ectomycorrhiza on sub-surface weathering environments and the geochemical carbon cycle. *American Journal of Science* 311(5), 369–403.

Trumbore, S. (2009) Radiocarbon and soil carbon dynamics. *Annual Review of Earth and Planetary Sciences* 37(1), 47–66.

van Oost, K., Quine, T.A., Govers, G., De Gryze, S., Six, J., Harden, J.W., Ritchie, J.C., McCarty, G.W., Heckrath, G., Kosmas, C. et al. (2007) The impact of agricultural soil erosion on the global carbon cycle. *Science* 318.

Vernadsky, V.I. (2005) Some words about the Noösphere. *21st Century Science and Technology* 18(1), 16–21. (Translated from Russian by Rachel Douglas (Executive Intelligence Review) from the original 1943 article and a 1945 translation by the author's son.)

Victoria, R., Banwart, S.A., Black, H., Ingram, H., Joosten, H., Milne, E. and Noellemeyer, E. (2012) The benefits of soil carbon. In: *UNEP Year Book 2012: Emerging Issues in Our Global Environment*. UNEP, Nairobi, pp. 19–33 (http://www.unep.org/yearbook/2012/pdfs/UYB_2012_CH_2.pdf, accessed 1 July 2014).

Vitousek, P.M., Chadwick, O.A., Crews, T.E., Fownes, J.H., Hendricks, D.M. and Herbert, D. (1997) Soil and ecosystem development across the Hawaiian islands. *GSA Today* 7(9), 1–8.

Vitousek, P.M., Porder, S., Houlton, B.Z. and Chadwick, O.A. (2010) Terrestrial phosphorous limitation: mechanisms, implications, and nitrogen–phosphorus interactions. *Ecological Applications* 20(1), 5–15.

von Blanckenburg, F. (2005) The control mechanisms of erosion and weathering at basin scale from cosmogenic nuclides in river sediment. *Earth and Planetary Science Letters* 237(3–4), 462–479.

Waisanen, P.J. and Bliss, N.B. (2002) Changes in population and agricultural land in conterminous United States counties, 1790–1997. *Global Biogeochemical Cycles* 16(4), 84-1–84-19.

West, A.J. (2012) Thickness of the chemical weathering zone and implications for erosional and climatic drivers of weathering and for carbon-cycle feedbacks. *Geology* 40, 811–814.

White, A.F. (2003) Natural weathering rates of silicate minerals. In: Drever, J.I. (ed.) *Surface and Ground Water, Weathering, and Soils*. Treatise on Geochemistry, Vol 5. Series editors H.D. Holland and K.K. Turekian. Elsevier-Pergamon, Oxford, UK, pp. 133–168.

White, A.F., Blum, A.E., Schultz, M.S., Huntington, T.G., Peters, N.E. and Stonestrom, D.A. (2002) Chemical weathering of the Panola Granite: solute and regolith elemental fluxes and the weathering rate of biotite. In: Hellmann, R. and Wood, S.A. (eds) *Water–Rock Interactions, Ore Deposits, and Environmental Geochemistry: A Tribute to David A. Crerar*. The Geochemical Society, St Louis, Missouri, pp. 37–59.

Wilkinson, B. (2005) Humans as geologic agents: a deep-time perspective. *Geology* 33(3), 161–164.

Wilkinson, B.H. and McElroy, B.J. (2007) The impact of humans on continental erosion and sedimentation. *Geological Society of America Bulletin* 119(1–2), 140–156.

Yoo, K. and Mudd, S.M. (2008) Toward process-based modeling of geochemical soil formation across diverse landforms: a new mathematical framework. *Geoderma* 146(1–2), 248–260.

Zalasiewicz, M.W. (2008) Are we now living in the Anthropocene? *GSA Today* 18, 4–8.

7 Soil Carbon Dynamics and Nutrient Cycling

David Powlson*, Zucong Cai and Philippe Lemanceau

Abstract

The quantity of organic carbon in soil and the quantity and type of organic inputs have profound impacts on the dynamics of nutrients. Soil organic matter itself represents a large reservoir of nutrients that are released gradually through the action of soil fauna and microorganisms: this is especially important for the supply of N, P and S to plants, whether agricultural crops or natural vegetation. Organic matter also modifies the behaviour and availability of nutrients through a range of mechanisms including increasing the cation exchange capacity of soil, thus leading to greater retention of positively charged nutrient ions such as Ca, Mg, K, Fe, Zn and many micronutrients. Carboxyl groups in organic matter, and in root exudates or microbial metabolites, form complexes with various metal ions, usually increasing their availability to plants. In some cases, the formation of stable complexes has a detoxifying effect, for example by making Al and Cu less available to plants or microorganisms. Organic matter influences soil physical conditions greatly, especially through the formation or stabilization of aggregates and pores; this indirectly influences the availability of water and dissolved nutrients to plant roots. Organic matter and organic inputs are the source of energy for heterotrophic soil organisms, variations in organic carbon content and composition, impacting biome size, diversity and activities. These complex interactions between organic carbon and the soil biome require additional research to be fully understood. The implications for nutrient dynamics differ between nutrient-rich situations such as agricultural topsoils and nutrient-poor environments such as subsoils or boreal forests. In agricultural soils, excessive inputs of organic matter in manures can lead to pollution problems associated with losses of N and P.

Introduction

Soil carbon (C) plays a major role in regulating the cycling of plant nutrients, especially, but not limited to, nitrogen (N), phosphorus (P) and sulfur (S). This is partly because the organic entities in soil contain these elements combined with C, and thus act as a source of nutrients as organic matter undergoes decomposition. But it is also because the C in soil organic matter is a source of energy for soil organisms, which are mostly heterotrophs, and thus acts as the driver for various biologically mediated processes involved in nutrient transformations. In addition to being a source of nutrients, and a controlling factor in nutrient transformations, organic C in soil contributes to soil cation exchange

*E-mail: david.powlson@rothamsted.ac.uk

capacity due to the action of carboxyl groups. Thus, soils with a higher C content generally have the ability to retain cations such as calcium (Ca), magnesium (Mg), zinc (Zn), iron (Fe) and many others that are important for plant growth. Organic C has an indirect effect on nutrient availability to agricultural crops through its influence on soil physical conditions, especially structural attributes such as the formation and stabilization of aggregates. This in turn has a positive influence on the infiltration and retention of water and the growth of plant roots – factors that are of central importance for agricultural productivity and food security. These factors, and others, are considered individually.

Organic Matter as a Reservoir of Plant Nutrients

In addition to C, the organic moieties that constitute soil organic matter contain substantial quantities of elements that are highly significant as plant nutrients, especially N, P and S. The C:N ratio of soils is surprisingly constant, with values almost always in the range between 10:1 and 12:1. Higher values in the range 15:1–20:1 occur in peat soils. In a global data set analysed by Kirkby *et al.* (2011), the mean C:N ratio was 11:1. The measured C:N ratio can vary somewhat according to the amount of fresh plant material in a soil at the time of sampling; this may come from recently added crop residues or, especially in the case of soils under grass, be due to dead plant residues. In one set of soils examined by these authors, the C:N ratio of the 'light fraction' organic matter (plant-like material separated by either sieving or density separation) had a C:N ratio ranging from 13:1 to 21:1, more similar to values typical of plants. The C:S ratio in soil is somewhat more variable than the C:N ratio, but generally fairly constant, and averaged 79:1 in a subset of soils where S had been analysed (Kirkby *et al.*, 2011). P content is linked less strongly to C, mainly because a substantial fraction of soil P is present in inorganic forms, in contrast to N or S. If only *organic* forms of P were considered, these authors found a general correlation between C and organic P, though much weaker than for N or S. The mean ratio C:organic P was 133:1.

The relatively constant values for the ratio C:N:S:organic P in the stable fraction of soils (i.e. excluding fresh plant material if this is present) across a very wide range of soils may be evidence that this material is derived largely from microbial sources, because the ratios in plant material are much more variable. The main input of organic matter to soil is plant material, but it is thought that it is the products of its biological transformations by soil fauna and microorganisms that constitute the more stabilized forms of organic matter accumulating in soil. There is evidence from spectroscopic studies that about 85% of the organic N in soil is in the form of amide groups, consistent with being in proteins. However, proteins are among the most labile natural macromolecules, and when added to soil under experimental conditions, normally persist for no more than 2 or 3 days, so it is surprising to find evidence of their dominant position within soil organic matter (SOM). It is thought that this unexpected persistence is a result of stabilization processes that include chemical entrapment by association with humic substances and physical adsorption on clay surfaces.

Table 7.1 shows that even a soil with the relatively low organic C content of 1% (at the low end of the range for arable soils in temperate regions) contains over 2000 kg N ha^{-1} in the cultivated layer. Even if only a few per cent of this stock of N becomes available to plants each year, it makes a significant contribution to plant nutrition. For example, crops growing in unfertilized treatments of

Table 7.1. Quantities of nitrogen (N), sulfur (S) and organic phosphorus (P) contained in the organic matter of a typical soil in arable agriculture in the temperate region. Values refer to the 0–20 cm depth and assume a soil bulk density of 1.3 g cm^{-3}.

Approximate elemental content in 1 ha of soil (0–20 cm depth) under arable cropping (kg ha^{-1})			
C	N	S	Organic P
26,000	2,360	330	200

long-term experiments typically remove at least 20 kg N ha^{-1} annually (i.e. equivalent to about 1% of the total stock of N in soil organic matter), and annual plant uptake in semi-natural ecosystems could also be of this order. Although part of this N uptake will be derived from N deposited in rain or through dry deposition from the atmosphere, in most cases the majority is from the mineralization of soil organic matter. In normal agricultural situations in the temperate region, it is common for N derived from this source to account for at least 30% of total N uptake by crops (Macdonald et al., 1997). In tropical and subtropical environments, the contribution of soil organic matter is even greater, with an average of 79% of total N uptake by crops being derived from soil organic N in one coordinated set of experiments with 13 sites in nine countries (Dourado-Neto et al., 2010).

Such data are derived from experiments using ^{15}N-labelled fertilizers; uptake of *unlabelled* N is taken as indicating that unlabelled soil organic matter is the source. However, this can be a slight overestimate, due to an artefact of the methodology termed 'pool substitution'. Even so, the results still show that soil organic matter is a major source of N for agricultural crops. Indeed, a major thrust of practically oriented research in soil science and agronomy is to identify ways of predicting more accurately the quantity of N that will become available for crop uptake during the growing season on a field-specific basis. The more N that can be derived from soil sources, the less is required from synthetic fertilizer, which has a large greenhouse gas footprint from the manufacturing process. So, any management practices that can maintain or increase the quantity of organic matter in a soil have multiple benefits for crop nutrition. These include: (i) contributing to food security; (ii) increasing the economic efficiency of agriculture through saving on the purchase of N synthetic fertilizer by the farmer; and (iii) decreasing the greenhouse gas emissions associated with N fertilizer. Although total soil organic C or N content gives some indication of potential N mineralization under field conditions, the precision or prediction is poor, and hence the need for ongoing studies on this.

Figure 7.1 shows an example of agricultural management influencing SOM content, and this in turn influencing crop yields through the supply of N. Two cereal crops, either winter wheat or spring barley, were grown following three previous cropping regimes. These were: (i) continuous arable cropping such that organic inputs to soil were small; (ii) 3 years of a grass pasture that received N fertilizer – organic inputs from

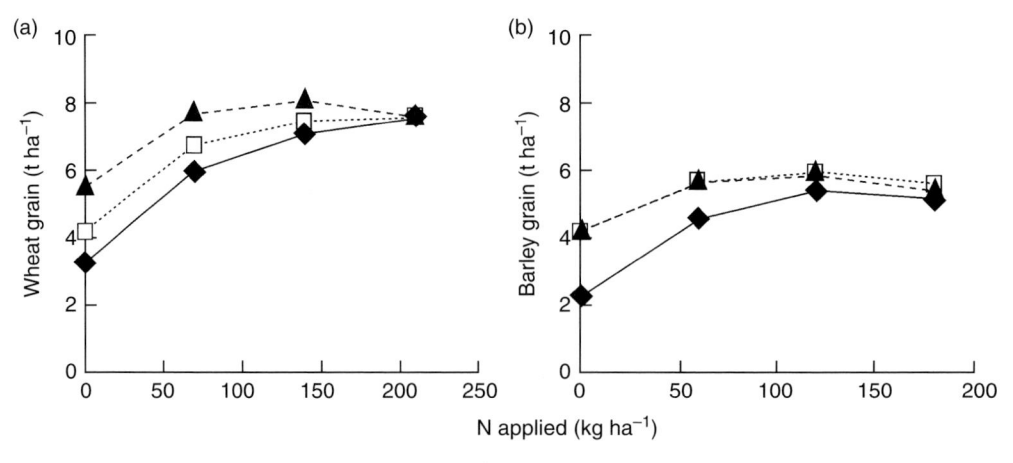

Fig. 7.1. Yields of (a) winter wheat and (b) spring barley in the Woburn Ley-Arable Experiment on a sandy soil in south-east UK. Graphs show yields of arable test crops (t ha^{-1}) following three different previous cropping sequences: 3 years arable followed by 2 years arable test crops, ♦; 3 years grass ley + N followed by 2 years arable test crops, □; 3 years grass/clover ley followed by 2 years test crops, ▲. (From Johnston et al., 2009.)

grass roots would have been larger than from roots of the arable crops in treatment (i); or (iii) 3 years of a pasture comprising a mixture of grass and clover – this treatment would lead to organic inputs at least as large as in treatment (ii) plus additional N from biological N fixation by the clover. Where no N fertilizer was applied to the spring barley crop (Fig. 7.1b), grain yield doubled from 2 to 4 t ha^{-1} when comparing treatments (ii) and (iii) with the lower organic matter treatment (i). That this was a result of additional N supply in the two treatments that included 3 years of pasture was shown because where N fertilizer was applied to spring barley, yield increased substantially in treatment (i), such that crop yields in all three treatments became approximately equal. The benefit to crop yield was proportionately greater than the increase in soil C content, indicating that the additional N was coming in large part from relatively fresh fractions of SOM – especially from legume residues in the case of the pasture that included clover. There were similar trends with winter wheat: grain yield in the absence of added N fertilizer increased from about 3 t ha^{-1} in the low SOM all-arable treatment to 5.5 t ha^{-1} where wheat followed the grass/clover pasture.

Role of Organic Carbon in Retention of Nutrients

Plant nutrient species that occur as positively charged ions (cations) are retained in soil through the action of negatively charged sites on soil constituents. Many such sites occur at the surfaces of clay minerals and are associated with non-crystalline oxides in soil, and are also due to carboxyl groups in organic matter. On an equal mass basis, organic matter has a much greater ability to retain cations (termed cation exchange capacity, CEC) than soil inorganic constituents. For example, the CEC of soil organic matter is typically around 300 cmol$_c$ g^{-1} compared to values for clays in the range <10–100 cmol$_c$ g^{-1}. Even though the organic matter content in soil is typically only a few per cent by mass, because of its large CEC, organic matter often contributes 50% or more of soil total CEC. Thus, changes in SOM content are highly significant for a soil's ability to retain nutrients such as Ca, Mg, potassium (K), Fe and Zn, plus a range of micronutrients that occur in cationic forms.

An additional and related benefit of increased SOM content is that it contributes buffering capacity to soil, such that a soil with higher SOM content can resist the natural tendency of soils to become more acid under the influence of rain which is slightly acidic or additions such as nitrogen fertilizer. This is important in soils used for arable crops. An extreme example was seen in a long-term fertilizer experiment in China (Zhao et al., 2010). In a subtropical soil in a high rainfall region that started at pH 5.7, soil pH in treatments receiving N fertilizer but no additional organic matter fell to the range 4.2–4.7 after 12 years and led to virtual crop failures of wheat and maize. By contrast, where manure was applied in addition to N fertilizer, pH increased slightly to 5.9, and where manure alone was applied, it rose to 6.6.

Impacts of Organic Carbon on Soil Physical Properties and Implications for Nutrient Availability to Plants

Organic matter in soil is highly influential in favouring the formation of aggregates at different scales. A soil with a pronounced aggregate structure, with stable pore spaces between the aggregates, is favourable for root growth and the uptake of water and dissolved nutrients. The strong relationship between the stability of aggregates and organic C content has been demonstrated in studies covering many hundreds of soils worldwide (Tisdall and Oades, 1982). Microaggregates (regarded as those <250 μm in diameter) are formed through direct interactions between clay particles and organic matter through the carboxyl or phenolic groups in organic matter, usually via cations such as Ca, Fe or aluminium (Al) associated with clay surfaces. Organic matter in such microaggregates is either composed of relatively stable forms and/or is further stabilized against decomposition through association with the clay particles. Microaggregates are thought to be assembled into larger units (macroaggregates) through the action of more transient organic entities, especially polysaccharides derived either from roots or as products of microbial decomposition.

It has been shown that, under appropriate conditions, resistance to a penetrometer being pushed into soil is a good predictor of resistance to penetration by a growing root, and that this is influenced strongly by soil C content (Bengough and McKenzie, 1997; Whalley et al., 2005). Organic C content also strongly influences soil bulk density and soil strength, and these have major impacts on root growth (Whalley et al., 2007). The influence of soil physical structure on root exploration in soil, and the resulting uptake of P, a relatively immobile nutrient in soil, is illustrated by the results in Table 7.2. Three crops (spring barley, potato and sugarbeet) were grown in a range of treatments in a field experiment in the UK where organic C and crop-available P (termed Olsen P) differed due to manure and P additions during the previous 12 years. Because a wide range of Olsen P levels were present, it was possible to determine the P level required to achieve maximum yield when N and other nutrients were not limiting growth. Table 7.2 shows that, for the three crops, the maximum yield that could be reached was similar in the low C and the high C soil. However, in the soil with lower organic C content (0.87%), the required Olsen P level to achieve this yield was 2–3 times greater than in the soil containing 1.40% C. This was suspected to be due to the less extensive root growth in the low C soil because of its visibly poorer soil structure; this soil was particularly difficult to plough. This conclusion was confirmed when samples of the soils were brought to the laboratory, air-dried and ground to pass through a 2 mm sieve, thus eliminating the differences in structure between the low C and the high C soils. The soils were then used in a pot experiment in which ryegrass was grown, and in this situation the level of Olsen P required to attain the maximum yield was equal in both soils – a strong indication that it was differences in soil structure that were influencing P uptake and crop growth under field conditions. Thus, maintaining soil structure that is conducive to root proliferation is important for the efficient use of nutrients; adopting management practices that maintain or increase the concentration of soil C is generally the most effective means of achieving this.

Organic C and the Abundance, Diversity and Activity of Soil Biota

It is very well established that soils having a higher organic matter content, or those receiving larger inputs of organic matter, have larger microbial biomass content and exhibit greater activities of the different groups that

Table 7.2. Effect of soil organic carbon (C) concentration on crop yield responses to available phosphorus (Olsen P) in soils having differing C and Olsen P concentrations due to past treatments. (From Johnston et al., 2009.)

Crop product	Soil organic C (%)	Maximum crop yield[a] (t ha^{-1})	Olsen P associated with maximum yield (mg kg^{-1})	Variance accounted for (%)
Field experiments				
Spring barley grain	1.40	5.00	16	83
	0.87	4.45	45	46
Potato tubers	1.40	44.7	17	89
	0.87	44.1	61	72
Sugarbeet sugar	1.40	6.58	18	87
	0.87	6.56	32	61
Pot experiment				
Ryegrass dry matter	1.40	6.46	23	96
	0.87	6.52	25	82

[a]Yield shown is slightly less than the absolute maximum. It is yield at 95% of the maximum assessed from curves of yields plotted against Olsen P concentration.

can be measured. It is also well established that soils are extremely heterogeneous and show very great spatial variation in their properties. One example of this is the existence of hotspots of SOM where biotic growth and activity are far greater than in the surrounding soil. Such sites include the rhizosphere, where a significant fraction of the photosynthates are released as rhizodeposits (Nguyen, 2003), and the so-called detritusphere, corresponding to the dead plant materials (Gaillard et al., 1999). Recently, Clemmensen et al. (2013) have showed that in boreal forest, a major part (50–70%) of the carbon stored by the forest derives from roots and root-associated microorganisms.

A significant pool of organic C, resulting from the humification process, is stabilized in soil into forms with turnover times measured in decades to centuries. In fact, there is strong evidence from research using a combination of physical fractionation and spectroscopic methods that the largest and most stable pool of organic C in soil is derived from microbial constituents and metabolites and is intimately associated with clay minerals (Courtier-Murias et al., 2013). Despite the great stability of this pool, it still contributes to plant nutrient supply because of its large size: modelling of soil N turnover indicates that slow turnover of a large pool probably contributes a similar amount of plant-available N as the fast turnover of smaller pools (Smith et al., 1996).

The stabilization of some fractions of organic matter in soil has been ascribed by some authors to the lack of energy required for microbial activity. Clearly, other mechanisms are involved, and will often be of much greater significance, as discussed by Courtier-Murias et al. (2013). However, the idea is supported by the observation that, in some circumstances, the addition of fresh plant material causes increased decomposition of existing SOM – a mechanism termed a priming effect. It has particularly been noted in subsoils where there is only a small quantity of fresh material (Fontaine et al., 2007). Evidence for plant regulation of this process has recently been provided by Fontaine et al. (2011). These authors have shown that plants may regulate the mineralization of recalcitrant fractions of SOM through their requirement for nutrients, especially nitrogen, by tuning activities of microorganisms and, more specifically, fungal communities.

The idea that microbial diversity impacts C mineralization has been supported by recent reports (Baumann et al., 2012; Pascault et al., 2013). By coupling molecular and isotopic tools (454 pyrosequencing in conjunction with stable isotope analyses), Pascault et al. (2013) showed that the intensity of the priming effect in soil might depend on the stimulation of particular functional groups of microorganisms. Combining these tools, Clemmensen et al. (2013) showed that root-associated fungi played a major role in carbon cycling in boreal forest. Baumann et al. (2012) evidenced that redundant functions such as sugar mineralization, but also specialized C transformations such as lignin decomposition, were affected by a decrease of soil microbial diversity. The role of the diversity of invertebrates in carbon cycling was also underlined in the meta-analysis made by Nielsen et al. (2011). This report stresses a positive relation between species richness and C cycling, especially in low-diversity conditions. Altogether, these data highlight the contribution of community diversity at each level of the food web chain, starting with fauna known to contribute to the fragmentation of organic matter and then followed by microorganisms involved in the degradation and mineralization processes.

However, there is also counterintuitive evidence that microbial biodiversity in soil is not necessarily associated with increased SOM content or plant inputs. Hirsch et al. (2009) reported results from a field experiment in the UK where adjacent plots had been under three managements for over 50 years: grass, arable crops or kept bare by frequent ploughing. Not surprisingly, the organic C concentration in the bare soil was much lower than in the others, and total microbial biomass C was only 7% of that in the soil under grass and 33% of that in the arable treatment. Fresh organic matter (as indicated by material of low density, termed light fraction) was also very low in the bare soil; as a consequence, both the abundance and diversity of mesofauna (mites and collembola), which survived mainly on fresh plant inputs, were much lower in the bare fallow than in either of the others. In the case of the bacterial communities, their *abundance*, as assessed by culturing on a low-nutrient agar, was much lower in the bare fallow soil than in soil with fresh plant inputs – consistent with the greatly decreased microbial biomass C content. By contrast, bacterial *diversity* was similar in the bare fallow soil to that in the soils receiving fresh plant inputs from grass or arable crops, whether assessed by phospholipid fatty acid (PLFA) analysis, Biolog or extracted DNA. Because DNA examined in this way is considered to represent the species present, though the definition of species in bacteria is a matter of debate (Prosser *et al.*, 2007), the distinct 16S rRNA gene PCR products revealed by denaturing gradient gel electrophoresis (DGGE) were referred to as *operational taxonomic units* (OTUs). Hence, the finding of a high bacterial diversity in the 50-year, bare-fallowed soil is a discovery that favours evidence for the stability of the soil bacterial community. The possible implications of these findings for the organisms involved in nutrient transformations are not known.

Besides their implication in the food web chain, Six *et al.* (2006) stressed the contribution of soil microbial communities to the development of well-structured soils through the production of a range of metabolites. Of particular importance is the role of fungal exocellular polysaccharides and fungal hyphae in the formation of macroaggregates. Soil microbial communities can derive energy and C substrate for growth from both native SOM and added organic materials such as plants and manures, but community sizes are invariably larger where there are added inputs. The improved structural attributes arising from a large and active microbial community is of great significance for the growth of plant roots and the effectiveness with which crop plants can access water and nutrients.

Modification of Availability of Plant Nutrients through Complexation

Soil organic matter carries carboxyl and phenolic groups that are able to compete strongly with inorganic ligands in complexing metal cations. Generally, metal–organic complexes have less positive charge than the original metal ion, so complexation enhances desorption from soil minerals. Thus, the process is highly significant in altering the solubility, mobilization and availability of metal cations: the overall result is dependent on the properties of the cation–organic complex, the type and quantities of organic matter and cations and the influence of plants (Ondrasek and Rengel, 2012). In the case of organic acids, the greater the number of carboxylic groups present, the stronger their metal-complexing ability: organic acids with only one carboxyl group (lactate, formate and acetate) have very little influence. For metals, there is generally a decline in their propensity to form metal–organic complexes in the order: trivalent > bivalent > monovalent (Jones, 1998).

In addition to the role of organic matter present in soil, organic acids released from plant roots are also able to form complexes with metal cations. Forming organic complexes is an important mechanism by which plants respond to nutrient deficiencies in soils (Jones, 1998). It has been well known that roots of graminaceous plants secrete so-called phytosiderophores consisting of organic acids such as citric acids to form Fe^{3+}–citrate complexes in soil solution, which can be taken up readily by plants (Von Wirén *et al.*, 2000). Microbial siderophores may also contribute to iron nutrition in both dicotyledonous plants (Vansuyt *et al.*, 2007) and monocotyledonous graminaceous species (Shirley *et al.*, 2011). Organic complexes may also be an uptake mechanism for plants responding to manganese (Mn), and is also observed in Mn-deficient soils. Soil organic matter components do not directly form complexes with anionic nutrients, such as P, but their availability can also be enhanced by forming Fe^{3+}– and Al^{3+}–organic complexes, which release phosphorus from Fe–P and Al–P minerals.

For some metals, forming organic complexes is a process of detoxification to plants. Typically, toxicity of Al can be eliminated substantially by forming Al–organic acid complexes in apoplast and soil solution.

Copper (Cu) also readily forms complexes with soil organic matter and becomes unavailable (Inaba and Takenaka, 2005). Amendment with organic manure is considered an effective countermeasure for detoxification of Cu in Cu-contaminated soils (Paradelo et al., 2011). The molecular weight of organic matter plays a role in the toxicity and bioavailability of heavy metals to plants: organic moieties with larger molecular weight form more stable complexes, and have greater detoxification ability, than those with smaller molecular weight.

On the other hand, since the mobility of cationic nutrients is enhanced by the formation of metal–organic complexes, the complexed metals may be lost through leaching, leading to plant deficiency. For instance, peatland soils are usually deficient in Cu and cobalt (Co). Cu deficiency can also occur in soils with a high content of total Cu and organic matter, because of the low bioavailability of Cu–organic complexes (Edwards et al., 2012).

Problems from Excess Organic Inputs

Although increased content of C is generally beneficial for the functioning of agricultural soils, an excess can be deleterious. The main example is the situation where large amounts of animal manure, often from intensive livestock enterprises, are applied to a limited area of land. If much of the organic matter in the manure is in readily mineralizable forms, the released N can become a pollution hazard to water (as nitrate or dissolved organic N) and air as ammonia or nitrous oxide. For example, surface-applied manures, especially those with a high concentration of readily decomposable N compounds, leads to large losses of N to the atmosphere as ammonia. This can be reduced by either rapid incorporation or injection, in the case of liquid slurries, but this can lead to increased production of nitrate in soil and a greater risk of nitrate leaching (Sagoo et al., 2007). All N loss processes from manure, whether ammonia volatilization, nitrous oxide emission or nitrate leaching, are influenced greatly by the management practices used as well as by the type of manure. Key management factors are timing of application, the quantity applied (ensuring it is in line with crop requirements) and attempting to predict the supply of crop-available N from manures so that fertilizer applications are reduced accordingly (Salazar et al., 2005). Zhou et al. (2010) gives examples of very large accumulations of nitrate in the soil profile, leading to a high risk of nitrate leaching to water, when this was not done in the case of vegetable crops grown under plastic in north-west China. Nitrate accumulations to a depth of 1 m were frequently in the range 100–500 kg nitrate-N ha^{-1}. Similarly, P released from manures can also accumulate in soil and become a water pollutant. The rapidly increasing number of 'confined animal feeding operations' (CAFOs) around the world means that there is a trend towards large quantities of manure to be concentrated at a limited number of locations. This manure is a resource for increasing soil organic C content and as a source of nutrients for crops, but its management presents a major challenge if pollution from the sites of application is to be avoided.

Under flooded conditions, such as lowland rice cultivation, excess labile organic C input leads to extreme anaerobic conditions, which stress the growth of rice plants and soil microbial diversity and activities. It also leads to increased emissions of methane, a greenhouse gas about 25 times more powerful than CO_2.

Interactions between N and C in Natural Ecosystems

In addition to the importance of C for nutrient cycling in agricultural soils, and hence for food security (one of the provisioning roles of ecosystems), soil C and nutrients interact within natural and semi-natural ecosystems. Excess N is released from agricultural systems as various gases (oxidized and reduced forms of N) and transferred elsewhere. One negative impact of this is soil acidification, as demonstrated in China (Guo et al., 2010). This can affect both agricultural and non-agricultural soils. N deposited on semi-natural ecosystems can have negative impacts through alterations to the diversity of vegetation, but can also have impacts that have a beneficial global influence. Tree growth in northern boreal forests is generally limited by shortage of N. Even small quantities of N deposited in these regions stimulates the growth of vegetation and can lead to increased accumulation of C – initially in vegetation but then in soil. This process may well be quantitatively significant within the global C cycle, with implications for climate change mitigation, though at the expense of alterations to ecosystem structure and biodiversity (Magnani et al., 2007).

Acknowledgements

This work was supported by the European Union within the project EcoFINDERS (FP7-264465). Rothamsted Research receives financial support from the UK Biotechnology and Biological Sciences Research Council (BBSRC) through Institute Strategic Programme Grants.

References

Baumann, K., Dignac, M.F., Rumpel, C., Bardoux, G., Sarr, A., Steffens, M. and Maron, P.A. (2012) Soil microbial diversity affects soil organic matter decomposition in a silty grassland soil. *Biogeochemistry* doi:10.1007/s10533-012-9800-6.

Bengough, A.G. and McKenzie, B.M. (1997) Sloughing of root cap cells decreases the frictional resistance to maize (*Zea mays* L.) root growth. *Journal of Experimental Botany* 48, 885–893.

Clemmensen, K.E., Bahr, A., Ovaskainen, O., Dahlberg, A., Ekblad, A., Wallander, H., Stenlid, J., Finlay, R.D., Wardle, D.A. and Lindahl, B.D. (2013) Roots and associated fungi drive long-term carbon sequestration in boreal forest. *Science* 339, 1615–1618.

Courtier-Murias, D., Simpson, A.J., Marzadori, C., Baldoni, G., Ciavatta, C., Fernandez, J.M., López-de-Sa, E.G. and Plaza, C. (2013) Unravelling the long-term stabilization mechanisms of organic materials in soils by physical fractionation and NMR spectroscopy. *Agriculture, Ecosystems and Environment* 171, 9–18.

Dourado-Neto, D., Powlson, D.S., Abu Bakar, R., Bacchi, O.O.S., Basanta, M.V., thi Cong, P., Keerthisinghe, G., Ismaili, M., Rahman, S.M., Reichardt, K. *et al*. (2010) Multiseason recoveries of organic and inorganic nitrogen-15 in tropical cropping systems. *Soil Science Society of America Journal* 74, 139–152.

Edwards, A.C., Coull, M., Sinclair, A.H., Walker, R.L. and Watson, C.A. (2012) Elemental status (Cu, Mo, Co, B, S and Zn) of Scottish agricultural soils compared with a soil-based risk assessment. *Soil Use and Management* 28, 167–176.

Fontaine, S., Barot, S., Barré, P., Bdioui, N., Mary, B. and Rumpel, C. (2007) Stability of organic carbon in deep soil layers controlled by fresh carbon supply. *Nature* 450, 277–280.

Fontaine, S., Henault, C., Aamor, A., Bdioui, N., Bloor, J.M.G., Maire, V., Mary, B., Revaillot, S. and Maron, P.A. (2011) Fungi mediate long-term sequestration of carbon and nitrogen in soil through their priming effect. *Soil Biology and Biochemistry* 43, 86–96.

Gaillard, V., Chenu, C., Recous, S. and Richard, G. (1999) Carbon, nitrogen and microbial gradients induced by plant residues decomposition in soil. *European Journal of Soil Science* 50, 567–578.

Guo, J.H., Liu, X.J., Zhang, Y., Shen, J.L., Han, W.X., Zhang, W.F., Christie, P., Goulding, K.W.T., Vitousek, P.M. and Zhang, F.S. (2010) Significant acidification in major Chinese croplands. *Science* 327, 1008–1010.

Hirsch, P.R., Gilliam, L.M., Sohi, S.P., Williams, J.K., Clark, I.M. and Murray, P.J. (2009) Starving the soil of plant inputs for 50 years reduces abundance but not diversity of soil bacterial communities. *Soil Biology and Biochemistry* 41, 2021–2024.

Inaba, S. and Takenaka, C. (2005) Effects of dissolved organic matter on toxicity and bioavailability of copper for lettuce sprouts. *Environment International* 31, 603–608.

Johnston, A.E., Poulton, P.R. and Coleman, K. (2009) Soil organic matter: its importance in sustainable agriculture and carbon dioxide fluxes. *Advances in Agronomy* 101, 1–57.

Jones, D.L. (1998) Organic acids in the rhizosphere – a critical review. *Plant and Soil* 205, 25–44.

Kirkby, C.A., Kirkegaard, J.A., Richardson, A.E., Wade, L.J., Blanchard, C. and Batten, G. (2011) Stable soil organic matter: a comparison of C:N:P:S ratios in Australian and other world soils. *Geoderma* 163, 197–208.

Macdonald, A.J., Poulton, P.R., Powlson, D.S. and Jenkinson, D.S. (1997) Effects of season, soil type and cropping on recoveries, residues and losses of 15N-labelled fertilizer applied to arable crops in spring. *Journal of Agricultural Science, Cambridge* 129, 125–154.

Magnani, F., Mencuccini, M., Borghetti, M., Berbigier, P., Berninger, F., Delzon, S., Grelle, A., Hari, P., Jarvis, P.G., Kolari, P. *et al.* (2007) The human footprint in the carbon cycle of temperate and boreal forests. *Nature* 447, 848–850.

Nguyen, C. (2003) Rhizodeposition of organic C by plants: mechanisms and controls. *Agronomie* 23, 375–396.

Nielsen, U.N., Ayres, E., Wall, D.H. and Bardgett, R.D. (2011) Soil biodiversity and carbon cycling: a review and synthesis of studies examining diversity–function relationships. *European Journal of Soil Science* 62, 105–116.

Ondrasek, G. and Rengel, Z. (2012) The role of soil organic matter in trace element bioavailability and toxicity. In: Ahmad, P. and Prasad, M.N.V. (eds) *Abiotic Stress Responses in Plants: Metabolism, Productivity and Sustainability*. Springer Science+Business Media, LLC, New York, Dordrecht, Heidelberg, London, pp. 403–424.

Paradelo, R., Villada, A. and Barral, M.T. (2011) Reduction of the short-term availability of copper, lead and zinc in a contaminated soil amended with municipal solid waste compost. *Journal of Hazardous Materials* 188, 98–104.

Pascault, N., Ranjard, L., Kaisermann, A., Bachar, D., Christen, R., Terrat, S., Mathieu, O., Lévêque, J., Mougel, C., Henault, C., *et al.* (2013) Stimulation of different functional groups of bacteria by various plant residues as a driver of soil priming effect. *Ecosystems* 16, 810–822.

Prosser, J.I., Bohannan, B.J.M., Curtis, T.P., Ellis, R.J., Firestone, M.K., Freckleton, R.P., Green, J.L., Green, L.E., Killham, K., Lennon, J.E. *et al.* (2007) The role of ecological theory in microbial ecology. *Nature Reviews Microbiology* 5, 384–392.

Sagoo, E., Williams, J.R., Chambers, B.J., Boyles, L.O., Matthews, R. and Chadwick, D.R. (2007) Integrated management practices to minimise losses and maximise the crop nitrogen value of broiler litter. *Biosystems Engineering* 97, 512–519.

Salazar, F.J., Chadwick, D., Pain, B.F., Hatch, D. and Owen, E. (2005) Nitrogen budgets for three cropping systems fertilised with cattle manure. *Bioresource Technology* 96, 235–245.

Shirley, M., Avoscan, L., Bernaud, E., Vansuyt, G. and Lemanceau, P. (2011) Comparison of iron acquisition from Fe-pyoverdine by strategy I and strategy II plants. *Botany* 89, 731–735.

Six, J., Frey, S.D., Thiet, R.K. and Batten, K.M. (2006) Bacterial and fungal contributions to carbon sequestration in agroecosystems. *Soil Science Society of America Journal* 70, 555–569.

Smith, J.U., Bradbury, N.J. and Addiscott, T.M. (1996) SUNDIAL: a PC-based system for simulating nitrogen dynamics in arable land. *Agronomy Journal* 88, 38–43.

Tisdall, J.M. and Oades, J.M. (1982) Organic matter and water-stable aggregates in soils. *Journal of Soil Science* 33, 141–163.

Vansuyt, G., Robin, A., Briat, J.F., Curie, C. and Lemanceau, P. (2007) Iron acquisition from Fe-pyoverdine by *Arabidopsis thaliana*. *Molecular Plant–Microbe Interactions* 20, 441–447.

Von Wirén, N., Khodr, H. and Hider, R.C. (2000) Hydroxylated phytosiderophore species possess an enhanced chelate stability and affinity for iron (III). *Plant Physiology* 124, 1149–1157.

Whalley, W.R., Leeds-Harrison, P.B., Clark, L.J. and Gowing, D.J.G. (2005) Use of effective stress to predict the penetrometer resistance of unsaturated agricultural soils. *Soil and Tillage Research* 84, 18–27.

Whalley, W.R., To, J., Kay, B.D. and Whitmore, A.P. (2007) Prediction of the penetrometer resistance of soils with models with few parameters. *Geoderma* 137, 370–377.

Zhao, B.-Q., Li, X.-Y., Li, X.-P., Shi, X.-J., Huang, S.-M., Wang, B.-R., Zhu, P., Yang, X.-Y., Liu, H., Chen, Y. *et al.* (2010) Long-term fertilizer experiment network in China: crop yields and soil nutrient trends. *Agronomy Journal* 102, 216–230.

Zhou, J.B., Chen, Z.J., Liu, X.J., Zhai, B.N. and Powlson, D.S. (2010) Nitrate accumulation in soil profiles under seasonally open 'sunlight greenhouses' in northwest China and potential for leaching loss during summer fallow. *Soil Use and Management* 26, 232–339.

8 Soil Hydrology and Reactive Transport of Carbon and Nitrogen in a Multi-scale Landscape

Christopher Duffy* and Nikolaos Nikolaidis

Abstract

This chapter examines the role of soil in water filtration, its impact on carbon and nitrogen in biogeochemical transformations and its relation to the larger landscape, where soil functions are key to clean water, essential to human sustenance. Soil composition and chemical weathering are essential factors in soil structure and formation, which affect the hydrologic properties of soil and chemical transport significantly. The role of clays on aggregation, carbon (C) sequestration, pH, etc., carbon/nitrogen/phosphorus (C/N/P) cycles and plant growth (plant exudates) on reactive transport are examined. Examples of water filtration and solute transformation of a functioning soil (producing clean water) and of failure to transform (producing toxicity and contamination) are presented. Special focus is given on the parameterization of hydrologic and reactive transport models that cover a range of scales from soil profile, to hill slopes and the catchment. A variety of modelling strategies presently exist for biogeochemical modelling, and typically each focuses on a particular scale, with scale-appropriate processes, mechanisms and states. They range from bottom-up approaches, where plot-scale studies use intensive monitoring and detailed local modelling of process-level biogeochemical cycles for C and N, to regional- and continental-scale approaches to simulating the C–N dynamics in atmospheric models that, by necessity, may neglect the details of the processes understanding obtained in the plot-scale research. Ecosystem approaches extend the plot-scale models for C–N and water to landscape scales maintaining systematic processes, but may not include detailed geospatial structure and coupled hydrodynamic processes of the larger catchment and river basin. A strategy for merging scales and concepts intrinsic to plot-, landscape- and catchment-scale carbon-based biogeochemical research is proposed. The approach will describe existing process models for each scale of research, including hydrologic impacts. We propose a strategy for an integrated hydrodynamic approach to numerical modelling that can resolve the geospatial characteristics of water, carbon and nitrogen cycles over entire catchments, including first- and second-order streams, which link plot and hill-slope studies within this framework. The approach integrates plot-to-catchment scales for mesoscale model application for scales that range from 100 m to 10^5 km^2. The approach has important implications for ongoing soils research at Critical Zone Observatories, which are advanced field research facilities, to scale up their science understanding to larger domains and will improve the prospect of carbon–nitrogen management greatly.

*E-mail: cxd11@psu.edu

Chemical Weathering, Reactive Transport and C–N Dynamics

The Thematic Strategy for Soil Protection prepared for the European Union (EU) has identified soil ecosystem functions and services as important to the global economy and human sustenance (European Commission, 2006). These services go beyond the most easily identified service of biomass (food and fibre) production, and includes carbon and nitrogen sequestration, preservation of terrestrial biodiversity and the gene pool, water filtration–transformation, provision of raw materials, landscape and heritage (Banwart and SoilTrEC Partners, 2011; Banwart et al., 2011). Water filtration and mass transformation of transported substances (collectively termed below as water filtration and transformation) such as nutrient ions or dissolved pollutants are an important soil service for the regulation and provision of clean water, essential to human sustenance.

Organic carbon addition to the soil, either through natural (above- and belowground vegetation inputs), allochtonous inputs (i.e. compost, green manure, manure) or manipulation of vegetation/crop (e.g. crop rotation, nitrogen-fixing plants, intercropping), is the starting point of a series of biogeochemical mechanisms and processes that all together comprise the abovementioned soil functions and soil services including water filtration and transformation. Soil carbon is the key driver of soil structure that provides water regulation and drainage, and is the main substance controlling soil biogeochemistry and thus affecting water transformation reactions. The physical role of organic carbon on soil hydraulic properties including macropore development and water-holding capacity is also a factor.

Water, as it moves along a hydrologic pathway, alters the chemical composition of not only the soil and rock that come in contact but also its own composition. A conceptual schematic of the predominant hydrologic transport processes and soil characteristics affecting solute transport is presented in Fig. 8.1 (adapted from Kohne et al., 2009a,b). The figure illustrates a vertical cut of the unsaturated soil along a hill slope. Hydrologic processes that control the drainage characteristics of soil, and in turn its soil moisture,

Fig. 8.1. Hydrologic transport processes and soil characteristics affecting solute transport. (From Kohne et al., 2009a,b.)

depend on the structure of the soil matrix as well as the development of preferential flow macropore structures such as root holes, fractures and earthworm burrows. The combination of the inherent characteristics of soil matrix (homogeneous versus mixed permeability) and the macropore structure are primary factors affecting the hydraulic retention of water in the soil (Gerke and van Genuchten, 1993a,b; Kohne et al., 2009a,b).

A critical element for the conceptual model is the role of soil in water filtration and transformation. The foundations of this are the soil-forming processes (Jenny, 1941, 1994; Drever, 1997; Langmuir, 1997). Soils are formed from parent rock as a function of climate conditions, chemical and physical weathering processes and microbial and plant action, leading to minerals with specific chemical composition and physical properties. The parent material can be classified broadly into igneous, metamorphic and sedimentary rock types containing primary minerals that determine the rate and chemical composition of soils. Climate conditions modulate physical erosion and chemical weathering that through time reduces particle sizes, leading to increased mineral reactive surface area. The most important minerals are feldspars, quartz, olivine, pyroxenes, amphibole (hornblend), micas (biotite and moskovite) and clays (montmorillonite, kaolinite, etc.). Other minerals found in nature are those composed of soluble salts (halite, gypsum, pyrite), carbonates (calcite and dolomite), phosphate rocks (apatite), aluminium (Al) oxides (gibbsite and diaspore) and iron (Fe) oxides (goethite and hematite). Understanding the soil-forming processes lays the foundation for soil characterization and model development.

Figure 8.2 presents a schematic of the biogeochemical reactions affecting solute transport in the soil. Soil and sediments have the capability to remove ions from water (adapted from Kohne et al., 2009a,b). The water–soil interface plays a significant role in water filtration and transformation. Clays due to the isomorphic substitution of their central metal ion have a permanent negative charge. In aquatic environments, the negative surface charge is balanced with cations, which are exchanged by other cations of higher charge. Ion exchange is a reversible reaction that can affect clay and non-clay minerals, as well as organic and oxide coatings.

Fig. 8.2. Biogeochemical reactions affecting solute transport. (From Kohne et al., 2009a,b.)

Where the mineral surface charge is a function of soil pH, soils have the ability to adsorb/desorb anions and cations. These two reactions are occurring simultaneously in the soil, and these processes are hard to separate. Ion exchange reduces the leaching of calcium (Ca^{2+}), magnesium (Mg^{2+}), potassium (K^+) and sodium (Na^+) from the soil, making them bioavailable to the plants, serving as macronutrients necessary for plant growth. On the other hand, adsorption controls the mobility of anions, cations and heavy metals in the subsurface. Chemical weathering dissolves the minerals and transforms their chemical composition into new, secondary minerals (Drever, 1997; Langmuir, 1997). Chemical weathering is a kinetic phenomenon. There are two types of chemical weathering: one that results in complete dissolution without further precipitation (congruent dissolution), and the weathering that results in dissolution, where new products are formed and dissolved ions partially precipitate out of solution (incongruent dissolution); for instance, the chemical weathering of kaolinite results in the production of gibbsite.

Surface soil structure also develops through aggregate formation (Banwart et al., 2012; Nikolaidis and Bidoglio, 2013). Clay minerals and oxide coatings bind with soil organic matter and microbial biomass, creating aggregate particles that are larger than the original components. Humic and fulvic substances bind strongly to clay minerals and oxide coatings, forming organic–metal surface complexes. Negatively charged clay surfaces of minerals bind organic compounds through cation bridges, and iron and aluminum oxide surfaces bind organic compounds by electrostatic forces (Nikolaidis and Bidoglio, 2013). pH, redox status, clay content and oxide coating content affect organic sorption and surface complexation to surfaces. Silt-clay size particles (<53 µm) form microaggregate size particles (53–250 µm), while the combination of the two fractions form macroaggregate size particles and a well-graded soil. The organic matter sequestered within the macroaggregates is less susceptible to degradation. Stamati et al. (2012b) have shown that the mean particle diameter of the soil can increase significantly in a set-aside (uncultivated) soil where there is active carbon addition compared to a cultivated soil, and it can be an order of magnitude higher than the mean particle diameter of the actual minerals within the soil. Soil aggregates sequester carbon and nutrients for plant growth and increase soil moisture within the aggregate micropores for plant and microbial growth (Stamati et al., 2012a). The developed soil structure allows oxygen diffusion from the atmosphere and water drainage necessary to prevent waterlogging. Microbial action degrades the sequestered organic material, decreasing the cohesion of the aggregate, which eventually breaks up, allowing for faster organic matter decomposition and associated depletion of nutrients and the release of CO_2 to the atmosphere. In addition to soil biota, aboveground plant communities drive carbon sequestration and nutrient turnover through organic matter input to the soil. How much carbon and nutrients are stored in the soil depends on the balance between microbial decomposition and plant fixation rates.

Plant roots play a significant role in the soil aggregation process by entangling soil particles with the mycorrhizal system and through root exudates that can prime microbial activity, change soil pH and impact local mineral weathering activity. Other soil terrestrial fauna, and especially earthworms, play a significant role in soil aggregation, due to bioturbation, excretion of casts and deposition of mucus on the walls of the burrow (Nikolaidis and Bidoglio, 2013).

Human activities have impacted soil ecosystem services significantly (Nikolaidis, 2011), and in particular the soil function to filter and transform water. The main ecosystem drivers that affect soil functions and services are climate change, land-use change and aboveground biodiversity change (Nikolaidis, 2011). In relatively undisturbed forest and grassland soil ecosystems, nutrient cycling (C, N, P, K) is regulated by the plant–soil–microorganism system, and nutrient leaching is controlled tightly by the hydrologic pathway and the reactivity of the host minerals (Stamati et al., 2011). Riparian forests have been used as natural and engineered systems to improve the water quality of drainage water in agricultural areas (Broadmeadow and Nisbet, 2004). Once the soil system is disturbed by converting forest or grassland to cultivated and managed soils, soil structure is lost and macroaggregates are mechanically broken down, exposing protected organic matter to microbial degradation and nutrient leaching. Within 5–10 years after the land-use conversion, significant quantities of organic carbon and nitrogen of the soil are depleted (Mann, 1986; Lal, 2004), decreasing the soil fertility and decreasing the ability for water filtration and transformation. Livestock grazing of pastures exerts significant impact on soils due to the decrease of aboveground biodiversity, soil compaction, increase in erosion and soil loss, as well as increase in the inputs of manure. Stamati et al. (2011) have shown that a nearly linear relationship exists between dissolved organic nitrogen (DON) export and livestock N load that operates at regional scales. De-vegetation of grazing lands coupled with increased carbon inputs in the form of manure caused decline in the biochemical quality of soil and leaching of soluble organic matter.

Modelling of Soil Hydrology and Coupled C–N Biogeochemical Transport Across Scales

Reactive transport modelling has been used as a way to gain better understanding of the coupled biogeochemical process in soils and groundwaters (Regnier et al., 2003) for the past two decades. Several recent articles are available that review the current state of the art of reactive transport modelling in fractured rock, investigating the design of nuclear depository facility (MacQuarrie and Mayer, 2005), contaminant transport in groundwaters (Steefel et al., 2005), tracer and pesticide transport in structure soils (Kohne et al., 2009a,b) and organic matter and soil structure dynamics (Nikolaidis and Bidoglio, 2013).

Even though significant advancement has been made, further improvement in our understanding of reactive transport modelling will require addressing the following issues: treatment of preferential flow and non-equilibrium (Kohne et al., 2009a,b), pore-scale structures and pore heterogeneity (Gharasoo et al., 2012), as well as two flow domain issues and microenvironments (Steefel et al., 2005). Upscaling of processes from the laboratory to the field has always been an obstacle to model parameter estimation. A critical review on upscaling sorption/desorption processes in reactive transport models concluded that inclusion of small-scale processes might not always lead to better prediction of the larger-scale behaviour of metal/radionuclide transport due to conceptual model errors, geochemical heterogeneities and incorrect model discretization compared to the scale of the processes (Miller et al., 2010). Steefel and co-workers (2005) have summarized the main challenges in reactive transport as follows:

- treatment of chemical microenvironments and assessment of how they affect the bulk geochemistry, including C–N dynamics;
- treatment of flow, chemistry and mechanical deformation of the matrix in a unified way;
- resolving the discrepancies between laboratory and field reaction rates; and
- upscaling reactive transport processes.

Strategies for Merging Scales and Concepts to Advance Hydrologic and C–N Reactive Transport

A wide range of modelling strategies exists for biogeochemical modelling with a range of scales, processes, mechanisms and states. These approaches range from bottom-up approaches, where plot-scale studies use intensive monitoring and detailed local modelling to improve our understanding of process-level biogeochemical cycles for C and N, to regional- and continental-scale approaches to simulating the C–N dynamics in atmospheric models that by necessity may neglect the details of processes understanding obtained in the plot-scale research. Ecosystem approaches extend the plot-scale models for C–N and water to landscape scales maintaining systematic processes, but may not include detailed geospatial structure and coupled hydrodynamic processes of the larger catchment and river basin. Other models are available that represent some variables of interest based on statistical inference, plot or hybrid approaches (Arnold et al., 1998, SWAT model; Stöckle et al., 2003, CropSyst model; Schwarz et al., 2011, Sparrow model). The intention is not to propose a new model, but rather to integrate process-based plot-scale models with a catchment-scale model that fills a gap for efficient modelling of lateral transport to first- and second-order streams while also allowing simulation scales of mesoscale catchments.

A strategy for merging scales and concepts intrinsic to plot-, landscape- and catchment-scale biogeochemical research is proposed. An integrated hydrodynamic approach to numerical modelling can resolve the geospatial characteristics of water and carbon and nitrogen cycles over entire catchments. The intention is that the mesoscale model will have application for scales that range from 10^{-2} to 10^5 km^2. The current status of modelling N and C dynamics can be described in terms of two types:

1. The soil profile model approach where detailed vertical distributions of water, energy and solutes are explicitly modelled (1-D continuum) while plant growth/decay, landscape features, soil are represented as a discrete domain defined for some characteristic of the landscape or an average over some area. At this scale, lateral spatial gradients are not explicitly resolved. A primitive version of this type uses vertical averaging to represent the soil in terms of the bulk water, energy and solute dynamics. The soil profile can include vertical macropore-matrix effects (Gerke and van Genuchten, 1993b) and complete kinetic descriptions of the coupled water and biogeochemical reactions.

2. A semi-distributed model approach combines the concept of hydrologic response units (HRUs) and the 1-D continuum approach from (1) to divide the catchment into characteristic landscape units that allow mass transfer to adjacent sub-areas based on the average response in the HRU. In this approach, the lateral mass transfer uses a one-way coupling strategy to fluid flow. This approach preserves the 1-D continuum characteristics including vertical macropore effects, but is limited to empirical lateral mass transport, which limits the use of distributed validation data sets (e.g. groundwater and baseflow data).

Ideally, the next generation of simulation models should expand on the experience gained in the development of these models and incorporate a modern scientific and software infrastructure. It should be biophysically based, with fully distributed hydrology, have a common object-oriented structure to simulate nutrient flow and vegetation processes, and transport across land units, and the ability to simulate the variations in flows generated by management. The modelling infrastructure outlined above follows a 'bottom-up', process-based approach to simulating landscape-level effects. Processes occurring in land segments, and landscape-level processes governed by the movement of water in watersheds, need to be represented to simulate water quality at any point in the landscape.

Technically, it is progress in this direction that will allow a seamless integration between stakeholder demands and science delivery in the form of model outputs. In this context, a new 2-D continuum approach for lateral surface runoff and groundwater flow with embedded and fully coupled 1-D soil profile processes is proposed. This approach is still not a fully 3-D continuum strategy, since these are limited to relatively small-scale computational domains and our practical goal here is a catchment-scale model. It recognizes the extreme computational requirements of the 3-D hydrobiogeochemical continuum by a suitable dimension reduction, but still makes progress at the catchment scale (10^{-2} to 10^5 km^2).

As an example of this strategy, one could use a catchment hydrodynamic model, the Penn State Integrated Hydrologic Model (PIHM; http://www.pihm.psu.edu) that is currently being implemented at Critical Zone Observatory (CZO) sites in the USA and Europe, with the 1-D Integrated Critical Zone (ICZ) model. The PIHM code is a multi-process hydrologic model and geographic information system (GIS) toolkit for building unstructured numerical mesh with initial estimates of parameters over the catchment domain. Major hydrological processes are fully coupled by the semi-discrete finite volume approach, simulating water, energy and solute transport at the catchment scale (Qu and Duffy, 2007; Kumar et al., 2009, 2010). PIHM can also solve for plant water uptake components, groundwater–stream exchange and recharge over the catchment stream network (Fig. 8.3). The transport equations in PIHM allow the simulation of residence time and age of water in soils, groundwater and streams.

The 1D-ICZ model simulates key soil ecosystem functions such as biomass production, carbon and nutrient sequestration, water transformation and filtration and belowground biomass production. The 1D-ICZ is a combination of four submodels: (i) a flow, transport and bioturbation model, HYDRUS 1D, which simulates flow, heat and solute transport in the unsaturated zone; (ii) a chemical equilibrium model and weathering to account for the

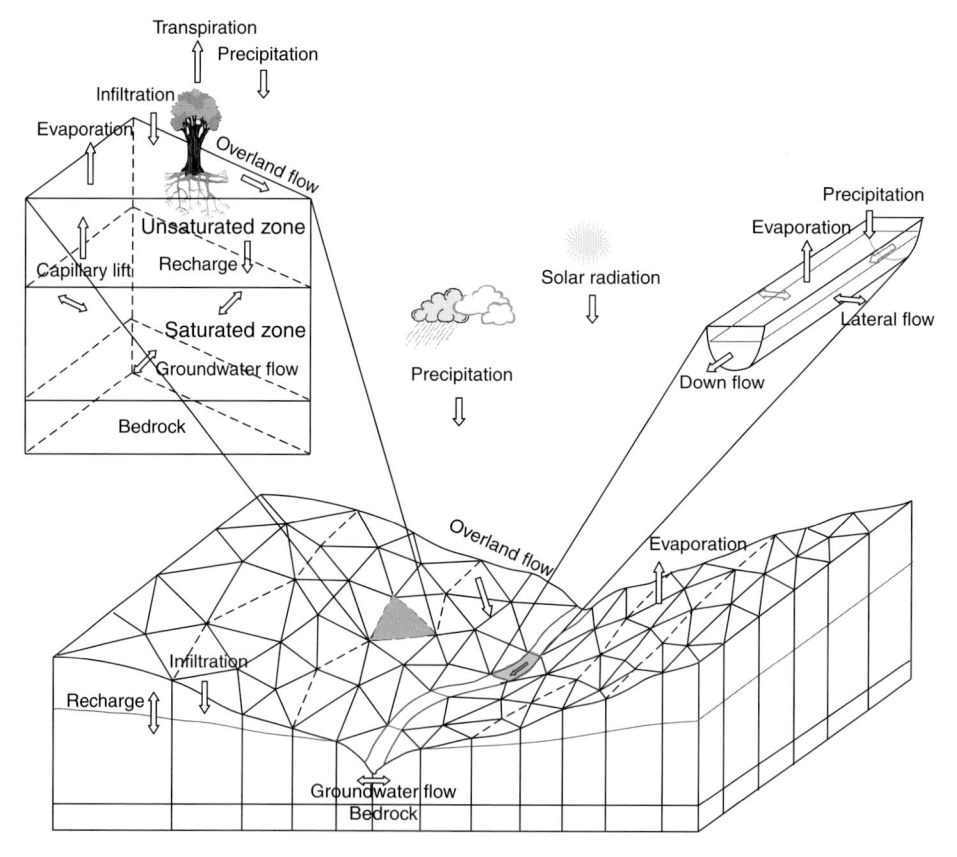

Fig. 8.3. Spatial discretization used for the numerical representation of the watershed in PIHM. Interacting hydrologic processes are assigned to each prismatic element (top left) and on each linear river element (top right). The local system of ordinary differential equations (ODEs) corresponding to the processes acting on a unit prismatic element is termed as 'model kernel'. (From Qu and Duffy, 2007.)

effect of water transformation; (iii) a C/N/P dynamics and structure model, CAST, that links the transformations of organic matter with a dynamic model of soil aggregation/disaggregation, a simplified terrestrial ecology model that is comprised of mycorrhizal fungi, microorganisms (BIO pool), consumers and predators (FAUNA pool) and a plant/root dynamics model; (iv) a plant dynamics model, PROSUM, that is based on biomass production ecological principles and predicts the dynamics of key variables (e.g. above- and belowground production of litter C and N; nutrient and water uptake) in response to key drivers (temperature; availability of light, water, CO_2 and the nutrient elements N, P, Ca, Mg and K; grazing and management events) for the wide range of vegetation types (outlined in Banwart et al., 2012).

Integrating a 1-D ecosystem function model into a fully distributed catchment model would provide a rigorous biophysically based model that computes rates of change and updates numerous state variables based on biophysical principles within the soil profile, while transporting the water and solutes to streams draining the catchment.

The conceptual strategy goes beyond a distributed C–N–biogeochemical model that integrates multiple data sources to include a database that links the model to the 'essential terrestrial variables', or ETVs, that allow dynamic simulation across spatial and temporal scales relevant to decision making for soil management and land use. As part of this model synthesis is the question of data and data support. A multi-scale model as proposed here must also have efficient access to data for topography, soils, geology and climate, land cover/land use, as well as reactive transport properties and parameters and open-source GIS tools to set up the model domain and parameters. Dynamic modelling of geospatial landscape processes requires a comprehensive database that can be linked to each land use, soil, geology and climate. This model–data architecture is a major challenge and, to our knowledge, existing databases lack the harmonized data products necessary for modelling that reflect current and past land-use and management practices, are geospatially and geotemporally consistent and are fully coupled to the model within an efficient cyberinfrastructure. Such a prototype is being developed at PennState (see http://www.hydroterre.psu.edu) and at the EC Joint Research Centre (see http://fate.jrc.ec.europa.eu/interactive-maps-and-data). Figure 8.4 presents a prototype service showing the soils, land parcel and wetland inventory in a central database used to automate the development of the PIHM model for a site in Pennsylvania, USA. Ideally, this prototype cyberinfrastructure would enable users to have efficient access to all necessary data products and to be able to carry out simulations for any catchment in a region from a consistent data source.

Clearly, coupling a catchment hydrodynamic model with an integrated carbon and nitrogen and reactive transport code that resolves processes at spatial scales from plot to catchments must overcome several obstacles:

- **Data support across scales**: the process of connecting an independent C–N reactive transport model with a spatially distributed hydrodynamic model will require a new approach to model data and parameterization. We propose a new strategy that integrates geospatial/temporal data, political or property boundaries, hydrographic, hydrologic, soil, climate, reactive transport properties and topographical and other community-derived geospatial data (http://www.hydroterre.psu.edu).
- **Multi-state parameter estimation**: a wide range of parameter estimation tools are presently available that can be adopted for automated parameter estimation.
- **Interoperability**: the lack of interoperability between water data, models and model simulations, and the lack of fast access between essential watershed data and computing resources,

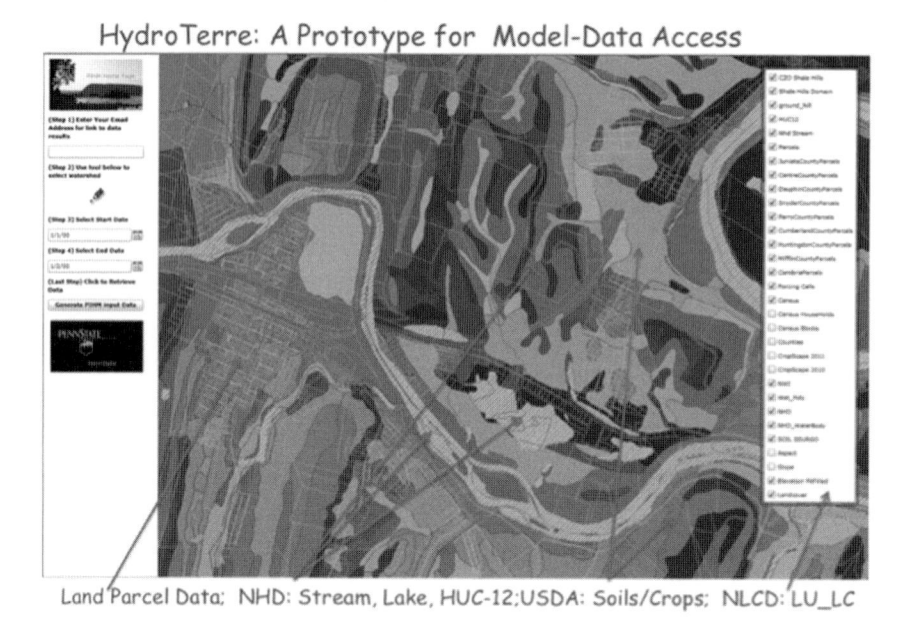

Fig. 8.4. Prototype services showing the soils, land parcel, wetland inventory in a central database used to automate the development of the PIHM model for a site in Pennsylvania, USA. (From Leonard and Duffy, 2013.)

impedes current efforts to predict and manage Earth's water and ecological resources. To date, databases for large-scale applications have been created and applied ad hoc to attend to a specific demand, rather than a growing set of new and future demands, with limited reuse of information due to lack of compatibility among seemingly similar efforts.

- **Scaleability of model data for decision making**: it is critical that the next-generation algorithms and technologies for data access, simulation and delivery should apply to all scales for plot, landscape and catchment scales, such that management practice can be tested across scales, from single fields to large watersheds, specifically targeting the impact of landscape practices and long-term outcomes.

The modelling strategy proposed has important implications for soils research at Critical Zone Observatories to scale up their science understanding to larger domains, which will improve the prospect of carbon–nitrogen management greatly.

The novelty of the proposed strategy is that for the first time a spatially explicit and physically based hydrological model will be coupled with the nutrient cycling and agronomic simulation model, a necessary step towards a complete virtual representation of the physical and biological matrix of a watershed. This is enabled largely by the steady progress in computing power, accessibility of high-resolution soils, weather and land-use data, and advances in necessary theoretical and empirical knowledge for predictions in terrestrial hydrologic and biogeochemical processes. The development of such a model will allow simulating virtual scenarios in agricultural catchments, support the analysis of emerging and fully coupled properties of the soil carbon system and implement realistic management optimization schemes that consider the variability of the landscape explicitly, so that management in one land unit can be evaluated transparently for its on-site and off-site effects.

References

Arnold, J.G., Srinivasan, R., Muttiah, R.S. and Williams, J.R. (1998) Large-area hydrologic modeling and assessment: Part I. Model development. *Journal of American Water Resources Association* 34, 73–89.

Banwart, S.A. and SoilTrEC Partners (2011) Save our Soils. Comment article. *Nature* 474, 151–152, 9 June.

Banwart, S.A., Bernasconi, S., Bloem, J., Blum, W., Brandao, M., Brantley, S., Chabaux, F., Duffy, C., Lundin, L., Kram, P. *et al*. (2011) Assessing soil processes and function across an international network of critical zone observatories: research hypotheses and experimental design. *Vadose Zone Journal* 10, 978–987.

Banwart, S.A., Bernasconi, S., Bloem, J., Blum, W., de Souza, D.M., Chabaux, F., Duffy, C., Lundin, L., Kram, P., Nikolaidis, N. *et al*. (2012) Soil processes and functions across an international network of critical zone observatories: introduction to experimental methods and initial results. *Comptes Rendu Geosciences* 344, 758–772.

Broadmeadow, S. and Nisbet, T.R. (2004) The effects of riparian forest management on the freshwater environment: a literature review of best management practice. *Hydrology and Earth System Sciences* 8, 286–305.

Drever, J.I. (1997) *The Geochemistry of Natural Waters: Surface and Groundwater Environments*, 3rd edn. Prentice-Hall, Upper Saddle River, New Jersey, 436 pp.

European Commission (2006) *Thematic Strategy for Soil Protection*, COM (2006) 231. European Commission, Brussels.

Gerke, H.H. and van Genuchten, M.T. (1993a) Evaluation of a first-order water transfer term for variably saturated dual-porosity flow models. *Water Resources Research* 29(4), 1225–1238.

Gerke, H. and van Genuchten, M.T. (1993b) A dual-porosity model for simulating the preferential movement of water and solutes in structured porous media. *Water Resources Research* 29(2), 305–319.

Gharasoo, M. Centler, F., Regnier, P., Harms, H. and Thullner, M. (2012) A reactive transport modeling approach to simulate biogeochemical processes in pore structures with pore-scale heterogeneities. *Environmental Modeling and Software* 30, 102–114.

Jenny, H. (1941, 1994) *Factors of Soil Formation*. McGraw-Hill, New York (1941) and Dover, New York (1994), 191 pp.

Kohne, J.M., Kohne, S. and Simunek, J. (2009a) A review of model applications for structured soils: (a) Water flow and tracer transport. *Journal of Contaminant Hydrology* 104, 4–35.

Kohne, J.M., Kohne, S. and Simunek, J. (2009b) A review of model applications for structured soils: (b) Pesticide transport. *Journal of Contaminant Hydrology* 104, 36–60.

Kumar, M., Bhatt, G. and Duffy, C.J. (2009) An efficient domain decomposition framework for accurate representation of geodata in distributed hydrologic models. *International Journal of Geographical Information Science* 23(12), 1569–1596.

Kumar, M., Bhatt, G. and Duffy, C.J. (2010) An object-oriented shared data model for GIS and distributed hydrologic models. *International Journal of Geographical Information Science* 24(7), 1061–1079.

Lal, R. (2004) Soil carbon sequestration impacts on global climate change and food security. *Science* 304, 1623–1627.

Langmuir, D. (1997) *Aqueous Environmental Geochemistry*. Prentice-Hall, Upper Saddle River, New Jersey, 600 pp.

Leonard, L. and Duffy, C.J. (2013) Cyber infrastructure for Distributed Water Resource Modelling: A National Prototype for Model-Data Web Services and Workflows. *Environmental Modelling and Software* 50, 85–96.

MacQuarrie, K.T.B. and Mayer, K.U. (2005) Reactive transport modelling in fractured rock: a state-of-the-science review. *Earth-Science Reviews* 72, 189–227.

Mann, L.K. (1986) Changes in soil carbon storage after cultivation. *Soil Science* 142(5), 279–288.

Miller, A.W., Rodriguez, D.R. and Honeyman, B.D. (2010) Upscaling sorption/desorption processes in reactive transport models to describe metal/radionuclide transport: a critical review. *Environmental Science and Technology* 44, 7996–8007.

Nikolaidis, N.P. (2011) Human impacts on soil: tipping points and knowledge gaps. *Applied Geochemistry* 26, S230–S233.

Nikolaidis, N.P. and Bidoglio, G. (2013) Soil organic matter dynamics and structure. *Sustainable Agriculture Reviews* 12, 175–200.

Qu, Y. and Duffy, C.J. (2007) A semidiscrete finite volume formulation for multiprocess watershed simulation. *Water Resources Research* 43, W08419, doi:10.1029/2006WR005752.

Regnier, P., Jourabchi, P. and Slomp, C.P. (2003) Reactive-transport modeling as a technique for understanding coupled biogeochemical processes in surface and subsurface environments. *Netherlands Journal of Geosciences* 82, 5–18.

Schwarz, G.E., Alexander, R.B., Smith, R.A. and Preston, S.D. (2011) The regionalization of national-scale SPARROW models for stream nutrients. *JAWRA Journal of the American Water Resources Association* 47, 1151–1172.

Stamati, F., Nikolaidis, N.P., Venieri, D., Psillakis, E. and Kalogerakis, N. (2011) Dissolved organic nitrogen as an indicator of livestock impacts on soil biochemical quality. *Applied Geochemistry* 26, S340–S343.

Stamati, F., Nikolaidis, N.P., Banwart, S.A. and Blum, W.E. (2012a) A coupled carbon, aggregation, and structure turnover (CAST) model for topsoils. *GeoDerma* 211–212, 51–64.

Stamati, F., Nikolaidis, N.P. and Schnoor, J.L. (2012b) The role of soil texture on carbon and nitrogen sequestration in agricultural soils of different climates. *Agriculture, Ecosystems and Environment* 165, 190–200.

Steefel, C.I., DePaolo, D.J. and Lichtner, P.C. (2005) Reactive transport modeling: an essential tool and a new research approach for the earth sciences. *Earth and Planetary Science Letters* 240, 539–558.

Stöckle, C.O., Donatelli, M. and Nelson, R. (2003) CropSyst, a cropping systems simulation model. *European Journal of Agronomy* 18(3), 289–307.

9 Climate Change Mitigation

Martial Bernoux* and Keith Paustian

Abstract

Terrestrial ecosystems play a major role in regulating the concentrations of three greenhouse gases (CO_2, CH_4 and N_2O), of which CO_2 is the most important in terms of the impact on the global radiative balance. Soils play a major role in the global carbon (C) cycle and CO_2 dynamics; thus, management of soil carbon appears essential and more and more inevitable.

The capacity of natural and managed agroecosystems to remove carbon dioxide from the atmosphere in a manner that is not immediately re-emitted into the atmosphere is known as carbon sequestration: carbon dioxide is absorbed by vegetation through photosynthesis and stored as carbon in biomass and soils, and released through autotrophic and heterotrophic respiration. Forests, croplands and grasslands can store large amounts of carbon in soils for relatively long periods. Soils are the larger terrestrial pool of organic carbon. Moreover, soil carbon sequestration is beneficial for soil quality, both over the short term and long term, and can be achieved through land management practices adapted to the specific site characteristics. The ability of soils to sequester carbon depends on climate, soil type, vegetation cover and land management practices.

According to the fourth assessment report of the Intergovernmental Panel on Climate Change (IPCC), the total technical greenhouse gas (GHG) mitigation potential of agriculture (considering all gases and sources) is estimated to be in the range 4.5–6 Gt CO_2-equivalent year^{-1} by 2030. Estimates indicate that many of these options are of relatively low cost and generate significant co-benefits in the form of improved agricultural production systems, resilience and other ecosystem services. Moreover, many of the technical options are readily available and could be deployed immediately. About 90% of this potential can be achieved by soil C sequestration through cropland management, grazing land management, restoration of organic soils and degraded lands, and water management in rainfed and irrigated croplands. In most cases, such management practices include the management of organic residues produced on site or coming from outside the field or the farm. It has been estimated that the global world production of residues in the agriculture sector is about 3.8 Pg C and, to date, the use of this resource has not been optimized; a large part is still being burned.

Over the past two decades, other practices have been tested and are still controversial, such as biochar or chipped ramial wood application in cultivated fields. Biochar is a stabile carbon amendment, produced from pyrolysis of biomass, which may increase biomass productivity as well as sequester C from the source biomass. The scientific validation of these practices is still incomplete.

*E-mail: martial.bernoux@ird.fr

Full participation of the agricultural sector in GHG mitigation still faces some challenges and barriers related to measurement, monitoring and reporting requirements in C offset markets. Further improvements are needed in methodologies and approaches that would help project designers and policy makers to integrate significant mitigation effects in agriculture development projects.

Introduction and Background

At global level, agriculture and forestry are major contributors to increased greenhouse gas (GHG) concentrations, from emissions of nitrous oxide (N_2O), methane (CH_4) and carbon dioxide (CO_2). Agricultural emissions of N_2O and CH_4 amount to c.10–12% of total global anthropogenic emissions of GHGs (IPCC, 2007), with emissions in 2005 estimated at 3.3 Pg CO_2-equivalent for CH_4 and 2.8 Pg CO_2-equivalent for N_2O. Soil represents the main C stock in agricultural systems, and decreases or increases of soil organic C (SOC) stocks, due to changes in land use and management, climate or other drivers, result in net emissions or removals of CO_2. Global estimates for the net CO_2 contribution from agricultural soils are difficult to quantify and have large uncertainty, but recent IPCC (2007) estimates are for a net emission of 40 Tg CO_2-eq year^{-1}. This relatively small net contribution is in contrast to a much larger historical loss of C from soils, and also potential losses induced by global warming, which could be in the order of several Pg CO_2 year^{-1} (Amundson, 2001). In a recent study, Shevliakova et al. (2009) estimated a net annual terrestrial carbon source due to all land-use activities ranging from 1.1 to 1.3 Pg C during the 1990s. Those authors also estimated separately, but with higher uncertainties, the contributions from grasslands (ranging from an emission of 0.37 Pg C year^{-1} to a sink of 0.15 Pg C year^{-1}) and for croplands (net emissions in the range 0.6–0.9 Pg C year^{-1}).

In any case, the contribution of agriculture to GHG emissions varies from country to country, depending mainly on the structure of the economy. Excluding land-use change and forestry (LUCF), agricultural emissions vary from a few per cent (e.g. 1% for Jordan in 2000, 6% for the USA, 10% for the 27 members of the European Union in 2010, 14% for China in 1994) to one-half or more of total emissions (e.g. 48% for Brazil in 2005, 68% for Benin in 2000, 81% for Ethiopia in 1995 and even 91% for Chad in 1993) (UNFCCC, 2012). For all Annex I countries together, the agricultural sector (excluding LUCF) contributes about 9% of total emissions.

This huge uncertainty applies also to soil C sinks at national and regional levels. In most cases, there is no agreement on the regional net contribution of SOC from agricultural soils. For instance, in Europe, some authors reported a recent net sink (Ciais et al., 2011a), whereas others reported net emissions (Bellamy et al., 2005; Ceschia et al., 2010). Most regions have even less available information (e.g. Africa; see Ciais et al., 2011b) and thus uncertainties are large and it is difficult to draw clear conclusions.

Agricultural activities are also directly or indirectly responsible for the largest part of the emissions, due to deforestation worldwide (Hosonuma et al., 2012). Collectively, agriculture, forestry and other land use (AFOLU) is recognized to contribute about one-third of global anthropogenic emissions (Baumert et al., 2005; IPCC, 2007), and thus agriculture should be strongly considered in policies and actions concerning GHG mitigation.

Although the AFOLU sector is a GHG source, it can also act as a sink for CO_2, through improved land use and management practices. Smith et al. (2007) estimated that the global potential for all mitigation options in agriculture, considering all gases and excluding fossil fuel replacement from bioenergy, were c.5500–6000 Tg CO_2-eq year^{-1} by 2030, with an associated 95% confidence interval from 3000 to 8700 Tg CO_2-eq year^{-1}. Moreover, the authors also estimated that about 90% of total mitigation potential was from soil carbon sequestration (SCS) including cropland and grassland management and the restoration of organic cultivated soil and degraded land. As with emissions, the mitigation potential and measures vary among regions; however, developing countries, which are responsible for about 75% of agricultural GHG emissions, also represent the largest share, about 70%, of the total mitigation potential (IPCC, 2007; UNFCCC, 2008).

Thus, managing soil C has implications both in terms of sources and sink of atmospheric CO_2. But the AFOLU sector must, above all, face the important challenge of supporting food production and food security. This imperative is recognized in the UNFCCC, where it is stated that stabilization of GHGs must be achieved while ensuring that food production is not threatened. Thus, particularly following the food crises of 2008 and 2009, the international debate has shifted somewhat away from GHG mitigation in the AFOLU sector towards food security and sustainability goals in other international agreements and initiatives (e.g. United Nations Convention to Combat Desertification (UNCCD), Convention on Biological Diversity (CBD), Global Soil Partnership, Rio+20, etc.), with a greater focus on adapting to climate changes and extreme climate events. However, adopting practices to achieve GHG mitigation and improving food security and soil sustainability are not, in most cases, mutually exclusive. Indeed, they can often be complementary in that SCS allows for the replenishment of soil organic matter, thereby providing several other benefits, including improved soil structure and stability that leads to reduced soil erosion, improved soil biodiversity, increased nutrient-holding capacity, increased nutrient-use efficiency, increased water-holding capacity and increased crop yields. Moreover, carbon in soil improves the resilience of cropping systems against both excess and lack of water, and accompanying improvements in soil biodiversity increases resilience to changing environmental conditions and stresses. Therefore, it strengthens the capacity to face extreme events (climate adaptation). Most mitigation and adaptation solutions are interrelated, and both must be planned together. In arid and semi-arid regions, soil degradation is widespread and most dryland soils are already degraded or are at high risk of degradation. Due to natural constraints, dryland soils contain a very small amount of carbon (typically below 1%). Thus, maintaining a minimum soil organic matter level is critical to maintain soil function. It is evident that in arid and semi-arid regions, SCS is more important for the non-GHG benefits, in terms of economic and social impacts, than the absolute amount of carbon sequestered.

The objectives of this chapter are thus to provide an updated overview of the different approaches for encouraging SCS, but also to give a focus on common erroneous shortcuts still present in the debate of mitigation and agriculture. Finally, this chapter will also discuss and propose solutions for removing barriers to the full implementation of SCS solutions into policies.

Options and Practices to Mitigate through Soil Carbon Sequestration at Field Scale

Many mitigation options focus on increasing SOC in the soil profile, but managing soils also influences fluxes of CH_4 and N_2O. Thus, to account for the overall GHG balance, Bernoux et al. (2006) proposed a definition of SCS: 'SCS for a specific agroecosystem, in comparison with a reference, should be considered as the result for a given period of time and portion of space of the net balance of all GHG expressed in $C–CO_2$-equivalent or CO_2-equivalent, computing all fluxes at the soil–plant–atmosphere interface, but also the indirect fluxes (gasoline, enteric fermentation, etc.).' A number of detailed reviews of agricultural management practices being promoted for SCS and GHG reductions have been published recently (e.g. Ogle et al., 2005; Smith et al., 2007; Eagle et al., 2010; Denef et al., 2011). Here, we present a brief overview of different types of activities from a global perspective.

The IPCC (2007) published global estimates of SCS (net change considering all direct GHGs, expressed as CO_2-eq) of broad sustainable land management categories, namely agronomy, nutrient management, tillage/residue management, water management and agroforestry (Table 9.1). Briefly, the 'agronomy' category corresponds to practices that may increase yields and thus generate

Table 9.1. Annual net mitigation potentials including non-CO_2 GHG (Mg CO_2-eq ha^{-1} year^{-1}) in each climate region for aggregate management categories.

Regions (moisture regime)	Sustainable land management categories	Net mitigation potential[a] (t CO_2-eq ha^{-1} year^{-1})		Average yield increase[b] (%)
		Cool	Warm	
Dry	Agronomy	0.39	0.39	116
	Nutrient management	0.33	0.33	72[c]
	Tillage/residues management	0.17	0.35	122
	Water management	1.14	1.14	92
	Agroforestry	0.17	0.35	81
Moist	Agronomy	0.98	0.98	122
	Nutrient management	0.62	0.62	118
	Tillage/residues management	0.53	0.72	55
	Water management	1.14	1.14	164
	Agroforestry	0.09	0.72	61

[a]Eggleston et al. (2006); [b]Branca et al. (2013b); [c]considers only organic nutrient management.

higher residues. Examples of such practices reported by Smith *et al.* (2007) include using improved crop varieties, extending crop rotations and rotations with legume crops. Nutrient management corresponds to the application of fertilizer, manure and biosolids, improving either the efficiency (adjusting application rate, improving timing, location, etc.) or diminishing the potential losses (slow-release fertilizer form or nitrification inhibitors). Tillage/residue management concerns the adoption of practices with less tillage intensity, ranging from minimum tillage to no-tillage, and with or without residue retention on the field. Water management brings together enhanced irrigation measures that can lead to an increase in productivity (and hence of the residues). Agroforestry encompasses a wide range of practices where woody perennials are integrated within agricultural crops.

Estimates of the mean net mitigation potential by these aggregated management categories are reported in Table 9.1. Due to the scarcity of data, only simplified categories were used in compiling mean estimates of C sequestration potential at the global scale, aggregated by major climate (Table 9.2; Plate 7).

Branca *et al.* (2013b) synthesized published data on the impact of the adoption of mitigation practices corresponding to these categories on crop productivity (average yield). Table 9.1 should be read with caution, because the values have been obtained from two different data sets, and also the main land management categories are too broad to allow a direct relationship between the mitigation potentials and average yield increases. But as a whole, these findings suggest that management options favourable for GHG mitigation have a positive effect on yields, and thus soil fertility.

Table 9.2. Correspondence between IPCC climate zones used in Eggleston *et al.* (2006) and simplified classification used by Smith *et al.* (2007) and IPCC (2007) for reporting GHG mitigation potentials.

IPCC climate zone	Simplified
Tropical montane dry	Warm dry
Tropical montane moist	Warm moist
Tropical wet	Warm moist
Tropical moist	Warm moist
Tropical dry	Warm dry
Warm temperate dry	Warm dry
Warm temperate moist	Warm moist
Cool temperate dry	Cool dry
Cool temperate moist	Cool moist
Boreal moist	Boreal moist
Boreal dry	Boreal dry

This finding has important implications for the potential and means of capturing synergies between mitigation and food security (Branca et al., 2013a).

Since the IPCC (2007) assessment, new practices have been considered, including the use of biochar as a soil amendment and also the application of 'chipped ramial wood' to cultivated fields.

Biochar is produced from pyrolysis of biomass heated between 300°C and up to 1000°C under zero (or extremely low) oxygen concentration. Biochar is considered to be constituted mostly of recalcitrant or stable organic C, and has been proposed as a mitigation solution that also improves soil properties and functionality. A recent report (Verheijen et al., 2010) concluded that the carbon sequestration potential of biochar was largely hypothesized and important uncertainties remained, particularly regarding the effects on CH_4 and N_2O fluxes. Biochar applications have also been found to improve the overall soil quality and increase yields. Jeferry et al. (2011) reported, in a quantitative review of the effect of biochar on crop productivity, a mean yield increase of 10%, although with a wide range in yield responses, from −28% to +39%. The highest increases were observed for acidic and neutral pH soils and medium- to coarse-texture soils. The efficacy of biochar for GHG mitigation is currently under debate, and there are still large unknowns and potential risks that need to be quantified.

Chipped ramial wood (CRW) consists of the twigs and branches of trees or shrubs used in cultivated land for mulching. Originally a by-product of hardwood logging and processing in Canada, some scientists, farmers and non-governmental organizations (NGOs) have advocated the use of CRW to improve soil fertility and rehabilitate degraded soils. With a high content of cellulose and lignin, CRM is considered to stimulate soil fungi. As is typically the case with new cultivation practices – to date, more attention has been paid by farmers, NGOs and extension services – scientific investigations are scarce. Barthès et al. (2010) performed a synthesis of available experimental results and concluded that CRW had a positive effect on crop yield (except for the first year in some cases), increased soil organic matter (and thus SOC) and improved medium-term nutrient availability (Fig. 9.1). However,

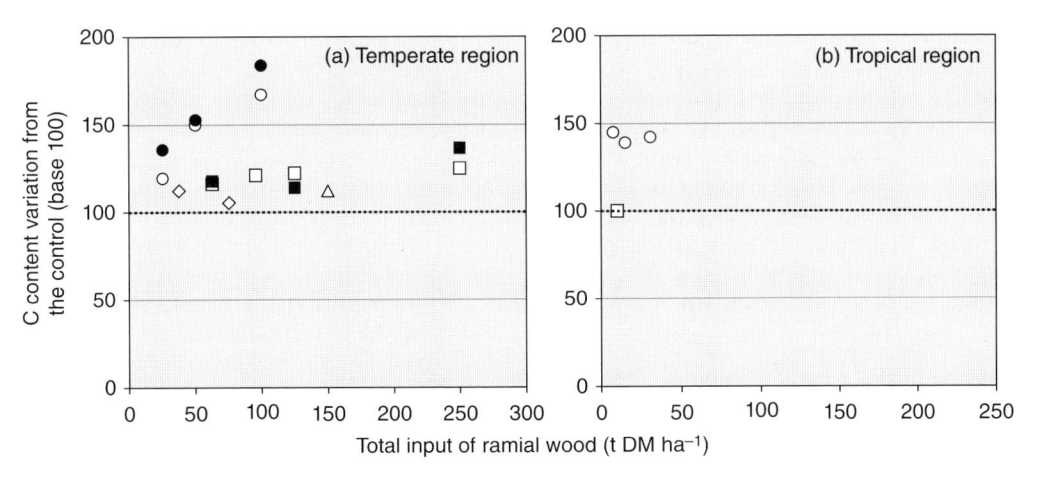

Fig. 9.1. Impact of the addition of chipped ramial wood (CRW, expressed in cumulative quantity in dry matter (DM) per hectare) on soil organic carbon in relation to a control without CRW. White symbols correspond to situation without additional inputs, black symbols to situation with CRW plus additional inputs. The same symbols correspond to different measurements during the experimentation. (From Barthès et al., 2010.)

it is recognized that the scarcity of available information does not allow for precise agronomic recommendation, and even more attention needs to be paid to the availability of the raw material and the time and energy necessary for preparing the material.

Moreover, the benefit of CRW as compared with non-woody amendments is poorly documented, and thus the comparative advantage in relation to more traditional inputs is not proved.

Recently, the World Bank (2012) published a report containing a meta-analysis that included more conventional agricultural practices aimed at GHG mitigation, as well as other technical solutions (e.g. biochars), and using more resolved land management categories and focusing in Latin America, Africa and Asia (Fig. 9.2).

Other options that have been evaluated recently for GHG mitigation potential include organic farming practices. Gattinger et al. (2012) performed a quantitative assessment based on pairwise comparisons of organic versus non-organic farming systems from 74 published studies. The authors found significant higher values for organic systems of 0.350 ± 0.108 kg C m^{-2} for stocks and 0.045 ± 0.021 kg C m^{-2} year^{-1} for sequestration rates. But they failed to identify clear drivers and when they used a more restrictive approach, excluding data sets with lowest data quality (e.g. excluding studies lacking direct measurements of soil bulk density), the sequestration rates were non-significant. Moreover, Leifeld et al. (2013) argued that the study by Gattinger et al. (2012) was biased and that their conclusion that practices central to organic farming could mitigate climate change through C sequestration in soils was misleading.

Issues and Barriers to Implementing Agricultural C Sequestration and GHG Reductions

While a variety of agronomic practices have been shown to reduce GHG emissions and/or sequester C in soils and biomass at the field scale, achieving widespread and broad improvements at regional/national scales – so that meaningful levels of mitigation can be achieved – faces a variety of challenges. The most obvious is that adopting best practices for GHG mitigation will, in most cases, incur additional costs to the land user, at least in the short-term, and therefore land users will need to be compensated to secure their participation. Agricultural soils have been largely excluded as mitigation options within the Clean Development Mechanism (CDM)[1] and the Kyoto Protocol (KP) system, although there is growing interest for inclusion of agricultural options in several voluntary market registries, as well as compliance markets in Australia and California. In order to realize a greater participation of agriculture in climate change mitigation, further progress is needed in emissions quantification, as well in meeting common GHG accounting standards that ensure that emission reduction and sequestration are real and verifiable. A brief discussion of these issues is given below, and the reader is referred to other chapters for a more in-depth treatment.

For SOC accounting, well-developed methods exist to measure SOC stocks and have been applied routinely for many decades. Automated dry combustion soil C analysers are highly accurate and, with proper laboratory techniques, the error in estimating the SOC contents of a soil sample is small. However, the relative scarcity of well-equipped laboratories in developing countries is an impediment. Near- or mid-infrared spectroscopy is another method under development that has considerable promise, particularly in developing countries (Brunet et al., 2007), and it has the potential for significant reductions in cost and increased sample throughput, if adequately calibrated (Kamau-Rewe et al., 2011).

The issue with relying on direct measurements for soil C accounting is not a lack of technology per se (as is sometimes believed), but rather cost and practicality. Because SOC exhibits significant spatial variability at field scales, and changes in stocks relative to background

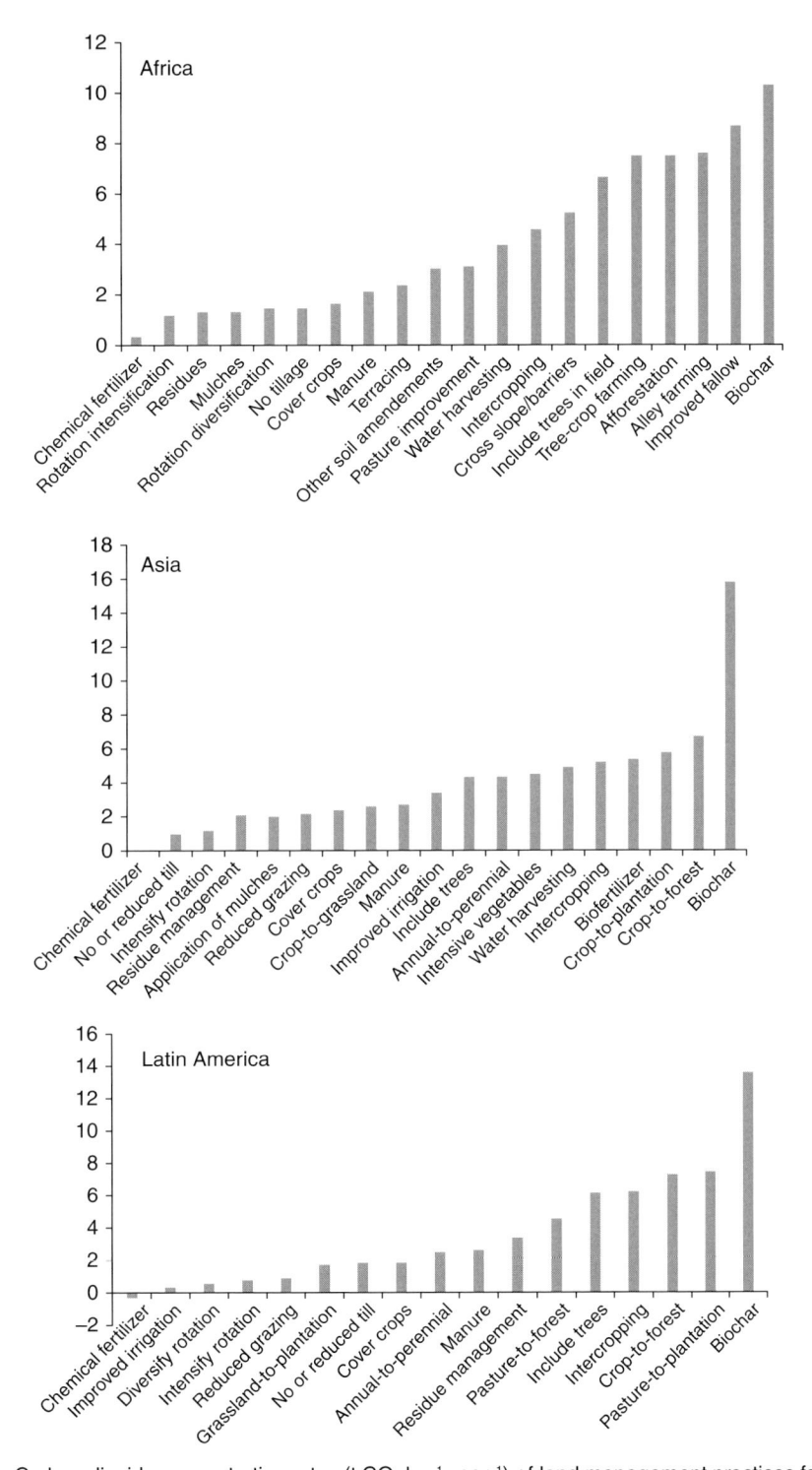

Fig. 9.2. Carbon dioxide sequestration rates (t CO_2 ha^{-1} year^{-1}) of land management practices for Africa, Asia and Latin America. (From The World Bank, 2012: Report No 67395-GLB, reproduced with authorization of The World Bank.)

SOC levels are often small, direct measurement of SOC changes for an individual field can require a large number of samples, and thus a high cost. Efficient sampling designs, which focus on repeated measurements over time at relocatable benchmark sites, can reduce greatly the sampling requirements (and cost) for detecting SOC change (Conant and Paustian, 2002; Lark, 2009; Spencer et al., 2011). However, for many smaller mitigation projects, particularly in areas with highly heterogeneous landscapes, quantifying SCS with direct soil measurements alone is likely to be too costly.

Using model-based quantification systems, supported by a coordinated network of 'on-farm' monitoring locations with direct SOC measurements for representative soils and management practices, has been proposed as an approach that can reduce costs significantly, while achieving acceptable accuracy and precision, for the measurement and monitoring required for agricultural options to be broadly accepted by C offset markets (Paustian et al., 2011; Spencer et al., 2011). This would enable monitoring at field and farm scales to focus on reporting of the practices that are occurring – when and where on the landscape – which can be accomplished using lower-cost methods such as remote sensing and self-reporting by the land users themselves (Paustian, 2012). This type of approach could reduce dramatically the transaction costs for agricultural mitigation projects and thus provide more net compensation to incentivize land users, potentially enabling a much broader participation in agricultural mitigation programmes.

Other C accounting issues, specifically additionality, permanence and leakage, pose challenges to a broad participation of agriculture in current and future mitigation policies. Further research and streamlined policy designs can reduce impediments to agricultural participation in GHG mitigation efforts.

Additionality refers to the condition that a mitigation project must produce results that would otherwise not be achieved in the absence of an explicit mitigation policy or effort – in other words, that the mitigation results are additional to what would have occurred with 'business as usual' (BAU). This is particularly important for agricultural activities that would not themselves be subject to mandatory GHG emission regulations, but rather functioning as an 'offset' to required emission reductions in an industrial or energy-sector business. Agricultural mitigation activities need to be demonstrably 'additional' in order to be viewed as of equal value to emission reductions in other sectors of the economy. Currently, most agriculturally related mitigation projects (indeed, projects of all kinds), primarily involving voluntary market systems, establish additionality on a project-by-project basis, in which projects are tested by different criteria (e.g. regulatory, financial, common practice) to establish that the project activities do not simply represent BAU.

Establishing additionality on a case-by-case basis, as in the CDM and many voluntary market standards, can entail development, evaluation and verification costs. Alternatives being explored are broader practice based or 'performance standards' that apply, for example, for an entire class of projects within a country or large region. However, setting and applying performance standards also faces significant challenges; first, in determining what constitutes an objective BAU baseline, particularly in developing countries where information on current and past land use and management practices is patchy at best for many countries. Second, there can be substantial variability for any set of agricultural systems and practices viewed at a regional or national scale, so setting a too liberal performance standard will increase the number of 'free riders', i.e. projects that are not really additional, whereas too strict a performance standard will preclude many projects with significant mitigation potential. Key to establishing workable yet rigorous performance standards for potential agricultural mitigation practices will be more data and understanding of the trends in the management practices that are actually occurring on the landscape and what the drivers are that influence management choices.

All biological C sinks are inherently non-permanent, in that organic carbon, whether in biomass or soils, is subject to re-emission as CO_2 to the atmosphere if the management practices that increase C stocks are reversed or somehow compromised. Soil C stock accumulations are, in general, less vulnerable than biomass C stocks to inadvertent stock losses due to natural disturbances, such as fire and pest outbreaks. Thus, intentional abandonment or reversal of the soil C sequestering practices represents the primary non-permanence risk for soils. Other challenges to long-term maintenance of prescribed practices include changes in land ownership and incentives to change cropping practices due to changes in commodity prices. To the extent that the adoption of improved practices leads to increased productivity and/or reduced costs over time (but require incentives to first establish), risks of non-permanence for many conservation management practices may be relatively low – i.e. farmers would have additional (non-mitigation related) incentives to continue to maintain the practices long term.

Various approaches have been proposed to adjust for non-permanence, including *ex ante* discounting and leasing or renting sequestration credits (Murray *et al.*, 2007), which assume that all additional stored carbon is re-emitted at the end of the project/contract period and the value of the credit is discounted accordingly. The general concept with rented or leased credits is that the buyer assumes liability for emissions of stored carbon after the end of the contract period, and thus would have either to renew the lease or to replace it with other credits or emission reductions. This decreases the economic value of SCS-based activities dramatically, and thus is a strong disincentive for many agricultural mitigation projects.

Another approach to reducing non-permanence risk includes the use of a permanence buffer, such as in the Verified Carbon Standard (VCS, 2013). The VCS uses a risk assessment approach to rate projects in terms of their relative risk for non-permanence in order to establish an amount of offset credits that are assigned to a pooled buffer account, which is maintained within the overall offset registry. Over time, if a project maintains its integrity and is able to mitigate the non-permanence risk successfully, the number of credits held in the buffer account is reduced. The advantage to the buffer approach is that the non-permanence risk is held within the registry so that activities that have a low risk of non-permanence are not penalized unduly and full-value 'permanent' credits are issued to the buyer. To do so, however, requires an increased monitoring and assessment effort within the programme to ensure the integrity of the permanence buffer.

Leakage refers to unintended emissions that occur outside the boundary of a mitigation project that arise due to the establishment of the project, thus nullifying all or part of the emission reductions (or sink enhancement) achieved by the project. For example, an effort to halt deforestation and the accompanying CO_2 emissions in one location could result in the forest-clearing activities moving to another location, so that emissions are not really reduced as a result of the project, unless proper safeguards are in place to prevent leakage. This type of displaced activity leakage is most likely to occur for agricultural activities that involve land-use changes, such as afforestation of agricultural land or conversion of cropland to grassland reserves, which reduce or displace the main productive use of the land, such as food or fibre production. Thus, effective agricultural mitigation projects promoting SCS will be those that maintain or enhance food and fibre production. This type of approach is also highly commensurate with objectives to increase food security and buffer against the effect of climate change, as discussed previously in the chapter.

Integrating the Spatial Dimension: A Mandatory Aspect

The spatial dimension has to be taken into account since a technical solution could increase emissions at the scale of an individual field level, but might at the same time represent a good mitigation strategy at the landscape, or even national, level. For example, while agronomic intensification at the field scale to increase production could increase emissions locally, it is evident that at a regional level, increasing production in areas capable of high crop productivity will reduce conversion pressure on other land (forest/savannah degradation) and reduce GHG emissions at the regional scale. Thus, at a regional level, increased access to fertilizers for farmers can be a candidate mitigation solution compared to other alternatives (reduced intensification, higher dependencies from importation, additional land degradation). This is particularly important in developing countries where, erroneously, nitrogen fertilization is in most cases considered only as an N_2O emissions source. Burney et al. (2010) demonstrated the importance of taking into account a landscape dimension when identifying cost-effective ways to avoid GHG emissions. Future agricultural productivity is critical, as it will shape emissions from the conversion of native landscapes to food and biofuel crops. However, the Burney et al. (2010) analysis noted that investment in agricultural research was rarely mentioned as a mitigation strategy. Based on their estimates of the net effect on GHG emissions of historical agricultural intensification between 1961 and 2005, they found that while emissions from factors such as fertilizer production and application had increased, the net effect of higher yields meant an overall decrease in GHG emissions. Their figures show that each dollar invested in agricultural yields has resulted in 249 fewer kg CO_2-eq emissions relative to 1961 technology, and that investment in yield improvements compares favourably with other commonly proposed mitigation strategies. This analysis is particularly important for countries that need to increase food productivity at the national level. However, the problem remains of how to quantify the impact of national policies or of the implementation of field plot technical solutions at the landscape level. The bioenergy debates have shown that land-use change (LUC) is critical for environmental assessment. LUC includes direct LUC (dLUC), occurring in the study area itself, and indirect LUC (iLUC), occurring outside the study area but resulting from changes within the study area. Lapola et al. (2010) reported that iLUC could, in certain cases, overcome carbon savings from biofuels. During recent years there has therefore been an increasing demand by project managers and policy makers for suitable GHG assessment tools to reap the benefits of a landscape-scale approach. The IPCC has published guidelines and good practices for GHG accounting (Eggleston et al., 2006), and various tools have been developed to help those performing GHG assessment within these guidelines. Denef et al. (2012) classified these tools into calculators, protocols, guidelines and models. Two recent publications (Colomb et al., 2013; Milne et al., 2013) showed that many calculators were available for landscape-scale assessment to meet different needs (and cover different parts of the world). The level of uncertainty of the appraisal performed with those tools remains high, but is acceptable so long as the aim is mainly to raise awareness, guide policy decisions and demonstrate synergy between development and mitigation (Branca et al., 2013b).

Conclusion

SCS should be seen not only in a mitigation perspective but also better in a wider approach considering other ecosystems benefits and optimization of the trade-offs (Fig. 9.3). Promoting better C management will be a multiple-win solution: for the global scale and the climate systems, for the local scales and the fight against soil degradation, and for the achievement of food security at all levels. The mobilization of local solutions for SCS will be possible at larger or even global scales, only if SCS is considered for all its co-benefits.

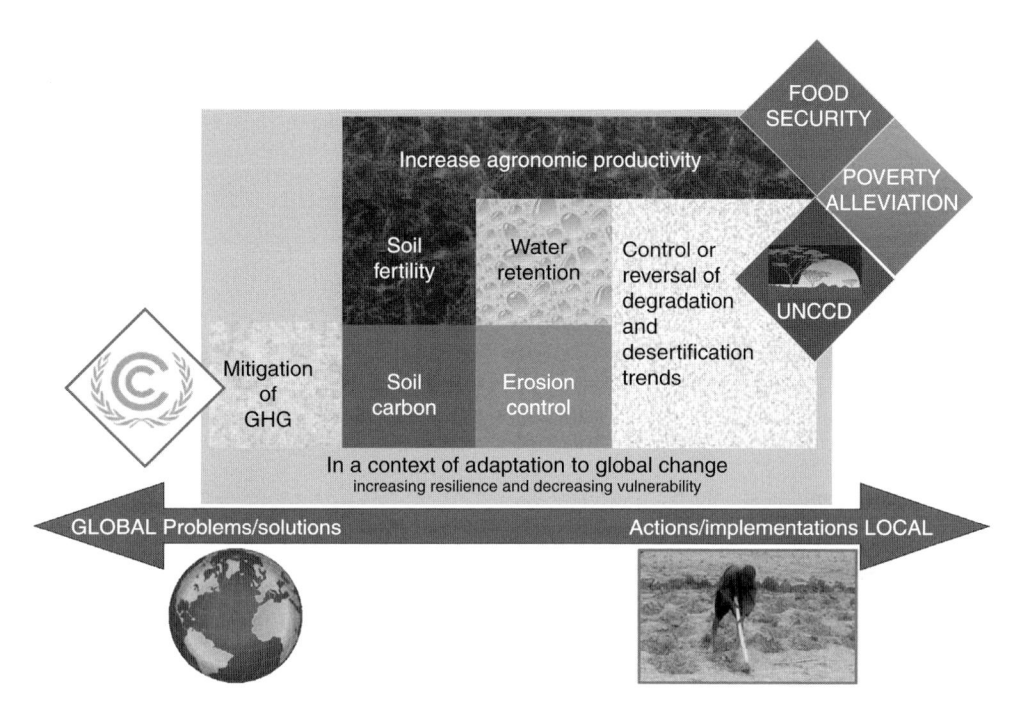

Fig. 9.3. Linking the global and local dimensions: soil C is related with the mitigation of GHG and thus the UNFCCC (on the left) but soil C is also related with soil fertility, erosion control and water retention, and thus indirectly with agronomic productivity and land degradation, and at the policy level with the UNCCD and food security (on the right). (From Bernoux and Chevallier, 2013.)

Note

[1]CDM recently adopted a practice for enhanced biological N fixation by legumes for soil N_2O reductions.

References

Amundson, R. (2001) The carbon budget in soils. *Annual Review of Earth and Planetary Sciences* 29, 535–562.
Barthès, B.G., Manlay, R.J. and Porte, O. (2010) Effets de l'apport de bois raméal sur la plante et le sol: une revue des résultats expérimentaux. *Cahiers Agricultures* 19(4), 280–287.
Baumert, K.A., Herzog, T. and Pershing, J. (2005) *Navigating the Numbers: Greenhouse Gas Data and International Climate Policy*. World Resources Institute, 132, Washington, DC.
Bellamy, P.H., Loveland, P.J., Bradley, R.I., Lark, R.M. and Kirk, G.J.D. (2005) Carbon losses from all soils across England and Wales 1978–2003. *Nature* 437, 245–248.
Bernoux, M. and Chevallier, T. (2013) Le carbone dans les sols des zones sèches. Des fonctions multiples indispensables. *Dossier thématique n°10 from the Comité Scientifique Français de la Désertification – CSFD. ISSN 1172–6964*. CSFD/Agropolis International, Montpellier, France, 44 pp. (http://www.csf-desertification.org/dossier/item/dossier-carbone-sols-zones-seches, accessed 4 July 2014).
Bernoux, M., Cerri, C.C., Cerri, C.E.P., Siqueira Neto, M., Metay, A., Perrin, A.S., Scopel, E., Razafimbelo, T., Blavet, D., Piccolo, M.C., Pavei, M. and Milne, E. (2006) Cropping systems, carbon sequestration and erosion in Brazil, a review. *Agronomy for Sustainable Development* 26(1), 1–8.
Branca, G., Hissa, H., Benez, M.C., Medeiros, K., Lipper, L., Tinlot, M., Bockel, L. and Bernoux, M. (2013a) Capturing synergies between rural development and agricultural mitigation in Brazil. *Land Use Policy* 30, 507–518.

Branca, G., Lipper L., McCarthy, N. and Jolejole, M.C. (2013b) Food security, climate change and sustainable land management: a review. *Agronomy for Sustainable Development* 33, 635–640.

Brunet, D., Barthes, B.G., Chotte, J.L. and Feller, C. (2007) Determination of carbon and nitrogen contents in Alfisols, Oxisols and Ultisols from Africa and Brazil using NIRS analysis: effects of sample grinding and set heterogeneity. *Geoderma* 139(1–2), 106–117.

Burney, J., Davis, S. and Lobella, B. (2010) Greenhouse gas mitigation by agricultural intensification. (http://www.pnas.org/cgi/doi/10.1073/pnas.0914216107, accessed 4 July 2014).

Ceschia, E., Béziat, P., Dejoux, J.F., Aubinet, M., Bernhofer, C., Bodson, B., Buchmann, N., Carrara, A., Cellier, P., Di Tommasi, P., et al. (2010) Management effects on net ecosystem carbon and GHG budgets at European crop sites. *Agriculture, Ecosystems and Environment* 139, 363–383.

Ciais, P., Gervois, S., Vuichard, N., Piao, S.L. and Viovy, N. (2011a) Effects of land use change and management on the European cropland carbon balance. *Global Change Biology* 17, 320–338, doi:10.1111/j.1365-2486.2010.02341.x.

Ciais, P., Bombelli, A., Williams, M., Piao, S.L., Chave, J., Ryan, C.M., Henry, M., Brender, P. and Valentini, R. (2011b) The carbon balance of Africa: synthesis of recent research studies. *Philosophical Transactions of the Royal Society A: Mathematical, Physical and Engineering Sciences* 369, 2038–2057, doi:10.1098/rsta.2010.0328.

Colomb, V., Touchemoulin, O., Bockel, L., Chotte, J.L., Martin, S., Tinlot, M. and Bernoux, M. (2013) Selection of appropriate calculators for landscape-scale greenhouse gas assessment for agriculture and forestry. *Environmental Research Letters* 8, 015029.

Conant, R.T. and Paustian, K. (2002) Spatial variability of soil organic carbon in grasslands: implications for detecting change at different scales. *Environmental Pollution* 116, 127–135.

Denef, K., Archibeque, S. and Paustian, K. (2011) Greenhouse gas emissions from U.S. agriculture and forestry: a review of emission sources, controlling factors, and mitigation potential. *Interim Report to USDA under Contract #GS23F8182H* (http://www.usda.gov/oce/climate_change/techguide/Denef_et_al_2011_Review_of_reviews_v1.0.pdf, accessed 4 July 2014).

Denef, K., Paustian, K., Archibeque, S., Biggar, S. and Pape, D. (2012) Report of Greenhouse Gas Accounting Tools for Agriculture and Forestry Sectors (Interim Report to USDA under Contract No GS23F8182H 140) (http://www.usda.gov/oce/climate_change/techguide/Denef_et_al_2012_GHG_Accounting_Tools_v1.pdf, accessed 4 July 2014).

Eagle, A.J., Henry, L.R., Olander, L.P., Haugen-Kozyra, K., Millar, N. and Robertson, G.P. (2010) Greenhouse gas mitigation potential of agricultural land management in the United States. A synthesis of the literature (http://nicholasinstitute.duke.edu/sites/default/files/publications/TAGGDLitRev-paper.pdf, accessed 1 March 2013).

Eggleston, H., Buendia, L., Miwa, K., Ngara, T. and Tanabe, K. (2006) The National Greenhouse Gas Inventories Programme, Intergovernmental Panel on Climate Change. 2006 IPCC Guidelines for National Greenhouse Gas Inventories (http://www.ipcc-nggip.iges.or.jp/public/2006gl/index.html, accesssed 4 July 2014).

Gattinger, A., Muller, A., Haeni, M., Skinner, C., Fliessbach, A., Buchmann, N., Mäder, P., Stolze, M., Smith, P., El-Hage Scialabba, N. and Niggli, U. (2012) Enhanced top soil carbon stocks under organic farming. *Proceedings of the National Academy of Sciences USA* 109(44), 18226–18231.

Hosonuma, N., Herold, M., Sy, V.D., Fries, R.S.D., Brockhaus, M., Verchot, L., Angelsen, A. and Romijn, E. (2012) An assessment of deforestation and forest degradation drivers in developing countries. *Environmental Research Letters* 7, 044009.

IPCC (Intergovernmental Panel on Climate Change) (2007) *Climate Change 2007: Mitigation of Climate Change. Contribution of Working Group III to the Fourth Assessment Report of the Intergovernmental Panel on Climate Change*. Cambridge University Press (http://www.ipcc.ch/publications_and_data/publications_ipcc_fourth_assessment_report_wg3_report_mitigation_of_climate_change.htm, accessed 4 July 2014).

Jeferry, S., Verhijen, F.G.A., van der Velde, M. and Bastos, A.C. (2011) A quantitative review of the effect of biochar application to soils on crop productivity using meta-analysis. *Agriculture, Ecosystems and Environment* 144, 175–187.

Kamau-Rewe, M., Rasche, F., Cobo, J.G., Dercon, G., Shepherd, K.D. and Cadish, G. (2011) Generic prediction of soil organic carbon in Alfisols using diffuse reflectance Fourier-transform mid-infrared spectroscopy. *Soil Science Society of America Journal* 75(6), 2358–2360.

Lapola, D.M., Schaldach, R., Alcamo, J., Bondeau, A., Koch, J., Koelking, C. and Priess, J.A. (2010) Indirect land-use changes can overcome carbon savings from biofuels in Brazil. *Proceedings of the National Academy of Sciences USA* 107, 3388–3393.

Lark, R.M. (2009) Estimating the regional mean status and change of soil properties: two distinct objectives for soil survey. *European Journal of Soil Science* 60, 748–756.

Leifeld, J., Angers, D.A., Chenu, C., Fuhrer, J., Kätterer, T. and Powlson, D.S. (2013) Organic farming gives no climate change benefit through soil carbon sequestration. *Proceedings of the National Academy of Sciences USA* 110(11) (http://www.pnas.org/content/110/11/E984, accessed 4 July 2014).

Milne, E., Neufeldt, H., Rosenstock, T., Smalligan, M., Cerri, C.E.P., Malin, D., Easter, M., Bernoux, M., Ogle, O., Casarim, F. *et al.* (2013) Methods for the quantification of GHG emissions at the landscape level for developing countries in smallholder contexts. *Environmental Research Letters* 8, 015019.

Murray, B.C., Sohngen, B. and Ross, M.T. (2007) Economic consequence of consideration of permanence, leakage and additionality for soil carbon sequestration projects. *Climatic Change* 80, 127–143.

Ogle, S.M., Breidt, F.J. and Paustian, K. (2005) Agricultural management impacts on soil organic carbon storage under moist and dry climatic conditions of temperate and tropical regions. *Biogeochemistry* 72, 87–121.

Paustian, K. (2012) Agriculture, farmers and GHG mitigation: a new social network? *Carbon Management* 3(3), 253–257.

Paustian, K., Ogle, S.M. and Conant, R.T. (2011) Quantification and decision support tools for US agricultural soil carbon sequestration. Chapter 16. In: Hillel, D. and Rosenzweig, C. (eds) *Handbook of Climate Change and Agroecosystems: Impact, Adaptation and Mitigation*. Imperial College Press, London, pp. 307–341.

Shevliakova, E., Pacala, S.W., Malyshev, S., Hurtt, G.C., Milly, P.C.D., Caspersen, J.P., Sentman, L.T., Fisk, J.P., Wirth, C. and Crevoisier, C. (2009) Carbon cycling under 300 years of land use change: importance of the secondary vegetation sink. *Global Biogeochemical Cycles* 23, GB2022, doi: 10.1029/2007GB003176.

Smith, P., Martino, D., Cai, Z., Gwary, D., Janzen, H.H., Kumar, P., McCarl, B., Ogle, S., O'Mara, F., Rice, C. *et al.* (2007) Greenhouse gas mitigation in agriculture. *Philosophical Transactions of the Royal Society, B* 363, 789–813, doi:10.1098/rstb.2007.2184.

Spencer, S., Ogle, S.M., Breidt, F.J., Goebel, J.J. and Paustian, K. (2011) Designing a national soil carbon monitoring network to support climate change policy: a case example for US agricultural lands. *Greenhouse Gas Measurement and Management* 1(3–4), 167–178.

UNFCCC (United Nations Framework Convention on Climate Change) (2008) *Challenges and Opportunities for Mitigation in the Agricultural Sector*. Technical Paper number FCCC/TP/2008/8. United Nations, Bonn, Germany, 101 pp.

UNFCCC (2012) GHG emission profiles. United Nations, Bonn, Germany (http://unfccc.int/ghg_data/ghg_data_unfccc/ghg_profiles/items/3954.php, accessed 4 July 2014).

VCS (Verified Carbon Standard) (2013) Agriculture, forestry and other land use (AFOLU) requirements. Verified Carbon Standard Version 3 (http://www.v-c-s.org/sites/v-c-s.org/files/AFOLU%20Requirements,%20v3.4.pdf, accessed 4 July 2014).

Verheijen, F., Jeffery, S., Bastos, A.C., van der Velde, M. and Diafas, I. (2010) *Biochar Application to Soils. A Critical Scientific Review of Effect on Soil Properties, Processes and Functions*. EUR 24099 EN, Office for the Official Publications of the European Communities, Luxembourg, 149 pp.

World Bank, The (2012) *Carbon sequestration in agricultural soils*. Report No 67395-GLB. International Bank for Reconstruction and Development/The World Bank, Economic and Sector Work. Washington, DC, 118 pp. (http://documents.worldbank.org/curated/en/2012/05/16274087/carbon-sequestration-agricultural-soils, accessed 4 July 2014).

10 Soil Carbon and Agricultural Productivity: Perspectives from Sub-Saharan Africa

Andre Bationo*, Boaz S. Waswa and Job Kihara

Abstract

Soil carbon plays a key role in maintaining crop productivity in the soils in sub-Saharan Africa (SSA). This is more so considering that most smallholder farmers cannot afford the use of adequate amounts of inorganic fertilizers to restore the proportion of nutrients lost through crop harvests, soil erosion and leaching. Complicating the situation is the huge proportion of land under threat of degradation in the form of soil erosion and nutrient decline. There are numerous opportunities for improving soil carbon as a basis of ensuing sustainable agriculture. This paper discusses the role of soil carbon in agricultural production, with special focus on sub-Saharan Africa. First, the paper presents a discussion on the functions of soil carbon (biological, chemical and physical). This is followed by a look at the causes of carbon variation across agroecosystems. Management of soil carbon and productivity is evaluated in the context of resource availability, quality and soil organic matter pools. Drawing from the integrated soil fertility management practices in Africa, the paper discusses various strategies for organic carbon management and the implication of the same on crop productivity and soil properties. A special focus is given to the lessons learned from long-term experiments across Africa.

Introduction

Managing soils so that carbon stocks are sustained and even enhanced is of crucial importance in ensuring sustainable crop production. Soil organic carbon (SOC) is considered as the most important indicator of soil quality and agronomic sustainability. It is the main constituent of soil organic matter (SOM). Organic matter impacts on the physical, chemical and biological properties of soils. The amount of carbon in a soil is influenced by the balance between inputs (plant residues) and losses, mainly microbial decomposition and associated mineralization. This amount will vary with factors such as the specific land use undergoing change, soil type and texture, soil depth, bulk density, management and climate. The decrease of organic matter in topsoils can have dramatic negative effects on the water-holding capacity of the soil, on structure stability and compactness, nutrient storage and supply, and on soil biological life such as mycorrhizas and nitrogen (N)-fixing bacteria (Sombroek et al., 1993).

---

*E-mail: abationo@outlook.com

For centuries, many farming systems have relied on the SOM to sustain production. However, with increasing intensification, land degradation and climate change, the quantity of SOM has declined rapidly, thereby threatening the capacity of the land to produce sustainably. In the past 25 years, one-quarter of the global land area has suffered a decline in productivity and in the ability to provide ecosystem services because of soil carbon losses (Bai et al., 2008). The situation is made worse in tropical soils, which are considered as more risky because cropping is synonymous to nutrient removal in the already impoverished soils with insufficient replenishment. There is considerable concern that, if SOM concentrations in soils are allowed to decrease too much, then the productive capacity of agriculture will be compromised. Soils exhibit different behaviour, and as such we would expect to have different SOC levels. However, there is a general consensus among scientists that a 2% soil carbon (3.5% SOM) is a critical level for temperate soils below which potentially serious decline in soil quality will occur (Loveland and Webb, 2003).

Functions of Soil Organic Carbon

Organic matter is of great importance in soil, because it impacts on the physical, chemical and biological properties of soils (Baldock and Skjemstad, 1999; Fig. 10.1). Physically, it promotes aggregate stability, and therefore water infiltration, percolation and retention. It impacts on soil chemistry by increasing cation exchange capacity, soil buffer capacity and nutrient supply. Biologically, it stimulates the activity and diversity of organisms in soil (Allison, 1973).

While SOM is primarily carbon, it also contains nutrients essential for plant growth, such as nitrogen, phosphorus, sulfur and micronutrients. Organisms in the soil food web decompose SOM and make these nutrients available (Brussaard et al., 2007). The rate of SOM decomposition and turnover depends mainly on the interplay between soil biota, temperature, moisture and a soil's chemical and physical composition (Taylor et al., 2009). In addition to soil's clay content, there is evidence to suggest that an increase in organic carbon content can increase the amount of water present in the soil. Managing soil water is critical in enhancing crop productivity, especially with the adverse effects of climate change and rainfall variability in most regions of the world. A marginal increase in the amount of plant available water of a soil can help maintain or enhance potential productivity by allowing the soil to retain more water, applied either as rain or irrigation.

Biological functions
- Provides energy to biological processes
- Provides nutrients (N, P and S)
- Contributes to resilience

Functions of soil organic matter

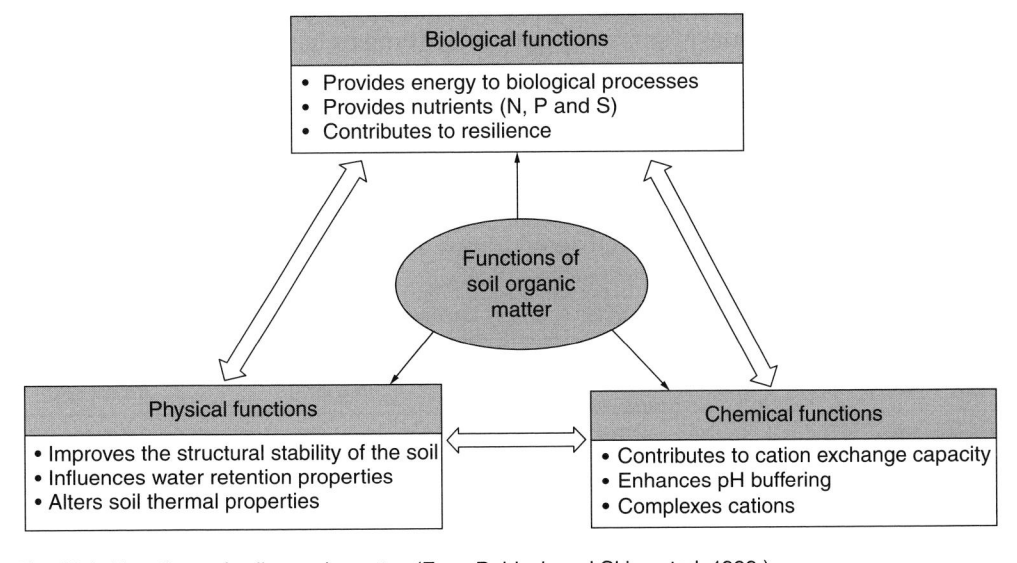

Physical functions
- Improves the structural stability of the soil
- Influences water retention properties
- Alters soil thermal properties

Chemical functions
- Contributes to cation exchange capacity
- Enhances pH buffering
- Complexes cations

Fig. 10.1. Functions of soil organic matter. (From Baldock and Skjemstad, 1999.)

(Palm et al., 2001). According to Palm et al. (2001), N, polyphenol and lignin contents are the major residue quality factors determining the decomposition rate of organic materials and the subsequent nutrient release. These factors influence the management options of organic materials. Organic resources can thus be classified as of high, low or intermediate quality based on the proportions of the above parameters. High-quality organic resources contain high proportions of N and low proportions of lignin and polyphenols. Materials with such characteristics decompose fast and release nutrients immediately for plant use, but contribute less to the build-up of SOC. At the other extreme, low-quality organic resources contain low amounts of N but high proportions of lignin and polyphenols. Such materials are slow in decomposition and nutrient release, and hence can contribute more to the build-up of the SOM pool. The contributions of these legumes to nutrient supply and SOC depends on the quality of the organics used. Legume residues have a low C:N ratio and a low lignin and polyphenol content, and are regarded as high quality (Swift et al., 1994). These therefore decompose relatively fast, releasing nutrients to the current crop. Their contribution to SOC build-up may be low relative to other organic resources such as maize stover and manure, which have a relatively high C:N ratio. As a result, manipulation of resource quality, particularly N and polyphenol content, or organic resources used as nutrient sources is a potentially important way of managing SOM (Mafongoya et al., 1998; Palm et al., 2001; Kimetu et al., 2004).

Soil Organic Matter Pools

Soil organic matter consists of diverse fractions ranging from young, biologically active pools to more recalcitrant, 'passive' pools (Schimel et al., 1985). These fractions and their dynamics are particularly useful in understanding the link between SOM dynamics and nutrient availability (Motavalli et al., 1994). SOC pools have different turnover rates (Buyanovsky et al., 1994; Six et al., 2002). With respect to soil carbon sequestration, it is most desirable to fix atmospheric C in those pools having long turnover times. Eswaran et al. (1995) defined four pools based on carbon dynamics: first, an 'active or labile pool' of readily oxidizable compounds. The formation of this pool is dictated largely by plant residue inputs (and hence management), and climate. Second, a 'slowly oxidized pool' associated with soil macroaggregates. The dynamics and size of this pool are affected by soil physical properties such as mineralogy and aggregation, as well as agronomic practices. Third, a 'very slowly oxidized pool' associated with microaggregates, where the main controlling factor is the water stability of aggregates and agronomic practices have only little effect. Fourth, a 'passive or recalcitrant pool', where clay mineralogy is the main controlling factor and there are probably no effects due to agronomic practices. Indicative residence times of the above pools are termed as 'labile', 'moderate', 'slow' and 'passive', respectively (Table 10.3).

Changes in total SOC proportions in the soil take a long time to manifest. Partitioning SOC into the different pools using fractionation procedures offers a better understanding of the dynamics of soil carbon otherwise masked by the native pool. Kapkiyai et al. (1999), in a study in Kenya, observed that organic matter and microbial biomass among treatments were proportionately larger than changes in total SOC. Similar observations have been reported by other studies employing SOM fractionation procedures (Cambardella and Elliot, 1992; Christensen, 1992; Woomer et al., 1994).

The recent development of models has improved understanding of the dynamics of SOM and the nutrients associated with it. Models can be used to provide a better understanding of decomposition and accumulation processes and to predict future conditions from previous experience of SOM. Soil carbon models such as the Rothamsted carbon model (Jenkinson, 1990) and the CENTURY model (Parton et al., 1988) have been developed for temperate regions and validated against data from long-term experiments across the tropics. These models have been quite successful in predicting changes in SOM for local soils, climatic conditions and agricultural practices. Model predictions are potentially valuable for improving management practices where experimental data are limited, as is the case of most regions in the tropics (Syers, 1997).

Using the models, it is now understood that the amount of carbon in any undisturbed agroecosystems is never constant, but oscillates around a steady state driven by the accumulation of dry matter and disturbances by fires or other drastic events such as droughts. Most of the carbon losses are experienced in the initial 5–10 years, affecting mainly the labile carbon fractions. After each disturbance, a period of constant management is required in order to reach a new equilibrium that may be lower, similar or higher than the original one (Johnson, 1995) (Fig. 10.2). Depending on the management, it will take at least 25–50 years before a new organic carbon steady state is reached in soils (Baldock and Skjemstad, 1999; Batjes, 2001).

Soils vary in their ability to resist change or to recover after disturbance (Greenland and Szabolcs, 1994; Syers, 1997). Eswaran (1994) defined soil resilience as the ability of the soil or system to revert to its original or near original performance subsequent to stress.

Table 10.3. Organic carbon (C) pools and estimated ranges in the quantities and turnover times of the different types of organic matter (OM) present in agricultural soils. (From Jastrow and Miller, 1997.)

Organic C pool	Type of organic matter	Turnover time (year)	Proportion of total OM (%)
Labile	Microbial biomass	0.1–0.4	2–5
Rapid	Litter	1–3	–
	Particulate OM	5–20	18–40
	Light fraction	1–15	10–30
Moderate to slow	Within macroaggregates		20–35
Passive	Within microaggregates		
	Physically sequestered	50–1000	20–40
	Chemically sequestered	1000–3000	20–40

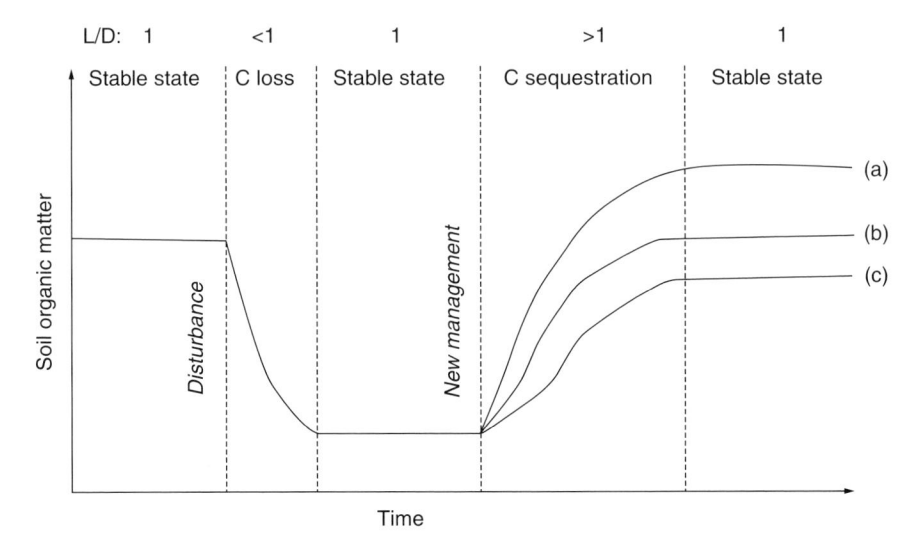

Fig. 10.2. Conceptual model of soil organic matter decomposition/accumulation following disturbance. Scenarios: (a) stabilization at above-original level; (b) stabilization at original level; (c) stabilization at lower than original level. L/D is the ratio of litter production over decomposition. (From Johnson, 1995.)

This definition emphasized the ability to recover with appropriate inputs rather than the ability to resist change. Characterizing the resilience capacity allows a differentiation of resistant, resilient, fragile and marginal soils, and a delineation of their response to stress (Lal et al., 1989). This is the basis for targeting management options.

Conclusion

There is no doubt that SOC plays a key role in maintaining the crop productivity of soils. Given this importance, the maintenance of an adequate level should be a guiding principle in developing management practices. However, just what constitutes an adequate level is likely to vary according to the soil type, environmental conditions and farming systems. There are numerous opportunities for improving soil carbon. Unfortunately, the amount of organics available at farm level for use in soil improvement is low, due to the many competing uses. The use of inorganic fertilizer will remain an important option for increasing crop productivity, and hence the amount of residues available for multiple uses, among them soil improvement. The contribution of organic resources to SOC will vary with the accompanying management and quality of the resources. Integration of different qualities of organics is needed if the production objective is to achieve both immediate soil fertility and maintenance and crop improvement in the long term.

References

Allison, F.E. (1973) *Soil Organic Matter and Its Role in Crop Production*. Elsevier Scientific Publishers Co, Amsterdam, 637 pp.

Bai, Z.G., Dent, D.L., Olsson, L. and Schaepman, M.E. (2008) Proxy global assessment of land degradation. *Soil Use and Management* 24, 223–234.

Baldock, J.A. and Skjemstad, J.O. (1999) Soil organic carbon/soil organic matter. In: Peverill, K.I., Sparrow, L.A. and Reuter, D.J. (eds) *Soil Analysis: An Interpretation Manual*. CSIRO Publishing, Collingwood, Victoria, Australia, pp. 159–170.

Batjes, N.H. (2001) Options for increasing carbon sequestration in West African soils: an exploratory study with special focus on Senegal. *Land Degradation and Development* 12, 131–142.

Bekunda, M.A., Nkonya, E., Mugendi, D. and Msaky, J.J. (2002) Soil fertility status, management, and research in East Africa. *East African Journal of Rural Development*, September 2002 (http://tinyurl.com/mws7vce, accessed 10 July 2014).

Brown, S. and Lugo, A.E. (1990) Effects of forest clearing and succession on the carbon and nitrogen contents of soil in Puerto Rico and U.S. *Plant and Soil* 124, 53–64.

Bruce, J.P., Frome, M., Haites, E., Janzen, H., Lal, R. and Paustian, K. (1999) Carbon sequestration in soils. *Journal of Soil and Water Conservation* 54, 382–389.

Brussaard, L., de Ruiter, P.C. and Brown, G.G. (2007) Soil biodiversity for agricultural sustainability. *Agriculture, Ecosystems and Environment* 121, 233–244.

Buyanovsky, G.A., Aslam, M. and Wagner, G.H. (1994) Carbon turnover in soil physical fractions. *Soil Science Society of American Journal* 58, 1167–1173.

Cambardella, C.A. and Elliot, E.T. (1992) Particulate soil organic matter across a grassland cultivation sequence. *Soil Science Society of America Journal* 56, 777–783.

Carsky, R.J., Jagtap, S., Tian, G., Sanginga, N. and Vanlauwe, B. (1998) Maintenance of soil organic matter and N supply in the moist savanna zone of West Africa. In: Lal, R. (ed.) *Soil Quality and Agricultural Sustainability*. Ann Arbor Press, Chelsea, Michigan, pp. 223–236.

Christensen, B.T. (1992) Physical fractionation of soil organic matter in primary particle size and density separates. *Advances in Soil Science* 20, 1–90.

Eswaran, H. (1994) Soil resilience and sustainable land management in the context of Agenda 21. In: Greenland, D.J. and Szabolcs, I. (eds) *Soil Resilience and Sustainable Land Use*. CAB International, Wallingford, UK, pp. 21–32.

Eswaran, H., van den Berg, E., Reich, P. and Kimble, J. (1995) Global soil carbon resources. In: Lal, R., Kimble, J., Levine, E. and Stewart, B.A. (eds) *Soils and Global Change*. Lewis Publishers, Boca Raton, Florida, pp. 27–43.

Greenland, D.J. and Szabolcs, I. (1994) *Soil Resilience and Sustainable Land Use*. CAB International, Wallingford, UK.

Jastrow, J.D. and Miller, R.M. (1997) Soil aggregate stabilization and carbon sequestration: feedbacks through organo-mineral associations. In: Lal, R., Kimble, J.M., Follet, R.F. and Stewart, B.A. (eds) *Soil Management and the Greenhouse Effect*. Lewis Publishers, Boca Raton, Florida, pp. 207–223.

Jenkinson, D.S. (1990) The turnover of organic carbon and nitrogen in soil. *Philosophical Transactions of the Royal Society B* 329, 361–368.

Johnson, M.G. (1995) The role of soil management in sequestering soil carbon. In: Lal, R., Kimble, J.M., Follet, R.F. and Stewart, B.A. (eds) *Soil Management and the Greenhouse Effect*. Lewis Publishers, Boca Raton, Florida, pp. 351–363.

Kapkiyai, J.J., Karanja, N.K., Qureshi, J.N., Smithson, P.C. and Woomer, P.L. (1999) Soil organic matter and nutrient dynamics in a Kenyan Nitisol under long-term fertilizer and organic input management. *Soil Biology and Biochemistry* 31, 1773–1782.

Kimetu, J.M., Mugendi, D.N., Palm, C.A., Mutuo, P.K., Gachengo, C.N., Bationo, A., Nandwa, S. and Kungu, J.B. (2004) Nitrogen fertilizer equivalencies of organics of differing quality and optimum combination with inorganic nitrogen source in Central Kenya. *Nutrient Cycling in Agroecosystems* 68, 127–135.

Lal, R. (2003) Soil erosion and the global carbon budget. *Environment International* 29(4), 437–450.

Lal, R., Hall, G.F. and Miller, F.P. (1989) Soil degradation: I. Basic processes. *Land Degradation and Rehabilitation* 1, 51–69.

Loveland, P. and Webb, J. (2003) Is there a critical level of organic matter in the agricultural soils of temperate regions: a review. *Soil and Tillage Research* 70, 1–18.

Mafongoya, P.L., Giller, K.E. and Palm, C.A. (1998) Decomposition and nitrogen release patterns of tree prunings and litter. *Agroforestry Systems* 38, 77–97.

Motavalli, P.P., Palm, C.A., Parton, W.J., Elliot, E.T. and Frey, S.D. (1994) Comparison of laboratory and modeling simulation methods for estimating soil carbon pools in tropical forest soils. *Soil Biology and Biochemistry* 26, 935–944.

Palm, C.A., Myers, R.J.K. and Nandwa, S.M. (1997) Combined use of organic and inorganic nutrient sources for soil fertility maintenance and replenishment. In: Buresh, R.J., Sanchez, P.A. and Calhoun, F.G. (eds) *Replenishing Soil Fertility in Africa. SSSA Special Publication, 51*. SSSA (Soil Science Society of America), Madison, Wisconsin, pp. 193–217.

Palm, C.A., Gachengo, C.N., Delve, R.J., Cadisch, G. and Giller, K.E. (2001) Organic inputs for soil fertility management in tropical agroecosystems: application of an organic resource database. *Agriculture Ecosystems and Environment* 83, 27–42.

Parton, W.J., Stewart, J.W.B. and Cole, C.V. (1988) Dynamics of C, N, P, and S in grassland soils: a model. *Biogeochemistry* 5, 109–131.

Prudencio, C.Y. (1993) Ring management of soils and crops in the West African semi-arid tropics: the case of the mossi farming system in Burkina Faso. *Agriculture, Ecosystems and Environment* 47, 237–264.

Schimel, D.S., Colman, D.C. and Horton, K.A. (1985) Soil organic matter dynamics in paired rangeland and cropland toposequences in North Dakota. *Geoderma* 36, 201–214.

Six, J., Conant, R.T., Paul, E.A. and Paustian, K. (2002) Stabilization mechanisms of soil organic matter: implications for C-saturation of soils. *Plant and Soil* 241(2), 155–176.

Sombroek, W.G., Nachtergaele, F.O. and Hebel, A. (1993) Amounts, dynamics and sequestrations of carbon in tropical and subtropical soils. *Ambio* 22, 417–426.

Swift, M.J., Seward, P.D., Frost, P.G.H., Qureshi, J.N. and Muchena, F.N. (1994) Long-term experiments in Africa: developing a data-base for sustainable land use under global change. In: Leigh, R.A. and Johnson, A.E. (eds) *Long-term Experiments in Agricultural and Ecological Sciences*. CAB International, Wallingford, UK, pp. 229–251.

Syers, J.K. (1997) Managing soils for long-term productivity. *Philosophical Transactions of the Royal Society B* 352, 1011–1021.

Taylor, L.L., Leake, J.R., Quirk, J., Hardy, K., Banwart, S.A. and Beerling, D.J. (2009) Biological weathering and the long-term carbon cycle: integrating mycorrhizal evolution and function into the current paradigm. *Geobiology* 7, 171–191.

Tiessen, H., Stewart, J.W.B. and Moir, J.O. (1983) Changes in organic and inorganic phosphorus composition of two grassland soils and their particle size fractions during 60–90 years of cultivation. *Journal of Soil Science* 34(4), 815–823.

Vanlauwe, B. (2003) Integrated soil fertility management research at TSBF: the framework, the principles, and their application. In: Gichuru, M.P., Bationo, A., Bekunda, M.A., Goma, P.C., Mafongoya, P.L., Mugendi, D.N., Murwira, H.M., Nandwa, S.M., Nyathi, P. and Swift, M.J. (eds) *Soil Fertility Management in Africa: A Regional Perspective*. Academy Science Publishers/TSBF-CIAT, Nairobi, 306 pp.

Woomer, P.L., Martin, A., Albrecht, A., Resck, D.V.S. and Sharpenseel, H. (1994) The importance and management of soil organic matter in the tropics. In: Woomer, P.L. and Swift, M.J. (eds) *The Biological Management of Tropical Soil Fertility*. Wiley, Chichester, UK, pp. 47–80.

Zach, A., Tiessen, H. and Noellemeyer, E. (2006) Carbon turnover and ^{13}C natural abundance under land-use change in the semiarid La Pampa, Argentina. *Soil Science Society of America Journal* 70, 1541–1546.

1

Low status; medium to strong degradation
High status; medium to strong degradation
Low status; weak degradation
Low status; improving
High status; stable to improving
Bare lands
Urban land
Water

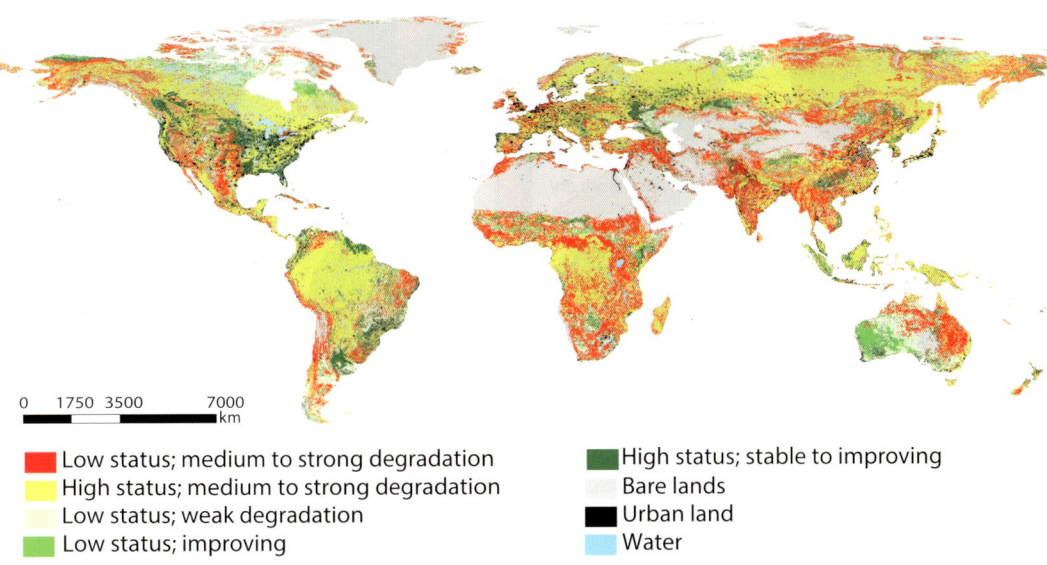

2

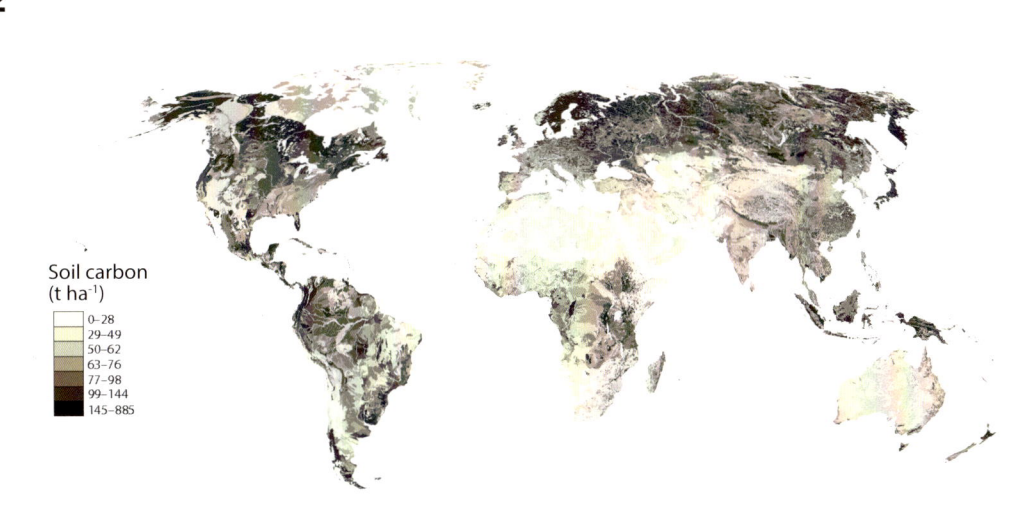

Soil carbon (t ha^{-1})
0–28
29–49
50–62
63–76
77–98
99–144
145–885

Plate 1. World map of soil status. Soil degradation is the decrease in the capacity of soil functions to supply ecosystem services and deliver benefits. Declining agricultural productivity is one example of loss of soil function and the benefits flowing from soil. The colour coding shows the status of land to deliver ecosystem services (low, high) and the state of soil functions (strong or weak degradation, improving). (From Nachtergaele et al., 2011.)

Plate 2. World map of soil carbon content in the top 1 m of the land surface. Higher carbon stocks are found in more humid regions with drylands coinciding more extensively with the land degradation shown in Plate 1. (From UNEP-WCMC, 2009.)

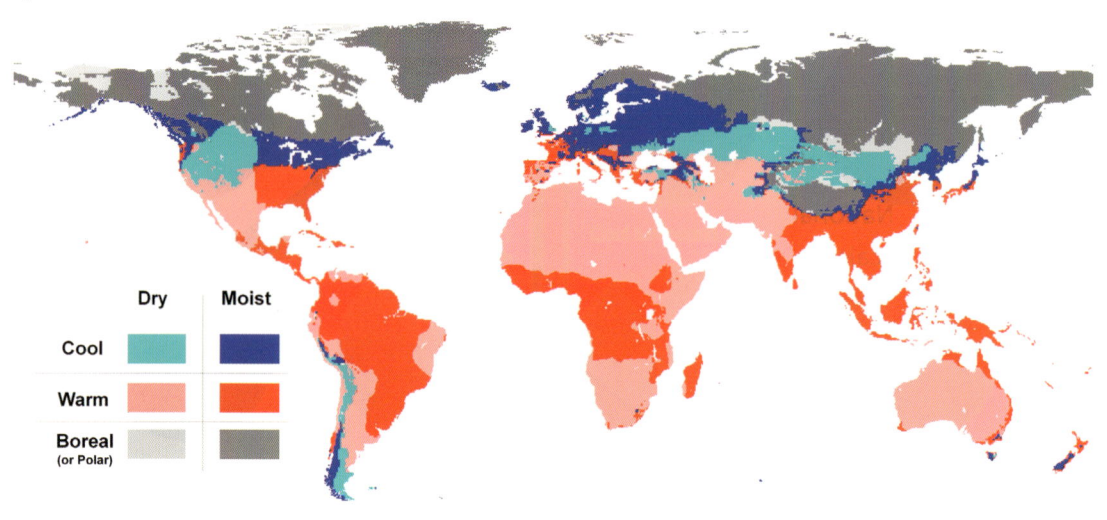

Plate 7. Geographical repartition of simplified climate categories.
Plate 8. Some examples of natural and anthropogenic organic matter and carbon particles occurring in soils (typical size approximately 100 μm); pictures by Bertrand Ligouis. (For methods and more examples, see Ligouis *et al.*, 2005).

9

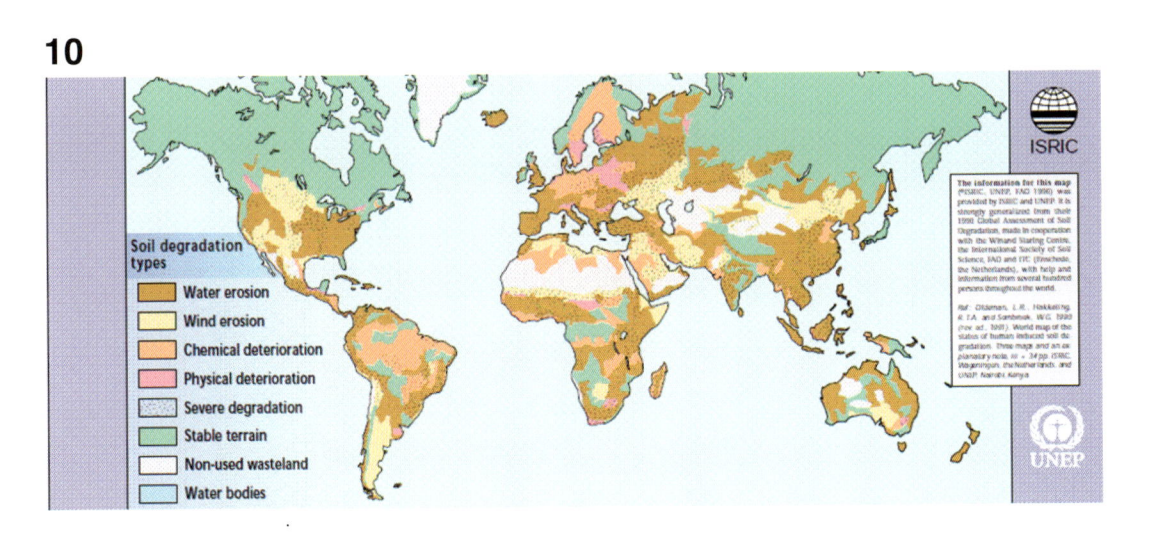

10

Plate 9. Highly coloured, foaming water in Black Burn, which drains organic carbon-rich soil on Cragside Estate, Northumberland, UK. (Pictures courtesy of Dr Paul Sallis.)
Plate 10. Global distribution of various soil degradation types. (From Oldeman et al., 1991.)

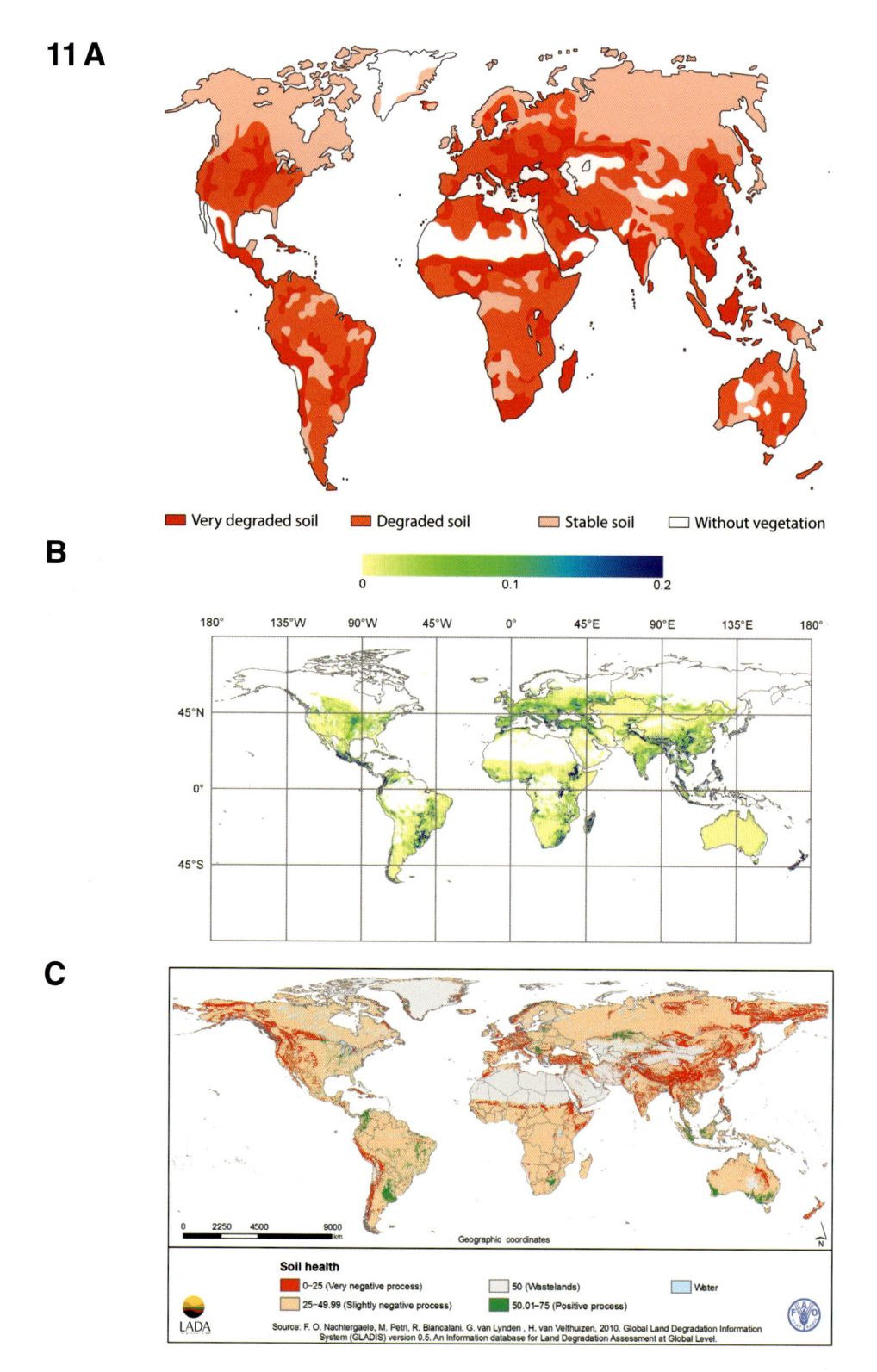

Plate 11. Global soil degradation according to various studies. (A) Middleton and Thomas, 1997. (B) Simulated global distribution of agricultural carbon erosion (cropland + pasture and rangeland) (Mg C ha^{-1} year^{-1}) (van Oost et al., 2007). (C) Geographical distribution of soil health trends according to LADA (Nachtergaele et al., 2011).

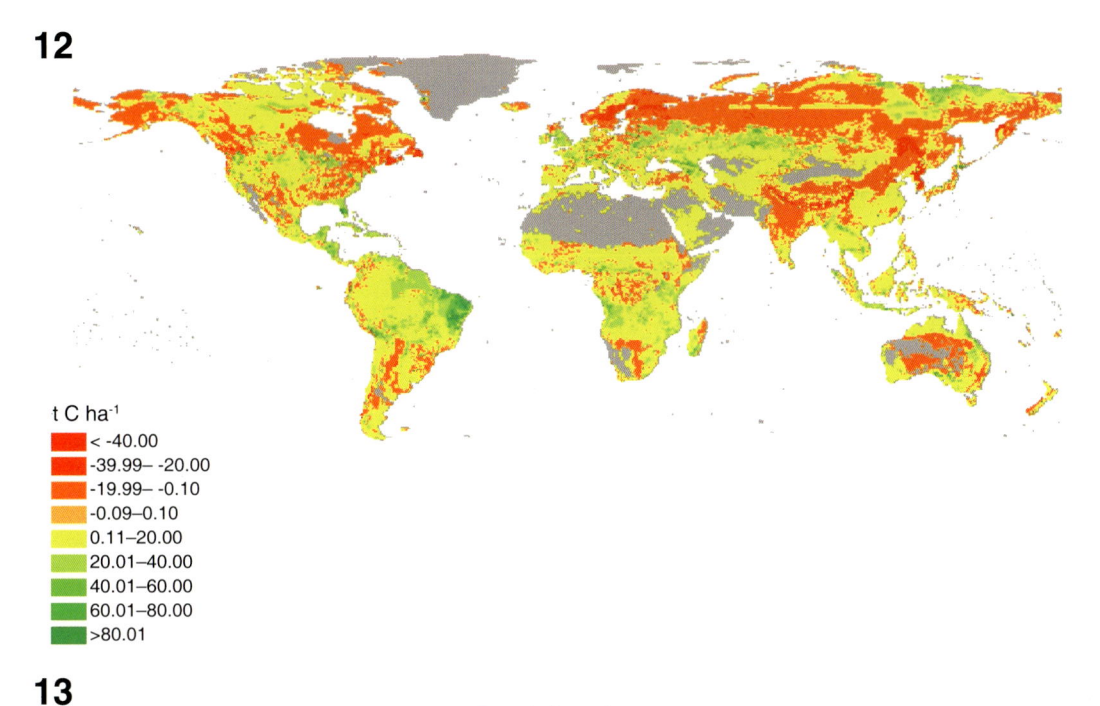

Plate 12. Average trend in SOC stock change 1971–2100 across ten climate scenarios. (From Gottschalk et al., 2012.)
Plate 13. Map of South America highlighting the area covered by the Cerrado, Atlantic Forest and Southern Grassland biomes.

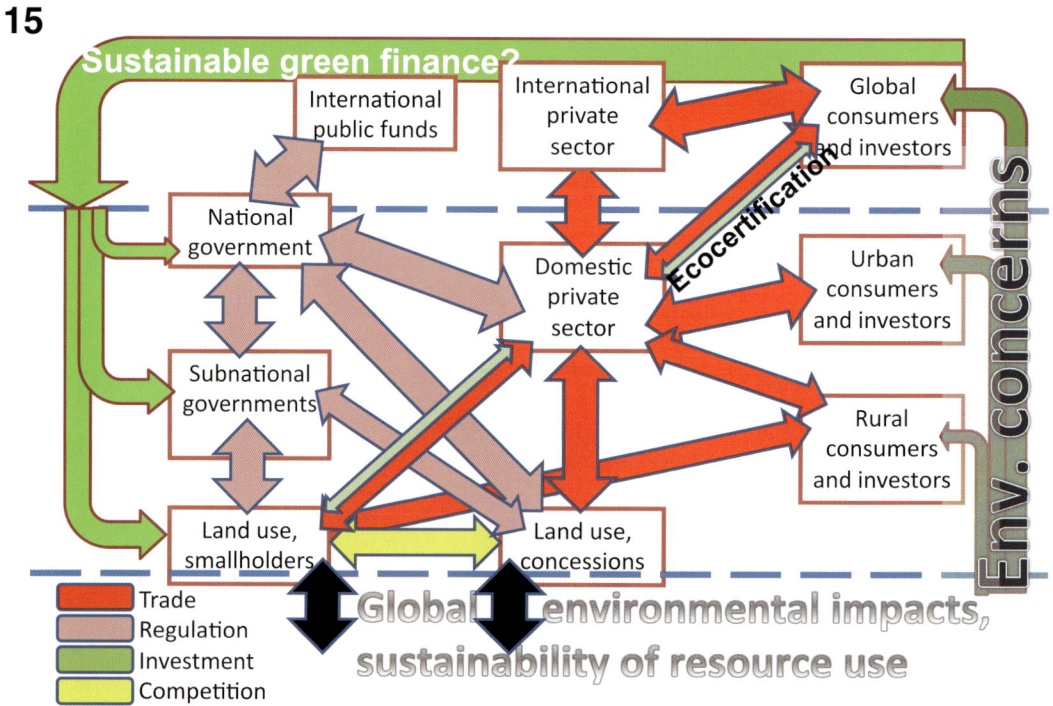

Plate 14. Abandoned, heavily degraded agricultural peatland in Chernigiv region, Ukraine, where the soil has become so hydrophobic and dry that flourishing vegetation can only be found in the ditches.
Plate 15. Institutional map of actors and relations involved in land-use decisions, investments and benefits from the associated value chains.

11 Soil as a Support of Biodiversity and Functions

Pierre-Alain Maron* and Philippe Lemanceau

Abstract

The soil is a major reservoir of biological diversity on our planet. It also shelters numerous biological and ecological processes and therefore contributes to the production of a considerable number of ecosystem services. Among the ecological, social and economic services identified, the role of soil as a reservoir of diversity has now been well established, along with its role in nutrient cycling, supporting primary productivity, pollution removal and storing carbon.

Since the development of industrialization, urbanization and agriculture, soils have been subjected to numerous variations in environmental conditions, which have resulted in modifications of the diversity of the indigenous microbial communities. As a consequence, the functional significance of these modifications of biodiversity, in terms of the capacity of ecosystems to maintain the functions and services on which humanity depends, is now of pivotal importance. The concerns emanating from the scientific community have been reiterated in the *Millennium Ecosystem Assessment* (MEA, 2005) published by the policy makers. This strategic document underlines the need to consider biodiversity as an essential component of ecosystems, not only because of its involvement in providing services essential to the well-being of human societies but also because of its intrinsic value in terms of a natural patrimony that needs to be preserved. This objective cannot be raised without the improvement of our ability to predict the effects of environmental changes on soil biodiversity, ecosystem functioning and the associated services; this requires a better quantification of soil biodiversity at different temporal and spatial scales, and its translation into biological functioning. Major advances in molecular biology since the mid-1990s have allowed the development of techniques to investigate and resolve the diversity of soil microbial communities (Maron *et al.*, 2007).

This chapter describes present and ongoing conceptual and methodological strategies employed to assess and understand better the distribution and evolution of soil microbial diversity at different spatial and temporal scales. It also presents actual knowledge about the link between soil microbial diversity and soil processes, with emphasis on C and N cycling, which are determinant for many of the ecosystem services.

Introduction

Soil represents one of the most important stocks of carbon (C) of the biosphere, with an estimated amount of 2000–3500 Gt of soil organic matter (SOM) (twice more than in the atmosphere). Thus, C is a major soil component, controlling its ability to deliver many

---

*E-mail: Pierre-alain.maron@dijon.inra.fr

services on which human well-being is dependent. Soil is also one of the most important reservoirs of biological diversity on our planet and, above all, one of the last bastions of such biodiversity (Swift et al., 1998). Since most soil organisms are heterotrophs, SOM is of major importance for the sustaining of this biodiversity. One gram of soil is reported to host up to 10 billion microorganisms and thousands of different species (Torsvik and Øvreås, 2002; Roesch et al., 2007). Soil is also the site of numerous biological and ecological processes, and therefore performs a considerable number of ecosystem services resulting from the complex taxonomic and functional assemblages of the indigenous communities and interactions between organisms (Coleman and Whitman, 2004). Pimentel et al. (1997) estimated the economic profit derived from soil biodiversity at US$1546 billion, although the relative value of the associated services remains to be determined (Huguenin et al., 2006).

Among the ecological, social and economic services identified, the role of soil as a reservoir of biodiversity has now been well established, along with its role in surface-water purification, recycling of mineral elements (soil fertility) and carbon storage (as a sink for atmospheric CO_2, soil fertility), the latter processes being related directly to climatic changes and plant productivity (Pimentel et al., 1997). However, the studies involving a characterization/quantification of soil biodiversity and its translation into biological functions are much fewer than those dealing with the biodiversity of organisms living on its surface (notably plants), and therefore our knowledge of the belowground diversity and functioning remains limited (Prosser et al., 2007). This holds even more for microbial communities, which are still considered as being ubiquitous on the basis of the statement made by Beijerinck (1913) that 'everything is everywhere' and showing a high functional redundancy. Thus, soil microbial communities are regarded as a functional 'black box', generating fluxes of different intensities that are solely dependent on abiotic factors such as temperature, moisture and pH. The hypothesis that the diversity and composition of microbial communities, as well as trophic interactions between populations, may play a functional role has so far basically been excluded (McGill, 1996; Gignoux et al., 2001). This has resulted in: (i) a limited effort in characterizing soil microbial diversity, its spatial distribution and contribution to soil functioning (Ranjard et al., 2010); and (ii) an insufficient, even non-existent, consideration of microbial diversity in the models currently used to quantify fluxes of matter and energy in soils (Ingwersen et al., 2008).

Technical difficulties account partly for these scientific lacunae in microbial ecology. Microorganisms, as their name implies, are of microscopic size (in the order of one micrometre for bacteria). Their diversity in soils is huge (Torsvik and Øvreås, 2002; Bates et al., 2011), and the very large range of environmental conditions (soil types, climatic zones, land uses) is likely to enhance this diversity even more. Most (more than 90%) of these microorganisms cannot be cultured on available media (Schloss and Handelsman, 2003; Rajendhran and Gunasekaran, 2008), which has meant that, until recently, they could not be studied. Soil microorganisms are also hidden within the soil, which is a heterogeneous but structured matrix, and thus hinders access to this biotic component. These various difficulties have, for a long time, permitted only a truncated vision of soil biodiversity.

However, thanks to major advances in molecular biology during the past 20 years, techniques have been developed to investigate and decipher the diversity of soil microbial communities in situ and without a priori knowledge (Ranjard et al., 2001). In this scientific context, this chapter aims at providing a better understanding of present and future prospects in soil microbial ecology and functioning, with specific attention on the C cycle. It is divided into three main sections. The first provides a survey of the development of molecular tools and how they represent a unique opportunity to progress in our knowledge of soil biodiversity and functioning. The second section focuses on the importance of SOM as a driver of soil abundance and biodiversity at different spatial scales. The third section presents actual developments aiming at linking soil biodiversity with ecosystem services, with a specific focus on C cycle.

Methods of Characterization of Soil Microbial Communities

The scientific domain described as 'microbial ecology' is about 50 years old, and is thus young. Its step-by-step evolution has been promoted mainly by methodological developments (Fig. 11.1; Maron *et al.*, 2007). Among these developments, numerous molecular tools are now available to characterize the microbial information contained in the nucleic acids extracted from environmental samples (Swift *et al.*, 1998; Qin *et al.*, 2010). They allow the routine characterization of variations in microbial community abundance, structure and diversity in multiple situations (for review, see Ranjard *et al.*, 2001). With the development of these so-called 'molecular ecology' approaches, the number of studies dedicated to the characterization of microbial communities in various environments or subjected to different perturbations has increased exponentially (Morris *et al.*, 2002). All these studies are highly promising and are providing insights into chronic or punctual modifications of microbial biodiversity in natural environments.

Total microbial communities and those determining particular functions or belonging to specific taxonomic groups of particular interest or presenting a danger can therefore be quantified. The abundance of total microbial communities can be measured from the microbial molecular biomass (Dequiedt *et al.*, 2011). The microbial biomass is a well-known marker of soil biological functioning (Horwath and Paul, 1994) and represents a sensitive and early indicator of changes of soil management (farming practices, contamination; Ranjard *et al.*, 2006). It is determined by quantifying the microbial DNA in soil extracts and is correlated with the microbial biomass (Marstorp *et al.*, 2000; Leckie *et al.*, 2004) measured after fumigation extraction (Vance *et al.*, 1987). The advantage of molecular biomass, compared with the biomass obtained after fumigation extraction, is that it can be measured on dry soil samples, which means that it can be determined at the same time as physico-chemical analyses, using a moderate-throughput system to establish reference values for interpreting the results.

Total community (bacteria or fungi), functional communities, or those belonging to a particular taxonomic group, can also be quantified from soil DNA extracts by applying quantitative PCR to determine the number of copies of the rRNA genes (whole communities), the functional gene shared by populations in the functional community or of a specific sequence in the targeted taxonomic group, respectively. One example of a functional group concerns bacteria with the ability to reduce nitrous oxides (Regan *et al.*, 2011). For example, Regan *et al.* (2011), by quantifying denitrifying genes (*nirK*, *nirS*, *nosZ*) in a permanent grassland under elevated atmospheric CO_2, showed that high N_2O emissions under elevated CO_2 correlated with lower nosZ to nirK ratios, suggesting that increased N_2O emissions under elevated CO_2 might be caused by a higher proportion of N_2O-producing, rather than N_2O-consuming (N_2-producing) denitrifiers.

From a qualitative point of view, communities can be characterized either by their structure, i.e. assemblage of the different constitutive populations, or their diversity, i.e. the different types of organisms present. The genetic structure is classically determined by the molecular fingerprints of the communities (for review, see Ranjard *et al.*, 2001). Until recently, comparisons between studies were hampered by the great variety of techniques employed, and an effort of standardization was clearly required. This has been undertaken by national and European programmes (Gardi *et al.*, 2009), such as those conducted in the UK (Countryside Survey, http://www.countrysidesurvey.org.uk), in France (Network for the Measurement of Soil Quality, Réseau de Mesure de la Qualité des Sols – RMQS, http://www.gissol.fr/programme/rmqs/rmqs.php) and in Europe (EcoFINDERS, http://ecofinders.dmu.dk/).

The recent progress achieved with new high-throughput sequencing technologies

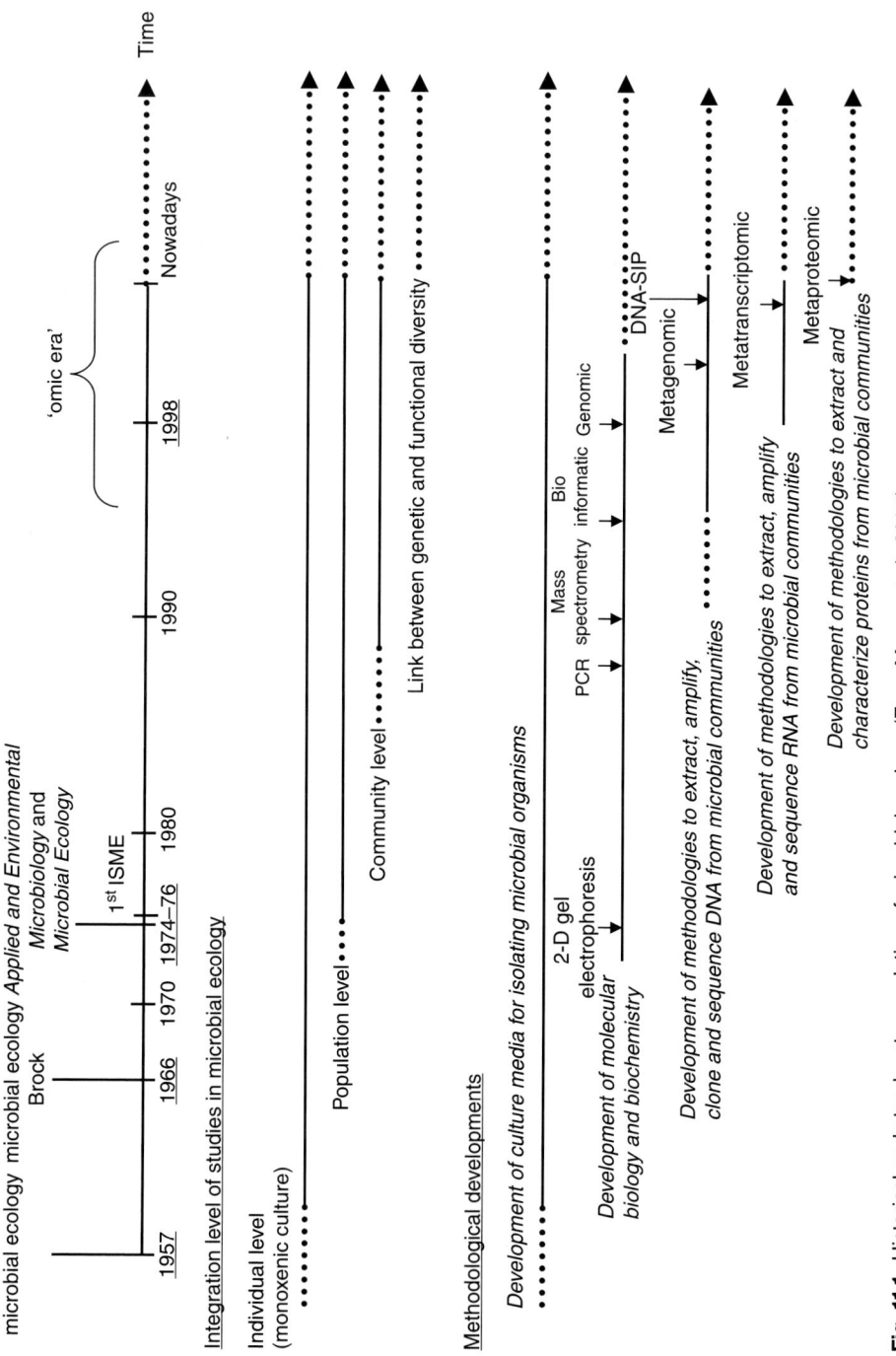

Fig. 11.1. Historical and step-by-step evolution of microbial ecology. (From Maron et al., 2007.)

together with their highly significant reduction in cost (the price of high-throughput sequencing has fallen from US$5292 down to US$0.09 per DNA megabase over the past decade; http://www.genome.gov/sequencingcosts/) corresponds to a major technical revolution for the characterization of the diversity of soil microbial communities. More precisely, pyrosequencing allows several tens of thousands, even hundreds of thousands, of DNA sequences to be obtained from a single metagenomic DNA. It represents, consequently, the most powerful method to date to quantify precisely and characterize microbial diversity through the identification, in a holistic way, of the microbial populations in complex environments (Roesch et al., 2007; Fulthorpe et al., 2008). Two complementary strategies can be distinguished for these metagenomics approaches. The first strategy consists of analysing all the DNA sequences extracted from the indigenous communities in a given environment by targeting either a gene providing particular taxonomic information (ribosomal genes) (Terrat et al., 2012) or functional data (functional genes) (Philippot et al., 2013). The second strategy is based on the mass sequencing of all DNA fragments without targeting specific genes (sequencing of all the metagenome). This approach is highly promising and has proved particularly useful in analysing metagenomes from the Sargasso Sea (Dalton, 2004), an acid mine (Tyson et al., 2004), the digestive tube (Qin et al., 2010) and a soil (Vogel et al., 2009). However, it is time- and cost-consuming and still requires important bioinformatics developments, which limits the number of samples that can be analysed and consequently the application of this strategy for ecological studies. In contrast, the first strategy allows a very large number of environmental situations to be examined, and thus appreciation of a variety of environments. In addition, by the targeting of a limited number of genes, it provides access to the taxonomic and functional diversities of communities involved in particular ecosystemic services. Roesch et al. (2007) used this strategy to characterize microbial diversity in four soils originating from distant geographical locations (USA, Canada and Brazil). Such studies represent the first exhaustive descriptions of the enormous richness of soil bacterial diversity. Fulthorpe et al. (2008) then produced an alternative analysis of these data, and demonstrated the weak similarity in community composition between the soils, thereby revealing that distantly sampled soils carried few species in common. These data support the hypothesis that various pedoclimatic characteristics, as well as land-use and soil management history, can lead to different indigenous microbial diversities. Pyrosequencing techniques were also used in other investigations to decipher soil microbial diversity and elucidate the distribution of diversity within particular taxonomic groups of soil bacteria. Jones et al. (2009) sampled and characterized soils from North to South America (from Alaska to Patagonia) to determine the influence of abiotic soil parameters on the abundance of Acidobacteria. They used this approach to define the ecological attributes of the targeted groups and to rank the environmental parameters that most explained their spatial distribution. Limited knowledge of the taxonomic and functional sequences is one of the main blocks restricting our ability to identify new species or new functional genes. Although few studies have been published as yet, the scientific community is unanimous in affirming the relevance and enormous potential of this type of approach for characterizing the diversity of soil microorganisms (Christen, 2008). Application of the methods described in this section on suitable samplings brings precious knowledge for our understanding of the importance of SOM as a driver of the microbial abundance and diversity in soil.

Organic Matter as a Support of Soil Biodiversity

Soil represents a highly complex and heterogeneous matrix. This property has strong repercussions on the distribution of both nutritive resources (Arrouays et al., 2001) and organisms in soil. Indeed, at the microscale, soil provides a heterogeneous habitat for microorganisms characterized by variations in pH values, pore size and substrate and nutrient, water and oxygen availabilities, all impacting microbial survival and development (Mummey et al., 2006). This has been highlighted by various studies based on direct observations (Chenu et al., 2001; Nunan et al., 2003), microbial biomass measurements (Jocteur Monrozier et al., 1991; Gestel and Merckx, 1996) and community structure and bacterial diversity characterizations (Ranjard et al., 2000; Mummey et al., 2006; Kong et al., 2011), which all evidenced a heterogeneous distribution of microbial communities.

At the scale of the microbial habitat, carbon substrates represent major drivers of the quantitative and qualitative distribution of microbial diversity, since they induce the formation of ecological niches or 'hotspots' that correspond to zones of high biological activity (Stotzky, 1997). Main carbon inputs to soil are represented by: (i) the release of photosynthates by living plants from the roots, called rhizodeposition (Nguyen, 2003); and (ii) plant biomass resulting from its decay after its death. In both cases, the provision of organic substrates to heterotrophic soil microorganisms in soils which are mostly oligotrophic leads to a stimulation of the density and activity of the microflora around the roots, corresponding to the so-called 'rhizosphere' hotspot (Hiltner, 1904), and around the plant residues, which corresponds to the 'detritusphere' hotspot (Gaillard et al., 1999). The quality of these trophic niches depends on plant genotypes which release different amounts and types of rhizodeposits during their development (Gransee and Wittenmayer, 2000) and in their residues after their death. In response to the resulting variations of soil organic status, microbial density, diversity and activity in the hotspots differ during plant development and plant degradation compared to those in bulk soil (Mougel et al., 2006; Pascault et al., 2010). Recently, Davinic et al. (2012) defined soil aggregates as trophic niches of bacterial diversity, highlighting specific bacterial assemblages associated with the quality of organic matter within these soil microhabitats, supporting previous studies (Ranjard and Richaume, 2001).

On larger scales, the distribution of microbial communities has also been demonstrated to be heterogeneous, but structured. The relative contribution of SOM to the microbial distribution of abundance and diversity varies according to the scale considered. Thus, SOM remains a major driver of microbial communities' abundance and diversity on the scale of a farm plot (Lejon et al., 2007), together with soil texture (Johnson et al., 2003; Lejon et al., 2007), pH (Bååth and Anderson, 2003), land use (Nicolardot et al., 2007) and plant cover (Lejon et al., 2005). However, on the landscape and up to the country scale, the physico-chemical properties (mainly soil pH) and land use, but not SOM, constitute the main drivers of soil microbial communities (Dequiedt et al., 2011). As an example, on the scale of France, the molecular biomass, which represents the microbial abundance, has been demonstrated to vary from 5 to 15 µg of DNA g^{-1}, depending on the soil. These variations were associated with the major soil types, and in particular their physico-chemical properties and land uses. More precisely, the molecular biomass values were linked positively to the clay and calcium (Ca) contents, pH value and cation exchange capacity of soils (Dequiedt et al., 2011). The mean molecular biomass values were highest in grassland and lowest in orchards and vineyards, maybe because of the poor plant diversity observed in orchards and vineyards due to perennial culture of the same plant genotypes and the frequent absence of grass between the rows in these crops. More precisely, low plant diversity in orchards and vineyards may lead to a low diversity of C substrates released into the soil, which in addition to the low soil covering by plant cover in these systems, may explain the low mean molecular biomass values observed. In the same way, analyses of the genetic structure of bacterial communities at the scale of France, based on Automated Ribosomal Intergenic Spacer Analysis (A-RISA) fingerprinting, showed that the distribution of microbial diversity was also heterogeneous but spatially structured (Dequiedt et al., 2009).

This distribution was again affected by soil type (physico-chemical characteristics, especially pH) and land use, but SOM represented a minor driver on this scale. The major influence of pH, on the distribution of the structure and diversity of microbial communities on large spatial scales, has also been demonstrated by analogous studies in the UK (Griffiths et al., 2011) and USA (Fierer and Jackson, 2006; Jones et al., 2009).

Linking Soil Microbial Diversity to Soil Processes

The question of the link between diversity and ecosystem functioning, fundamental in functional ecology, has, historically and up to now, been addressed mainly by ecologists studying macroorganisms, and particularly plants, through experimentations involving manipulation of diversity (taxonomic diversity or diversity of functional groups) (Balvanera et al., 2006). First studies were reported in 1843, represented by the experimentation set up on the English experimental station located at Rothamsted (Lawes et al., 1882). This long history explains the abundant literature available on this topic at present. These different works were able to instigate a very rich theoretical and conceptual framework (i.e. ecological insurance, complementarities between niches, functional redundancy), allowing a better understanding of how biodiversity could influence ecosystem functioning (Loreau, 2000). In spite of sometimes contradictory results, it is globally clear that plant diversity has positive effects on ecosystem functioning, performance and stability, and thereby their capacity to provide ecosystem services (Naeem and Li, 1997; Tilman et al., 1997).

Regarding microorganisms, this question remains open, since very few studies have really examined this aspect. As a consequence, the microbial component of soil is still roughly taken into account in mathematical models designed to predict the fate of major elements in the environment (Ingwersen et al., 2008). Indigenous microbial communities are considered in these models as a functional 'black box', with a very high level of functional redundancy. As a result, only the size of the microbiota (biomass), but not the diversity or the composition of this pool, is integrated into such models. Until recently, two main reasons were put forward to justify this pragmatic approach: (i) these 'black box' models were able to provide accurate simulations of SOM dynamics for a variety of land uses with a minimal set of parameters; and (ii) understanding of the diversity of the microbial communities involved in ecosystem processes had been restricted, mainly because of methodological limitations.

Thanks to progress in modelling and new methods in molecular microbial ecology (see above), these reasons are no longer justified. Recent works on modelling the priming effect (i.e. the increase in soil organic C mineralization following the input of a fresh organic C compound) suggest that the density of microbial communities, together with the competition between different microbial functional groups, may control the rate of mineralization of native SOM (Fontaine and Barot, 2005; Neill and Gignoux, 2006). These theoretical studies have confirmed that microbial diversity plays a crucial functional role in organic matter turnover in soil, and clearly suggest that this parameter must be included in the models. They have also evidenced the need for an empirical demonstration of the functional role of microbial diversity in ecosystem processes.

Experiments performed on the grassland systems of Jasper Ridge (California, USA) by Horz et al. (2004) showed that modifications of the diversity of the nitrating community in soil in response to global changes (increased atmospheric CO_2, temperature, nitrogen (N) deposition and soil moisture) could lead to an increase in nitrification, thereby illustrating that microbial diversity plays an important role in the nitrogen cycle. This is in agreement with Philippot et al. (2013), who recently demonstrated that a reduction of the diversity of the denitrifying community strongly decreased N_2O emission from soil.

In contrast, studies to date on the involvement of microbial diversity in the carbon cycle have not provided clear evidence to support such a link (Griffiths et al., 2000; Wertz et al., 2006), even if it has been suggested in some studies (Zogg et al., 1997; Adrén et al., 1999; Strickland et al., 2009). One reason often proposed is that, within the soil microbial community, the decomposition of organic matter in soil would be a highly redundant function. However, analysis of the enzymatic capacities required for SOM degradation has highlighted the non-uniform distribution of these capacities within the soil microbial component. In particular, those involved in the final degradation steps are carried by only a small subset of the soil microbial community (Hu and van Bruggen, 1997; Schimel and Gulledge, 1998). This could explain the successions of microbial populations observed during the degradation of plant residues added to soil (Bernard et al., 2007; Nicolardot et al., 2007; Pascault et al., 2010), suggesting that different populations are required for the different steps of SOM decomposition.

In agreement with this hypothesis, Bell et al. (2005) and Liebich et al. (2007) showed that microbial diversity was playing a major role in the degradation of plant residues. However, these studies were based on the use of consortia of microbial species that had previously been isolated on culture media, which therefore induced a strong selective bias (only 1–10% of soil microorganisms can be cultured) and consequently precluded the possibility of considering the resulting microbial consortium as representative of the indigenous communities (Griffiths et al., 2001). To overcome this bias, Baumann et al. (2012) developed an experimental strategy relying on creating a diversity gradient by inoculating sterile soil microcosms with different dilutions of a soil suspension. This strategy allowed the characterization of SOM degradation along a microbial diversity gradient under controlled laboratory conditions. The microbial consortia that developed in the corresponding microcosms contained several hundreds of different populations, even for the less diverse treatment. Results obtained clearly demonstrated that microbial diversity altered bulk chemical structure and the decomposition of plant litter sugars, and influenced the microbial oxidation of particular lignin compounds, thus changing SOM composition.

Regarding the importance of the C cycle for ecosystem dynamics, and the contradictory results of the insufficiently numerous studies available in the literature, it appears that further investigation and new fundamental studies need to be carried out to elucidate more the role of microbial diversity in C transformations in soil. In contrast to earlier investigations on this topic, methodological developments offer a unique opportunity to decipher relations between the C cycle and active microbial populations. As an example, the recently developed DNA stable-isotope probing (DNA-SIP) method allows specific characterization of the communities actively involved in the decomposition of C substrates labelled with stable isotopes (e.g. ^{13}C) (Neufeld et al., 2006; Bernard et al., 2007; Chen and Murrell, 2010). Another advantage of using labelled compounds is the possibility to monitor not only mineralization of the labelled C substrate but also that of native SOM at the same time as the dynamics of the degrading communities responsible for the decomposition of each C pool. It thus provides an opportunity to evaluate the importance of microbial diversity in ecosystem processes such as the priming effect that may play an important role in soil carbon balance (Fontaine et al., 2003; Bernard et al., 2007; Kuzyakov, 2010; Pascault et al., 2013). Use of such methods, including for manipulating microbial diversity, may constitute a decisive step towards experimental demonstration of the functional significance of microbial diversity in C cycling in soil.

Nevertheless, extrapolation of the conclusions derived from investigations performed under simplified controlled conditions will obviously be limited, and results will need to be generally applicable in order to acquire a truly predictive dimension. A promising complementary strategy to achieve this goal in future will be to combine the powerful and robust tools used to characterize microbial biodiversity (i.e. pyrosequencing of ribosomal genes from soil samples) and functioning, with extensive sampling on a massive spatial scale (landscape, region, country or continent). With such a strategy, it should be possible to establish: (i) a statistical link between the different parameters of soil microbial diversity (i.e. synthetic diversity index/species richness/evenness) and the intensity of the microbial processes involved in the major biogeochemical cycles, this being a prerequisite to determining the importance of microbial diversity in the provision of ecosystem services by the soil environment; and (ii) the relative importance of the environmental filters controlling the functionality of soil microbial communities. Large-scale sampling, as in national soil survey networks, offers unique opportunities to implement this strategy (Gardi et al., 2009).

Conclusions

Organic matter is essential for soil properties and functioning. It contributes to soil aggregation, structure stability and porosity (Duchicela et al., 2012). These impact water regulation, and therefore plant productivity. More generally, organic matter supports life in soils, since most soilborne organisms are heterotrophic. Their activities lead to the decomposition and mineralization of SOM, making nutrients available to plants and again contributing to primary productivity. Studies reported in this chapter report that the rate of SOM mineralization is promoted when biodiversity increases, which therefore would be expected to be favourable to plant productivity (Baumann et al., 2012). However, a possible trade-off has to be considered. Indeed, the enhancement of SOM mineralization contributes to decreasing soil C stock and increasing CO_2 emission, when C storage in soils represents a major issue in the context of global change. In contrast, decreased SOM mineralization would be favourable to C storage and would contribute to decreased CO_2 emission, which would favour the control of global change. However, as indicated above, this would impair plant productivity and therefore the incorporation of organic matter from primary production, and thus would decrease C storage in soils. Thus, proper agroecological practices have to be developed in order to minimize these trade-offs by monitoring soil biodiversity. One of them, proposed by Fontaine et al. (2011), would be to modulate organic mineralization to match the nutritional requirement of the plants in order to avoid the flush of mineralization which would lead to C loss and possible leaching of N. Further research needs to be carried out in order to test hypotheses on the relations between soil biodiversity and functioning with regards to the C status in soil. This research represents a major challenge not only in microbial ecology but also in agroecology in the context of global change.

References

Adrén, O., Brussaard, L. and Clarholm, M. (1999) Soil organism influence on ecosystem-level process – bypassing the ecological hierarchy? *Applied Soil Ecology* 11, 177–188.

Arrouays, D., Deslais, W. and Badeau, V. (2001) The carbon content of topsoil and its geographical distribution in France. *Soil Use and Management* 17, 7–11.

Bååth, E. and Anderson, T.H. (2003) Comparisons of soil fungal/bacterial ratios in a pH gradient using physiological and PLFA-based techniques. *Soil Biology and Biochemistry* 35, 955–963.

Balvanera, P., Pfisterer, A.B., Buchmann, N., He, J.S., Nakashizuka, T., Raffaelli, D. and Schmid, B. (2006) Quantifying the evidence for biodiversity effects on ecosystem functioning and services. *Ecology Letters* 9, 1146–1156.

Bates, S.T., Berg-Lyons, D., Caporaso, J.G., Walters, W.A., Knight, R. and Fierer, N. (2011) Examining the global distribution of dominant archaeal populations in soil. *The ISME Journal* 5, 908–917.

Baumann, K., Dignac, M.F., Rumpel, C., Bardoux, G., Sarr, A., Steffens, M. and Maron, P.A. (2012) Soil microbial diversity affects soil organic matter decomposition in a silty grassland soil. *Biogeochemistry* doi:10.1007/s10533-012-9800-6.

Beijerinck, M.W. (1913) De infusies en de ontdekking der backterien. In: *Jaarboek van de Knoniklijke Akademie van Wetenschappen*. Muller, Amsterdam.

Bell, T., Newman, J.A., Silverman, B.W., Turner, S.L. and Lilley, A.K. (2005) The contribution of species richness and composition to bacterial services. *Nature* 436, 1157–1160.

Bernard, L., Mougel, C., Maron, P.A., Nowak, V., Henault, C., Lévêque, J., Haichar, F., Berge, O., Marol, C., Balesdent, J. et al. (2007) Dynamics and identification of microbial populations involved in the decomposition of ^{13}C labeled wheat residue as estimated by DNA- and RNA-SIP techniques. *Environmental Microbiology* 9, 752–764.

Chen, Y. and Murrell, J.C. (2010) When metagenomics meets stable-isotope probing: progress and perspectives. *Trends in Microbiology* 18, 157–163.

Chenu, C., Hassink, J. and Bloem, J. (2001) Short-term changes in the spatial distribution of microorganisms in soil aggregates as affected by glucose addition. *Biology and Fertility of Soils* 34, 349–356.

Christen, R. (2008) Global, sequencing: a review of current molecular data and new methods available to assess microbial diversity. *Microbes and Environments* 23, 252–272.

Coleman, D.C. and Whitman, W.B. (2004) Linking species richness, biodiversity and ecosystem function in soil systems. *Pedobiologia* 49, 479–497.

Dalton, R. (2004) Natural resources: bioprospects less than golden. *Nature* 429, 598–600.

Davinic, M., Fultz, L.M., Acosta-Martinez, V., Calderón, F.J., Cox, S.B., Dowd, S.E., Allen, V.G., Zak, J.C. and Moore-Kucera, J. (2012) Pyrosequencing and mid-infrared spectroscopy reveal distinct aggregate stratification of soil bacterial communities and organic matter composition. *Soil Biology and Biochemistry* 46, 63–72.

Dequiedt, S., Thioulouse, J., Jolivet, C., Saby, N.P.A., Lelievre, M., Maron, P.A., Martin, M.P., Chemidlin-Prévost-Bouré, N., Toutain, B., Arrouays, D., Lemanceau, P. and Ranjard, L. (2009) Biogeographical patterns of soil bacterial communities. *Environmental Microbiology Reports* 1, 251–255.

Dequiedt, S., Saby, N.P.A., Lelievre, M., Jolivet, C., Thioulouse, J., Toutain, B., Arrouays, D., Bispo, A., Lemanceau, P. and Ranjard, L. (2011) Biogeographical patterns of soil molecular microbial biomass as influenced by soil characteristics and management. *Global Ecology and Biogeography* 20, 641–652.

Duchicela, J., Vogelsang, K.M., Schultz, P.A., Kaonongbua, W., Middleton, E.L. and Bever, J.D. (2012) Non-native plants and soil microbes: potential contributors to the consistent reduction in soil aggregate stability caused by the disturbance of North American grasslands. *New Phytologist* 196, 212–222.

Fierer, N. and Jackson, R.B. (2006) The diversity and biogeography of soil bacterial communities. *Proceedings of the National Academy of Sciences USA* 103, 626–631.

Fontaine, S. and Barot, S. (2005) Size and functional diversity of microbe populations control plant persistence and carbon accumulation. *Ecology Letters* 8, 1075–1087.

Fontaine, S., Mariotti, A. and Abbadie, L. (2003) The priming effect of organic matter: a question of microbial competition? *Soil Biology and Biochemistry* 35, 837–843.

Fontaine, S., Henault, C., Aamor, A., Bdioui, N., Bloor, J.M.G., Maire, V., Mary, B., Revaillot, S. and Maron, P.A. (2011) Fungi mediate long term sequestration of carbon and nitrogen in soil through their priming effect. *Soil Biology and Biochemistry* 43, 86–96.

Fulthorpe, R.R., Roesch, L.F.W., Riva, A. and Triplett, E.W. (2008) Distantly sampled soils carry few species in common. *The IMSE Journal* 2, 901–910.

Gaillard, V., Chenu, C., Recous, S. and Richard, G. (1999) Carbon, nitrogen and microbial gradients induced by plant residues decomposition in soil. *European Journal of Soil Science* 50, 567–578.

Gardi, C., Montanarella, L., Arrouay, D., Bispo, A., Lemanceau, P., Mulder, C., Ranjard, L., Rombke, L., Rutger, M. and Menta, C. (2009) Soil biodiversity monitoring in Europe: ongoing activities and challenges. *European Journal of Soil Science* 60, 807–819.

Gestel, M. and Merckx, R. (1996) Spatial distribution of microbial biomass in microaggregates of a silty-loam soil and the relation with the resistance of microorganisms to soil drying. *Soil Biology and Biochemistry* 28, 503–510.

Gignoux, J., House, J.I., Hall, D., Masse, D, Nacro, H.B. and Abbadie, L. (2001) Design and test of a generic cohort model of soil organic matter decomposition: the SOMKO model. *Global Ecology Biogeography* 10, 639–660.

Gransee, A. and Wittenmayer, L. (2000) Qualitative and quantitative analysis of water-soluble root exudates in relation to plant species and development. *Journal of Plant Nutrition and Soil Sciences* 163, 381–385.

Griffiths, B.S., Ritz, K., Bardgett, R., Cook, R., Chritensen, S., Ekelund, F., Sørensen, S.J., Bååth, E., de Ruiter, P.C., Dolfing, J. and Nicolardot, B. (2000) Ecosystem response of pasture soil communities to fumigation-induced microbial diversity reductions: an examination of the biodiversity-ecosystem function relationship. *OIKOS* 90, 279–294.

Griffiths, B.S., Ritz, K., Wheatley, R., Kuan, H.L., Boag, B., Chritensen, S., Ekelund, F., Sørensen, S.J., Muller, S. and Bloem, J. (2001) An examination of the biodiversity–ecosystem function relationship in arable soil microbial communities. *Soil Biology and Biochemistry* 33, 1713–1722.

Griffiths, R.I., Thomson, B., James, P., Bell, T., Bailey, M. and Whiteley, A.S. (2011) The bacterial biogeography of British soils. *Environmental Microbiology* 13, 1642–1654.

Hiltner, L. (1904) Über neuere Erfahrungen und Probleme auf dem Gebiete der Bodenbakteriologie unter besonderer Berücksichtigung der Gründüngung und Brache. *Arbeiten der Deutschen Landwirtschaftlichen Gesellschaft*, 98, 59–78.

Horwath, W.R. and Paul, E.A. (1994) Microbial biomass. In: Weaver, R.W. (ed.) *Methods of Soil Analysis*, SSSA Book Series 5. Soil Science Society of America Inc, Madison, Wisconsin, pp. 727–752.

Horz, H.P., Barbrook, A., Field, C.B. and Bohannan, B.J.M. (2004) Ammonia oxidizing bacteria respond to multifactorial global change. *Proceedings of the National Academy of Sciences USA* 101, 15136–15141.

Hu, S. and van Bruggen, A.H.C. (1997) Microbial dynamics associated with multiphasic decomposition of 14C-labeled cellulose in soil. *Microbial Ecology* 33, 134–143.

Huguenin, M.T., Leggett, C.G. and Paterson, R.W. (2006) Economic valuation of soil fauna. *European Journal of Soil Biology* 42, S16–S22.

Ingwersen, J., Poll, C., Streck, T. and Kandeler, E. (2008) Micro-scale modelling of carbon turnover driven by microbial succession at a biogeochemical interface. *Soil Biology and Biochemistry* 40, 872–886.

Jocteur Monrozier, L., Ladd, J., Fitzpatrick, R., Foster, R. and Rapauch, M. (1991) Components and microbial biomass content of size fractions in soils of contrasting aggregation. *Geoderma* 50, 37–62.

Johnson, M.J., Lee, K.Y. and Scow, K.M. (2003) DNA fingerprinting reveals links among agricultural crops, soil properties, and the composition of soil microbial communities. *Geoderma* 114, 279–303.

Jones, R.T., Robeson, M.S., Lauber, C.L., Hamady, M., Knight, R. and Fierer, N.A. (2009) Comprehensive survey of soil acidobacterial diversity using pyrosequencing and clone library analyses. *The IMSE Journal* 7, 1–12.

Kong, A.Y.Y., Scow, K.M., Córdova-Kreylos, A.L., Holmes, W.E. and Six, J. (2011) Microbial community composition and carbon cycling within soil microenvironments of conventional, low-input, and organic cropping systems. *Soil Biology and Biochemistry* 43, 20–30.

Kuzyakov, Y. (2010) Priming effects: interactions between living and dead organic matter. *Soil Biology and Biochemistry* doi:10.1016/j.soilbio.2010.074.003.

Lawes, J.B., Gilbert, J.H. and Master, M.T. (1882) Agricultural, chemical, and botanical, results of experiments on the mixed herbage of permanent grasslands, conducted for more than twenty years in succession on the same land. Part II. The botanical results. *Philosophical Transactions of the Royal Society, A and B* 173, 1181–1423.

Leckie, S.E., Prescott, C.E., Grayston, S.J., Neufeld, J.D. and Mohn, W.W. (2004) Comparison of chloroform fumigation-extraction, phospholipid fatty acid, and DNA methods to determine microbial biomass in forest humus. *Soil Biology and Biochemistry* 36, 529–532.

Lejon, D.P.H., Chaussod, R., Ranger, J. and Ranjard, L. (2005) Microbial community structure and density under different tree species in an acid forest soil (Morvan, France). *Microbial Ecology* 50, 614–625.

Lejon, D.P.H., Sebastia, J., Lamy, I., Nowak, V., Chaussod, R. and Ranjard, L. (2007) Microbial density and genetic structure in two agricultural soils submitted to various organic managements. *Microbial Ecology* 53, 650–653.

Liebich, J., Schloter, M., Schäffer, A., Vereecken, H. and Burauel, P. (2007) Degradation and humification of maize straw in soil microcosms inoculated with simple and complex microbial communities. *European Journal of Soil Science* 58, 141–151.

Loreau, M. (2000) Biodiversity and ecosystem functioning: recent theoretical advances. *OIKOS* 91, 3–17.

Maron, P.A., Ranjard, L., Mougel, C. and Lemanceau, P. (2007) Metaproteomics: a new approach for studying functional microbial ecology. *Microbial Ecology* 53, 486–493.

Marstorp, H., Guan, X. and Gong, P. (2000) Relationship between dsDNA, chloroformlabile C and ergosterol in soils of different organic matter contents and pH. *Soil Biology and Biochemistry* 32, 879–882.

McGill, W.B. (1996) Review and classification of ten soil organic matter (SOM) models. In: Powlson, D.S., Smith, P. and Smith, J.U. (eds) *Evaluation of Soil Organic Matter Models*. Springer, Rothamsted, UK, pp. 111–132.

MEA (Millennium Ecosystem Assessment) (2005) *Ecosystem and Human Well-Being*. World Resources Institute, Washington, DC.

Morris, E.C., Bardin, M., Berge, O., Frey-Klett, P., Fromin, N., Girardin, H., Guinebretière, M.H., Lebaron, P., Thiery, J.M. and Troussellier, M. (2002) Microbial biodiversity: approaches to experimental design and hypothesis testing in primary scientific literature from 1975 to 1999. *Microbiology and Molecular Biology Reviews* 66, 592–616.

Mougel, C., Offre, P., Ranjard, L., Corberand, T., Gamalero, E., Robin, C. and Lemanceau, P. (2006) Dynamic of the genetic structure of bacterial and fungal communities at different developmental stages of *Medicago truncatula* Gaertn. cv. Jemalong line J5. *New Phytologist* 170, 165–175.

Mummey, D.L., Holben, W., Six, J. and Stahl, P. (2006) Spatial stratification of soil bacterial populations in aggregates of diverse soils. *Microbial Ecology* 51, 404–411.

Naeem, S. and Li, S.B. (1997) Biodiversity enhances ecosystem reliability. *Nature* 390, 507–509.

Neill, C. and Gignoux, J. (2006) Soil organic matter decomposition driven by microbial growth: a simple model for a complex network of interactions. *Soil Biology and Biochemistry* 38, 803–811.

Neufeld, J.D., Dumont, M.G., Vohra, J. and Murrell, J.C. (2006) Methodological considerations for the use of stable isotope probing in microbial ecology. *Microbial Ecology* 55, 435–442.

Nguyen, C. (2003) Rhizodeposition of organic C by plants: mechanisms and controls. *Agronomie* 23, 375–396.

Nicolardot, B., Bouziri, L., Bastian, F. and Ranjard, L. (2007) Influence of location and quality of plant residues on residue decomposition and genetic structure of soil microbial communities. *Soil Biology and Biochemistry* 39, 1631–1644.

Nunan, N., Wu, K., Young, I.M., Crawford, J.W. and Ritz, K. (2003) Spatial distribution of bacterial communities and their relationships with the micro-architecture of soil. *FEMS Microbiology Ecology* 44, 203–215.

Pascault, N., Nicolardot, B., Bastian, F., Thiebeau, P., Ranjard, L. and Maron, P.A. (2010) In situ dynamics and spatial heterogeneity of soil bacterial communities under different crop residue management. *Microbial Ecology* 60, 291–303.

Pascault, N., Ranjard, L., Kaisermann, A., Bachar, D., Christen, R., Terrat, S., Mathieu, O., Lévêque, J., Mougel, C., Henault, C. *et al.* (2013) Stimulation of different functional groups of bacteria by various plant residues as a driver of soil priming effect. *Ecosystems* 16(5), 810–822.

Philippot, L., Spor, A., Hénault, C., Bru, D., Bizouard, F., Jones, C.M., Sarr, A. and Maron, P.A. (2013) Loss in microbial diversity affects nitrogen cycling in soil. *The ISME Journal* 7(8), 1609–1619.

Pimentel, D.C., Wilson, C., McCullum, C., Hunag, R., Dwen, P., Flack, J., Tran, Q., Saltman, T. and Cliff, B. (1997) Economic and environmental benefits of biodiversity. *Bioscience*, 47, 747–757.

Prosser, J.I., Bohannan, B.J.M., Curtis, T.P., Ellis, R.J., Firestone, M.K., Freckleton, R.P., Green, J.L., Green, L.E., Killham, K., Lennon, J.J. *et al.* (2007) Essay – The role of ecological theory in microbial ecology. *Nature Reviews Microbiology* 5, 384–392.

Qin, J., Li, R., Raes, J., Arumugam, M., Burgdorf, K.S., Manichanh, C., Nielsen, T., Pons, N., Levenez, F., Yamada, T. *et al.* (2010) MetaHITConsortium, a human gut microbial gene catalogue established by metagenomic sequencing. *Nature* 464, 59–65.

Rajendhran, J. and Gunasekaran, P. (2008) Strategies for accessing soil metagenome for desired applications. *Biotechnology Advances* 26, 576–590.

Ranjard, L. and Richaume, A.S. (2001) Quantitative and qualitative microscale distribution of bacteria in soil. *Research in Microbiology* 152, 707–716.

Ranjard, L., Poly, F., Combrisson, J., Richaume, A., Gourbiere, F., Thioulouse, J. and Nazaret, S. (2000) Heterogeneous cell density and genetic structure of bacterial pools associated with various soil microenvironments as determined by enumeration and DNA fingerprinting approach (RISA). *Microbial Ecology* 39, 263–272.

Ranjard, L., Poly, F. and Nazaret, S. (2001) Monitoring complex bacterial communities using culture-independent molecular techniques: application to soil environment. *Research in Microbiology* 151, 167–177.

Ranjard, L., Echairi, A., Nowak, V., Lejon, D.P.H., Nouaïm, R. and Chaussod, R. (2006) Field and microcosm experiments to evaluate the effects of agricultural copper treatment on the density and genetic structure of microbial communities in two different soils. *FEMS Microbiology Ecology* 58, 303–315.

Ranjard, L., Dequiedt, S., Jolivet, C., Saby, N.P.A., Thioulouse, J., Harmand, J., Loisel, P., Rapaport, A., Fall, S., Simonet, P. *et al.* (2010) Biogeography of soil microbial communities: a review and a description of the ongoing French national initiative. *Agronomy for Sustainable Development* 30, 359–365.

Regan, K., Kammann, C., Hartung, K., Lenhart, K., Muller, C., Philippot, L., Kandeler, E. and Marhan, S. (2011) Can differences in microbial abundances help explain enhanced N_2O emissions in a permanent grassland under elevated atmospheric CO_2? *Global Change Biology* 17, 3176–3186.

Roesch, L.F.W., Fulthorpe, R.R., Riva, A., Casella, G., Hadwin, A.K.M., Kent, A.G., Daroub, S.H., Camargo, F.A.O., Farmerie, W.G. and Triplett E.W. (2007) Pyrosequencing enumerates and contrasts soil microbial diversity. *The ISME Journal* 1, 283–290.

Schimel, J.P. and Gulledge, J. (1998) Microbial community structure and global trace gases. *Global Change Biology* 4, 745–758.

Schloss, P.D. and Handelsman, J. (2003) Biotechnological prospects from metagenomics. *Current Opinion in Biotechnology* 14, 303–310.

Stotzky, G. (1997) Soil as an environment for microbial life. In: Van Elsas, J.D., Trevors, J.T. and Wellington, E.M.H. (eds) *Modern Soil Microbiology*. Marcel Dekker, New York, pp. 1–20.

Strickland, M.S., Lauber, C., Fierer, N. and Bradford, M.A. (2009) Testing the functional significance of microbial community composition. *Ecology* 90, 441–451.

Swift, M.J., Andren, O., Brussaard, L., Briones, M., Couteaux, M., Ekschmitt, K., Kjoller, A., Loiseau, P. and Smith, P. (1998) Global change, soil biodiversity, and nitrogen cycling in terrestrial ecosystems: three case studies. *Global Change Biology* 4, 729–743.

Terrat, S., Christen, R., Dequiedt, S., Lelievre, M., Nowak, V., Regnier, T., Bachar, D., Plassart, P., Wincker, P., Jolivet, C. et al. (2012) Molecular biomass and MetaTaxogenomic assessment of soil microbial communities as influenced by soil DNA extraction procedure. *Microbial Biotechnology* 5, 135–141.

Tilman, D., Knops, J., Wedin, D., Reich, P., Ritchie, M. and Siemann, E. (1997) The influence of functional diversity and composition on ecosystem processes. *Science* 277, 1300–1302.

Torsvik, V. and Øvreås, L. (2002) Microbial diversity and function in soil: from genes to ecosystems. *Current Opinion in Microbiology* 5, 240–245.

Tyson, G.W., Chapman, J., Hugenholtz, P., Allen, E.E., Ram, R.J., Richardson, P.M., Solovyev, V.V., Rubin, E.M., Rokhsar, D.S. and Banfield, J.F. (2004) Community structure and metabolism through reconstruction of microbial genomes from the environment. *Nature* 428, 37–43.

Vance, E.D., Brookes, P.C. and Jenkinson, D.S. (1987) An extraction method for measuring soil microbial biomass C. *Soil Biology and Biochemistry* 19, 703–707.

Vogel, T.M., Hirsch, P.R., Simonet, P., Jansson, J.K., Tiedje, J.M., van Elsas, J.D., Nalin, R., Philippot, L. and Bailey, M.J. (2009) TerraGenome: a consortium for the sequencing of a soil metagenome. *Nature Reviews Microbiology* 7, 252.

Wertz, S., Degrange, V., Prosser, J.I., Poly, F., Commeaux, C., Freitag, T., Guillaumaud, N. and Le Roux, X. (2006) Maintenance of soil functioning following erosion of microbial diversity. *Environmental Microbiology* 8, 2162–2169.

Zogg, G.P., Zak, D.R., Ringelberg, D.B., MacDonald, N.W., Pregitzer, K.S. and White, D.C. (1997) Compositional and functional shifts in microbial communities due to soil warming. *Soil Science Society of America Journal* 6, 475–481.

12 Water Supply and Quality

David Werner* and Peter Grathwohl

Abstract

Water is filtered while passing through soil, and soil organic carbon plays an important role in water purification through the retention of organic and inorganic pollutants. But no filter lasts forever, and no matter how strongly pollutants adsorb to soil organic matter, they will not be retained indefinitely and will eventually break through organic carbon-rich soil horizons and may reach the groundwater. Therefore, the organic pollutant biodegradation in soil by microorganisms is a most important complementary process to pollutant retention by sorption. Soil organic carbon also shapes soil microbial communities and activities as an important substrate and habitat, and forms the metabolic capabilities and activities leading to pollutant breakdown in soils. If released into soil pore water, dissolved or colloidal organic matter may cause problems for drinking water supply as a carrier of associated pollutants, by giving taste, odour or colour to water and through the formation of disinfection by-products. Furthermore, the release of dissolved organic carbon from soils may lead to oxygen depletion in seepage water and subsequent mobilization of metals.

Introduction

Soils are the most important buffer and filter for water. Water tends to get cleaned and filtered while passing through soil. When comparing different raw water sources as potential intakes for drinking water treatment and supply purposes, groundwater usually provides the better raw water quality than surface water in terms of colour, turbidity and counts of coliform bacteria. The soil's water purification capacity depends on dissolved contaminant sorption by soil particles, the entrapment of suspended particles, including bacteria and viruses in the soil's pore space and the mineralization of biodegradable matter by soil microorganisms. Soil organic carbon is an important sorbent matrix for dissolved water contaminants, and also an important substrate for soil biota and a soil habitat-shaping factor. It is thus an important determinant of a soil's water filtration capacity. The release of soil organic matter into soil pore water may, however, also complicate drinking water treatment or mobilize pollutants.

*E-mail: david.werner@ncl.ac.uk

Water Filtration

The role of soil organic carbon in the sorption and retention of pollutants

Soils are the most important filters for rainwater on its way down to groundwater. Almost all drinking waters at some point have been filtered by soils. The organic pollutant retention capacity of soils typically increases with organic carbon content, and organic carbon normalized sorption coefficients play an important role in predicting the organic pollutant fate. Soil organic matter also contains functional groups such as carboxyl, hydroxyl and amine groups, which, depending on pH, may contribute substantially to the soil cation exchange capacity, and thus the retention of cationic inorganic pollutants. Increased soil organic carbon contents are thus beneficial for the purification of seepage water and the retention not only of, for example, pesticides in agriculture (Werner et al., 2013) but also of the many compounds from urban space or atmospheric pollution.

Soil organic matter comes in various forms (Plate 8) that interact distinctly with contaminants (Grathwohl, 1990). More mature, thermally altered forms of organic carbon such as coals, or combustion-derived carbon such as charcoals and soot adsorb hydrophobic organic chemicals much more strongly than normal soil organic matter. On the other hand, more mature, thermally altered forms of organic carbon tend to have lower elemental oxygen-to-carbon ratios, and hence a lower abundance of functional groups. The extreme heterogeneity in terms of the nature of particulate organic matter in soils is largely unknown, although fluorescence microscopy techniques exist to identify and quantify a large variety of organic matter and black carbon particles (Ligouis et al., 2005) (some examples are shown in Plate 8). These different types of organic matter particles in soils have different sorption characteristics that depend on organic matter facies (Kleineidam et al., 1999; Karapanagioti et al., 2000). Thermally altered organic matter (e.g. kerogens, coals, chars) contains micropores, which strongly adsorb hydrophobic organic compounds already at low concentrations (Kleineidam et al., 2002). With the suggested application of biochars in agricultural soils, consideration of carbon-type specific contaminant sorption becomes ever more relevant (Bushnaf et al., 2011; Khan et al., 2013). Urban soils, but also soils in flood plains of industrial countries, also contain various forms of tars, coke, chars and soot, for example from sealcoats, coal-based steel and electric power production, which are highly loaded with polycyclic aromatic hydrocarbons (Yang et al., 2008, 2010). How stable these different forms of carbon are in soils and whether carbon-sorbed contaminants would eventually be released is unclear. 'Normal' soil organic matter derived from modern plant residues may be much less stable than the thermally altered form, and degradation of organic carbon may lead to water-soluble fractions of organic carbon.

The water purification capacity of soils has come under increasing pressure since the industrial and agricultural revolution. In industrial soils, organic carbon of natural or anthropogenic origin was frequently loaded with legacy contaminants, to such an extent that soils became a long-lasting secondary source of pollution of seepage water or air. Similarly, rural soils have been, and still are, sinks for atmospheric pollutants or pesticides, and may become sources in the future. Whether or not soil pollutants finally end up in groundwater depends on the pollution attenuation by sorption and biodegradation. Both processes are crucially influenced by soil organic carbon. This applies especially for organic compounds, but also for many inorganic species.

Combined Effects of Sorption and Biodegradation on Water Filtration

Contaminant sorption by soil organic carbon is beneficial in the short term, as it removes pollutants from seepage water, and may reduce toxic effects significantly by reducing the contaminant availability for bio-uptake. This reduction in bioavailability, however, also slows down biodegradation, which leads to increased persistence of organic compounds, especially in topsoils. The effect of sorption on the kinetics of biodegradation and advective pollutant transport in infiltrating soil pore water can be derived from a simple mass balance equation that assumes the aqueous concentration as the rate-limiting factor for biodegradation (first-order kinetics):

$$\theta \frac{dC_w}{dt} + \rho \frac{dC_s}{dt} = -\lambda \theta C_w - u_z \theta \frac{\partial C_w}{\partial z} \quad (12.1)$$

θ (–) and ρ (g cm^{-3}) represent the volumetric water content and the dry bulk density in soil. C_w (g cm^{-3}) is the pollutant concentration in soil pore water, and C_s (g g^{-1}) the pollutant concentration associated with the solid soil matrix. The parameter λ is the first-order rate constant for biodegradation (s^{-1}), u_z (cm s^{-1}) is the soil pore water flow velocity in the z direction, and z (cm) is the vertical distance from the soil surface towards the groundwater table, t (s) the time. Assuming local equilibrium conditions between soil and aqueous concentrations characterized by a distribution coefficient K_d (= C_s/C_w) [cm^3 g^{-1}] results in:

$$\frac{dC_w}{dt} = -\frac{\lambda}{\left(1+\frac{\rho}{\theta}K_d\right)} C_w - \frac{u_z}{\left(1+\frac{\rho}{\theta}K_d\right)} \frac{\partial C_w}{\partial z}$$

$$(12.2)$$

Thus, the term $\left(1+K_d \frac{\rho}{\theta}\right)$ slows down the organic pollutant biodegradation and also advective pollutant transport. In this model, rates are slowed down by sorption (i.e. K_d), but also depend on the solid-to-liquid ratio (here: ρ/θ), i.e. decrease with decreasing water content. For many organic pollutants, the magnitude of K_d increases with soil organic matter content. Therefore, water contents and organic matter may be major reasons for the persistence and retention of strongly hydrophobic organic pollutants by organic matter in soils.

In terms of the soil's water filtration capacity, the most important long-term consideration is the attenuation of pollution over a certain distance (i.e. between the soil surface and the groundwater table). If one assumes, as above, that adsorbed pollutants are not only immobile but also inaccessible to soil microorganisms, and describes the removal of water-dissolved, mobile pollutants by first-order rate biodegradation kinetics, then the sorption will have no net long-term benefit in terms of pollutant attenuation during soil passage. This can, for instance, be seen, by considering a hypothetical scenario with a fixed, constant pollutant concentration in soil pore water, $C_{w,top}$ (g cm^{-3}), at the soil–atmosphere interface and a constant pore-water flow velocity, u_z. For this hypothetical scenario, the pollutant concentration in soil pore water will tend towards a steady state ($dC_w/dt = 0$), in which the concentration profile no longer changes with time but is simply a function of distance from the soil surface, described by the following exponential decay:

$$C_w(z) = C_{w,top} \cdot e^{-\left(\frac{\lambda}{u_z} \cdot z\right)} \quad (12.3)$$

The exponential decay of the pollutant concentration with increasing downward distance from the soil surface can be interpreted as a measure for the sustainable, long-term pollutant attenuation in soil. The sustainable pollution attenuation depends on the ratio between the first-order biodegradation rate, λ, and the soil pore-water flow velocity, u_z, but not on the model parameter, K_d, which accounts for the pollutant sorption. Enhanced sorption will delay the migration of the pollutant, and hence the time needed for the system to reach the steady state described by Eqn 12.3. Equally, sorption would delay the disappearance of the pollutant from the soil once concentrations at the soil surface return to zero. If we assume that sorbed pollutants are inaccessible to soil microorganisms, then the water-filtration benefits of pollutant sorption by soil organic carbon will be only temporary. Sorption initially reduces pollutant leaching when soil is freshly exposed to a new source of contamination, but this apparent benefit is annulled by the disadvantage of prolonged pollutant leaching once the contamination input has ceased. This analysis is only valid for the model assumptions outlined above, which paint a rather simple picture of the interactions between pollutants, the soil matrix and soil microorganisms. Soil organic carbon affects the pollutant fate, not only as a sorbent matrix (i.e. via K_d) but also indirectly through its profound effects on soil microorganisms, and hence the biodegradation term in Eqn 12.1.

The Role of Soil Organic Carbon in the Biodegradation of Organic Pollutants

If water purification relied on pollutant sorption by organic matter in soil only, it could not be effective. No filter lasts forever, and no matter how strongly pollutants adsorb to soil organic matter, they will not be retained indefinitely and will eventually break through organic carbon-rich soil horizons and leach into surface water and groundwater. Hence, the pollutant breakdown by soil organisms is a critical complementary component of long-term sustainable water purification in soil. Soil organic carbon effects on the organic pollutant biodegradation process can be synergistic or antagonistic. Many soil microorganisms subsist on plant organic matter such as leaf litter or root exudates and live in the root zone, which is also the organic matter-rich topsoil layer. Much of the soil organic carbon is humified and poorly biodegradable, but soil organic carbon content nevertheless correlates with the abundance and activity of soil microbes (Schnürer et al., 1985; Ritz et al., 2004), and microbial carbon may contribute several per cent of total organic carbon in soil (Ananyeva et al., 2008). To what extent natural organic carbon mineralization may stimulate directly or indirectly the biodegradation of xenobiotic organic compounds depends on which mechanism controls the microbial breakdown of pollutants.

Metabolic capability

Capability to biodegrade xenobiotic organic compounds is indispensable for their breakdown by soil microbial communities. Soil organic carbon is a key factor in shaping this capability. A diverse array of catabolic enzymes has been developed through the evolutionary adaptation of soil microorganisms to the diversity of complex natural organic molecules present in soil organic matter, and many of these microbial enzymes are perfectly suited to metabolizing structurally related xenobiotic compounds (Singer et al., 2003). Furthermore, existing genes coding for enzymes involved in natural organic matter biodegradation may provide suitable templates for the evolution of novel genes involved in xenobiotic compound biodegradation. For instance, the microbial biodegradation of xenobiotic s-triazine ring compounds, a class which includes many pesticides, dyes and explosives, is related to the metabolism of pyrimidine and purine rings, which are structural components of essential biomolecules such as deoxyribonucleic acid (DNA) and ribonucleic acid (RNA), adenosine triphosphate (ATP) and nicotinamide adenine dinucleotide (NAD) (Wackett et al., 2002).

Co-metabolism versus substrate–substrate inhibition

Co-metabolism refers to the ability of microorganisms to biodegrade a xenobiotic compound together with co-substrates that may occur naturally and at much higher concentration in soil. Co-metabolism can be explained by non-specific enzymes which transform a wide range of different organic compounds. For instance, extracellularly excreted peroxidases and laccases involved in the oxidation of lignin also enable ligninolytic fungi to biodegrade polycyclic aromatic hydrocarbons (Bamforth and Singleton, 2005). Monooxygenases and dioxygenases play an important role in the removal of both natural and xenobiotic organic compounds (Bamforth and Singleton, 2005; Meynet et al., 2012). Co-metabolism enables the biodegradation of xenobiotic organic compounds present at very low concentrations, since microorganisms may use the co-substrate(s) to meet their carbon and energy needs. Since soil organic matter is a rich source of potential co-substrates, it facilitates the co-metabolic biodegradation of xenobiotic organic compounds. Indeed, the addition of organic carbon-rich substrates to contaminated soils is used as a bioremediation strategy to stimulate microbial activity and facilitate the co-metabolic removal of pollutants (Wagner and Zablotowicz, 1997).

Some versatile degraders of xenobiotic organic compounds, for instance *Pseudomonas putida* strains, are known to regulate the expression of their degradation pathways in response to substrate availability (Rojo, 2010), and also to maintain the carbon–nitrogen balance (Amador *et al.*, 2010). For instance, expression of the genes of the alkane degradation pathway encoded in the *P. putida* octane (OCT) plasmid are subject to negative and dominant global control, depending on the carbon source used and on the physiological status of the cell (Dinamarca *et al.*, 2002). Effective regulation of metabolic pathways in response to carbon substrate availability and carbon–nitrogen ratios will increase the competitiveness of the microorganisms that can grow on their preferred substrate. Substrate–substrate inhibition refers to the situation where expression of the metabolic pathway leading to the breakdown of one substrate, for instance a xenobiotic compound, is inhibited in the presence of another, more preferred substrate, for instance more readily biodegradable natural organic compounds. This provides a potential mechanism by which the presence of soil organic matter could impede the biodegradation of xenobiotic compounds in soil.

Redox conditions

Redox conditions define the thermodynamic landscape of xenobiotic compound biodegradation in soils. While most xenobiotic compounds are biodegraded most rapidly by aerobic microorganisms using oxygen as their ultimate electron acceptor, some important classes of xenobiotic compounds are transformed more effectively under anaerobic conditions. For instance, reductive dechlorination is often an important first step in the metabolism of persistent, heavily chlorinated pesticides, and many nitro-aromatic compounds are susceptible to nitro reduction under anaerobic conditions. Inorganic pollutant speciation will also depend on the prevailing redox conditions. Because it is the most abundant organic substrate in soil, the decomposition of natural organic matter will set the soil redox conditions. Especially in very wet conditions and waterlogged soils, the available oxygen in soil pore water may be consumed rapidly by natural organic matter decomposition, which results in anaerobic conditions and the accumulation of partially decomposed natural organic matter in organic carbon-rich soils such as peat.

Nutrient availability

An adequate supply of essential nutrients is important for the biodegradation of xenobiotic organic compounds in soil (Atlas and Philp, 2005). Microbial degradation of soil organic matter releases nutrients into soil solution, which can then potentially support the growth of xenobiotic compound-degrading bacteria. Soil organic matter decomposition releases nitrogen, phosphorus, potassium, calcium, magnesium and other essential growth elements assimilated (Tian *et al.*, 1992). Organic matter also contributes to the cation exchange capacity of soils, and thus it helps with the retention of positively charged inorganic nutrients released by the decomposition of biomolecules. Soil organic matter decomposition also releases chelating organic chemicals into soil solution, which may enhance the solubility and bioavailability of essential micronutrients such as copper, iron, manganese and zinc, which is particularly important in alkaline soils. Dissolved organic matter may also prevent phosphate precipitation.

Xenobiotic compound availability

As outlined above, soil organic matter is an important sorbent matrix for xenobiotic compounds. While association with the organic matter within soil aggregates reduces the mobility and toxicity of xenobiotic compounds (Luthy et al., 1997), sorption may also reduce the xenobiotic compound's accessibility to pollutant-degrading microorganisms. Experimental evidence shows that organic soil amendments reduce the aerobic mineralization of xenobiotic compounds by reducing their readily bioaccessible concentrations in soil (Marchal et al., 2013), but soil organic matter bound compounds apparently remain to some extent susceptible to microbial breakdown (Yang et al., 2012). If soil organic matter bound compounds remain susceptible to biodegradation, the benefits of the pollutant retention and subsequent breakdown of immobilized pollutants would be lasting.

Generation of Colloidal and Dissolved Organic Carbon

Soil organic carbon has many benefits for water filtration, as outlined above, but the dissolution of smaller and more polar natural organic molecules into water, for instance the release of fulvic and humic acids by organic carbon-rich soils, can also create problems for water supply. Dissolved organic carbon facilitates the transport of metals and hydrophobic organic pollutants in surface water and groundwater, and may thus enable entry of these pollutants into reservoirs and aquifers used for water supply. High contents of naturally dissolved organic carbon give colour to water (see Plate 9), which is aesthetically undesirable in water used for consumption and may also add unpleasant tastes and odours. Dissolved organic carbon is a growth substrate for microorganisms, and it may cause high cell counts and biofouling problems in drinking water treatment facilities, for instance the clogging of filters and membranes. Natural dissolved organic matter is also an important precursor for the formation of undesirable by-products from water disinfection; for instance, the formation of trihalomethanes if chlorination is used. Dissolved organic matter is not released continuously into seepage water, but depends on the 'pre-history' of the soil. Freezing/thawing events, snow melts/floods in winter have been observed to release peak concentrations of dissolved organic matter that is often associated with the transport of pollutants. Furthermore, the digging of trenches for laying cables to wind farms through organic carbon-rich soil is speculated to have led to increased dissolved organic matter leaching into surface waters and drinking water reservoirs.

References

Amador, C.I., Canosa, I., Govantes, F. and Santero, E. (2010) Lack of CbrB in *Pseudomonas putida* affects not only amino acids metabolism but also different stress responses and biofilm development. *Environmental Microbiology* 12(6), 1748–1761.

Ananyeva, N.D., Polyanskaya, L.M., Susyan, E.A., Vasenkina, I.V., Wirth, S. and Zvyagintsev, D.G. (2008) Comparative assessment of soil microbial biomass determined by the methods of direct microscopy and substrate-induced respiration. *Microbiology* 77(3), 356–364.

Atlas, R.M. and Philp, J. (2005) *Bioremediation: Applied Microbial Solutions for Real-World Environmental Cleanup*. ASM Press, Washington, DC.

Bamforth, S.M. and Singleton, I. (2005) Bioremediation of polycyclic aromatic hydrocarbons: current knowledge and future directions. *Journal of Chemical Technology and Biotechnology* 80(7), 723–736.

Bushnaf, K.M., Puricelli, S., Saponaro, S. and Werner, D. (2011) Effect of biochar on the fate of volatile petroleum hydrocarbons in an aerobic sandy soil. *Journal of Contaminant Hydrology* 126(3–4), 208–215.

Dinamarca, M.A., Ruiz-Manzano, A. and Rojo, F. (2002) Inactivation of cytochrome o ubiquinol oxidase relieves catabolic repression of the *Pseudomonas putida* GPo1 alkane degradation pathway. *Journal of Bacteriology* 184(14), 3785–3793.

Grathwohl, P. (1990) Influence of organic-matter from soils and sediments from various origins on the sorption of some chlorinated aliphatic hydrocarbons – implications on Koc correlations. *Environmental Science and Technology* 24(11), 1687–1693.

Karapanagioti, H.K., Kleineidam, S., Sabatini, D.A., Grathwohl, P. and Ligouis, B. (2000) Impacts of heterogeneous organic matter on phenanthrene sorption: equilibrium and kinetic studies with aquifer material. *Environmental Science and Technology* 34, 406–414.

Khan, S., Wang, N., Reid, B.J., Freddo, A. and Cai, C. (2013) Reduced bioaccumulation of PAHs by *Lactuca satuva* L. grown in contaminated soil amended with sewage sludge and sewage sludge derived biochar. *Environmental Pollution* 175, 64–68.

Kleineidam, S., Rugner, H., Ligouis, B. and Grathwohl, P. (1999) Organic matter facies and equilibrium sorption of phenanthrene. *Environmental Science and Technology* 33, 1637–1644.

Kleineidam, S., Schuth, C. and Grathwohl, P. (2002) Solubility-normalized combined adsorption-partitioning sorption isotherms for organic pollutants. *Environmental Science and Technology* 36(21), 4689–4697.

Ligouis, B., Kleineidam, S., Karapanagioti, H.K., Kiem, R., Grathwohl, P. and Niemz, C. (2005) Organic petrology: a new tool to study contaminants in soils and sediments. In: Lichtfouse, E., Dudd, S. and Robert, D. (eds) *Environmental Chemistry. Green Chemistry and Pollutants in Ecosystems. Part I.* Springer, Berlin, pp. 89–98.

Luthy, R.G., Aiken, G.R., Brusseau, M.L., Cunningham, S.D., Gschwend, P.M., Pignatello, J.J., Reinhard, M., Traina, S.J., Weber, W.J. and Westall, J.C. (1997) Sequestration of hydrophobic organic contaminants by geosorbents. *Environmental Science and Technology* 31(12), 3341–3347.

Marchal, G., Smith, K.E.C., Rein, A., Winding, A., Trapp, S. and Karlson, U.G. (2013) Comparing the desorption and biodegradation of low concentrations of phenanthrene sorbed to activated carbon, biochar and compost. *Chemosphere* 90(6), 1767–1778.

Meynet, P., Hale, S.E., Davenport, R.J., Cornelissen, G., Breedveld, G.D. and Werner, D. (2012) Effect of activated carbon amendment on bacterial community structure and functions in a PAH impacted urban soil. *Environmental Science and Technology* 46(9), 5057–5066.

Ritz, K., McNicol, W., Nunan, N., Grayston, S., Millard, P., Atkinson, D., Gollotte, A., Habeshaw, D., Boag, B., Clegg, C.D. *et al.* (2004) Spatial structure in soil chemical and microbiological properties in an upland grassland. *FEMS Microbiology Ecology* 49(2), 191–205.

Rojo, F. (2010) Carbon catabolite repression in Pseudomonas: optimizing metabolic versatility and interactions with the environment. *FEMS Microbiology Reviews* 34(5), 658–684.

Schnürer, J., Clarholm, M. and Rosswall, T. (1985) Microbial biomass and activity in an agricultural soil with different organic matter contents. *Soil Biology and Biochemistry* 17(5), 611–618.

Singer, A.C., Crowley, D.E. and Thompson, I.P. (2003) Secondary plant metabolites in phytoremediation and biotransformation. *Trends in Biotechnology* 21(3), 123–130.

Tian, G., Kang, B.T. and Brussaard, L. (1992) Biological effects of plant residues with contrasting chemical compositions under humid tropical conditions – decomposition and nutrient release. *Soil Biology and Biochemistry* 24(10), 1051–1060.

Wackett, L.P., Sadowsky, M.J., Martinez, B. and Shapir, N. (2002) Biodegradation of atrazine and related s-triazine compounds: from enzymes to field studies. *Applied Microbiology and Biotechnology* 58(1), 39–45.

Wagner, S.C. and Zablotowicz, R.M. (1997) Effect of organic amendments on the bioremediation of cyanazine and fluometuron in soil. *Journal of Environmental Science and Health Part B-Pesticides Food Contaminants and Agricultural Wastes* 32(1), 37–54.

Werner, D., Garratt, J. and Pigott, G. (2013) Sorption of 2,4-D and other phenoxy herbicides to soil, organic matter, and minerals. *Journal of Soils and Sediments* 13(1), 129–139.

Yang, Y., Ligouis, B., Pies, C., Grathwohl, P. and Hofmann, T. (2008) Occurrence of coal and coal-derived particle-bound polycyclic aromatic hydrocarbons (PAHs) in a river floodplain soil. *Environmental Pollution* 151(1), 121–129.

Yang, Y., Metre, P.C.V., Mahler, B.J., Wilson, J.T., Ligouis, B., Razzaque, M.M., Schaeffer, D.J. and Werth, C.J. (2010) Influence of coal-tar sealcoat and other carbonaceous materials on polycyclic aromatic hydrocarbon loading in an urban watershed. *Environmental Science and Technology* 44(4), 1217–1223.

Yang, Y., Shu, L., Wang, X.L., Xing, B.S. and Tao, S. (2012) Mechanisms regulating bioavailability of phenanthrene sorbed on a peat soil-origin humic substance. *Environmental Toxicology and Chemistry* 31(7), 1431–1437.

13 Wind Erosion of Agricultural Soils and the Carbon Cycle

Daniel E. Buschiazzo* and Roger Funk

Abstract

Wind erosion is an important process of both progressive and regressive pedogenesis in arid and semi-arid environments around the world. In semi-arid regions, which are influenced by carbon-poor dust depositions from deserts, the properties as a sink area should be maintained to enable C enrichment by continued soil formation. On agricultural land, wind erosion is a soil-degrading process, resulting mainly from the very effective sorting processes. Coarse particles remain in the field, whereas the finest and most valuable parts of the soil get lost, like particles of the silt and clay fractions and soil organic matter. The latter is not regarded in most carbon balances, although this particulate loss can reach considerable amounts. The processes of wind erosion are subject to a great spatial and temporal variability, making its quantification difficult. In this chapter, we expose wind erosion in the context of its influence on soil organic carbon and prove considerable losses by first measurements.

Introduction

Wind erosion is an important process of both progressive and regressive pedogenesis in arid and semi-arid environments around the world. Soils, which have been formed by aeolian processes over centuries, are now endangered to be destroyed by the same processes within a short time. Associated dust emissions influence physical and chemical processes in the atmosphere, have negative effects on air quality and affect other ecosystems far away from the source areas. About 2000 Mt dust are emitted each year into the atmosphere, of which three-quarters are deposited on land surfaces and one-quarter on the oceans (Shao et al., 2011). Dust depositions may have an important role in the nutrient cycle of natural ecosystems or low-input agriculture, as shown from the Amazonas forests (Swap et al., 1992; Kaufman et al., 2005; Koren et al., 2006) and the Sudan–Sahelian zone (Jahn, 1995; Herrmann, 1996; Stahr et al., 1996; Goudie and Middleton, 2001). Dust depositions on the oceans activate phytoplankton growth, which has a direct impact on the global carbon cycle and on carbon sequestration as well, as stated by the 'iron hypothesis' of Martin and Fitzwater (1988) and the 'silica hypothesis' of Harrison (2000). Consequently, there is a close relationship between the cycles of dust, carbon

*E-mail: buschiazzo@agro.unlpam.edu.ar/debuschiazzo@yahoo.com

and energy in the global context (Shao et al., 2011). In longer timescales, dust also has modifying influences on the climate (Martínez-Garzia et al., 2011). Deserts are the major global source of dust, but also the surrounding semi-arid steppe ecosystems have changed in the last decades from former sinks into substantial source regions. Causes have always been inappropriate land use such as the destruction of vegetation by overgrazing or the conversion from grazing land to cropland. Because steppe soils represent one of the largest carbon stocks, its degradation and the loss of this area as a sink have a significant influence on the global carbon balance.

On agricultural land, wind erosion is a soil-degrading process, but quantitative data are rather approximate estimates. Heavy sand and dust storms attract attention by disturbing the public once in a while, but in general, the processes mostly happen unnoticed. Chepil (1960) has already pointed out that annual average soil losses of up to 40 t ha^{-1} are possible without any visible indications of soil movement. Wind erosion has been recognized as a gradual soil-degradation process, which removes predominantly the finest and most valuable particles of a soil, such as silt and clay particles, as well as the soil organic matter. In addition to the creeping degradation, single wind-erosion events may result in soil losses of more than 100 t ha^{-1} and cause considerable on-site and off-site damages (Funk et al., 2004; Goossens and Riksen, 2004; Hoffmann et al., 2011). In the last decades, the spatial extent of wind erosion has increased, caused mainly by changes of agricultural land use and inappropriate farming practices. Increasing demands on food production have expanded arable land use to marginal sites such as natural grassland or forest. Some more factors favouring wind erosion on arable land are:

- A higher level of mechanization has led to larger fields and, in return, to the removal of hedges and other landscape structures.
- The drainage of arable land causes faster drying of the soil surface, resulting in organic matter decomposition and decreasing soil aggregate stability.
- Overgrazing is a significant causative factor in the semi-arid regions, where no other type of land use is possible (Frielinghaus and Schmidt, 1993; van Lynden, 1995; Riksen et al., 2003; Hoffmann et al., 2008a).

Wind erosion has been regarded mainly as a soil-removing process, but also soil-particle sorting, long-distance nutrient transport, fertilizing aquatic and terrestrial ecosystems far away from the origin or the increase of soil heterogeneity are parts of the problem, with local to global consequences on soil properties and the carbon cycle as well. Erosion and deposition processes both take place on large areas and are therefore difficult to identify. In contrast to water erosion, where the eroded material follows determined paths, wind-eroded material is widely dispersed over the landscape. These direct and indirect effects are difficult to evaluate entirely and the current ongoing debate is whether soils, affected by erosion, act as a source or a sink for carbon. In agricultural soils, organic carbon is concentrated in the top layer (0–30 cm) and therefore vulnerable to wind erosion in its particulate form (particulate organic matter: POM) because of its lower density compared to the mineral particles (Lal, 2003). Cultivated organic soils are most susceptible to erosion and to massive decomposition processes. Mineral soils cover the wide range from high-erodible sandy soils to non-erodible loam and clay soils. The erodibility of a soil is attributed mainly to the texture and organic matter content, which in turn influence the water-holding capacity and the ability of the soil to produce aggregates or crusts (Chepil and Woodruff, 1961). In general, sandy soils are highly erodible because they dry quickly, form only a few, weak aggregates and contain a great part of particles in the most erodible fraction of 80–200 μm in diameter. Loamy soils are more resistant against wind erosion but have a greater potential for dust production (particles <60 μm) if they erode.

Due to the low net primary production of most regions affected by wind erosion, removed carbon (C) can be regarded as an irretrievable loss at the eroded site (Yan et al., 2005).

Even at the landscape scale, C losses by wind erosion are not balanced, because sorting effects cause the transport of the finer and lighter fractions over long distances and its deposition at much larger areas than the eroded sites. Buschiazzo et al. (1991) found that the organic carbon contents of cultivated soils were significantly different from those of virgin soils and the A-horizon's thickness decreased on average by 7 cm. Carbon losses of 33–57% of the original soil were measured by Zach et al. (2006) within 12–18 years of continuous cultivation. The aggregate size distribution, as well as the particle composition, is also affected by land use. Aggregates are destroyed by tillage, and fine particles are blown out by wind erosion, which lowers the carbon and nitrogen contents in soils (Mendez et al., 2006). Concomitant reduction in aggregate stability might further lead to a positive feedback for wind erosion by increasing the soil erodibility.

Difficulties arise from balancing the contribution of wind erosion to other soil organic carbon (SOC)-reducing processes. The easiest way is to trap and analyse the eroded material directly at the field and to calculate the soil and SOC losses. The two most popular systems to sample aeolian sediment are the so-called Big Spring number height (BSNE; Fryrear, 1986) and the modified Wilson and Cook samplers (MWAC; Kuntze et al., 1990). Both samplers are highly effective to quantify wind erosion in the field (Mendez et al., 2011). For balancing soil losses or gains on a measuring field of about 1 ha, a certain number of samplers are needed. Sterk and Stein (1997) used 21 samplers on a field of 0.24 ha, Funk et al. (2004) 15 samplers on 2.25 ha and Visser et al. (2004) 17 samplers on a 1.6 ha field in regular and irregular grids. This sampler density is necessary to derive important sediment transportation parameters as a function of the distance, like the vertical flux density, the particle composition and the SOC content, and to calculate the final soil loss (Funk et al., 2004). Based on 15 years of wind erosion measurements on sandy soils (SOC 0.9%) in north-eastern Germany, Funk (2013, unpublished) calculated an average SOC loss of 117 kg ha^{-1}. This is far above the limit value of 75 kg ha^{-1} according to good management practice in the 'Cross Compliance' regulations of the EU (EU, 2009).

The nutrient content of the emitted dust is greater than the original soil, with significantly higher available phosphorus (P), nitrogen (N) and organic matter contents (Ramsperger et al., 1998; Hoffmann et al., 2008a). Buschiazzo et al. (2007) have measured enrichment ratios of 2–5 for N and of 1.5–8 for P for dust transported at heights of 0.13 and 1.5 m, respectively. Funk (1995) estimated an enrichment ratio of about 8 for the SOC content of eroded soil material from a sandy soil, measured at heights of 1 and 6 m. Measurements in Niger by Sterk et al. (1996) showed 17× higher contents of potassium (K), C, N and P in the dust, trapped at a height of 2 m, compared to the topsoil. The remaining sediments on the eroded field or at its leeward boundary have a distinctly lower content of nutrients and SOC, easily detectable by the much lighter colour.

Additional information on the effect of wind erosion on SOC stocks is available from indirect measurements, but methods balancing the SOC stocks and turnover rates are more complex and have to consider periods of more than 10 years (Kolbe, 2010). For example, Buschiazzo and Taylor (1993) compared, in the semi-arid Pampas of Argentina, heavy and lightly eroded soils after many years of contrasting management systems and concluded that eroded soils lost 25–35% of their original SOC contents. At a long-term monitoring site in northern Germany, SOC losses of 25 t ha^{-1} could be assigned to wind erosion, reducing the SOC content from 2.5% in 1990 to 1.8% in 2009 (−28%). Differences in the SOC stocks that could not be explained by the balance method of Kolbe (2010) were most likely caused by wind erosion.

Physics of the Processes

Generally, in comparison to other soil-degrading processes, the availability of reliable data of wind erosion effects on the SOC cycle is rare. But, nevertheless, some inferences can be drawn based on the current physical knowledge of aeolian transport. Soils in the broader sense are a mixture of grains of different sizes, shapes and properties. Natural soils have a large variety of grain sizes of mineral and organic components, which are in most cases aggregated, and the mobilization of the finest particles of a soil by wind erosion results mainly from the impacts of sand particles. As shown in Fig. 13.1, the wind transports particles in three main forms: creeping, saltation (mainly sand) and suspension (mainly dust). Creeping and saltation include relatively coarse and heavy particles with relevance for local processes, while suspension includes fine and light particles affecting regional and global processes (Shao, 2000). The way that wind erosion further affects the SOC losses has to be linked closely with the spatial distribution of the eroded material and its temporal variability.

not only the highest soil losses but also the highest SOC concentrations in the eroded material at the beginning of an erosion event (Fig. 13.2). The availability of erodible material has a limitation due to the enrichment of coarser fractions or stable aggregates from deeper and moister layers at the surface during erosion, lowering the total amounts and SOC concentrations of the dust following erosion events.

The increasing SOC content of the emitted dust can be explained with separation processes by density differences. Figure 13.3 shows dust on a filter sampled at a height of 3 m. All particles can be regarded to have the same aerodynamic characteristics, well-sorted mineral particles of a uniform size and much larger organic particles (upper left side), representing a larger volume/mass compared to the original composition. The aerodynamic separation therefore causes a relative increase of SOC in the emitted dust, which increases in height.

Temporal Variability

On agricultural land, the susceptibility to wind erosion is subject to permanent changes. The highest susceptibility occurs always after tillage operations that have loosened, crumbled and levelled the surface. That causes

Spatial Variability

Wind erosion processes are subject to a high spatial variability, at both the small and the large scale. At the field site, the limited transport capacity of the wind causes closely adjacent areas of soil losses and gains.

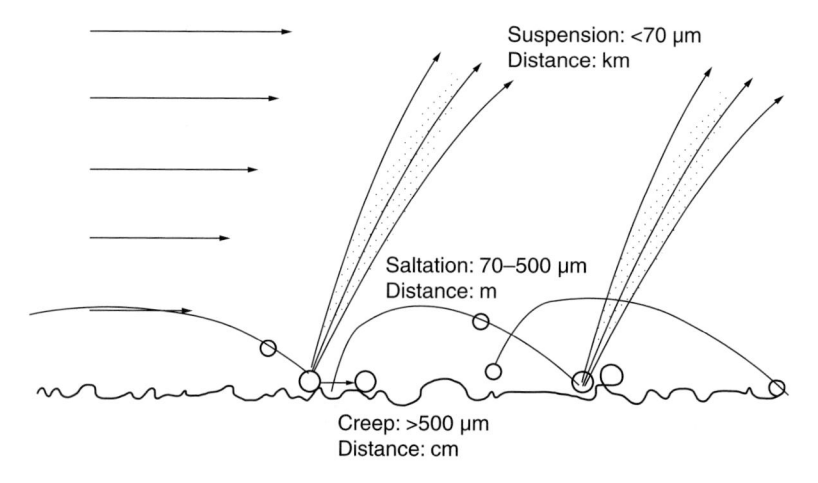

Fig. 13.1. Main transport modes of wind erosion depending on particle diameter and transported distances by wind.

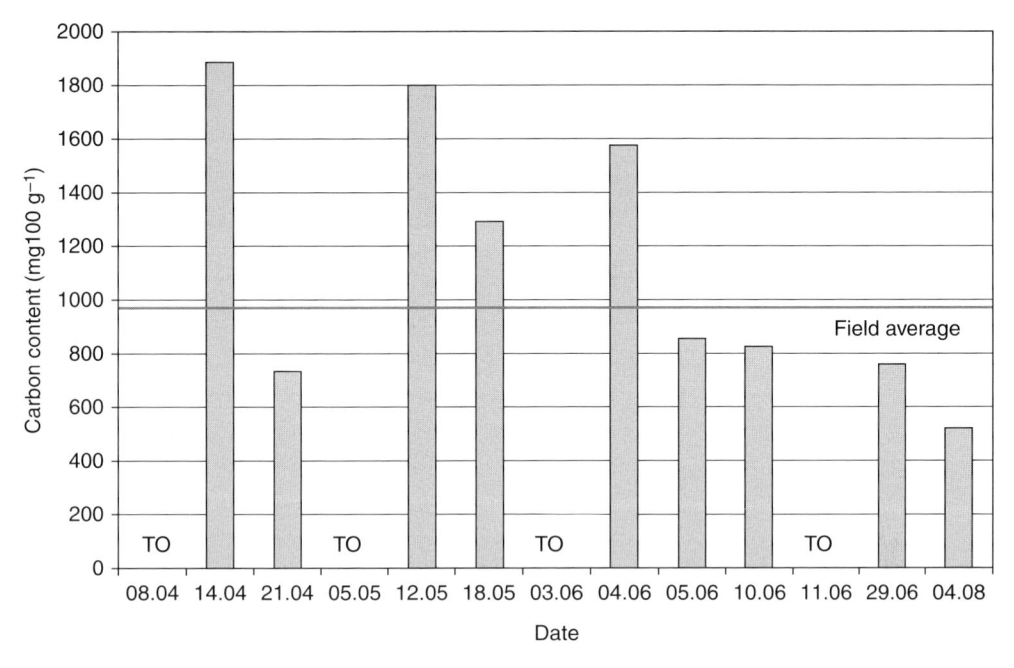

Fig. 13.2. Carbon content of wind-eroded material during 4 months of measurements; each TO marks a tillage operation with a chisel of about 10 cm depth; after that, the highest concentration of SOC was always measured.

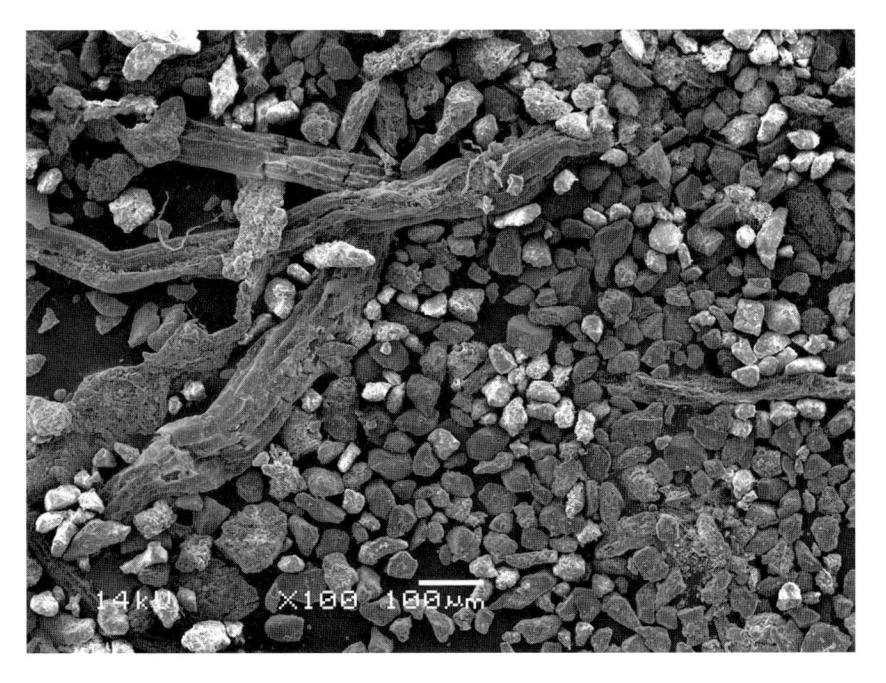

Fig. 13.3. Scanning electron microcopy capture of dust sampled at a height of 3 m; mineral particles are uniform in size of about 50 μm, organic particles with the same aerodynamic characteristics are much larger, resulting in a relative increase of SOC content.

On highly erodible sandy soils, the transport capacity can be reached after very short distances (10–20 m), and despite very high transport rates, the soil loss after the saturation is quite low (Fig. 13.4, left). Soil and SOC losses affect the first part of a field much more than the downstream parts, where SOC-poor material is already deposited. So, wind erosion increases soil heterogeneity, with consequences on soil physical properties and eventually crop production.

At the large scale, topographical effects are relevant if other surface characteristics are homogeneous and/or deposition processes are considered over a long period with a predominant transport direction (Goossens, 1997, 2006). Topography influences the wind velocity close to the ground, which may result in the typical deposition patterns on summits, windward- or leeward-orientated slopes related to the prevailing transport direction (Goossens and Offer, 1997; Zufall et al., 1999; Hoffmann et al., 2008b). The lowest deposition could always be found on summits and the upper part of windward slopes, the highest on leeward slopes and in depressions.

Conclusions

Wind erosion and dust depositions are important processes of soil formation, and soil destruction as well. In semi-arid regions that are influenced by carbon-poor dust depositions from deserts, the properties as a sink area have to be maintained to enable C enrichment by continued soil formation. On arable land, measures are necessary to prevent wind erosion and the preferred emission of carbon-rich dust. The particulate losses of SOC by wind erosion are not included in most balances, but the available data prove a considerable part of these losses from the soils affected.

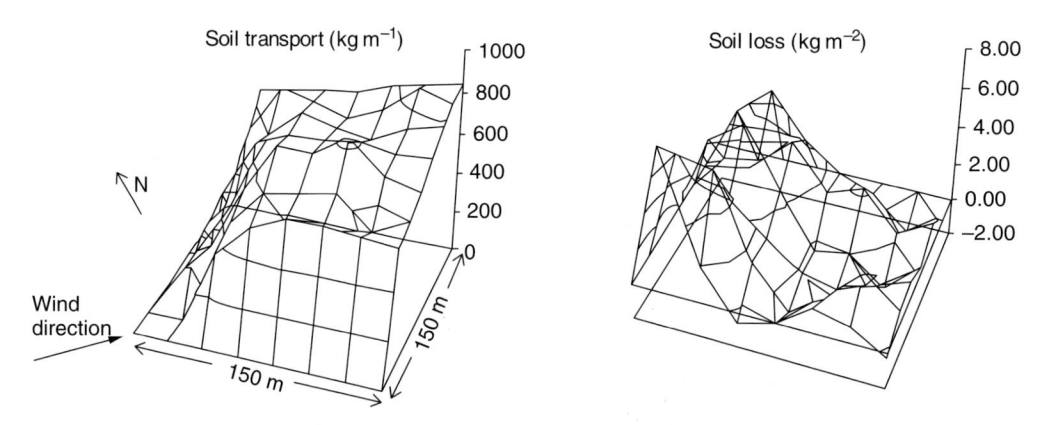

Fig. 13.4. Spatial distribution of wind erosion on a field measuring 150 × 150 m; left side: total transported soil per metre width; right side: appropriate soil loss per m^2 (negative values = deposition).

References

Buschiazzo, D.E. and Taylor, V. (1993) Efectos de la erosión eólica sobre algunas propiedades de suelos de la region Semiárida Pampeana Central. *Ciencia del Suelo* 10–11, 46–53.

Buschiazzo, D.E., Quiroga, A.R. and Stahr, K. (1991) Patterns of organic matter accumulation in soils of the semiarid Argentinean Pampas. *Zeitschrift für Pflanzenernährung und Bodenkunde* 154, 437–441.

Buschiazzo, D.E., Zobeck, T.M. and Abascal, S.A. (2007) Wind erosion quantity and quality of an Entic Haplustoll of the semi-arid pampas of Argentina. *Journal of Arid Environments* 69, 29–39.

Chepil, W.S. (1960) Conversion of relative field erodibility to annual soil losses by wind. *Soil Science Society of America Proceedings* 24, 143–145.

Chepil, W.S. and Woodruff, N.P. (1961) The physics of wind erosion and its control. *Advances in Agronomy* 15, 211–302.

EU (European Union) (2009) Council Regulation (EC) No 73/2009. Common rules for direct support schemes for farmers under the common agricultural policy and establishing certain support schemes for farmers (http://ec.europa.eu/agriculture/direct-support/legal-basis/index_en.htm, accessed 15 July 2014).

Frielinghaus, M. and Schmidt, R. (1993) Onsite and offsite damages by erosion in landscapes of East Germany. In: Wicherek, S. (ed.) *Farm Land Erosion. Template Plains Environment and Hills*. Elsevier Science Publishers BV, Amsterdam, pp. 47–49.

Fryrear, D.W. (1986) A field dust sampler. *Journal of Soil and Water Conservation* 41/42, 117–120.

Funk, R. (1995) *Quantifizierung der Winderosion auf einem Sandstandort unter besonderer Berücksichtigung der Vegetationswirkung*. ZALF-Bericht, No. 16, Müncheberg, Germany.

Funk, R., Skidmore, E.L. and Hagen, L.J. (2004) Comparison of wind erosion measurements in Germany with simulated soil losses by WEPS. *Environmental Modelling and Software* 19(2), 177–183.

Goossens, D. (1997) Long-term aeolian loess accumulation modelled in the wind tunnel: the Molenberg case (central loess belt, Belgium). *Zeitschrift für Geomorphologie* 41, 115–129.

Goossens, D. (2006) Aeolian deposition of dust over hills: the effect of dust grain size on the deposition pattern. *EarthSurf Process Landforms* 31, 762–776.

Goossens, D. and Offer, Z.Y. (1997) Aeolian dust erosion on different types of hills in a rocky desert: wind tunnel simulations and field measurements. *Journal of Arid Environments* 37, 209–229.

Goossens, D. and Riksen, M. (2004) Wind erosion and dust dynamics at the commencement of the 21th century. In: Goossens, D. and Riksen, M. (eds) *Wind Erosion and Dust Dynamics, Simulations, Modelling*. ESW Publications, Wageningen, the Netherlands, pp. 7–13.

Goudie, A. and Middleton, N. (2001) Saharan dust storms: nature and consequences. *Earth-Science Reviews* 56, 179–204.

Harrison, K.G. (2000) Role of increased marine silica input on paleo-pCO_2 levels. *Paleoceanography* 15, 292–298.

Herrmann, L. (1996) *Staubdepositionen auf Böden West-Afrikas*. Hohenheimer Bodenkundliche Hefte 36, Universität Hohenheim, Germany.

Hoffmann, C., Funk, R., Li, Y. and Sommer, M. (2008a) Effect of grazing on wind driven carbon and nitrogen ratios in the grasslands of Inner Mongolia. *Catena* 75, 182–190.

Hoffmann, C., Funk, R., Wieland, R., Li, Y. and Sommer, M. (2008b) Effects of grazing and topography on dust flux and deposition in the Xilingele grassland, Inner Mongolia. *Journal of Arid Environments* 72(5), 792–807.

Hoffmann, C., Funk, R., Reiche, M. and Li, Y. (2011) Assessment of extreme wind erosion in Inner Mongolia, China. *Aeolian Research* 3, 343–351.

Jahn, R. (1995) *Ausmaß äolischer Einträge in circumsaharischen Böden und ihre Auswirkungen auf Bodenentwicklung und Standorteigenschaften*. Hohenheimer Bodenkundliche Hefte 23, Universität Hohenheim, Germany.

Kaufman, Y.J., Koren, I., Remer, L.A., Tanré, D., Ginoux, P. and Fan, S. (2005) Dust transport and deposition observed from the Terra-Moderate Resolution Imaging Spectroradiometer (MODIS) spacecraft over the Atlantic Ocean. *Journal of Geophysical Research* 110, D10S12, doi:10.1029/2003JD004436.

Kolbe, H. (2010) Site-adjusted organic matter-balance method for use in arable farming systems. *Journal of Plant Nutrition and Soil Science* 173, 778–787.

Koren, I., Kaufman, Y., Washington, R., Todd, M., Rudich, Y.J., Vanderlei, M. and Rosenfeld, D. (2006) The Bodele depression: a single spot in the Sahara that provides most of the mineral dust to the Amazon forest. *Environmental Research Letters* 1(014005), 1–5, doi:10.1088/1748-9326/1/1/014005.

Kuntze, H., Schäfer, W. and Frielinghaus, M. (1990) Quantifizierung der Bodenerosion durch Wind. Final Report of the BMBF-Project. Federal Agency of Soil Research of Lower Saxony, Bremen, Germany.

Lal, R. (2003) Soil erosion and the global carbon budget. *Environment International* 29, 437–450.

Martin, J.H. and Fitzwater, S.E. (1988) Iron deficiency limits phytoplankton growth in the north-east Pacific subarctic. *Nature* 331, 341–343.

Martínez-Garzia, A., Rosell-Mele, A., Jaccard, S.L., Geibert, W., Sigman, D.M. and Haug, G.H. (2011) Southern ocean dust–climate coupling over the past four million years. *Nature* 476, 312–316.

Mendez, M.J., Funk, R. and Buschiazzo, D.E. (2011) Field wind erosion measurements with Big Spring Number Eight (BSNE) and Modified Wilson and Cook (MWAC) samplers. *Geomorphology* 129, 43–48.

Mendez, M.J., Oro, L.D., Panebianco, J.E., Colazo, J.C. and Buschiazzo, D.E. (2006) Organic carbon and nitrogen in soils of semiarid Argentina. *Journal of Soil and Water Conservation* 61, 230–235.

Ramsperger, B., Peinemann, N. and Stahr, K. (1998) Deposition rates and characteristics of aeolian dust in the semi-arid and sub-humid regions of the Argentinean Pampa. *Journal of Arid Environments* 39, 467–476.

Riksen, M., Brouwer, F., Spaan, W., Arrue, J.L. and Lopez, M.V. (2003) What to do about wind erosion. In: Warren, A. (ed.) *Wind Erosion on Agricultural Land in Europe*. European Commission, EUR 20370, Brussels, pp. 39–54.

Shao, Y. (2000) *Physics and Modelling of Wind Erosion*. Kluwer Academic Publisher, Dordrecht, the Netherlands.

Shao, Y., Wyrwoll, K.H., Chappell, A., Huang, J., Lin, Z., McTainsh, G.H., Mikami Masao, Tanaka, T.Y., Wang, X. and Yoon, S. (2011) Dust cycle: an emerging core in Earth system science. *Aeolian Research* 2, 181–204.

Stahr, K., Herrmann, L. and Jahn, R. (1996) Long distance dust transport in the Sudano-Sahelian Zone and the consequences for soil properties. In: Buerkert, B., Allison, B.E. and von Oppen, M. (eds) *Proceedings of the International Symposium 'Wind Erosion in West Africa: The Problem and Its Control'*. University of Hohenheim, Germany, 5–7 December 1994. Markgraf Verlag, Weikersheim, Germany, pp. 23–33.

Sterk, G. and Stein, A. (1997) Mapping wind-blown mass transport by modelling variability in space and time. *Soil Science Society of America Journal* 61, 232–239.

Sterk, G., Herrmann, L. and Bationo, A. (1996) Wind-blown nutrient transport and soil productivity changes in southwest Niger. *Land Degradation and Development* 7, 325–335.

Swap, R., Garstang, M., Greco, S., Talbot, R. and Kallberg, P. (1992) Saharan dust in the Amazon basin. *Tellus B* 44, 133–149.

van Lynden, G.W.J. (1995) European soil resources. Current status of soil degradation, causes impacts and need for action. *Nature and Environment* 71, Council of Europe, Strasbourg, France.

Visser, S.M., Sterk, G. and Snepvangers, J. (2004) Spatial variation in wind-blown sediment transport in geomorphic units in Burkina Faso using geostatistical mapping. *Geoderma* 120, 95–107.

Yan, H., Wang, S.Q., Wang, C.Y., Zhang, G.P. and Patels, N. (2005) Losses of soil organic carbon under wind erosion in China. *Global Change Biology* 11, 828–840.

Zach, A., Tiessen, H. and Noellemeyer, E. (2006) Carbon turnover and carbon-13 natural abundance under land use change in semiarid savanna soils of La Pampa, Argentina. *Soil Science Society of America Journal* 70, 1541–1546.

Zufall, M.J., Dai, W.P. and Davidson, C.I. (1999) Dry deposition of particles to wave surfaces: II. Wind tunnel experiments. *Atmospheric Environment* 33, 4283–4290.

14 Historical and Sociocultural Aspects of Soil Organic Matter and Soil Organic Carbon Benefits

Christian Feller*, Claude Compagnone, Frédéric Goulet and Annie Sigwalt

Abstract

In this chapter, soil organic matter (SOM) benefits will be considered from two different perspectives: (i) the scientific perception of 'SOM benefits' between the 18th century and today; and (ii) how various contemporary religions and societies, including farmers of Western cultures, perceive soil and SOM benefits.

Perceptions of the benefits of SOM (or humus) varied greatly in Western culture according to changes in historical scientific theories. Different periods can be considered. In the first part of 19th century, the 'theory of humus' by Thaer, dealing with a large popularity of SOM management for soil humus, was considered as the main nutrient for plants. In 1840, the new 'theory of the mineral nutrition of plants by Liebig demonstrated that humus was not the main source of nutrients for plants, with the consequence that there was no important need to manage organic fertilization: the popularity of humus was largely decreasing. With the emergence of environmental problems due to bad SOM management, the popularity of OM management is newly increasing. The best example is the concept that soil could be a large reservoir for atmospheric carbon sequestration, and this confers special attention to plant residue management.

In addition to scientific knowledge or economic considerations, the practices of farmers around the world are also highly dependent on their own culture (religious and cult aspects). To illustrate this point, this chapter gives as examples not only the beliefs of the Buryat (Lake Baikal) and the Dogon people (Mali) but also the opinions of three groups of French farmers towards soil and the benefits of SOM, dealing with completely different attitudes vis-à-vis the adoption of different agricultural alternatives.

Introduction

The concept of 'benefit' (arising from an idea, a discovery, an invention, a procedure, etc.) implicitly suggests its possible application. The sociological impacts of such applications may involve a whole society (e.g. benefits from agriculture), a particular sociocultural group (appropriation of new techniques by agronomists) or a single person.

Hence, this chapter will deal essentially with the perception of the 'soil organic matter (SOM) benefits' at different sociocultural levels. As carbon (C) is a part of SOM, SOM benefits include 'soil organic carbon (SOC) benefits', and these will be considered from three different perspectives, i.e.:

---

*E-mail: christian.feller@ird.fr

- the historical perception of 'SOM benefits' by the scientific community, mainly between the 18th century (at the time when a distinction was drawn between organic and mineral components) and today;
- some examples of how various societies and religions perceive both soil and SOM; and
- how farmers practising various types of agricultural interventions refer to SOM and to its attributed properties.

Historical Perception of 'Soil OM Benefits'[1]

One of the first questions to be clarified in relation to likely 'soil OM benefits' is the assessment of the origin of plant nutrients that have the greatest effects on soil fertility. This discussion has been the subject of heated scientific controversy since the end of the 18th century, up to 1840. The importance of SOM in sustaining other soil functions beyond the provision of nutritive elements to the plant was identified progressively thereafter. The role of SOM regarding the capacity of the world's soil resources to deliver agricultural and environmental services and to sustain human societies at both local (e.g. fertility maintenance) and global (e.g. mitigation of atmospheric C emissions) scales has only more recently been established (Dewar and Cannell, 1992).

Humus as a source of plant carbon nutrition (1770–1840)

From ancient times until the 18th century, many hypotheses were formulated about the source of nutrients for plants, i.e. air, water or soil. It was only after a distinction had been made between mineral and organic components in the second half of 18th century that 'humus' (or SOM) was considered a plant nutrient, in addition to other sources.

Hassenfratz (1792a,b) – without referring to experimental facts – argued that a fraction of humus in the form of soluble carbon was assimilated directly by plants (carbon heterotrophy) and was the almost unique source for plant carbon nutrition. At the same time as cited by Bourde (1967), several authors, for example Priestley (1777), Ingen-Housz (1779), Senebier (1782) and de Saussure (1804), partially refuted these assumptions by demonstrating both the gaseous origin of carbon and the role of light in photosynthesis. None the less, de Saussure (1804) still considered that a small part of plant material was possibly derived from soluble humus. Contradictory debates arose on the topic, but many agricultural scientists shared an intermediate point of view and assigned functions to both SOM and the air in plant nutrition. In particular, this was the case for the famous German agronomist, Albrecht Daniel Thaer (1752–1828), known for the 'theory of humus' developed in his seminal book, *Principles of Rational Agriculture* (1809).

Thaer's principles contained some unverified theoretical concepts on plant nutrition that served as a basis for the first rational and systematic approach to fertilization within the context of sustainable cropping practices (de Wit, 1974; Feller et al., 2003). Thaer's 'theory of humus' incorporated an analysis of the management of soil fertility as well as the concept of sustainability that deserve particular attention.

Thaer's book was published during the controversy about whether soil or atmosphere was the actual source of carbon used by plants. Thaer did not deny that atmospheric CO_2 could be a carbon source for plants, but since this source seemed unlimited, he considered soil humus and its management as the main limiting factor of plant carbon nutrition. According to Thaer: (i) most plant dry matter derives from the 'soil nutritive juices' contained in the portion of soil humus that is soluble in hot water; and (ii) plant demand for 'juices' is selective and varies according to the species cultivated. Management of soil fertility must therefore be based both on the management of soil humic balance and on crop succession.

Although erroneous, these assertions encompassed the whole soil–plant system and were used to support the first quantified, complex, but integrated system of analysis for the diagnosis and prediction of fertility (Feller et al., 2003). This was certainly the first example of real concern with farming sustainability, and what is more, it was based on organic practices. In terms of plant production, this period certainly marked the golden age of 'soil OM benefits' until the emergence of the mineral theory of plant nutrition.

The mineralist period (1840–1940)

Although Liebig takes many of his ideas from the work of Sprengel (1838; cf. van der Ploeg et al., 1999), his authoritative text, *Die organische Chemie in ihrer Anwendung auf Agrikultur und Physiologie* (1840), is often considered as the first reflection on the origin of plant dry matter from mineral compounds based on scientific experiments. It was the birth of the 'mineral theory' of plant nutrition and the beginning of the 'NPK era'. Carbon was considered to be derived from carbon dioxide, hydrogen from water and other nutrients from the soluble salts present in soil and water. Since Liebig's synthesis accounted rather satisfactorily for the fertilizing effect of mineral inputs, it provided the basis for modern agricultural sciences. Liebig promoted the use of mineral fertilizers to compensate for soil mineral depletion, and his work paved the way for recommendations on the massive use of chemical fertilizers in cropping patterns and the abandonment of organic or organomineral fertilization. None the less, Liebig, as 'one of the last "complete" men among the Great Europeans' (Hyams, 1976), was himself an advocate of mixed fertilization. As a result, until World War I, organic or organomineral inputs to soils were no longer considered to benefit plant production, even if scientific investigations had demonstrated the indirect importance of SOM for properties playing a role in soil fertility.

The potential of chemical fertilization to increase crop yield was widely recognized at the end of the 19th century, the industrial synthesis of nitrogen (N) and the processing of phosphorus (P) being mastered by the early 20th century. The mineralist approach reached its peak in the 30-year period following World War II, with the development of high-input, heavily subsidized agriculture in Europe and North America. During the first half of the 20th century, the generalization of agricultural practices without any return of SOM to the soil led to intense land degradation. The clearest examples are the spectacular water and wind erosions of the Dust Bowl and Black Sunday in the USA (14 April 1935).

From agronomy to ecology (1940–1992)

Societal and scientific doubts concerning the sustainability of intensive farming arose as early as in the 1930s, when a connection was suspected between the decline in soil fertility, the quality of the human diet and human health (Balfour, 1944). This promoted the definition of the 'healthy' function of SOM and the creation of the organic farming movement.

Steiner (1924) provided the incentive for the first organic farming movement, the so-called 'biodynamic agriculture'. Steiner's scientific bases and that of his followers (e.g. Pfeiffer, 1938) were shallow, and they referred to both holistic and cosmogonic concepts (i.e. interrelationships between the stars, soil and geochemical elements, plants, animals and humans) to propose a new kind of agriculture that excluded the use of any chemical input. The most influential and rational approaches to modern organic farming were those by Howard (1940, 1952), Balfour (1944) and Rodale (1945); for detailed reviews of the history of organic farming see Scofield (1986) and Lotter (2003). They shared one main objective, which was to improve soil, plant, animal and human health by biologically managing soil fertility. Two fundamental aspects of organic farming philosophy put SOM at the heart of cropping sustainability: the holistic paradigm and the Law of Return.

The philosophy of organic farming is fundamentally holistic and considers 'all life, all creation as being inextricably interrelated, such that something done or not done to one member, part or facet will have an effect on everything else' (Merrill, 1983). This is illustrated best by the biotic pyramid of Albrecht (1975; cited in Merrill, 1983). This pyramid is composed of several layers, with soil at the base and humans at its apex. According to this schematic, any degradation of soil quality threatens the whole of civilization, and even mankind itself; hence the need for careful soil husbandry.

The Law of Return stems from the concept of the 'living substances cycle', which originated in ancient times and reappeared in treatises on agriculture in the 16th and 17th centuries. A rupture of this principle is one of the factors that have been suggested in several historical research works to explain the collapse of civilizations, attributed to failures of their agriculture. The question is still at the heart of critical issues in terms of urban waste recycling (Magid et al., 2001). According to this principle, life can only be maintained provided living beings, or at least the residues of their activities and their bodies, are recycled at each step of the biotic pyramid. A crucial process is thus the establishment of organic flows to the soil to maintain its fertility. Since this return is SOM mediated, Balfour (1944), and above all, Rusch (1972), adopted a sceptical position towards what they termed Liebig's 'rather naive theory', and developed a partly rigorous (Balfour and Howard), partly ideological (Rusch) analysis of the agroecological role of SOM. Howard's opinion, as expressed in *The Soil and Health* (1952), matches Balfour's holism, his causal interpretation of the relationship between soil, plant, animal and human health being anchored in the idea of a cycle of the proteins and their quality between living beings. Even if his opinion was to some extent ideological, Howard (1940, 1952) wrote rigorous technical handbooks on the production of compost, which he termed 'manufactured humus'.

Over the past few years, there has been a renewed interest in the scientific community for holistic approaches to soil management, as evidenced by the proliferation of scientific meetings, research programmes (and consequently publications) on the topic of 'soil health'. An International Federation of Organic Agriculture Movements was even created in 1972; its first international conference was held in 1977.

Towards ecological agriculture (1992)

The era of an 'ecological agriculture' may be seen to result symbolically from the 1992 Earth Summit in Rio de Janeiro. 'Sustainable development' came to global attention with the publication of a report of the World Commission on Environment and Development (WCED, 1987) in which it was defined as 'development that meets the need of present generations without compromising the ability of future generations to meet their own need'. This clear congruence with environmental concerns about the impact of intensive, high-input agriculture, coupled with the failure to achieve persistent and consistent results in many parts of the world, notably in Africa, stimulated a substantial effort to find sustainable means of agricultural production (Conway and Barbier, 1990), focusing obviously on the use of renewable natural resources. As regards the management of soil fertility, this new approach generated substantial attention on the manner in which organic matter and biological processes were being dealt with (Scholes et al., 1994).

One of the key features in terms of sustainable practices consists of managing soil fertility through a combination of organic matter (crop residues, compost or manure) and mineral nutrient inputs (Pieri, 1992). This rediscovery of the benefits of the ancient concept of the integrated provision of nutrients became the mainstay of soil fertility at the turn of the 20th century (Mokwunye and Hammond, 1992; Palm et al., 1997). Maintaining and/or improving SOM status is central to its philosophy. The scientific challenge remains unchanged and aims at extending the ecological principles beyond the manipulation of the plants' components (with indirect influence on the soil biota, decomposition processes and humus dynamics as a consequence), to include a more direct manipulation of the soil biota (Swift, 1998). Some success was obtained with

N-fixing bacteria (Giller, 2001), which still have to be matched with other groups of microorganisms.

In managed sustainable agriculture, the modern concept of SOM as a dynamic, biologically regulated pool of energy, carbon and nutrients is congruent with the concept of fertility as defined by Balfour for organic agriculture: 'the capacity of soil to receive, store and transmit energy' (Balfour, 1976, in Merrill, 1983). This has the effect of enhancing the status of SOM management as an essential component of the design of new cropping schemes. In the Western world, for instance, research stations devoted to organic farming were created as early as 1939 (Haughley Research Trust in the UK by Balfour), 1945 (Rodale Institute in the USA), 1950 (Germany) or the mid-seventies (Switzerland and the Netherlands). At first, they were privately funded, but they are now financed, at least partly, by governments (Krell, 1997; Lotter, 2003).

The worldwide adoption of precision agriculture, along with the increased promotion of agroforestry systems (Steppler and Nair, 1987; Ewel, 1999), of composting, mulching and direct-sowing techniques (CIRAD, 1999) demonstrate the scientific value of integrated SOM management in terms of sustainable cropping schemes that were widely used before the mineral era but that had only been preserved in smallholders' agriculture (Altieri, 2002; Jackson, 2002; Tilman et al., 2002). On the other hand, and slow though it has been to grow, introducing ecological concepts into modern agriculture represents a return to principles that had been derived empirically from observation, many of which had been retained in traditional indigenous knowledge in various parts of the world. This development has been documented recently by McNeely and Scherr (2002), who celebrate the achievement of what they call 'ecological agriculture'.

Beyond its role in regulating global climate change (soil's capacity for C sequestration), and as a key compartment in nutrient cycles, SOM has also come to be valued for its influence on a wide range of so-called 'ecosystem services' in the sense of the Millennium Ecosystem Assessment (MEA, 2005); for example, food production, water management, climate protection/regulation, provision of energy, biodiversity preservation, transformation and dynamics, landscape management, cultural services, etc. Such services also include water availability and quality, soil erodibility and SOM as a source of energy for soil biota acting as biological control of plant, livestock and even human pathogens and diseases.

The numerous ecosystem services provided by soil through appropriate SOM management represent both 'soil OM benefits' and 'soil C benefits'. In that perspective, many new agricultural systems are nowadays being proposed under the generic terms of 'agroecology'; for example, 'organic farming', 'conservation agriculture' (with reduced or no tillage), 'agroforestry', etc. All these alternatives are based on the use of locally available natural resources (leguminous plants, rock phosphate, etc.) and the restitution of any plant or animal residue to the soils.

Religious and Sociocultural Perception of 'Soil OM Benefits'

The concept of SOM is linked strongly to scientific development in the western hemisphere, while the very notion of organic matter never really emerged among practitioners in other past or present civilizations. Nevertheless, the latter had, by and large, a holistic perception of the environment. The ideas of earth, soil and fertility were universal. One of the archetypal myths is that of the Goddess Earth, or Mother Goddess, found in almost all cultures as a symbol of fertility and fecundity, both of fields and of women. This deity was a woman: Demeter for the ancient Greeks, Ceres for the Romans.

The denomination of Mother Goddess has been adopted not only solely by priests and religions but also by scientists, as shown by Patzel (2010) in an article devoted to what belonged (in the psychoanalytical sense) to the 'unconscious' in the discourse of the pioneers in soil science at the end of the 19th century, like Fallou (1862) and Senft (1888), or Steiner (1924), Howard (1940,

1952), Balfour (1944) and Rusch (1972), among the supporters of organic farming.

Very different approaches to soil management can be observed, according to the cultural context, Western or not, and perceptions of the Mother Goddess. Two examples described by Lahmar and Ribaut (2001), among numerous others, will serve as illustrations:

- The religion of the Buryats, living around Lake Baikal, is halfway between Shamanism and Buddhism (according to Intigrinova, 2001). In Shamanism, the most important worship is that of the Mother Goddess, the body of the deity being the actual soil. It is hence strictly prohibited to offend or hurt the soil by working with sharp tools. Buddhism, however, allows some labouring of the soil, but only if absolutely necessary, and above all, in the process, one has to avoid killing any organism living within. This plainly shows the importance of beliefs and cults in managing fields.
- For the Dogons and other African cultures, Earth is a divinity and as such is sacred (according to Laleye, 2001). Earth is the Creator's spouse and is therefore at the origin of mankind. Among the marvellous reported legends, the creation of the world is called 'Amma's work'. Amma moves only arms and fingers to create the planets, whereas for Earth, he throws a roll of loam. Earth is created in the form of a female body, the sex of which is an ant-hill and the clitoris a termite mound. Amma tries to copulate with Earth, but the termite mound prevents the passage. To consummate the union, Amma needs to destroy the mound, excising Earth. The history goes on, and following an incestuous act, Earth is left sullied and soiled, a soiling that can only be mitigated through ploughing. As opposed to the foregoing example, this is how a cultural attitude may contribute to encourage physical labour.

These two examples show how important beliefs and cults may be with regard to the appropriation and adoption of particular agricultural practices like the minimal or no-tillage systems, a topic that is presently of concern not only for research scientists but also for practitioners in the USA, Brazil, Argentina and, more recently, in Europe. And how is the situation evolving nowadays from that point of view in developed or emerging countries where farmers are relatively close to the science and the extension services when it comes to accepting innovations?

The Place Held by OM in Farmers' Conceptions of Soil

Very little is known about farmer's perceptions of the role and importance of organic matter. Ongoing research by the authors with groups of farmers from the Vendée, western France, who use different production systems (conventional, no-till and organic farming) is trying to elucidate their opinions about SOM, and the preliminary results are presented here.

In the framework of sustainable agricultural development, the importance of carbon (C) for soil quality has now become central. It is crucial to grasp how farmers understand and characterize their soils and the function of C in order to enhance the implementation of farming practices that favour C sequestration in the soil and to understand how technical advisory services and research can reinforce these practices. The notion of understanding, as derived from cognitive sociology (Darré, 1985; Bouvier and Conein, 2007), refers to the idea that farmers have opinions with regard to their own practices, only partially explicit, which can be given verbal expression within the framework of reflective work (Giddens, 1984).

Research has considered farmers' conceptions, representations and perceptions of their soils over several years. Various studies have already been carried out in Europe (Ingram, 2008; Marie et al., 2008; Ingram et al., 2010; Coll et al., 2012; Goulet, 2013), as well as outside Europe (Messing and Fagerstrom, 2001; Desbiez et al., 2004;

Okoba and De Graaff, 2005; Moges and Holdenn, 2007). However, these studies most often identify how farmers see their soils in order to make agricultural advisory services more relevant, and thus more efficient in advising farmers on specific points such as soil erosion or loss of soil fertility. The present approach is original in that it does not consider a predefined problem but is concerned with farmers' overall understanding of their own soils in order to identify the role SOM or SOC represents for them. This comparative approach also makes it possible to highlight how these conceptions are specific to particular modes of agricultural production. It further allows us to observe how farmers develop their own way of thinking about soils, as well as to gauge the degree of the homogeneity of these ideas and approaches.

The results (Compagnone et al., forthcoming) showed that these farmers were only marginally concerned with C sequestration by soils as a form to mitigate climate change; rather, they expressed themselves in a practical way, useful to their peers in similar production schemes. In the same way, these farmers did not mention 'soil health'. Soil can be 'good' or 'bad', 'well' or 'badly maintained', or 'worked', but it is never described as healthy or sick, unlike the crops or residues that are on its surface.

Strictly speaking, different farmers have different conceptions of the importance of OM to the soil, depending on which production method they employ. Its interest for no-till farmers is in stimulating the biological activity of the soil and improving and regulating its structure. The aim of these farmers is, for instance, to increase the OM content of their soils from 4 to 5%. The problem is different for organic farmers: since they are not able to use synthetic fertilizers, OM inputs constitute their primary source of nutrient input. The major question for them relates to how essential nutrients are made available to the plant as effectively as possible. Lastly, for conventional agriculture farmers, this question of OM has no special meaning, but ranks on the same level as mineral fertilization or calcic amendments. Although these farmers recognize that the OM content of soils has to be increased so that the soil can continue 'to give', and questions of soil structure and life have been mentioned, they consider that OM does not determine soil productivity.

Finally, although all three groups mentioned the importance of the 'life of the soil', this concept was most present for farmers practicing no-till. For these farmers, the term 'biological' concerns the natural dynamism particular to the life of the soil, complex and multiple in its form, whereas farmers in organic farming use the same term to refer to a practice which prohibits the use of agrochemicals. The statement of the need to produce new technical references that are consistent with their production methods is associated with the assertion of the central role of OM in the soil by these two types of farmers. They generally criticize standard technical and scientific frameworks and their relevance to their particular production systems. However, this criticism leads no-till farmers to a production process of knowledge and expertise, while organic farmers call on agronomic research, at the same time underlining their difficulty in integrating and making use of technical and scientific language. The organic farmers in this study thus differed from farmers converted to organic practices at an earlier stage, who were reported as being strongly involved in the generation of local knowledge (Hellec and Blouet, 2012).

Acknowledgements

The authors thank Dr Jean-Paul Aeschlimann, Dr Jane C. Williams, Dr Jean-Luc Denizieux, Dr Susan Greenwood Etienne and Dr Eleanor Milne for their help with the English translation and editing. They are also very grateful to Prof Elke Noellemeyer, Prof Steve A. Banwart and Dr Eleanor Milne for their helpful comments and advice.

Note

[1] For more information on this section and detailed references see Feller *et al.* (2006, 2012) and Manlay *et al.* (2007).

References

Albrecht, W.A. (1975) *The Albrecht Papers*. ACRES USA, Raytown, Missouri.
Altieri, M.A. (2002) Agroecology: the science of natural resource management for poor farmers in marginal environments. *Agriculture, Ecosystems and Environment* 93(1–3), 1–24.
Balfour, E.B. (1944) *The Living Soil*. Faber and Faber, London.
Balfour, E.B. (1976) *The Living Soil and the Haughley Experiment*. Faber and Faber, London.
Bourde, A. (1967) *Agronomies et Agronomes en France au XVIIIème siècle (Agronomy and agronomists in France in the 18th Century)*. SEVPEN, Paris.
Bouvier, A. and Conein, B. (2007) *L'épistémologie Sociale. Une Théorie Sociale de la Connaissance*. Les éditions de l'EHESS, Paris.
CIRAD (1999) *Ecosystèmes Cultivés: l'approche Agro-Ecologique (The Agro-Ecological Approach to Cultivated Ecosystems)*. Agriculture et Développement, Cirad, France.
Coll, P., Le Velly, R., Le Cadre, E. and Villenave, C. (2012) La qualité des sols : associer perceptions et analyses des scientifiques et des viticulteurs. *Étude et Gestion des Sols* 19(2), 79–89.
Compagnone, C., Sigwalt, A. and Pribetich, J. (2013) Les sols dans la tête. Conceptions et modes de production des agriculteurs. *Étude et Gestion des Sols* 20(2), 81–95.
Conway, G.R. and Barbier, E.B. (1990) *After the Green Revolution: Sustainable Agriculture for Development*. Earthscan Publications, London.
Darré, J.P. (1985) La parole et la technique. *L'univers de Pensée des Eleveurs du Ternois*. L'Harmattan, Paris.
Desbiez, A., Matthews, R., Tripathi, B. and Ellis-Jones, J. (2004) Perceptions and assessment of soil fertility by farmers in the mid-hills of Nepal. *Agriculture, Ecosystems and Environment* 103, 191–206.
de Saussure, T. (1804) *Recherches Chimiques Sur la Végétation*. Nyon (Gauthiers-Villars, Paris 1957, Facsimile), Paris.
de Wit, C.T. (1974) Early theoretical concepts in soil fertility. *Netherlands Journal of Agricultural Science* 22, 319–324.
Dewar, R.C. and Cannell, M.G.R. (1992) Carbon sequestration in the trees, products and soils of forest plantations: an analysis using UK examples. *Tree Physiology* 11(1), 49–71.
Ewel, J.J. (1999) Natural systems as models for the design of sustainable systems of land use. *Agroforestry Systems* 45(1–3), 1–21.
Fallou, F.A. (1862) *Pedologie oder allgemeine und besondere Bodenkunde*. G. Schönfeld's Buchhandlung, Dresden, Germany.
Feller, C., Thuriès, L., Manlay, R., Robin, P. and Frossard, E. (2003) 'The principles of rational agriculture' by Albrecht Daniel Thaër (1752–1828). An approach to the sustainability of cropping systems at the beginning of the 19th century. *Journal of Plant Nutrition and Soil Science* 166, 687–698.
Feller, C., Manlay, R., Swift, M.J. and Bernoux, M. (2006) Functions, services and value of soil organic matter for human societies and the environment: a historical perspective. In: Frossard, E., Blum, W.E.H. and Warkentin, B.P. (eds) *Function of Soils for Human Societies and the Environment*. Special Publication 266. Geological Society of London, London, pp. 9–22.
Feller, C., Blanchart, E., Bernoux, M., Lal, R. and Manlay, R. (2012) Soil fertility concepts over the past two centuries. *Archives of Agronomy and Soil Science* 58(1), S3–S21.
Giddens, A. (1984) *The Constitution of Society: Outline of the Theory of Structuration*. Polity, Cambridge, UK.
Giller, K.E. (2001) *Nitrogen Fixation in Tropical Cropping Systems*, 2nd edn. CAB International, Wallingford, UK.
Goulet, F. (2013) Narratives of experience and production of knowledge within farmers' groups. *Journal of Rural Studies* 32, 439–447.
Hassenfratz, J.U. (1792a) Sur la nutrition des végétaux. *Annales de Chimie* 13, 178–192 and 318–380.
Hassenfratz, J.U. (1792b) Sur la nutrition des végétaux. *Annales de Chimie* 14, 55–64.
Hellec, F. and Blouet, A. (2012) Autonomie versus technicité. Deux conceptions de l'élevage laitier biologique dans l'est de la France. *Terrains et Travaux* 20, 157–172.

Howard, A. (1940) *An Agricultural Testament*. Oxford University Press, Oxford, UK.
Howard, A. (1952) *The Soil and Health: A Study of Organic Agriculture*, 2nd edn. The Devin-Adam Co, New York.
Hyams, E. (1976) *Soil and Civilization*. Harper Colophon Books, New York.
Ingen-Housz, J. (1779) *Experiments upon Vegetables: discovering their great power of purifying the common air in the sun-shine, and of injuring it in the shade and at night. To which is joined a new method of examining the accurate degree of salubrity of the atmosphere*. Elmsly and Payne, London.
Ingram, J. (2008) Agronomist-farmer knowledge encounters: an analysis of knowledge exchange in the context of best management practices in England. *Agriculture and Human Values* 25, 405–418.
Ingram, J., Fry, P. and Mathie, A. (2010) Revealing different understanding of soil held by scientists and farmers in the context of soil protection and management. *Land Use Policy* 27, 51–60.
Intigrinova, T. (2001) Les Bouriates et le sol: des rapports en évolution. In: Lahmar, R. and Ribaut, J.P. (eds), *Sols et Sociétés. Regards Pluriculturels*. Editions Charles Léopold Mayer, Paris, pp. 57–63.
Jackson, W. (2002) Natural systems agriculture: a truly radical alternative. *Agriculture, Ecosystems and Environment* 88, 111–117.
Krell, R. (ed.) (1997) Biological Farming Research in Europe. REU Technical Series No 54. *Proceedings of an an External Roundtable Held in Braunschweig*, 28 June 1997, Gemany. FAO, Rome.
Lahmar, R. and Ribaut, J.P. (eds) (2001) *Sols et Sociétés. Regards Pluriculturels*. Editions Charles Léopold Mayer, Paris.
Laleye, I.P. (2001) Divinité et sacralité de la terre chez les Négro-africains. In: Lahmar, R. and Ribaut, J.P. (eds) *Sols et Sociétés. Regards Pluriculturels*. Editions Charles Léopold Mayer, Paris, pp. 179–185.
Liebig, J. (1840) *Die organische Chemie in ihrer Anwendung auf Agrikultur und Physiologie*. Vieweg, Braunschweig, Germany.
Lotter, D.W. (2003) Organic agriculture. *Journal of Sustainable Agriculture* 21, 59–128.
McNeely, J.A. and Scherr, S.J. (2002) *Ecoagriculture: Strategies to Feed the World and Save Wild Biodiversity*. Island Press, Covelo, California.
Magid, J., Granstedt, A., Dýrmundsson, O., Kahiluoto, H. and Ruissen, T. (eds) (2001) Urban areas – rural areas and recycling – the organic way forward? Proceedings from NJF-Seminar No 327, Copenhagen, Denmark 20–21/08/2001. Danish Research Centre for Organic Farming, Copenhagen.
Manlay, R.J., Feller, C. and Swift, M.J. (2007) Historical evolution of soil organic matter concepts and their relationships with fertility and sustainability of cropping systems. *Agriculture, Ecosystems and Environment* 119, 217–233.
Marie, M., Le Gouée, P. and Bermond, M. (2008) De la terre au sol: des logiques de représentations individuelles aux pratiques agricoles. Étude de cas en Pays d'Auge. *Étude et Gestion des Sols* 15(1), 19–34.
MEA (Millenium Ecosystem Assesment) (2005) *Ecosystems and Human Well-being: Synthesis*. Island Press, Washington, DC.
Merrill, M.C. (1983) Eco-agriculture: a review of its history and philosophy. *Biological Agriculture and Horticulture* 1, 181–210.
Messing, I. and Fagerstrom, M.H.H. (2001) Using farmers' knowledge for defining criteria for land qualities in biophysical land evaluation. *Land Degradation and Development* 12, 541–553.
Moges, A. and Holdenn, N.M. (2007) Farmers' perceptions of soil erosion and soil fertility loss in Southern Ethiopa. *Land Degradation and Development* 18, 543–554.
Mokwunye, U. and Hammond, L. (1992) Myths and science of fertilizer use in the tropics. In: Lal, R. and Sanchez, P.A. (eds) *Myths and Science of Soils of the Tropics*. Soil Science Society of America, Madison, Wisconsin, pp. 121–134.
Okoba, B.O. and De Graaff, J. (2005) Farmers' knowledge and perceptions of soil erosion and conservation measure in the central highlands, Kenya. *Land Degradation and Development* 16, 475–487.
Palm, C.A., Myers, J.K. and Nandwa, S.M. (1997) Combined use of organic and inorganic nutrient sources for soil fertility maintenance and replenishment. In: Buresh, R.J., Sanchez, P.A. and Calhoun, F. (eds) *Replenishing Soil Fertility in Africa*. American Society of Agronomy and Soil Science Society of America, Madison, Wisconsin.
Patzel, N. (2010) The soil scientist's hidden beloved: archetypal images and emotions in the scientist's relationship with soil. Chapter 13. In: Edward, L. and Feller, C. (eds) *Soil and Culture*. Springer, Dordrecht, Heidelberg, London, New York, pp. 205–226.
Pfeiffer, E.E. (1938) *Die Fruchtbarkeit der Erde, ihre Erhaltung und Erneuerung; Das Biologisch-Dynamische Prinzip in der Natur*. Zbinden and Hügin, Basel.
Pieri, C. (1992) *Fertility of Soils: A Future for Farming in the West African Savannah*. Springer Series in Physical Environment. Springer-Verlag, Berlin.

Priestley, J. (1777) *Expériences et Observations sur Différentes Espèces D'air (Experiments and Observations on Different Kinds of Air)*. Translated from English to French by Gibbelin. Nyon, Paris.

Rodale, J.I. (1945) *Pay Dirt*. Rodale Press, Emmaus, Pennsylvania.

Rusch, H.P. (1972) *La Fécondité Du Sol* (original title: 'Bodenfruchtbarkeit', 1968, K.F. HaugVerlag; translated from German to French by C. Aubert). Le Courrier du Livre, Paris.

Scholes, M.C., Swift, J., Heal, O.W., Sanchez, P.A., Ingram, J.S.I., Dalal, R. and Woomer P.L. (1994) Soil fertility research in response to the demand for sustainability. In: Woomer, P. and Swift, M.J. (eds) *Biological Management of Tropical Soil Fertility*. Wiley, Chichester, UK.

Scofield, A.M. (1986) Organic farming. The origin of the name. *Biological Agriculture and Horticulture* 4, 1–5.

Senebier, J. (1782) *Mémoires physico-chimiques sur l'influence de la lumière solaire pour modifier les êtres des trois règnes de la nature et surtout ceux du règne vegetal*, Vol 3. B. Chirol, Geneva.

Senft, K. (1888) *Der Erdboden nach Entstehung, Eigenschaften und Verhalten zur Pflanzenwelt. Ein Lehrbuch für alle Freunde des Pflanzenreiches, namentlich für Forst- und Landwirthe*. HahnischeBuchhandlung, Hannover, Germany.

Sprengel, C. (1838) *Die Lehre von den Urbarmachungen and Grundverbesserungen*. Immanuel Müller Publishing Co, Leipzig, Germany.

Steiner, R. (1924) *Agriculture: A Course of Eight Lectures*. Biodynamic Agriculture Association, London.

Steppler, H.A. and Nair, P.K.R. (1987) *Agroforestry: A Decade of Development*. International Council for Research in Agroforestry (ICRAF), Nairobi.

Swift, M.J. (1998) Towards the second paradigm: integrated biological management of soil. In: Siqueira, J.O., Moreira, F.M.S., Lopes, A.S., Guilherme, L.G.R., Faquin, V., Furtini Neto, A.E. and Carvalho, J.G. (eds) *Soil Fertility, Soil Biology and Plant Nutrition Interrelationships*. SBCS/UFLA/DCS, Lavras, Brazil, pp. 11–24.

Thaer, A. (1809–1812) *Grundsätze der Rationnellen Landwirtschaft*, Vol 4. Realschulbuch, Berlin.

Tilman, D., Cassman, K.G., Matson, P.A., Naylor, R. and Polasky S. (2002) Agricultural sustainability and intensive production practices. *Nature* 418(6898), 671–677.

van der Ploeg, R.R., Bohm, W. and Kirkham, B. (1999) On the origin of the theory of mineral nutrition of plants and the Law of the Minimum. *Soil Science Society of America Journal* 63, 1055–1062.

WCED (World Commission on Environment and Development) (1987) *Our Common Future*. Oxford University Press, Oxford, UK.

15 The Economic Value of Soil Carbon

Unai Pascual*, Mette Termansen and David J. Abson

Abstract

Soil carbon has an economic value insofar as it is associated with an asset that provides benefits for humans. Demonstrating and measuring the economic value of soil carbon can provide valuable information for policy making. It makes explicit that soil carbon is not freely available. It signals the scarcity of the resource from a social point of view and also the extent to which investment in soil carbon should be prioritized relative to other investments. It also helps policy makers determine what type of economic instruments or incentives are necessary to align privately and socially optimal soil conservation decisions. In other words, the economic valuation of soil carbon provides information to help assess how efficiently a particular land management can reallocate the goods and services from soil to different and often competing uses. The chapter stresses the importance of the context of value formation by linking human preferences, knowledge and institutions to soil carbon. Then, by means of a conceptual framework, the chapter links the types of ecosystem services (supporting, regulating, provisioning and cultural services) derived from soil carbon to economic value components using the total economic value approach. Mapping the ecosystem service values of soil carbon needs to account for who appropriates the different values (private versus social values), whether the values are direct or indirect, so as to avoid double counting. Emphasis is given to the natural insurance value of soil carbon.

Introduction

Economics is about choice. Every economic decision is based on the weighing of values among different competing alternatives (Bingham et al., 1995). Economics can also be instrumental in understanding the benefits and costs to society of alternative decisions on soil carbon allocation. First of all, economics can provide land users and society at large with information about the level of scarcity of soil carbon. Economic valuation of soil carbon can be used to demonstrate that soil carbon is scarce, and thus not freely available, and that its deaccumulation or depreciation has societal costs. If these costs are not taken into account, the decisions of social actors, including individual land managers and policy makers, would be misguided. As a result, suboptimal decisions on soil carbon would be most likely, which would make society worse off compared to the outcome of better-informed decisions.

---

*E-mail: unai.pascual@bc3research.org

The foremost important aspect of understanding the economic value of soil carbon hinges on the idea that, with this information, society can prioritize to what extent we are willing to trade off soil carbon for other goods and services that enhance human well-being. This is related to the notion of opportunity costs at both the individual and the social level. For example, in order to accumulate soil carbon in agriculture, a farmer may have to give up a certain amount of something else; for example, biomass production for energy consumption. However, this may come as a benefit to society; for instance, in terms of reducing carbon emissions and increased carbon stocks (Lal, 2004, 2010; Sanderman and Baldock, 2010; see also Chapter 31, this volume).

At a social level, in certain circumstances, there may be opportunity costs of accumulating soil carbon, in terms of food security (Antle and Stoorvogel, 2008). Conversely, when society decides to convert natural ecosystems to agriculture, this results in the depletion of soil carbon (and terrestrial carbon more generally) by as much as 60% in temperate regions and at least 75% in the tropics (Lal, 2004). This comes at a cost to society; for instance, by decreasing soil quality, which translates into reduced land productivity, potential impacts on water quality and increased climate change.

Valuing soil carbon from an economic point of view, however, is not straightforward, since most of the benefits derived from its conservation and sustainable use are not reflected by markets. This is mostly because markets only reveal sufficient information about the scarcity of a small subset of goods and services from nature. Most natural resources, ecosystem processes and functional components are not incorporated in transactions as commodities or services, and their economic value is not reflected by market prices. One overarching reason is that soil carbon is a public good, which poses the fundamental limitations of a market system to provide comprehensive values involved in decision processes. In this respect, economics can be helpful to diagnose such a market failure problem and identify socially optimal investments in soil carbon.

The notion of optimality in the allocation of soil carbon is associated intrinsically with the notion of *externality* in economics. *Externalities* are non-intentional collateral effects that an action has on the welfare of third parties (originally non-target individuals). These could be either positive or negative. A positive externality or co-benefit, for instance, may exist if a programme to increase soil carbon sequestration by a target farmer group by reducing tillage intensity might also reduce soil erosion, and thus affect other farmers positively. Negative externalities can also occur, though. For example, the adoption of reduced tillage might be associated with increased pesticide use, which in turn might increase pesticide runoff and negatively impact the quality of water used by consumers at large.

The stock–flow natural capital framework (Costanza and Daly, 1992) much popularized by global initiatives such as the Millennium Ecosystem Assessment (MEA, 2005) can be used in this context. Here, soil carbon is the natural capital asset (Dominati *et al.*, 2010; Robinson *et al.*, 2013) and the flow of benefits it creates is the 'interest' on that capital that society receives. It is akin to the idea of private investors choosing a portfolio of capital to manage risky returns. Society needs to choose the amount of soil carbon that maintains future flows of benefits by accounting for the opportunity costs (trade-offs) of doing so. This economic approach to the value of soil carbon thus is anthropocentric, to the extent that the objective is an instrumental one; that is, fulfil needs or confer satisfaction to humans either directly or indirectly. In this way, valuing soil carbon is based on the intensity of changes in people's preferences with respect to land-use change and associated changes in soil carbon.

It follows that, instead of being an inherent property of soil carbon, its value is attributed by people through their willingness to pay for the goods and services that flow from it, which depends greatly on the socio-economic context in which valuation takes place – on human preferences, institutions, culture and so on (Pascual et al., 2010). Thus, it is important to be reminded that estimates of the economic value of soil carbon would reflect the current choice pattern of individuals, given multiple conditioning factors that determine and are determined by the distribution of income and wealth in society, the state of soil carbon, production technologies and expectations about the future. A change in any of these variables affects the estimated economic value of soil carbon.

The stock–flow framework can be useful to value natural assets such as soil carbon. The stock–flow model embedded in the ecosystem services framework lends itself directly to valuation and cost–benefit analysis (Wegner and Pascual, 2011). Here, we will employ this framework of analysis. However, it is also important to note the risks of such framework, as over-reliance on the ecological stock–flow framework obscures the depth and richness of natural sciences, and specifically ecological knowledge, as well as social sciences and economics in particular (Norgaard, 2010).

The Total Economic Value of Soil Carbon

Soil carbon shares the characteristics of a private good when managed by an individual agent, as it delivers benefits that can be appropriated privately, and of a public good when that is appropriated by society at large; for example, in terms of its role in stabilizing climate. It is important to consider this when demonstrating the value of soil carbon through the standard total economic value (TEV) framework. TEV is defined as the sum of the values of all benefit flows that soil carbon generates, both now and in the future – appropriately discounted (Pascual et al., 2010). It encompasses all components of preference satisfaction (or utility) derived from soil carbon using a common unit of account such as money or any other unit of measurement that allows comparisons of the benefits of the various goods and services derived from soil carbon. Normally, TEV is translated in monetary terms, since in most societies people are already familiar with money as a unit of account and because expressing relative preferences through TEV in monetary units can be useful to policy makers. There is a taxonomy and classification of the components of TEV for soil carbon and valuation tools that can be used to estimate such components (see Chapter 18, this volume). TEV is the aggregated value of soil carbon in a given state (Table 15.1).

The total output value of soil carbon

The components of the output value within the TEV framework can be mapped using the ecosystem services framework. Robinson et al. (2013) note that this framework is used more in the agricultural context than in soil science per se, probably because of the emphasis on final services like the production of food and fibre.

The main value of soil carbon stems from its intermediate role as a supporting service that underpins regulating services such as promoting resistance to the erosion of soils and regulating flooding by increasing infiltration, reducing runoff and slowing water movement from upland to lowland areas and reducing releases of agrochemicals, pathogens and contaminants to the environment by aiding their retention and decomposition (Victoria et al., 2012).

The insurance value of soil carbon

Information about the total mean output value can be enhanced by understanding the system's capacity to maintain the production of values in the face of variability and disturbance. This implies that besides an 'output' (yield) value, there also exists an 'insurance' value determined by risk preferences and the level of uncertainty (Turner et al., 2003).

The insurance value of soil carbon can be related to the soil's self-organizing capacity and resilience, sensu Holling (1973), which is related to the soil's capacity to absorb

Table 15.1. Output value (use and non-use values) components of TEV of soil carbon and associated significance.

Output value components of soil carbon	Value subtype	Meaning	Significance and examples
Use values	Direct use value	Results from direct human use of soil carbon	Medium – production of peat for energy production
	Indirect use value	Derived from the regulation services provided by soil carbon	High – sequestration of carbon, regulation of agricultural production
	Option value	Relates to the importance that people give to the future availability of soil carbon for personal benefit	Medium to high when information on indirect use values is available (e.g. climate regulation, etc.)
Non-use values	Bequest value	Value attached by individuals to the fact that in the future the individual will also have access to the benefits from soil carbon (intergenerational equity concerns)	High under sustainability policies
	Altruist value	Value attached by individuals to the fact that other people of the present generation have access to the benefits provided by soil carbon (intragenerational equity concerns)	Medium to high in non-individualistic societies
	Existence value	Value related to the satisfaction that individuals derive merely from the knowledge that soil carbon is protected	N/A

shocks and reorganize so as to maintain its essential structure and functions, in a given ecological state. Securing soil resilience thus is associated with maintaining a minimum amount of soil carbon that allows 'healthy' functioning of the soil. This minimum level of soil carbon can be approached through the concept of 'critical natural capital' (Brand, 2009). Thus, we contend that TEV of soil carbon is the sum of the total output value (its components being identified in Table 15.1) and the insurance value under risk and uncertainty. The insurance value exists in as much that people are risk averse and are willing to pay to secure a stable stream of the TEV in the face of variability and disturbance.

The literature on the natural insurance value of natural assets defines natural insurance in different ways. For example, Perrings (1995, p. 70) points out that the insurance value of an asset concerns its role in the amelioration of fundamental uncertainty associated with catastrophic events that cannot be reducible to the realm of actuarial risk, which is appropriate when we can allocate probabilities to the different states of nature. Similarly, Baumgärtner and Strunz (2014, p. 22) interpret the value of natural insurance 'as the value of one very specific function of resilience: to reduce an ecosystem user's income risk from using ecosystem services under uncertainty'. Following these ideas of the value of resilience-based natural insurance, we posit that the value of natural insurance (NIV) of soil carbon is composed of two related but differentiated value components: (i) the value of soil carbon for *self-protection* under uncertainty (the value of lowering the risk or probability of being hit by a disturbance, for example flooding or drought, causing a decrease in the total output value; and (ii) the value of *self-insurance* (SI) (lowering the size of the loss due to such an event occurring).

It can be argued that under risk and uncertainty to the variability and disturbance of soil carbon, both output and insurance values can be aggregated. The NIV of soil carbon measures how much a land manager would be willing to pay, say by investing in soil carbon-enhancing land management activities, in order to hedge against risk from exogenous shocks and stresses, assuming the land manager is risk averse. When the farmer is risk neutral, the NIV would be zero, and if, instead, he is risk loving, his willingness to pay (WTP) would be negative. In a context of uncertainty and/or ignorance, and assuming risk aversion towards shocks and stresses due, for example, to climate change, separating the total output value from the insurance value would imply that double counting would not occur.

Discounting the value of soil carbon

An additional complicating factor related to the economic regulation of soil carbon is the long timescale involved with depletion and accumulation in soils. This implies that the natural capital is slowly decreasing in value if its value is not fully recognized in decision making; further, it also implies that mismanagement of the soil carbon cannot be rectified in the short term. Decision making with long-term consequences is notoriously difficult. This has been studied and demonstrated in relation to many environmental concerns, most prominently in the climate change debate (Arrow et al., 1996). It is well known that balancing the concerns of present versus future generations has profound implications for investment in climate change mitigation, and equally for the case for investment in soil carbon.

In economics, this debate is portrayed technically through discounting as a means of capturing the role of considerations over equity and efficiency in natural resource use and investment (Dasgupta, 2006). Investment in soil carbon has all the same characteristics as the climate change mitigation debate, but investment in soil carbon is also an investment in the productive capacity of private assets, for example through the role in agricultural productivity. This raises a challenge as the beneficiaries from the investments become less clear and optimal regulation should account for the diversity with which different groups in society are impacted from alternative policy options. On the other hand, the fact that investment in soil carbon leads to tangible, local private benefits (although long term) also suggests that it might be possible to design policy initiatives which would reduce the gap between the current investment in soil carbon and the optimal level.

Capturing the Economic Value of Soil Carbon

Valuation of soil carbon goes beyond demonstrating the preference intensity of society with respect to possible trade-offs (externalities). Properly used, valuation can be used to develop policy instruments to appropriate and distribute or share the societal value of soil carbon. Appropriation refers to the process of capturing some or all of the demonstrated and measured values of soil carbon so as to provide economic incentives for its sustainable and optimal management. It aims at internalizing, through a regulatory framework based on either command-and-control or alternative market-based instruments, the demonstrated values of soil carbon so that those values affect investment decisions about soil carbon. Internalization is achieved either by setting quantity targets directly, creating markets when they are altogether missing, or correcting markets when they are 'incomplete'. In the benefit-sharing phase, appropriation mechanisms can be designed so that the demonstrated and captured soil carbon values are distributed based on social equity and fairness principles.

But, in addition to market failures, there are also two other failures that need to be solved for socially optimal soil carbon decisions: information and policy failures. An information failure for soil carbon might exist due to the difficulty of quantifying the level of goods and services that derive from it in terms that are comparable with services from human-made assets. Hence, soil carbon valuation can be used to unravel the complexities of social ecological relationships in relation to soil carbon, to make explicit how economic decisions would affect such values, and to express such value changes in some unifying metric such as money in order to be incorporated in policy decision processes (Mooney et al., 2005).

Policy failure stems from intentional policies that bias values associated with land-use decisions. For example, economic policy environments for agriculture and rural development in most countries are driven by heavy subsidized agriculture, which is in part responsible for a regulatory environment that creates conditions for farmers to mine soil carbon as an economic rational response (Antle et al., 2003). This is associated, for instance, with the creation of the economic conditions for certain land management not to be sufficiently attractive to farmers, such as soil restoration, woodland regeneration, no-till farming, using cover crops, nutrient management and manure and sludge application, among others (Lal, 2004). Since soil carbon accumulation is a function of past land-use and management practices, policy failures need to be addressed with a long-term perspective.

One way to solve the market failure of soil carbon is to create a market for it. This requires, one way or the other, commodifying of soil carbon. That is, bringing it to the market sphere by converting it into a good that has exchange value. There are various approaches that could be used to allow internalizing the economic value of soil carbon through a market-based approach. This can take different forms, and several examples have been reported in the literature for different natural assets (Ribaudo et al., 2010). One approach is based on directly creating a market for soil carbon credits that can be traded. Soil carbon credits for sequestration are a type of offset trade, and the stored carbon may be leased or sold (Jones, 2007).

A well-known example is an emissions trading market developed by a regulatory agency. This agency would stimulate demand for soil carbon by requiring regulated agents, farmers for example, to have enough allowances to meet a regulatory requirement and enforce property rights. Farmers would be allowed to trade the discharge allowances (water-quality trading and carbon cap-and-trade programmes). Members in this programme can meet their obligations by purchasing offsets from qualifying emissions reductions projects, including carbon sequestration in agricultural soils. However, the effectiveness of such a mechanism needs a solid regulatory framework to support purely voluntary exchanges. For instance, the Chicago Climate Exchange (CCX) in North America, which began in 2003 including over 350 companies and the public sector in the USA, Australia, Brazil, China, India and Costa Rica, traded 35 million tonnes (Mt) of CO_2 equivalents, from which sequestration in soils contributed to 46% of these. But for trading of soil carbon to be an efficient mechanism, the price of soil carbon needs to be attractive to farmers, and currently voluntary markets are undervaluing soil C. In fact, in 2010, the exchange of carbon credits ceased due to their low price (Holderieath et al., 2012).

A second possibility is using direct payment schemes to incentivize individuals to accumulate soil carbon. These are so-called soil carbon incentive payment (SCIP) schemes and are akin to payments for environmental services (Ferraro and Kiss, 2002; Kinzig et al., 2011). An emergent body of the theoretical literature (Antle and Diagana, 2003; Antle et al., 2003; Antle and Stoorvogel, 2008) has analysed the potential effects of soil carbon contracts with small farmers in developing countries to solve the reason for market failure. Such contracts imply direct payments to farmers and require them to adopt certain land-use or management practices. Such contracts would allow a flexible opt-in and opt-out approach by farmers without compromising the long-term permanence of soil carbon stocks, and thus without imposing excessive opportunity costs on farmers in the form of uncertainty about the long-term benefits of changed land management, commodity price uncertainty and other inherent risks associated with small, semi-subsistence agricultural systems.

Based on three case studies, they focused on regions with extensive soil degradation. In Machakos (Kenya), the carbon contracts required farmers to utilize minimum amounts of organic fertilizer (600 kg ha^{-1} per season) and mineral fertilizer (60 kg ha^{-1} year^{-1}). In Cajamarca (Peru), carbon sequestration was based on the adoption of terraces and agroforestry; in the Southern Groundnut Basin (Senegal), carbon contracts were based on the incorporation of crop residues and the application of mineral fertilizer.

A carbon payment scheme requires ways of measuring how a change in land management practices will change the pool of soil carbon, and how this would affect the various services associated with it, such as climate regulation. The carbon payments each season could be based either on the number of hectares on which these practices were adopted (a *per hectare* payment) or on the expected amount of carbon sequestered (a *per tonne* payment mechanism), the latter being economically more efficient because it better guarantees conditionality in areas of a high degree of spatial heterogeneity (Antle *et al*., 2003). That is, the contract pays farmers per unit of environmental service provided. Assuming that after the incentive scheme is in place farmers adopt those land-use and management practices that maximize economic returns (adjusted for risk if farmers are risk averse), Antle and Stoorvogel (2008) simulated the effects of such contracts on soil carbon stocks. The results showed that carbon contracts would increase the adoption of carbon sequestering practices significantly, with the adoption of soil carbon-enhancing agricultural practices being rather sensitive to the payment per expected soil carbon enhancement, or price of carbon, and other factors such as transaction costs. The simulation output also suggested that sufficiently high carbon payments could enhance the sustainability of small farmers' production systems, since increased use of organic fertilizers and the incorporation of crop residues and other organic matter was assumed to stabilize production by, for example, improving the soil's water-holding capacity. However, such enhanced sustainability would not occur in contexts associated with initially high rates of land degradation unless the payments were exceedingly high.

Similarly, The Australian Soil Carbon Accreditation Scheme (ASCAS) pays farmers for agricultural practices such as no-till farming in order to enhance soil carbon (Jones, 2007). The payments are determined by validated soil carbon increases above initial baseline levels determined for each so-called 'defined sequestration area' in cropping and grazing land, so vary according to rates of carbon capture. SCIPs are calculated at one-hundredth of the 100-year rate, which is usually the time frame for timber carbon contracts in Australia, at AUS\$25 t^{-1} CO$_2$ equivalent. For example, for a 0.15% increase in soil carbon, approximately 23.1 t C ha^{-1} (84.78 t CO$_2$ equivalent ha^{-1}), the price per hectare is AUS\$21.19 ha^{-1}. For an increase of 0.30% in a second year, the price paid in year 2 is AUS\$42.39 ha^{-1}, and for an increase of 0.45% in year 3, the price paid at that point would be AUS\$63.58 ha^{-1}, summing 127.16 ha^{-1} over 3 years (Jones, 2007).

Other examples around the world include the World Bank's BioCarbon Fund that provides payments to Kenyan smallholder farmers to improve their agricultural practices through the Kenya Agricultural Carbon Project, in order to increase both food security and soil carbon sequestration (World Bank, 2010). Similarly, in 2009, the Portuguese government introduced a soil carbon offset scheme based on dryland pasture improvement, involving around 400 farmers in about 42,000 ha by establishing perennial pasture (Watson, 2010).

Conclusion

First, this chapter outlines the main rationale for identifying the economic value of soil carbon. Second, the different components of values are described and their relationship to human well-being is classified. Finally, suggestions are made as to how potential soil carbon values can be realized.

Soil carbon relates in different ways to human well-being. Most notably, first, soil carbon has a significant role in the maintenance of regulating ecosystem services, such as climate regulation. Second, soil carbon also has a role in the production of private goods as agricultural food production. In a classification of the role of soil carbon in agricultural production, it is important to distinguish between impacts on mean yield and the impact on the variability of yield in response to environmental and climate variability. The different ways in which soil carbon is valued has important implications for policy design. The value of soil carbon in climate regulation relates to the benefits to current and future society as a whole. Therefore, policy schemes need to align benefits at temporal and spatial scales quite different to the scales that influence potential private opportunity costs in agriculture. Incentives to invest in soil carbon for robust future agricultural production have a very different distribution of costs and benefit. In this case, both benefit and costs accrue largely to the private agents, however much uncertainty and ignorance about costs and benefit may be involved. The chapter reviews current payment for ecosystem services schemes to give an overview of the experiences of constructing incentive schemes for the conservation and restoration of soil carbon, which is to capture some part of its wider societal value and distribute it to those who may forego the private benefits for doing so. Designing schemes to encourage private agents to take long-term perspectives on agricultural management and for the prudent handling of risks is important to encourage private self-protection. Payments for ecosystem services schemes are still in their infancy. Future research needs to acknowledge that, in most contexts, multiple services and multiple beneficiaries are in place. Therefore, targeting one form of value generation is likely to under-perform in terms of overall welfare generation, and equity issues are likely to become a major concern in future scheme developments. On the other hand, when valuing soil carbon, it is also important to recognize the intermediate role of carbon in the production of goods and services, leading to a risk of double counting and an overestimation of values.

References

Antle, J.M. and Diagana, B. (2003) Creating incentives for the adoption of sustainable agricultural practices in developing countries: the role of soil carbon sequestration. *American Journal of Agricultural Economics* 85, 1178–1184.

Antle, J.M. and Stoorvogel, J.J. (2008) Agricultural carbon sequestration, poverty and sustainability. *Environment and Development Economics* 13(3), 327–352.

Antle, J.M., Capalbo, S.M., Mooney, S., Elliott, E.T. and Paustian, K.H. (2003) Spatial heterogeneity, contract design, and the efficiency of carbon sequestration policies for agriculture. *Journal of Environmental Economics and Management* 46, 231–250.

Arrow, K.J., Cline, W., Maler, K.G., Munasinghe, M., Squitieri, R. and Stiglitz, J. (1996) Intertemporal equity, discounting and economic efficiency. In: Bruce, J., Lee, H. and Haites, E. (eds) *Climate Change 1995 – Economic and Social Dimensions of Climate Change*. Cambridge University Press, Cambridge, UK, pp. 125–144.

Baumgärtner, S. and Strunz, S. (2014) The economic insurance value of ecosystem resilience. *Ecological Economics* 101, 21–32.

Bingham, G., Bishop, R., Brody, M., Bromley, D., Clark, E.T., Cooper, W., Costanza, R., Hale, T., Hayden, G., Kellert, S., *et al*. (1995) Issues in ecosystem valuation: improving information for decision-making. *Ecological Economics* 14, 73–90.

Brand, F. (2009) Critical natural capital revisited: ecological resilience and sustainable development. *Landscape Ecology* 68, 605–612.

Costanza, R. and Daly, H.E. (1992) Natural capital and sustainable development. *Conservation Biology* 6, 37–46.

Dasgupta, P. (2006) *Comments on the Stern Review's Economics of Climate Change*. 11 November 2006 (revised: 12 December). University of Cambridge, Cambridge, UK.

Dominati, E., Patterson, M. and Mackay, A. (2010) A framework for classifying and quantifying the natural capital and ecosystem services of soils. *Ecological Economics* 69, 1858–1868.

Ferraro, P.J. and Kiss, A. (2002) Direct payments to conserve biodiversity. *Science* 298, 1718–1719.

Holderieath, J., Valdivia, C., Godsey, L. and Barbieri, C. (2012) The potential for carbon offset trading to provide added incentive to adopt silvopasture and alley cropping in Missouri. *Agroforestry Systems* 86(3), 345–353.

Holling, C.S. (1973) Resilience and stability of ecological systems. *Annual Review of Ecology and Systematics* 4, 1–23.

Jones, C. (2007) Australian soil carbon Accreditation Scheme (ASCAS). Mimeo. Managing the Carbon Cycle. Katanning Workshop, 21–22 March 2007 (http://www.amazingcarbon.com/What%20are%20Soil%20Credits.pdf, accessed 16 July 2014).

Kinzig, A.P., Perrings, C., Chapin, F.S., Polansky, S., Smith, V.K., Tilman, D. and Turner, B.L. (2011) Paying for ecosystem services – promise and peril. *Science* 334, 603–604.

Lal, R. (2004) Soil carbon sequestration impacts on global climate change and food security. *Science* 304, 1623–1627.

Lal, R. (2010) Beyond Copenhagen: mitigating climate change and achieving food security through soil carbon sequestration. *Food Security* 2, 169.

MEA (Millennium Ecosystem Assessment) (2005) *Ecosystems and Human Well-being: Synthesis*. Island Press, Washington, DC.

Mooney, H., Cooper, A. and Reid, W. (2005) Confronting the human dilemma: how can ecosystems provide sustainable services to benefit society? *Nature* 434, 561–562.

Norgaard, R.B. (2010) Ecosystem services: from eye-opening metaphor to complexity blinder. *Ecological Economics* 69(6), 1219–1227.

Pascual, U., Muradian, R., Brander, L., Gómez-Baggethun, E., Martín-López, M., Verman, M., Armsworth, P., Christie, M., Cornelissen, H., Eppink, F. et al. (2010) The economics of valuing ecosystem services and biodiversity. In: Kumar, P. (ed.) *The Economics of Ecosystems and Biodiversity Ecological and Economic Foundations*, Chapter 5. Earthscan, London, pp. 183–256.

Perrings, C. (1995) Biodiversity conservation as insurance. In: Swanson, T. (ed.) *The Economics and Ecology of Biodiversity Decline*. Cambridge University Press, Cambridge, UK, pp. 69–77.

Ribaudo, M., Greene, C., Hansen, L. and Hellerstein, D. (2010) Ecosystem services from agriculture: steps for expanding markets. *Ecological Economics* 69(11), 2085–2092.

Robinson, D.A., Hockley, N., Cooper, D., Emmett, B.A., Keith, A.M., Lebron, I., Reynolds, B., Tipping, E., Tye, A.M., Watts, C.W. et al. (2013) Natural capital and ecosystem services, developing an appropriate soils framework as a basis for valuation. *Soil Biology and Biochemistry* 57, 1023–1033.

Sanderman, J. and Baldock, J.A. (2010) Accounting for soil carbon sequestration in national inventories: a soil scientist's perspective. *Environmental Research Letters* 5, 034003.

Turner, K.T., Paavola, J., Cooper, P., Farber, S., Jessamy, V. and Georgiu, S. (2003) Valuing nature: lessons learned and future research directions. *Ecological Economics* 46, 493–510.

Victoria, R., Banwart, S.A., Black, H., Ingram, H., Joosten, H., Milne, E. and Noellemeyer, E. (2012) The benefits of soil carbon. In: *UNEP Year Book 2012: Emerging Issues in Our Global Environment*. UNEP, Nairobi, pp. 19–33 (http://www.unep.org/yearbook/2012/pdfs/UYB_2012_CH_2.pdf, accessed 16 July 2014).

Watson, L. (2010) Portugal gives green light to pasture carbon farming as a recognised offset. *Australian Farm Journal* January, 44–47.

Wegner, G. and Pascual, U. (2011) Cost–benefit analysis in the context of ecosystem services for human well-being: a multidisciplinary critique. *Global Environmental Change* 21(2), 492–504.

World Bank (2010) *Project Information Document: Kenya Agricultural Carbon Project* (http://www.worldbank.org/projects/P107798/kenya-agricultural-carbon-project?lang=en, accessed 16 July 2014).

16 Measuring and Monitoring Soil Carbon

Niels H. Batjes* and Bas van Wesemael

Abstract

Soils are the largest terrestrial reservoir of organic carbon, yet great uncertainty remains in estimates of soil organic carbon (SOC) at global, continental, regional and local scales. Compared with biomass carbon, changes in SOC associated with changes in land use and management, or climate change, must be monitored over longer periods. The changes are small relative to the very large stocks present in the soil, as is their inherent variability. This requires sensitive measurement techniques and due consideration for the minimum detectable difference (MDD). Relationships between environmental and management factors and SOC dynamics can be established using experimental field trials, chronosequence studies and monitoring networks. Soil monitoring networks (SMNs), for example, can provide information on direct changes of SOC stocks through repeated measurements at a given site, as well as data to parameterize and test biophysical models at plot scale. Further, they can provide a set of point observations that represent the (mapped) variation in climate/soil/land use and management at national scale, allowing for upscaling. SMNs must be designed to detect changes in soil properties over relevant spatial and temporal scales, with adequate precision and statistical power. Most SMNs, however, are in the planning or early stages of implementation; few networks are located in developing countries, where most deforestation and land-use change is occurring. Within these monitoring networks, sites may be organized according to different sampling schemes, for example regular grid, stratified approach or randomized; different statistical methods should be associated with each of these sampling designs. Overall, there is a need for globally consistent protocols and tools to measure, monitor and model SOC and greenhouse gas emission changes to allow funding agencies and other organizations to assess uniformly the possible effects of the impacts of land-use interventions, and the associated uncertainties, across the range of world climate, soils and land uses.

Introduction

Soils are needed for the production of food, fibre and timber, and they provide many ecosystem services, largely through the beneficiary functions of soil organic matter (Victoria et al., 2012). Although soils are the largest terrestrial reservoir of organic carbon, great uncertainty remains in the estimates of soil organic carbon (SOC) stocks and their changes at global, continental, regional and local scales (Kogel-Knabner et al., 2005; Milne et al., 2010; Paustian, 2012; Smith et al., 2012). Hence the need for

*E-mail: niels.batjes@wur.nl

a better understanding and quantification of the role of soils and the vegetation they support as natural regulators of greenhouse gas (GHG) emissions and climate change. Documenting such changes requires methodologies for monitoring, reporting and verification (MRV) of C stocks and GHG emissions that follow the principles of the United Nations Framework Convention on Climate Change (UNFCCC): transparency, consistency, comparability, completeness and accuracy (Bottcher et al., 2009). The relative importance of accurate measurement of the different carbon pools will vary across land-cover types (IPCC, 2006; Ravindranath and Ostwald, 2008; GOFC-GOLD, 2009; de Brogniez et al., 2011). In this respect, the IPCC National Accounting Guidelines recommend prioritizing the measurement and monitoring of the most significant carbon pools and those with the greatest potential to change (IPCC, 2006).

Soil properties vary in space, with depth and over time, with different measurement errors attached to them (Burrough, 1993). Soil monitoring involves the systematic measurement of soil properties to record their spatial and temporal changes. The assessment of SOC stock (changes), at a given site or for a given region, will require analyses of OC concentration, bulk density, content of coarse fragments (>2 mm) and soil depth. To be most effective, however, monitoring activities should consider a larger set of soil variables (e.g. Morvan et al., 2007), as well as information on the main biophysical (climate, terrain, soils and land use/vegetation or 'activity' data) and socio-economic drivers of change (e.g. Lambin, 1997; Ravindranath and Ostwald, 2008; de Brogniez et al., 2011). These variables may be recorded within the framework of a larger soil monitoring network (SMN) or a specific land use and management project. They are also needed for modelling possible changes in SOC stocks and GHG emissions using a range of empirical and process-based tools (see Chapter 17, this volume).

Several steps are needed in any measurement protocol to produce credible and transparent estimates of net changes in carbon stocks. These include: (i) designing a monitoring plan, including delineation/mapping of (project) boundaries, stratification of the project area, determining the type and number of sample plots, selection of C pools to be considered and appropriate procedures for their measurement, and frequency of monitoring; (ii) sampling procedures for field data collection for estimating above- and belowground carbon stocks (mainly, above- and belowground tree biomass; dead wood in standing and downed trees; non-tree vegetation; litter and duff; and SOC). In addition to this, consistent procedures are needed to analyse the results; for soil properties, these should include a quality assurance and quality control plan (IPCC, 2006; Ravindranath and Ostwald, 2008; de Brogniez et al., 2011). The focus of this chapter is on monitoring soil carbon changes.

Key issues on SOC monitoring and sampling/analysis approaches are discussed in this paper, based on a review of the recent literature. Drawing from this, main methodological requirements and general recommendations are proposed for 'good soil monitoring practice'.

Soil Monitoring Networks

Purpose

The primary reasons for collecting most forms of natural resources data are to reduce risks in decision making and to improve the understanding of biophysical processes (McKenzie et al., 2002; Ravindranath and Ostwald, 2008; de Brogniez et al., 2011). Soil monitoring may be defined as the systematic measurement of soil properties with a view to recording their temporal and spatial variations. SMNs generally comprise a set of sites/areas where changes in soil characteristics are documented through periodic assessment of an extended set of properties (Morvan et al., 2008; Arrouays et al., 2012). To be most effective, SMNs have to be integrated with other activities that generate essential knowledge for natural resource management, including land resource surveys, environmental/land-use history, field experiments and simulation modelling (Morvan et al., 2008; Ravindranath and Ostwald, 2008; Desaules et al., 2010).

The complementary benefits of mapping, monitoring and modelling are summarized in Table 16.1 (McKenzie *et al.*, 2002). Mapping, for example, is needed to stratify landscapes according to climate, soil type and land use to forecast possible GHG changes for the main source and sink categories (i.e. forest land, cropland, grassland, wetlands and other lands), as considered in the IPCC guidelines (IPCC, 2006; Milne *et al.*, 2012).

Methodological considerations

General

There are three main approaches (experimental field trials, chronosequence studies and monitoring networks) to determine the relationships between environmental and management factors and SOC dynamics (van Wesemael *et al.*, 2011). A wide range of studies have been published on statistical and other methods of environmental monitoring indicating that it is often cumbersome for a practitioner to extract the relevant information from these diverse materials. For example, the sampling area (block support), number and kind of (sub)samples, depth of sampling, range of parameters to be measured and analytical methods for their measurement often differ from one SMN to another (e.g. Morvan *et al.*, 2007; GOFC-GOLD, 2009; Batjes, 2011b; Lark, 2012). The objective of the SMN and its complexity will largely determine which statistical methodology should be used, as there are trade-offs between the different classes of design. A detailed discussion of the different statistical approaches needed to analyse random, grid-based or stratified monitoring schemes, however, is beyond the scope of this chapter (see, for example, De Gruijter *et al.*, 2006; Allen *et al.*, 2010).

Temporal and spatial scales versus detection limits

Some soil properties can be monitored easily; this includes those properties that vary least spatially, are responsive to management intervention and are the easiest to measure. Compared with biomass carbon, changes in SOC associated with changes in land use and management or climate change must be monitored over longer periods. The changes in SOC are small relative to the very large stocks present in the soil, as well as the inherent variability, which requires sensitive measurement techniques and due consideration for the minimum detectable

Table 16.1. Complementary benefits of mapping, monitoring and modelling. (From McKenzie *et al.*, 2002, with permission from CSIRO.)

Complementary relationship	Benefits
Mapping → Monitoring	• Spatial framework for selecting representative sites • System for spatial extrapolation of monitoring results • Broad assessment of resource condition
Monitoring → Mapping	• Quantifies and defines important resource variables for mapping • Provides temporal dimension to land suitability assessment (including risk assessments for recommended land management practices)
Modelling → Monitoring	• Determines whether trends in specific land attributes can be detected successfully with monitoring • Identifies key components of system behaviour that can be measured in a monitoring programme
Monitoring → Modelling	• Provides validation of model results • Provides input data for modelling
Modelling → Mapping	• Allows spatial and temporal prediction of landscape processes
Mapping → Modelling	• Provides input data for modelling • Provides spatial association of input variables

difference. Further, monitoring protocols must be designed to detect changes in soil properties over relevant spatial and temporal scales, with adequate precision and statistical power. For example, the effect of climate change on SOC is observed more readily at a broad scale than at a smaller spatial scale (Wang et al., 2010). Alternatively, the precision required for reporting possible avoided emissions, expressed as metric tonnes CO_2-equivalent, will be different (i.e. higher) for a strict 'compliance' (e.g. Kyoto type) than for a 'voluntary' (e.g. Chicago Carbon Exchange (CCX)) project on the carbon-offset market (Kollmuss et al., 2008). For the latter, C sequestration is generally seen as an additional benefit next to improving food security, resilience, biodiversity and human well-being/livelihood (Milne et al., 2010).

Sampling design and statistical methods

The main difficulty in assessing changes in SOC level at the field, landscape and regional scale is not linked to the accuracy of SOC analysis in the laboratory but to the design of an efficient sampling system (De Gruijter et al., 2006; Conant et al., 2010). Garten and Wullschleger (1999) were among the first to introduce the concept of minimum detectable difference (MDD). This concept is based on the Central Limit Theorem and provides the smallest difference between two sampling campaigns that can be detected, taking into account the number of samples and the variance of the SOC. According to Spencer et al. (2011), a national SMN should enable detection of broad-scale changes in SOC related to multiple drivers such as land use, management and climate change. Such an SMN should contribute to the understanding of the regional C cycle by detecting changes in soil fertility and fluxes of CO_2 between the soil and the atmosphere. Saby et al. (2008) applied the MDD concept to evaluate the statistical power of SMNs to detect a certain change in SOC stock. The parameters of the European SMNs such as sampling design, the coordinates of the monitoring sites and the number of sampling campaigns were obtained from questionnaires completed within the framework of the Environmental Assessment of Soil for Monitoring (ENVASO) project (Morvan et al., 2007), while the variance of the SOC within European countries was estimated based on the SOC map of Jones et al. (2005). Overall, the variance of SOC concentrations increases from Mediterranean countries to colder and more humid north European countries where extensive areas of organic soils occur. As the density of the monitoring sites varies widely between European countries, the MDD is generally high, ranging from 10 to 30 g C kg^{-1} in Nordic countries (Estonia, Luxembourg, Latvia, Northern Ireland and Finland) to less than 5 g C kg^{-1} in southern countries or countries with a dense SMN (England and Wales, Bulgaria, Italy, Greece, Hungary, Romania, France, Portugal, Spain, Belgium, Austria and Malta) (Saby et al., 2008). Given an expected mean rate of SOC change of 0.6% C year^{-1} (Bellamy et al., 2005), the relatively dense SMNs in most countries will detect such a change in close to 10 years. Similarly, Schrumpf et al. (2011) recommend continuous soil monitoring for SOC at time intervals of 10 years as a compromise between detectability of changes and temporal shifts in trends. It should be noted here, however, that this is longer than the duration of many land use and management projects that involve the measurement of SOC stock changes (i.e. for the baseline and at the end of the project). Alternatively, some countries use an interval of 5 years (see Table 16.2).

Spencer et al. (2011) evaluated the potential of an SMN on US agricultural lands to detect changes in SOC. They combined model-based changes in SOC, as produced by the Century model for the UNFCCC reporting, with the variance of the estimates of the model runs at a subset of the 186,000 National Resources Inventory sites (NRI). The results of these model runs indicate that the slope of the standard error against the sample size declines after model runs at 6000 NRI sites. They argue that an a priori knowledge of the variance per strata allows an optimal allocation of sampling units, resulting in an efficient use of resources for establishing an SMN.

Table 16.2. Summary of soil monitoring networks and sample design. (From van Wesemael et al., 2011.)

	Belgium	Germany	Mexico	New Zealand	Sweden	USA	Australia	Brazil	Canada	China
Objective	National SOC monitoring	National SOC monitoring	National SOC monitoring	National SOC monitoring	National SOC monitoring	National SOC monitoring	Baseline SOC for land use/soil combinations	SOC response to land use/ management change	SOC response to land use/ management change	Regional SOC monitoring
Region covered	Cropland and grassland in southern Belgium	Cropland and grazing land	Forest and non-forest land in particular pasture and shrubs	All regions and land uses	Cropland ~3 Mha	Cropland and grazing land in the USA	Cropland and grazing land[b]	Rodônia and Mato Grosso	No-till sites in Saskatchewan province	North-east (120 sites), north (241), east (356), south (119), north-west (148), south-west (97)
Starting date	National Soil Survey 1950–1970; resampled 2004–2007	November 2010	Started in 2003; each year 1/5 of the sites will be resampled	National soils database from 1938. Land use and carbon analysis system (LUCAS) started in 1996[a]	Full scale in 1995, some data from 1988	Planned	July 2009	2007	1997	78% started before 1985 and 87.5% continued until at least 1996
Site density (km² per site)	18 km²	64 km²	78 km²	202 km²	10 km²	Croplands: 438 km²; grazing lands: 1040 km²	Total number of locations is not known at this time.	N/A	N/A	N/A
Site selection	Stratified	Grid	Grid	Stratified	Grid	Stratified	Stratified	Stratified	Stratified	Stratified

Soil sampling										
Sub-samples	Composite	Composite	Composite	Single	Composite	Composite	Composite	Composite	Composite	Composite
Depth	0–30 cm and 0–100 cm	10 cm slices until 100 cm	0–30 cm and 30–60 cm	Variable, sampled by soil horizon; in 2009, 1235 samples to 30 cm	0–20 cm; 40–60 cm	10 cm slices until 75 cm	0–10, 10–20 and 20–30 cm	0–10; 10–20; 20–30; 30–40 cm;	0–10; 10–20; 20–30; 30–40 cm;	0–20 cm
Frequency	Once, but can be resampled in future if funding is available	Every 10 years	Every 5 years	Resampling is ongoing	1995 and 2005 done, will be repeated every 10 years	Each point will be sampled every 5–10 years	Once, but can be resampled in future if funding is available	Once	Sampled in 1997, 1999, 2005 and 2010	Annual sampling from 2010

[a]LUCAS network: http://www.mfe.govt.nz/issues/climate/lucas/.
[b]For consistency across the programme the Australian Land Use and Management (ALUM) Classification is used (http://www.daff.gov.au/abares/aclump/pages/land-use/alum-classification-version-7-may-2010/default.aspx).

Following up from the review of European SMNs (Morvan et al., 2007), the Joint Research Centre of the European Commission has launched an initiative to sample the topsoil at 22,000 points of the Land Use/Cover Area Survey (LUCAS project, see Montanarella et al., 2011). LUCAS is based on the visual assessment of land-use and land-cover parameters that are deemed relevant for agricultural policy. The soil sampling at the LUCAS points carried out in 2009 will produce the first coherent pan-European physical and chemical topsoil database. This topsoil survey resulted in a consistent spatial database of the soil cover across Europe, based on standard sampling and analytical procedures. A stratified sampling design was implemented to produce representative soil samples for major landforms and types of land cover of the participating countries.

Further initiatives for national SMNs were retrieved from a questionnaire that was sent to participants of a special session at the SOM 2009 conference in Colorado Springs, USA (Table 16.2). Some SMNs are designed to estimate country-specific land-use or management effects on SOC stocks, while others collect soil carbon and ancillary data to provide a nationally consistent assessment of SOC conditions across the major land-use types. The SMNs in Brazil and Canada use a paired-site approach in order to detect the SOC response to specific land management (no-till in Canada and conversion from forest to agriculture in Brazil). These are stratified by ecoregion or typical farming system. Three out of eight national inventories have a grid design, and the remainder are stratified according to land use, soil type and climate regions.

Spencer et al. (2011) give a comprehensive overview of the statistical considerations to be taken into account for an SMN. The three possibilities are simple random sampling, stratified sampling or grid-based sampling. Although random sampling is conceptually the simplest option, it can be difficult to implement and carries the risk of leaving aside some regions. Grid-based sampling is a practical and efficient technique and generally results in a better estimation of the variable of interest and an even distribution across the whole domain. A stratified approach allows for allocation of a greater number of samples in strata with a higher variability in SOC stocks. Generally, samples are allocated randomly within strata. Strata can be defined according to major climate, land-use and soil-type combinations. Such an approach has the advantage that SOC stock changes can be linked directly to the categories used for reporting by the UNFCCC (see Ravindranath and Ostwald, 2008). Using the results of Century model runs, Spencer et al. (2011) discuss the statistical power of different attribution approaches for sampling points to the strata.

It has been shown that marking individual sampling sites with either a physical marker (e.g. ball marker 3M, Austin, Texas) or precise positioning using a Differential Global Positioning System (DGPS) is the most efficient in order to decrease the MDD for eventual re-sampling in the future (Fig. 16.1). Generally, a composite sample, which involves taking subsamples and bulking them, is taken according to a fixed spatial pattern (Table 16.2). Studies of subsampling error of monitoring sites are crucial for interpretation of results and changes (Arrouays et al., 2012).

Error propagation

Error propagation methods have been used to estimate the contribution of the different variables required to calculate SOC stocks (C concentration, bulk density, stoniness and soil depth) (Goidts et al., 2009). Overall, the spatial variability of topsoil SOC stocks is larger in grasslands than in croplands (Schrumpf et al., 2011). Although the main source of uncertainty in the topsoil SOC stock varied according to scale, the variability of SOC concentration and of the stone content were the largest. When assessing SOC stock at the landscape scale, one should focus on the precision of SOC analyses from the laboratory, reducing the spatial variation of SOC, and use equivalent masses for SOC stock comparison (Goidts et al., 2009).

Sampling depth

Organic layers at the soil's surface need to be sampled separately from the underlying

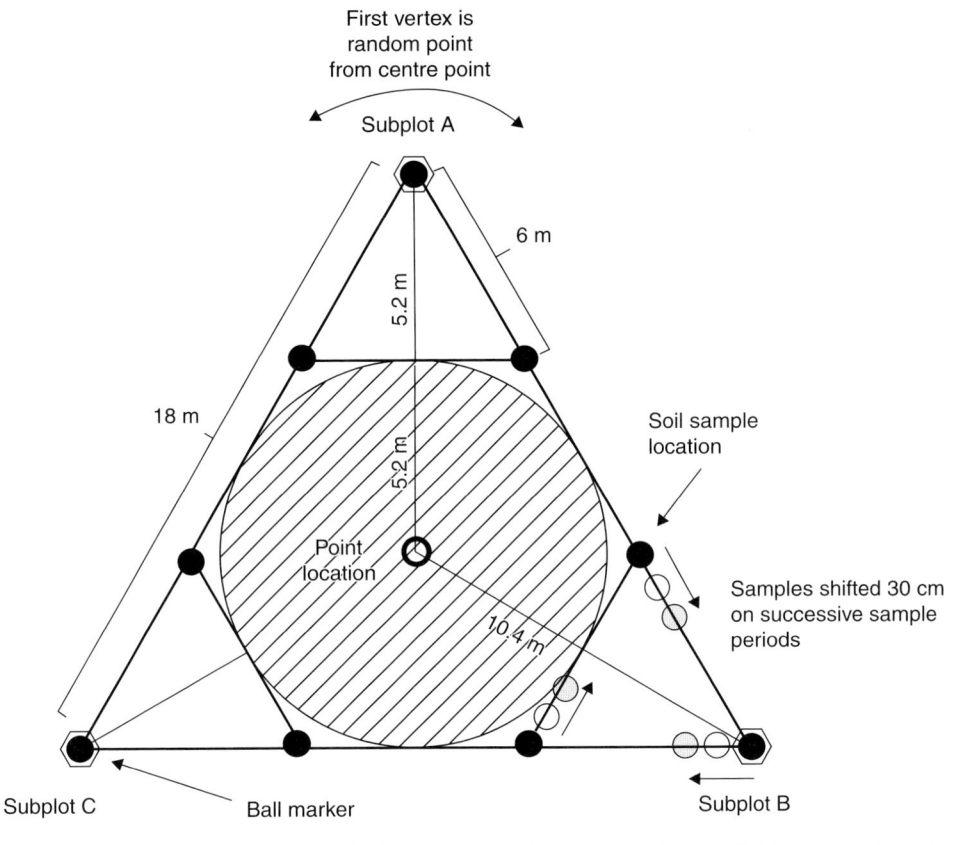

Fig. 16.1. Possible layout for a soil-monitoring site. (From Spencer *et al.*, 2011, with kind permission of Taylor & Francis.)

organo-mineral soil. A common practice is to sample mineral layers both by depth increment in the site as well as by pedogenetic horizons in a soil pit, located in close proximity to the monitoring site (Arrouays *et al.*, 2012) or in the centre of the monitoring plot (Fig. 16.2). Sampling to depths greater than 20–30 cm is recommended (Table 16.2, also: Batlle-Bayer *et al.*, 2010; Schrumpf *et al.*, 2011), as adoption of too shallow a sampling depth may preclude the conversion to SOC stock on an equivalent mass per area basis (Ellert and Bettany, 1995), which is the recommended practice (Wendt and Hauser, 2013). Consideration of a fixed depth for reporting possible SOC changes will further ignore the effect of land-use change induced modifications in bulk density (e.g. compaction). Estimation of SOC stocks based on a definite soil mass per area, however, accentuates differences in sites and soil cores as compared to the fixed depth method, due to the negative relationship between SOC concentration and bulk density (Schrumpf *et al.*, 2011). For pragmatic reasons, the IPCC method for national inventories considers SOC stock changes expressed on a volume basis to 30 cm depth (IPCC, 2006; Ravindranath and Ostwald, 2008), which is a simplification (Batjes, 2011a; Wendt and Hauser, 2013).

Analytical procedures

Most organizations implementing SMNs use long-established laboratory methods, and these can vary greatly between and within countries (Pleijsier, 1989; De Vos and Cools, 2011). This diversity can pose problems when using results from existing SMNs (Cools *et al.*, 2006;

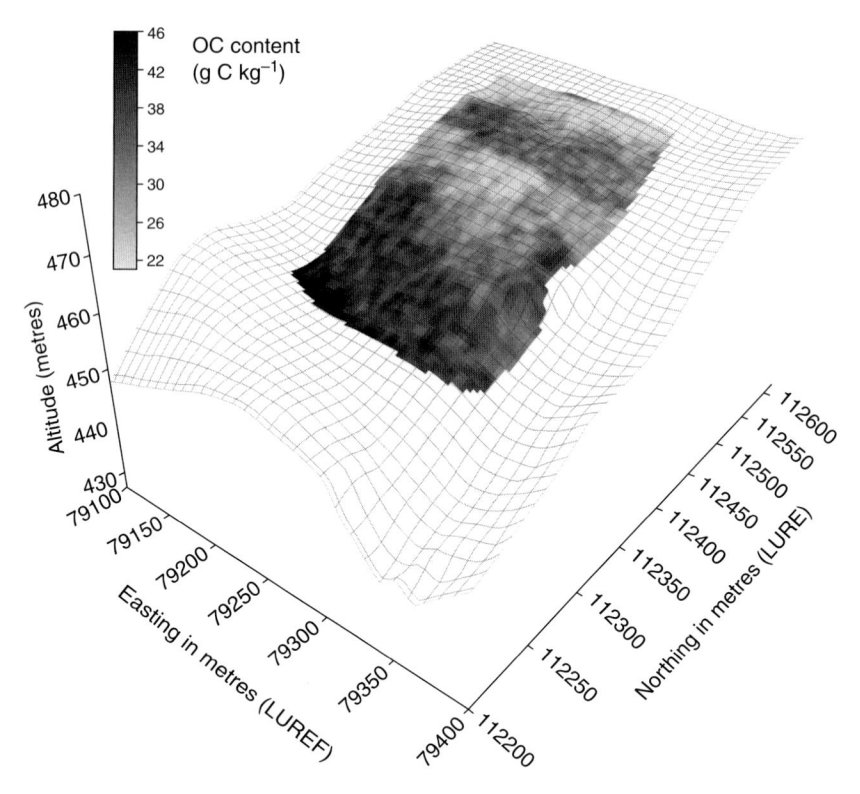

Fig. 16.2. Example of 'proximal sensing' approach for a field in Luxembourg. (From Stevens et al., 2010.)

Morvan et al., 2008). Adoption of a common standard (e.g. ISO TC 190 Soil Quality), however, would preclude comparison with earlier historic records, unless a set of reference samples could be analysed using both the new and 'old' methods. For example, dry combustion for SOC may be advocated rather than the more commonly used and cheaper Walkley and Black method, requiring the development of correction factors for incomplete oxidation (Lettens et al., 2007; Matus et al., 2009; Meersmans et al., 2009).

Harmonization may be defined as the minimization of systematic differences between different sources of environmental measures (Keune et al., 1991). There are some opportunities for harmonizing historic data obtained using different analytical methods, for example using regression analysis, but these are limited (Vogel, 1994; Cools et al., 2006; Panagos et al., 2013). Generally, a comprehensive comparison will require the establishment of benchmark sites devoted to harmonization and intercalibration of conventional soil analytical methods (Wagner et al., 2001; Kibblewhite et al., 2008; Morvan et al., 2008; Gardi et al., 2009). For each of these sites, key soil properties should be measured according to several (commonly used) analytical procedures, as well as the new standard reference method (e.g. ISO TC 190). In principle, this would allow comparison with results from earlier campaigns through the use of pedotransfer functions, which, inherently, will add uncertainty to the predicted values. There are no studies yet to assess how many calibration sites would be necessary for the world and how these sites may best be geo-located (Arrouays et al., 2012); interlaboratory comparisons remain critical here (van Reeuwijk, 1998; Cools et al., 2006).

Novel measurement techniques

Cost-effective techniques are needed to process the bulk of data derived from large soil-monitoring programmes. Visible and near infrared (Vis-NIR) and mid-infrared (MIR) reflectance spectroscopy have produced good results for the prediction of SOC content (McBratney et al., 2006; Shepherd and Walsh, 2007; Viscarra et al., 2010). Airborne imaging spectroscopy has been used for mapping topsoil properties (Ben-Dor et al., 2008). This technique provides good results for mapping the SOC content in the plough layer of bare soils (i.e. in seedbed condition).

The prediction of topsoil SOC content from remotely acquired spectral data is generally based on an empirical approach. Reference soil analyses of samples collected in the field are related to the spectral information through a multivariate calibration model used to predict the SOC values at locations for which there are no measured SOC data (Stevens et al., 2012). Several multivariate calibration models were developed to predict the SOC content in the plough layer of bare cropland fields in an airborne imaging spectroscopy scene of 420 km^2 in Luxembourg. For such large areas, the model performance depends strongly on the validation technique. The root mean square error of the most stringent validation procedure (excluding the fields used in the calibration) was equal to 4.7 g C kg^{-1}. Although this uncertainty is probably not good enough for the estimation of SOC stocks in individual fields, it can be used for regional mapping of SOC content and provides a unique insight into the spatial pattern of SOC content within fields (Fig. 16.2; Stevens et al., 2010). In all cases, however, conventionally measured SOC (dry combustion) in reference laboratories is necessary to calibrate the new techniques, and to build spectral libraries needed for the extension of spectral measurements in unsampled areas (Shepherd and Walsh, 2007; Bartholomeus et al., 2008; Terhoeven-Urselmans et al., 2010). Further, as techniques and standards for soil analyses are evolving continuously, it is good practice to preserve soil samples from SMNs so that they may be re-analysed in the future (McKenzie et al., 2002; Shepherd and Walsh, 2002; Arrouays et al., 2012).

Upscaling and Modelling

Typically, soil-monitoring activities encompass several decades of measurements (which implies long-term commitment from funding agencies and researchers). Appropriate data management tools are required to store the various data, check for errors and retrieve selected data for sharing and analysis (Cools et al., 2006; Lacarce et al., 2009; Batjes et al., 2013). The range of soil and ancillary data collated through SMNs and similar field sampling programmes should be stored in a (freely accessible) information system to support geostatistical analyses and modelling (see Chapter 17, this volume). At present, however, external access to SMN data is often restricted to the metadata (Morvan et al., 2008; Panagos et al., 2013), thereby greatly reducing their value to the scientific community and society.

In particular, there is a need for globally consistent protocols and tools to measure, monitor and model SOC changes and GHG emissions so that funding agencies and other organizations can assess uniformly the possible impacts of land-use interventions and climate change, as well as the associated uncertainties, across the range of world climates, soils and land uses. An example of such an integrated facility is the online Carbon Benefits Project tool developed for the Global Environmental Facility (GEF). It includes both empirical as well as process-based modelling approaches, which can be chosen based on the user requirements and available data through a user guidance module (Milne et al., 2010, 2012). Whether monitoring or modelling SOC dynamics/processes, the key issue is how to address complex issues of spatial and/or temporal variability at the scale of interest (Cerri et al., 2004; Maia et al., 2010; Smith et al., 2012). Opportunities include the use of ancillary data, scale-specific methods, development of spectral libraries, digital mapping of soil carbon and better integration of remote-sensing technologies into empirical and simulation SOC models (Grunwald et al., 2011; Croft et al., 2012; Minasny et al., 2013).

Conclusions

Based on the materials reviewed here, some basic methodological requirements and recommendations can be proposed for 'good SOC-monitoring practice' to support scientific and policy decisions. These include: (i) the provision of long-term continuity and consistency under changing boundary conditions, such as biophysical site conditions, climate change, methodologies, socio-economic setting and policy context; (ii) adoption of a scientifically and politically (e.g. for UNFCCC) appropriate spatial and temporal resolution for the measurements; (iii) ensuring continuous quality assurance at all stages of the measurement and monitoring process; (iv) measurement/observation and documentation of all potential drivers of SOC and GHG change; and (v) georeferenced samples, collated through SMNs, archived and the associated (harmonized) data made accessible through distributed databases to enhance the value of the collated data for multiple uses. In addition to this, SMNs should be included in a broader cross-method validation programme, ultimately to permit spatially and temporally validated comparisons both within and between countries.

The most common sampling design of SMNs aimed at monitoring regional/national SOC stocks is either stratified (according to soil/land use/climate) or grid based. Large countries with a low sampling density (<1 site per 100 km^2) generally prefer a stratified design so as to include all important units. The (expected) variability within these units should be determined to assess the optimal number of samples for each stratum. Such an approach will allow a statistical analysis of trends in SOC stocks for the soil/land use/climate units under consideration as an alternative or test for process-based models.

The establishment of SMNs poses various scientific, technical and operational challenges. The former are being addressed by various groups, such as the Soil Monitoring Working Group established in 2010 by the International Union of Soil Sciences (IUSS). From an operational point of view, to implement an integrated monitoring system it will be crucial to overcome initialization costs and unequal access to monitoring technologies. For the developing countries, this will require international cooperation, capacity building and technology transfer.

Acknowledgement

The authors gratefully acknowledge the invitation to the SCOPE Rapid Assessment Workshop held at the Joint Research Centre (JRC-EU), Ispra, Italy (18–22 March 2013).

References

Allen, D.E., Pringle, M.J., Page, K.L. and Dalal, R.C. (2010) A review of sampling designs for the measurement of soil organic carbon in Australian grazing lands. *The Rangeland Journal* 32, 227–246.

Arrouays, D., Marchant, B.P., Saby, N.P.A., Meersmans, J., Orton, T.G., Martin, M.P., Bellamy, P.H., Lark, R.M. and Kibblewhite, M. (2012) Generic issues on broad-scale soil monitoring schemes: a review. *Pedosphere* 22, 456–469.

Bartholomeus, H.M., Schaepman, M.E., Kooistra, L., Stevens, A., Hoogmoed, W.B. and Spaargaren, O.S.P. (2008) Spectral reflectance based indices for soil organic carbon quantification. *Geoderma* 145, 28–36.

Batjes, N.H. (2011a) Soil organic carbon stocks under native vegetation – revised estimates for use with the simple assessment option of the carbon benefits project system. *Agriculture, Ecosystems and Environment* 142, 365–373.

Batjes, N.H. (2011b) Research needs for monitoring, reporting and verifying soil carbon benefits in sustainable land management and GHG mitigation projects. In: De Brogniez, D., Mayaux, P. and Montanarella, L. (eds) *Monitoring, Reporting and Verification Systems for Carbon in Soils and Vegetation in African, Caribbean and Pacific Countries*. European Commission, Joint Research Center, Brussels, pp. 27–39.

Batjes, N.H., Reuter, H.I., Tempel, P., Leenaars, J.G.B., Hengl, T. and Bindraban, P.S. (2013) Development of global soil information facilities. *CODATA Data Science Journal* 12, 70–74 (https://www.jstage.jst.go.jp/article/dsj/12/0/12_WDS-009/_pdf, accessed 16 July 2014).

Batlle-Bayer, L., Batjes, N.H. and Bindraban, P.S. (2010) Changes in organic carbon stocks upon land use conversion in the Brazilian Cerrado: a review. *Agriculture, Ecosystems and Environment* 137, 47–58.

Ben-Dor, E., Taylor, R.G., Hill, J., Demattê, J.A.M., Whiting, M.L., Chabrillat, S., Sommer, S. and Donald, L.S. (2008) Imaging spectrometry for soil applications. *Advances in Agronomy* 97, 321–392.

Bellamy, P.H., Loveland, P.J., Bradley, R.I., Lark, R.M. and Kirk, G.J.D. (2005) Carbon losses from all soils across England and Wales 1978–2003. *Nature* 437, 245–248.

Bottcher, H., Eisbrenner, K., Fritz, S., Kindermann, G., Kraxner, F., McCallum, I. and Obersteiner, M. (2009) An assessment of monitoring requirements and costs of 'Reduced Emissions from Deforestation and Degradation'. *Carbon Balance and Management* 4, 7, doi:10.1186/1750-0680-4-7.

Burrough, P.A. (1993) Soil variability: a late 20th century view. *Soils and Fertilizers* 56, 529–562.

Cerri, C.E.P., Cerri, C.C., Paustian, K., Bernoux, M. and Mellilo, J.M. (2004) Combining soil C and N spatial variability and modeling approaches for measuring and monitoring soil carbon sequestration. *Environmental Management* 33, S274–S288, doi:10.1007/s00267-003-9137-y.

Conant, R.T., Ogle, S.M., Paul, E.A. and Paustian, K. (2010) Measuring and monitoring soil organic carbon stocks in agricultural lands for climate mitigation. *Frontiers in Ecology and the Environment* 9, 169–173.

Cools, N., Verschelde, P., Quataert, P., Mikkelsen, J. and De Vos, B. (2006) *Quality Assurance and Quality Control in Forest Soil Analysis: 4th FSCC Interlaboratory Comparison*. Forest Soil Coordinating Centre, Research Institute for Nature and Forest, Geraardsbegen (BE), 66 + annexes (on CD-Rom) (http://www.inbo.be/files/bibliotheek/52/167252.pdf, accessed 16 July 2014).

Croft, H., Kuhn, N.J. and Anderson, K. (2012) On the use of remote sensing techniques for monitoring spatio-temporal soil organic carbon dynamics in agricultural systems. *CATENA* 94, 64–74.

de Brogniez, D., Mayaux, P. and Montanarella, L. (2011) *Monitoring, Reporting and Verification Systems for Carbon in Soils and Vegetation in African, Caribbean and Pacific Countries*. Publications Office of the European Union, Luxembourg, 99 pp. (http://eusoils.jrc.ec.europa.eu/ESDB_Archive/eusoils_docs/other/EUR24932.pdf, accessed 16 July 2014).

De Gruijter, J.J., Brus, D.J., Bierkens, M.F.P. and Knottters, M. (eds) (2006) *Sampling for Natural Resource Monitoring*. Springer, Heidelberg, Germany, 338 pp.

De Vos, B. and Cools, N. (2011) *Second European Forest Soil Condition Report. Volume I: Results of the BioSoil Soil Survey*. Research Institute for Nature and Forest, Brussel, 369 pp. (http://www.inbo.be/docupload/4701.pdf, accessed 16 July 2014).

Desaules, A., Ammann, S. and Schwab, P. (2010) Advances in long-term soil-pollution monitoring of Switzerland. *Journal of Plant Nutrition and Soil Science* 173, 525–535.

Ellert, B.H. and Bettany, J.R. (1995) Calculation of organic matter and nutrients stored in soils under contrasting management regimes. *Canadian Journal of Soil Science* 75, 529–538.

Gardi, C., Montanarella, L., Arrouays, D., Bispo, A., Lemanceau, P., Jolivet, C., Mulder, C., Ranjard, L., Römbke, J., Rutgers, M. and Menta, C. (2009) Soil biodiversity monitoring in Europe: ongoing activities and challenges. *European Journal of Soil Science* 60, 807–819.

Garten, C.T. and Wullschleger, S.D. (1999) Soil carbon inventories under a bioenergy crop (Switchgrass): measurement limitations. *Journal of Environmental Quality* 28, 1359–1365.

GOFC-GOLD (Global Observation of Forest Cover and Land Dynamics) (2009) *A Sourcebook of Methods and Procedures for Monitoring and Reporting Anthropogenic Gas Emissions and Removals Caused by Deforestation, Gains and Losses of Carbon Stocks Remaining Forests, and Forestation*. GOFC-GOLD Project Office, Natural Resources Canada, Alberta, Canada, 197 pp. (http://www.gofc-gold.uni-jena.de/redd/sourcebook/Sourcebook_Version_Nov_2009_cop15-1.pdf, accessed 16 July 2014).

Goidts, E., Van Wesemael, B. and Crucifix, M. (2009) Magnitude and sources of uncertainties in soil organic carbon (SOC) stock assessments at various scales. *European Journal of Soil Science* 60, 723–739.

Grunwald, S., Thompson, J.A. and Boettinger, J.L. (2011) Digital soil mapping and modeling at continental scales: finding solutions for global issues. *Soil Science Society of America Journal* 75, 1201–1213.

IPCC (Intergovernmental Panel on Climate Change) (2006) *IPCC Guidelines for National Greenhouse Gas Inventories Volume 4: Agriculture, Forestry and other Land Use*. IPCC National Greenhouse Gas Inventories Programme, Hayama, Japan.

Jones, R.J.A., Hiederer, R., Rusco, E. and Montanarella, L. (2005) Estimating organic carbon in the soils of Europe for policy support. *European Journal of Soil Science* 56, 665–671.

Keune, H., Murray, A.B. and Benking, H. (1991) Harmonization of environmental measurement. *GeoJournal* 23, 249–255.

Kibblewhite, M.G., Jones, R.J.A., Montanarella, L., Baritz, R., Huber, S., Arrouays, D., Micheli, E. and Stephens, M. (2008) *Environmental Assessment of Soil for Monitoring. Volume VI: Soil Monitoring*

System for Europe. Office for Official Publications of the European Communities, Luxembourg, 72 pp. (http://eusoils.jrc.ec.europa.eu/projects/envasso/documents/ENV_Vol-VI_Final2_web.pdf, accessed 16 July 2014).

Kogel-Knabner, I., Lutzow, M.V., Guggenberger, G., Flessa, H., Marschner, B., Matzner, E. and Ekschmitt, K. (2005) Mechanisms and regulation of organic matter stabilisation in soils. *Geoderma* 128, 1–2.

Kollmuss, A., Lazarus, M., Lee, C. and Polycarp, C. (2008) *A Review of Offset Programs: Trading Systems, Funds, Protocols, Standards and Retailers (Version 1.1).* Stockholm Environment Institute Stockholm, 209 pp. (http://sei-us.org/Publications_PDF/SEI-ReviewOffsetPrograms1.1-08.pdf, accessed 17 July 2014).

Lacarce, E., Le Bas, C., Cousin, J.L., Pesty, B., Toutain, B., Houston Durrant, T. and Montanarella, L. (2009) Data management for monitoring forest soils in Europe for the Biosoil project. *Soil Use and Management* 25, 57–65.

Lambin, E.F. (1997) Modellling and monitoring land-cover change processes in tropical regions. *Progress in Physical Geography* 21, 375–393.

Lark, R.M. (2012) Some considerations on aggregate sample supports for soil inventory and monitoring. *European Journal of Soil Science* 63, 86–95.

Lettens, S., Vos, B.D., Quataert, P., van Wesemael, B., Muys, B. and van Orshoven, J. (2007) Variable carbon recovery of Walkley–Black analysis and implications for national soil organic carbon accounting. *European Journal of Soil Science* 58, 1244–1253.

McBratney, A.B., Minasny, B. and Viscarra Rossel, R. (2006) Spectral soil analysis and inference systems: a powerful combination for solving the soil data crisis. *Geoderma* 136, 272–278.

McKenzie, N., Henderson, B. and McDonald, W. (2002) *Monitoring Soil Change: Principles and Practices for Australian Conditions.* CSIRO Land and Water, CSIRO Mathematical and Information Sciences, National Land and Water Resources Audit, 112 pp. (http://www.clw.csiro.au/publications/technical2002/tr18-02.pdf, accessed 17 July 2014).

Maia, S.M.F., Ogle, S.M., Cerri, C.E.P. and Cerri, C.C. (2010) Soil organic carbon stock change due to land use activity along the agricultural frontier of the southwestern Amazon, Brazil, between 1970 and 2002. *Global Change Biology* 16, 2775–2788.

Matus, F.J., Escudey, M., Förster, J.E., Gutiérrez, M. and Chang, A.C. (2009) Is the Walkley–Black method suitable for organic carbon determination in chilean volcanic soils? *Communications in Soil Science and Plant Analysis* 40, 1862–1872.

Meersmans, J., van Wesemael, B. and van Molle, M. (2009) Determining soil organic carbon for agricultural soils: a comparison between the Walkley & Black and the dry combustion methods (north Belgium). *Soil Use and Management* 25, 346–353.

Milne, E., Sessay, M., Paustian, K., Easter, M., Batjes, N.H., Cerri, C.E.P., Kamoni, P., Gicheru, P., Oladipo, E.O., Minxia, M. *et al.* (2010) Towards a standardized system for the reporting of carbon benefits in sustainable land management projects. In: Abberton, M., Conant, R. and Batello, C. (eds) *Grassland Carbon Sequestration: Management, Policy and Economics (Proceedings of the Workshop on the Role of Grassland Carbon Sequestration in the Mitigation of Climate Change) (April 2009).* FAO, Rome, pp. 105–117 (http://www.fao.org/docrep/013/i1880e/i1880e00.htm, accessed 16 July 2014).

Milne, E., Neufeldt, H., Smalligan, M., Rosenstock, T., Bernoux, M., Bird, N., Casarim, F., Denef, K., Easter, M., Malin, D. *et al.* (2012) *Methods for the Quantification of Emissions at the Landscape Level for Developing Countries in Smallholder Contexts.* CGIAR Research Program on Climate Change, Agriculture and Food Security (CCAFS), Copenhagen, 59 pp. (http://www.focali.se/filer/CCAFS9_2012%20-3.pdf, accessed 16 July 2014).

Minasny, B., McBratney, A.B., Malone, B.P. and Wheeler, I. (2013) Digital mapping of soil carbon. *Advances in Agronomy* 118, 1–47.

Montanarella, L., Tóth, G. and Jones, A. (2011) Soil component in the 2009 LUCAS survey. In: Tóth, G. and Németh, T. (eds) *Land Quality and Land Use Information in the European Union.* Publication Office of the European Union, Luxembourg, pp. 209–219.

Morvan, X., Richer de Forges, A., Arrouays, D., Bas, C.L., Saby, N., Jones, R.J.A., Verheijen, F.G.A., Bellamy, P., Kibblewhite, M., Stephens, M. *et al.* (2007) Une analyse des stratégies d'échantillonnage des réseaux de surveillance de la qualité des sols en Europe. *Étude et Gestion des Sols* 14, 317–325.

Morvan, X., Saby, N.P.A., Arrouays, D., Le Bas, C., Jones, R.J.A., Verheijen, F.G.A., Bellamy, P.H., Stephens, M. and Kibblewhite, M.G. (2008) Soil monitoring in Europe: a review of existing systems and requirements for harmonisation. *Science of The Total Environment* 391, 1–12.

Panagos, P., Hiederer, R., van Liedekerke, M. and Bampa, F. (2013) Estimating soil organic carbon in Europe based on data collected through an European network. *Ecological Indicators* 24, 439–450.

Paustian, K. (2012) Agriculture, farmers and GHG mitigation: a new social network? *Carbon Management* 3, 253–257.

Pleijsier, K. (1989) Variability in soil data. In: Bouma, J. and Bregt, A.K. (eds) *Land Qualities in Space and Time*. PUDOC, Wageningen, the Netherlands, pp. 89–98.

Ravindranath, N.H. and Ostwald, M. (2008) *Carbon Inventory Methods – Handbook for Greenhouse Gas Inventory, Carbon Mitigation and Roundwood Production Projects*. Advances in Global Change Research, Volume 29. Springer, Heidelberg, Germany, 304 pp.

Saby, N.P.A., Bellamy, P.H., Morvan, X., Arrouays, D., Jones, R.J.A., Verheijen, F.G.A., Kibblewhite, M.G., Verdoodt, A., ÜVeges, J.B., Freudenschuß, A. and Simota, C. (2008) Will European soil-monitoring networks be able to detect changes in topsoil organic carbon content? *Global Change Biology* 14, 2432–2442.

Schrumpf, M., Schulze, E.D., Kaiser, K. and Schumacher, J. (2011) How accurately can soil organic carbon stocks and stock changes be quantified by soil inventories? *Biogeosciences* 8, 1193–1212.

Shepherd, K. and Walsh, M. (2002) Development of reflectance spectral libraries for characterization of soil properties. *Soil Science Society of America Journal* 66, 988–998.

Shepherd, K. and Walsh, M. (2007) Infrared spectroscopy—enabling an evidence-based diagnostic surveillance approach to agricultural and environmental management in developing countries. *Journal of Near Infrared Spectroscopy* 15, 1–20.

Smith, P., Davies, C.A., Ogle, S., Zanchi, G., Bellarby, J., Bird, N., Boddey, R.M., McNamara, N.P., Powlson, D., Cowie, A. *et al.* (2012) Towards an integrated global framework to assess the impacts of land use and management change on soil carbon: current capability and future vision. *Global Change Biology* 18, 2089–2101.

Spencer, S., Ogle, S.M., Breidt, F.J., Goebel, J.J. and Paustian, K. (2011) Designing a national soil carbon monitoring network to support climate change policy: a case example for US agricultural lands. *Greenhouse Gas Measurement and Management* 1, 167–178.

Stevens, A., Udelhoven, T., Denis, A., Tychon, B., Lioy, R., Hoffmann, L. and van Wesemael, B. (2010) Measuring soil organic carbon in croplands at regional scale using airborne imaging spectroscopy. *Geoderma* 158, 32–45.

Stevens, A., Miralles, I. and van Wesemael, B. (2012) Soil organic carbon predictions by airborne imaging spectroscopy: comparing cross-validation and validation. *Soil Science Society of America Journal* 76, 2174–2183.

Terhoeven-Urselmans, T., Shepherd, K.D., Chabrillat, S. and Ben-Dor, E. (2010) Application of a global soil spectral library as tool for soil quality assessment in sub-Saharan Africa. *A EUFAR Workshop on Quantitative Applications of Soil Spectroscopy (5–16 April 2010)*. Book of Abstracts, GFZ Potsdam, Germany, pp. 15.

van Reeuwijk, L.P. (1998) *Guidelines for Quality Management in Soil and Plant Laboratories*. FAO, Rome, 143 pp. (http://www.fao.org/docrep/W7295E/W7295E00.htm, accessed 16 July 2014).

van Wesemael, B., Paustian, K., Andrén, O., Cerri, C.E.P., Dodd, M., Etchevers, J., Goidts, E., Grace, P., Kätterer, T., McConkey, B.G. *et al.* (2011) How can soil monitoring networks be used to improve predictions of organic carbon pool dynamics and CO_2 fluxes in agricultural soils? *Plant and Soil* 338, 247–259.

Victoria, R., Banwart, S.A., Black, H., Ingram, H., Joosten, H., Milne, E. and Noellemeyer, E. (2012) The benefits of soil carbon. In: *UNEP Year Book 2012: Emerging Issues in Our Global Environment*. Nairobi, pp. 19–33 (http://www.unep.org/yearbook/2012/pdfs/UYB_2012_CH_2.pdf, accessed 16 July 2014).

Viscarra, R., Raphael, A., McBratney, A.B. and Minasny, B. (2010) *Proximal Soil Sensing*. Springer, Heidelberg.

Vogel, A.W. (1994) *Compatibility of Soil Analytical Data: Determinations of Cation Exchange Capacity, Organic Carbon, Soil Reaction, Bulk Density, and Volume Percent of Water at Selected pF Values by Different Methods*. Working Paper 94/07. ISRIC, Wageningen, the Netherlands (http://www.isric.org/isric/webdocs/Docs/ISRIC_Report_1994-07.pdf, accessed 16 July 2014).

Wagner, G., Desaules, A., Muntau, H., Theocharopoulos, S. and Quevauviller, P. (2001) Harmonisation and quality assurance in pre-analytical steps of soil contamination studies – conclusions and recommendations of the CEEM Soil project. *Science of the Total Environment* 264, 103–118.

Wang, D.D., Shi, X.Z., Wang, H.J., Weindorf, D.C., Yu, D.S., Sun, W.X., Ren, H.Y. and Zhao, Y.C. (2010) Scale effect of climate on soil organic carbon in the Uplands of Northeast China. *Journal of Soils and Sediments* 10, 1007–1017.

Wendt, J.W. and Hauser, S. (2013) An equivalent soil mass procedure for monitoring soil organic carbon in multiple soil layers. *European Journal of Soil Science* 64, 58–65.

17 Modelling Soil Carbon

Eleanor Milne* and Jo Smith

Abstract

Models that describe the dynamics of soil organic carbon (SOC) can be useful tools when estimating the impacts of land cover, land management and climate change on ecosystems. The development of SOC models started with single-compartment models that assumed a constant decomposition rate. As understanding of SOC dynamics improved, these were replaced by models with different compartments with varying decomposition rate constants. Models that deal with the decomposition of SOC as a continuum have been developed, but they require complex mathematics and are therefore less popular. Compartmentalized soil carbon models are at the core of complex models such as CENTURY and DNDC, which describe nutrient turnover in the entire ecosystem both above and below ground. The majority of such models have been developed using data from temperate ecosystems as studies on SOC stock change in temperate areas outnumber those from tropical areas. Application to tropical and subtropical areas therefore requires substantial parameterization and testing, and the availability of appropriate data sets remains a challenge.

The majority of SOC models were originally developed to model dynamics at the plot scale. Empirical methods to estimate SOC stock changes at the landscape to regional scale have been used successfully but are limited in their ability to predict future change. Linking dynamic SOC models to large-scale data sets using geographic information systems offers a way of estimating SOC change across large areas. However, account needs to be taken of the uncertainties associated with using multiple data sets with different scales. Uncertainties arise due to errors in process descriptions and imprecision in input data, and it is important that these are quantified against independent experimental measurements of the appropriate output values. As the scale of simulation increases, the precision of input data tends to decrease. It also becomes more difficult to quantify accurately the uncertainty in the simulations. These issues need to be addressed if the results of simulations are to be applied in a rigorous way.

Introduction

Soil organic matter (SOM, usually measured and expressed as content of soil organic carbon (SOC)) is a major determinant of the physical, chemical and biological properties of soils (Chapter 1, this volume). As such, gaining an understanding of how SOC levels will change under different environmental and anthropogenic circumstances is important. Models that simulate the turnover of organic carbon (C) in the soil are

*E-mail: eleanor.milne@colostate.edu

well established. They use mathematical relationships to describe the transformation of plant and animal material into organic matter under varying climatic and geographical circumstances, which in turn affect conditions in the soil (texture, moisture content and temperature). Application of SOC models is wide and varied. They can be used at the small scale to give farmers an indication of how much organic C to expect following different land management practices. They can be used to estimate soil C stocks across areas of similar land-use categories or land management activities where measured data are limited. In addition, they are also used in national greenhouse gas (GHG) inventories and to make projections about future soil C stocks under different land use and climate scenarios in climate change mitigation projects. In this chapter, we explore the history of SOC models and how they can be applied at different scales. We then consider the uncertainties associated with their use and research gaps in our current understanding.

Brief History of SOC Models

In 1941, Jenny developed a model for the turnover of nitrogen in soil (Jenny, 1941). This was then elaborated by Henin and Dupuis (1945), to describe the decomposition of organic C. They realized that during the initial decomposition of plant material, nitrogen was retained in the soil, whereas C was lost rapidly as CO_2, and therefore a modification of Jenny's model was needed (Jenkinson, 1990). By the end of the 1950s, Henin and Dupius's model had been fitted to varying data sets and modified further (Bartholomew and Kirkham, 1960; Nye and Greenland, 1960). In the model, the soil contains a pool of humus (fA_c), and yearly inputs of C from plant material are assumed to decompose, according to first-order kinetics, to add to this pool. The model can be described by Eqn 17.1, where C is the organic C content in the soil, t is the time in years and k is the fraction of this C decomposing each year (Jenkinson et al., 1994).

$$C = fA_c/k + (Co - fA_c/k)e^{-kt} \qquad (17.1)$$

Plant material decomposes rapidly in the first few months of addition to the soil, and more slowly from then on. This leads to the deduction that at least two different fractions of organic matter (and therefore organic C) exist, a labile fraction with a fast turnover time and a more resistant, stable fraction with a slow turnover time (Sauerbeck and Gonzalez, 1977). In addition, the C to nitrogen ratio in plant material varies, making it unlikely that all plant material has the same decomposition rate. Sugars and proteins break down first, followed by cellulose and finally lignin (van Keulen, 2001). These two facts spurred the development of models with multiple compartments, each with different decomposition rates.

Many multicompartmental models were developed to describe the decay of varying types of plant material in the first few days to months of addition to the soil (Hunt, 1977; Smith, 1979; McGill et al., 1981). Models of the longer-term turnover of organic C (10s to 100s of years) were fewer, as long-term data sets were needed to develop and parameterize models, and few of these existed at the time (a problem which persists today). The best-known early multicompartmental model is RothC, which was first devised in the 1970s (Jenkinson and Rayner, 1977) and is still in wide use today, be it in a modified form (Coleman and Jenkinson, 1996). RothC was developed using the long-term fertilizer experiments at Rothamsted Agricultural Research Station (UK), which provided unique data on the turnover of C in soils over a period of nearly 150 years (Jenkinson and Rayner, 1977). A challenge encountered in the development of RothC, universal to all models of the longer-term turnover of SOC, was finding data sets to test the model which were independent of those that had been used to develop it. The long-term experiments at Rothamsted provided a means of doing this. In essence, RothC has two compartments for incoming plant C, decomposable material (DPM) and resistant material (RPM). These then decay at different rates, according to first-order processes, to two more compartments; biomass or humified organic matter with CO_2 evolved during the process. In addition, there is a small pool of inert organic matter with a turnover time too slow to detect. Soil temperature and moisture modify the rates of decomposition. The partition of C to either CO_2, biomass or stabilized humus material is adjusted for soil texture (Coleman and Jenkinson, 1996).

This multicompartmental approach has been used in the majority of models that deal with the long-term turnover of C in soils, the main difference between models being the number of pools and associated rate constants. Short-term turnover rates can be determined by decomposition studies. For longer-term turnover, techniques have been employed that use natural labelling of SOM using stable ^{13}C tracers, labelling of SOM with 'bomb' ^{14}C and/or ^{14}C-dating techniques (von Lutzow et al., 2007). In reality, these pools are hypothetical, with decay being a continuous process. However, representing this in a model involves developing different equations for numerous individual types of plant material (Agren and Bossata, 1987) and has therefore proved less popular.

RothC requires the user to provide data on plant material inputs to the soil. In the 1980s and 1990s, several models were developed that linked plant growth models to SOM models, the purpose being to describe nutrient turnover both above and below ground. One example is CENTURY, developed using data from long-term experiments in the American Great Plains. CENTURY has a three-pool SOM model, two litter pools (often counted as SOM in conventional measurements) – metabolic and structural, which are roughly analogous to the DPM and RPM in RothC, three plant productivity submodels and a water budget model (Parton et al., 1988). Throughout the 1980s and 1990s, many models were developed using different data sets in different climate and soil regions. In 1996 (updated in 2001), model descriptions and metadata were brought together in a report by the SOM network (Smith et al., 1996a). In addition, Smith et al. (1997) carried out a comparison of nine of the most frequently used SOM models by testing them using data sets from a range of land uses. They found six of the models (RothC, CANDY, DNDC, CENTURY, DAISY and NCSOIL) performed better than the rest across a range of conditions.

The majority of SOC models were developed primarily to describe the dynamics of C in aerobic conditions in mineral soils. Fewer have been developed specifically for flooded soils and peatlands. By the 2000s, the need for models that could describe C dynamics in the large global C stock represented by peatlands had been recognized. The ECOSSE model was developed in response to this need, taking a multicompartmental approach in the same way as RothC (on which it is partially based), but including calculations that account for the domination of CH_4 evolution under anaerobic conditions (Smith et al., 2010). In recent years, the SOC modelling community has started to explore the use of different approaches such as bioclimatic envelope models to estimate C stocks and changes in peatlands at the regional to global scale with models such as PEATSTASH (Gallego-Sala et al., 2010).

Application to Different Climates and Land Covers/Uses

The application of SOM models to different environments, soils, land uses and climates is dependent on the data available for their development and evaluation. Models can be used to interpret observations at a particular site, to predict what might happen at another site or to predict SOM turnover in the future. If a model is used only to interpret what is occurring in a particular laboratory or field experiment, it can be run using just the data available in that experiment. However, if the application is predictive, simulations of the inherently long-term processes of SOM turnover must be tested against independent data. Such applications cannot be developed adequately by testing simulations against laboratory or short-term field data alone. Easy access to long-term experimental data has therefore controlled the development of such models, as long-term experiments (of at least

25 years duration) have not been established for all environments, soils, land uses and climates.

The early developments of SOM models, therefore, focused on sites where long-term experimental data were easily available. For example, the CENTURY model was developed first for grassland systems, against data from long-term sites in the northern Great Plains of the USA (Parton et al., 1988); the RothC model was developed for arable, grassland and forestry systems using data from the Rothamsted classical experiments (Jenkinson and Rayner, 1977). Because of this constraint, following the prevalence of long-term experiments, more SOM models have been developed in temperate regions and on mineral soils than in tropical systems or on highly organic soils.

If we aim to use SOM models to determine global changes in soil C stocks, it is important that models be expanded to include the wide range of soils and conditions found around the world. Jenkinson et al. (1991) used RothC to simulate changes in global C stocks and noted that RothC was not suitable for use in anaerobic conditions. Smith et al. (2007) also noted that RothC could not be used to simulate SOM turnover in highly organic soils. The establishment of SOMNET, a network of long-term experiments and models (Powlson et al., 1998), opened up the way to further developments of SOM models by providing access to a wider range of long-term experiments to allow adequate testing of long-term simulations.

Models have been developed to include the impact of nutrient limitations on SOM turnover. For example, RothC was expanded to include nitrogen (N) limitation on decomposition if the organic and mineral N content of the soil was inadequate to maintain a stable C:N ratio in the decomposed material (SUNDIAL, Bradbury et al., 1993). The CENTURY model includes N, phosphorus (P) and sulfur (S) limitation in a similar way, but uses a more complex approach that assumes immobilization of nutrients by decomposition of structural plant material and allows mineralization of nutrients from metabolic plant residue and the SOM pools according to different stable C to nutrient ratios for each organic matter pool and using a range of values for the stable ratios (Parton et al., 1988).

The impact of anaerobic conditions on decomposition has been described very simply in the ECOSSE model (developed from RothC and SUNDIAL by Smith et al., 2010) using a moisture rate modifier for decomposition that shows a linear decline in its value above field capacity. Similarly, anaerobic conditions are described in CENTURY as a linear decline in the rate of decomposition above a fitted parameter defining partial anaerobic conditions, up to another fitted parameter defining the soil as fully anaerobic (Parton et al., 1993; Chimner et al., 2002). In flooded anaerobic wetlands, particularly those undergoing cycles of wetting and drying, the effect of alternative electron acceptors in the soil (i.e. NO_3^-, Mn^{4+}, Fe^{3+} and SO_4^{2-} ions) becomes important and influences strongly the production of methane during anaerobic decomposition (van Bodegom and Stams, 1999). In some models, this effect is accounted for by simulating the pattern of decline of reduction (redox) potential (e.g. Cao et al., 1995). The Wetland-DNDC model estimates the redox potential of the soil layers in the saturated zone from observed patterns of variation in soils under fluctuating water table or continuously submerged conditions (Zhang et al., 2002). Other models simulate the behaviour of alternative electron acceptors directly, so avoiding the need to estimate redox potential (Matthews et al., 2000).

In order to simulate highly organic soils, some authors have developed new models, specifically designed to simulate the different processes occurring in organic soils (e.g. Frolking et al., 2001). An alternative approach modifies existing models for mineral soils for use in organic soils; this has the advantage that the adapted model can be applied seamlessly across the full range of soil types found in the landscape, but the models may be less well adapted to describe processes in organic soils than a purpose-built model. The ECOSSE model uses the description of anaerobic conditions and nutrient limitation, with added rate modifiers describing the impact of low pH, to extend simulations to highly organic as well as mineral soils (Smith et al., 2010). The CENTURY model uses a similar approach (Zhang et al., 2002). While CENTURY simulates plant growth and senescence, and therefore detrital C entering the soil (subject to decomposition), ECOSSE is initialized using an equilibrium run to determine the size of SOM pools and plant inputs needed to achieve steady state at the measured C content. This is a more robust approach for soils in steady state, but organic soils usually either accumulate C, due to the slow rate of decomposition compared to the plant inputs, or lose C, due to a land-use change that has disturbed the previously accumulating system. The size of soil C pools can be estimated if the rate of accumulation or degradation is known. However, the rate of accumulation or degradation is often unknown in larger-scale simulations. Alternative methods are needed for dealing with soils that are not in steady state, especially as the scale of simulations increases.

Simulation of SOM turnover under tropical conditions has to date been restricted by the lack of available data due to the complex and often inaccessible nature of tropical ecosystems, especially peatlands (Farmer et al., 2012). The effect of the higher temperatures found in tropical systems on the long-term dynamics of SOM has not been widely tested. Simulations of the decomposition of the large fragments of woody materials often found in tropical peat swamps have also not been well tested against long-term data.

The impact of salinity on the rate of decomposition and plant inputs has also not been widely incorporated into SOM models. This was simulated by Setia et al. (2012) using a rate modifier derived from laboratory and field experiments and a reduction factor for the plant inputs. This has not, as yet, been widely tested against long-term field experiments due to the lack of suitable data on a range of saline soils.

Volcanic soils pose another set of challenges, associated with adequately simulating the protection of SOM by the inorganic minerals in the soil. Shirato et al. (2004) modified the decomposition rate constant of the humus pool in RothC to account for protection of the humus by aluminium (Al) complexation. The rate constant was modified according to the content of pyrophosphate extractable aluminium in the soil. This modified version of RothC has been shown to improve simulations of four long-term experiments on andosols in Japan. Further evaluation of this approach against long-term experiments on other volcanic soils is needed.

Modelling SOC at Different Scales

The majority of SOC models were originally developed to model dynamics at the plot scale, as this is the scale at which experimental data are available. There are therefore numerous examples of application of different SOC models to plots under cultivated and native land use and following land-use transitions (Paustian et al., 1992; Li et al., 1994; Powlson et al., 1996; Leite et al., 2004). Many models have been tested and parameterized at the plot scale for different climate and soil conditions across the globe. Plot-scale studies typically represent homogeneous conditions, and are therefore limited in their applicability to other sites.

At the other end of the scale, global estimates of SOC change have been made using models that work on large grids ($0.5° \times 0.5°$) of the global surface. These are then linked to soil, climate and ecosystem type and use either a single pool or very generalized models to describe SOC turnover (King et al., 1997). Problems arise when we try to use such approaches at a smaller scale due to generalizations.

With growing interest in the role of soil C in the mitigation of climate change, a need to estimate SOC stocks and changes at the national to subnational scale has been recognized. Large-scale estimates require modelling in some form, both to scale up estimates of SOC stocks and to make projections of stock changes through time. The IPCC computational method was devised for national scale GHG accounting including SOC stock change (IPCC, 2004). It computes changes over time in a one-step process assuming a linear rate of change over a given period (the default being 20 years). It has been used in regional-scale assessments, showing most success when incorporating detailed regional data (Grace et al., 2004). Other studies have used regression approaches to estimate SOC stock changes at the regional scale, extrapolating several long-term experimental data sets into the future (Gupta and Rao, 1994; Smith et al., 2000). Others still have used more complicated regressions based on spatially explicit soil databases (Kern and Johnson, 1993; Kotto-Same et al., 1997).

As pointed out by Falloon et al. (2002), such regression approaches fail to capture the dynamic nature of SOC change. This is particularly relevant when looking at SOC change following a change in land use, as SOC tends to deplete rapidly at first, then more slowly. In order to capture these dynamics, studies have linked dynamic SOC models (originally developed at the plot scale) to large-scale data sets using geographic information systems (GIS) to make estimates at the regional to subnational scale. This allows many of the site-scale observations of SOC dynamics to inform larger-scale assessments (Paustian et al., 1997). Paustian et al. (1995) outlined the approach needed to link spatial data sets of soils, climate and land-use information to the CENTURY model, and this was subsequently used to make state and regional assessments of SOC stock change in the USA (Paustian et al., 2001, 2002). Other examples of this approach include Falloon et al. (1998), who linked the RothC model to GIS layers of soils, climate and land-use data for a 25,000 km² area of central Hungary. The methods developed by Paustian et al. (1995) were used to develop the GEFSOC Modelling System, a tool that allows the user to make large-scale assessments of SOC stocks and changes (Easter et al., 2007; Milne et al., 2007). The tools have been used to make national-scale (Al-Adamat et al., 2007; Kamoni et al., 2007) and subnational-scale (Bhattacharryya et al., 2007; Cerri et al., 2007) assessments, but could be applied at any scale providing data are available to parameterize and run the models.

The scale that perhaps presents the most challenges for SOC modelling, and any estimation of SOC stock change, is the landscape scale. Issues arise with data collation and adequate representation of heterogeneous land covers/uses. Also, at this scale, SOC stocks under fragmented land covers such as hedgerows or urban green spaces gain in significance (Viaud et al., 2010), and these are difficult to model. Another area where SOC modelling is in need of further development is in the horizontal transport and deposition of C, the relevance of which changes with scale. Some SOC turnover models include estimates of C loss from erosion, but few deal with deposition to another area or loss of dissolved C (Tipping et al., 2007).

Uncertainties in Modelling SOC

Uncertainty in large-scale simulations has two components: uncertainty due to inadequacies in the model (referred to by Beven, 2002, as structural errors), and uncertainty due to reduced detail and precision in data available at large scale (referred to here as input errors). If a model is to be applied at large scale, uncertainty in the simulations is likely to be greater than at field scale, due to the reduced *detail* of input data available. For example, detailed management factors of crops, such as sowing date and timing of fertilizer applications, cannot usually be specified when the resolution of the simulations is larger than the size of the management unit; the resolution of the simulation might be a 1 km² grid cell, whereas the size of a management unit might be a 5 ha field, so there will be many different values for the management factors within each 1 km² cell. Uncertainty in simulations at large scale is also greater than at field scale, due to the reduced *precision* of the input values; for example, the C content of the soil in a 10 ha field can be measured precisely, and the error in the measurement defined using replicates, whereas for applications at larger scale, the soil C content is often determined for 1 km² grid cells, with the C content estimated from typical or averaged soil C values for the major soil types identified in the cell (Batjes, 2009).

The uncertainty due to the structural and input errors can be quantified by evaluating the model at field scale, but using only input drivers that are available at the larger scale. In order to represent the uncertainty, a range of sites across the whole area to be simulated should be included in this field-scale evaluation. Good model performance is indicated statistically by simulations and measurements that are both coincident (indicating a close fit) and associated (indicating the trends in measurements are replicated) (Smith and Smith, 2007). The degree of coincidence can be used to represent the size of the uncertainty in the simulations.

Where measurements are replicated, the coincidence between simulated and measured values can be expressed as the 'lack of fit' statistic, and the significance of the coincidence determined using an F-test (Whitmore, 1991). If a data set is not replicated, the degree of coincidence can instead be determined by calculating the total error as the root mean squared error and the bias in the error as the relative error (Loague and Green, 1991; Smith et al., 1996b, 1997). The structural and input errors should be calculated separately to allow the source of errors to be understood and reduced, but the combined errors are then used to determine the accuracy of the model simulations at large scale.

SOC Models in Practical Application Today

Soil organic C models are in use today in many areas, including research of SOC dynamics and agricultural studies to estimate the long-term impacts of fertilizer and manure additions. They are also used increasingly in the reporting of GHG emissions. For national GHG inventories compiled for the United Nations Framework Convention on Climate Change (UNFCCC), reporters are required to estimate emissions of GHGs from soils under different land use and land-use change situations. The IPCC provide their computational model for this (IPCC, 2006). In addition, tools such as the Agriculture and Land Use National GHG Inventory Software (http://www.nrel.colostate.edu/projects/ALUsoftware) have been developed to make application of the method easier; however, the IPCC also encourage countries to use more sophisticated models where possible (IPCC, 2006).

With increasing interest in climate change mitigation potential from the agriculture, forest and land-use sector, agencies that fund land management activities are looking for ways to estimate the climate change impact of their activities. The US government has funded development of an online tool, COMET-VR, which allows farmers to estimate C sequestration and net GHG emissions from soils for farms in the USA. This allows farmers to take part in the US government's voluntary reporting of GHGs scheme. COMET-VR links a large set of databases containing information on soils, climate and management practices to run the CENTURY ecosystem simulation model dynamically to estimate soil C stocks and changes (COMET-VR, 2012). The user supplies basic land management information to the online system essentially to run a highly complex ecosystem model.

There are also examples of SOC models being used in the voluntary C market. The Verified Carbon Standard (VCS) recently approved a protocol that uses the RothC model in the estimation of SOC stock changes under sustainable land management projects – SALM (VCS, 2011). It is expected that more protocols involving models will be approved by the voluntary market as user-friendly interfaces are developed, making models more accessible. To date, several international funding agencies have developed their own tools for the reporting of C changes (including SOC). In the main, these are based on the IPCC method, examples being the Carbon Benefits Project's Simple and Detailed Assessments (Milne et al., 2010), FAO's EX-Act and the Cool Farm Tool. New reporting tools are likely to emerge in the near future, with continuing advances in graphical user interfaces that allow 'non-experts' to run SOC models and advances in web-based technologies for the management of spatial data. However, there are limitations, and more investment is still needed in the development of simple tools that can be applied in those areas of the globe most likely to experience dramatic change in SOC over the coming decades, e.g. the tropics and areas dominated by peatland.

Conclusions

Some researchers have recently suggested that, in order to make any further progress, a paradigm shift is needed in SOM modelling (Schmidt *et al.*, 2011). They argue that because the biotic and abiotic environment of the organic matter are as important in determining the C residence times of organic matter as the molecular structure of the material itself, organic matter should be described not by decay rate, pool stability or level of 'recalcitrance', as in the current models, but instead by quantifiable environmental characteristics governing stabilization, such as solubility, molecular size and 'functionalization'. While it is still unclear how such factors would be carried forward or used in dynamic simulations, it should also be noted that the characterization of pools in the current approaches already encompass both the biotic and abiotic factors impacting decomposability. We should guard against denying the huge amount of progress that has already been made in SOM modelling using the current approaches.

In the above discussion, we have shown how models can be used to quantify and report changes in soil C. We have discussed how they can be used to determine the impacts of land cover, land management and climate change on ecosystems in a range of different environments and at scales from plot to global. There are some environments to which the models are not yet well suited, but little by little, we are progressing our description of SOM dynamics to one that is able to cover the majority of soil environments given the appropriate input data to drive the model.

As the scale of simulation increases, the precision of input data tends to decrease, and it becomes more difficult to quantify accurately the uncertainty in the simulations. Reducing and quantifying the uncertainty in large-scale simulations is a key issue in SOM modelling if we are to have confidence in the figures they report and if they are to be of real value in informing policy decisions. However, most progress in reducing uncertainty is likely to be made by improving the quality of data available to drive models, in combination with developments in the models themselves.

The accuracy of large-scale simulations could be improved by developing better methods to initialize SOM models. A range of different methods has been used to initialize models. An approach that is robust at large scales uses the assumption of steady state to interpolate national soil C databases and to initialize the pools in the model in a way that characterizes the activity of the SOM (Smith *et al.*, 2005). This approach can reduce errors in simulations compared to the alternative approaches that estimate the initial pool size using typical values for a particular soil type.

However, one problem associated with the approach is the availability of accurate data on which to base the initialization. Although soil C is a standard soil parameter, reported in most soils databases, it is often estimated from a statistical sample of measurements for the particular soil type; the soil C value is then given for the major soil classifications in the grid cell (e.g. Batjes, 2009). More accurate estimates of soil C for each grid cell would improve the accuracy of simulations greatly, and it is in this area that a paradigm shift might generate most advances.

The assumption of steady state does not hold for some soil types, and development of mathematical approaches to initialize soils that are not in steady state would be a huge step forward. The development of methods to measure the soil pools rapidly would provide another way of solving this problem. While a range of fractionation techniques encompassing physical, biological and chemical fractionations have shown some success (Zimmermann *et al.*, 2007b), a more rapid method of characterizing soil pool sizes is needed for larger-scale simulations. Zimmermann *et al.* (2007a) have devised a method that may have potential to solve this problem using MIR spectroscopy as a rapid characterization of pool size; these methods need further development, especially in non-mineral soils.

For many years, modellers and experimentalists have worked separately on a problem that is reliant on both measurements and simulations for a solution. Future work requires modellers and experimentalists to work together to provide predictions of SOM decomposition that will be reliable, widespread and universal, and so provide their full potential value in informing policy.

References

Agren, G.I. and Bossata, E. (1987) Theoretical analysis of the long-term dynamics of carbon and nitrogen in soils. *Ecology* 68, 1181–1189.
Al-Adamat, R., Rawajfih, Z., Easter, M., Paustian, K., Coleman, K., Milne, E., Falloon, P., Powlson, D.S. and Batjes, N.H. (2007) Predicted soil organic carbon stocks and changes in Jordan between 2000 and 2030. *Agriculture, Ecosystems and Environment* 122, 35–45.
Bartholomew, W.V. and Kirkham, D. (1960) Mathematical descriptions and interpretations of culture-induced soil nitrogen changes. *Transactions of the 7th International Congress of Soil Science* 2, 471–477.
Batjes, N.H. (2009) Harmonized soil profile data for applications at global and continental scales: updates to the WISE database. *Soil Use and Management* 25, 124–127.
Beven, K. (2002) Towards a coherent philosophy for modelling the environment. *Proceedings of the Royal Society A* 458, 2465–2484.
Bhattacharyya, T., Pal, D.K., Easter, M., Batjes, N.H., Milne, E., Gajbhiye, K.S., Chandran, P., Ray, S.K., Mandal, C., Paustian, K. *et al.* (2007) Modelled soil organic carbon stocks and changes in the Indo-Gangetic Plains, India from 1980 to 2030. *Agriculture, Ecosystems and Environment* 122, 84–94.
Bradbury, N.J., Whitmore, A.P., Hart, P.B.S. and Jenkinson, D.S. (1993) Modelling the fate of nitrogen in crop and soil in the years following application of 15N-labelled fertilizer to winter wheat. *Journal of Agricultural Science* 121, 363–379.
Cao, M., Dent, J. and Heal, O. (1995) Modelling methane emissions from rice paddies. *Global Biogeochemical Cycles* 9, 183–195.
Cerri, C.E.P., Easter, M., Paustian, K., Killian, K., Coleman, K., Bernoux, M., Falloon, P., Powlson, D.S., Batjes, N.H., Milne, E. and Cerri, C. (2007) Predicted soil organic carbon stocks and changes in the Brazilian Amazon between 2000 and 2030. *Agriculture, Ecosystems and Environment* 122, 58–72.
Chimner, R.A., Cooper, D.J. and Parton, W.J. (2002) Modeling carbon accumulation in Rocky Mountain fens. *Wetlands* 22, 100–110.
Coleman, K. and Jenkinson, D.S. (1996) RothC-26.3—a model for the turnover of carbon in soil. In: Powlson, D.S., Smith, P. and Smith, J.U. (eds) *Evaluation of Soil Organic Matter Models*. NATO ASI Series, Vol. 138, Springer, Heidelberg, pp. 237–246.
COMET-VR (2012) USDA Voluntary Reporting Carbon Management Tool. http://www.nrcs.usda.gov/Internet/FSE_DOCUMENTS/nrcs144p2_053923.pdf (accessed 18 July 2014).
Easter, M., Paustian, K., Killian, K., Williams, S., Feng, T., Al Adamat, R., Batjes, N.H., Bernoux, M., Bhattacharyya, T., Cerri, C.C. *et al.* (2007) The GEFSOC soil carbon modelling system: a tool for conducting regional-scale soil carbon inventories and assessing the impacts of land use change on soil carbon. *Agriculture, Ecosystems and Environment* 122, 13–25.
Falloon, P.D., Smith, P., Smith, J.U., Szabo´, J., Coleman, K. and Marshall, S. (1998) Regional estimates of carbon sequestration potential; linking the Rothamsted carbon model to GIS databases. *Biology and Fertility of Soils* 27, 236–241.
Falloon, P., Smith, P., Szabo´, J. and Pasztor, L. (2002) Comparison of approaches for estimating carbon sequestration at the regional scale. *Soil Use Management* 18, 164–174.
Farmer, J., Matthews, R., Smith, J.U., Smith, P. and Singh, B.K. (2012) Assessing existing peatland models for their applicability for modelling greenhouse gas emissions from tropical peat soils. *Current Opinion in Environmental Sustainability* 3, 339–349.

Frolking, S., Roulet, N.T., Moore, T.R., Richard, P.J.H., Lavoie, M. and Muller, S.D. (2001) Modeling Northern peatland decomposition and peat accumulation. *Ecosystems* 4(5), 479–498.

Gallego-Sala, A.V., Clark, J.M., House, J.I., Orr, H.G., Prentice, C., Smith, P., Farewell, T. and Chapman, S.J. (2010) Bioclimatic envelope model of climate change impacts on blanket peatland distribution in Great Britain. Contribution to CR Special 24 'Climate change and the British Uplands'. *Climate Research* 45, 151–162.

Grace, P.R., Antle, J., Aggarwal, P.K., Andren, O., Ogle, S., Conant, R., Longmire, J., Ashkalov, K., Baethgen W.E., Valdivia, R., Paustian, K. and Davison, J. (2004) *Assessment of the Costs and Enhanced Global Potential of Carbon Sequestration in Soil*. International Energy Agency, IEA/CON/03/95, 130 pp.

Gupta, R.K. and Rao, D.L.N. (1994) Potential of wastelands for sequestering carbon by reforestation. *Current Science* 66, 378–380.

Henin, S. and Dupuis, M. (1945) Essai de bilan de la matiere organique du sol. *Annales Agronomiques* 15, 17–29.

Hunt, H.W. (1977) A simulation model for decomposition in grasssland. *Ecology* 58, 469–484.

IPCC (2004) Good practice guidance for land use, land-use change and forestry. In: Penman, J., Gytarsky, M., Hiraishi, T., Krug, T., Kruger, D., Pipatti, R., Buendia, L., Miwa, K., Ngara, T., Tanabe, K. and Wagner, F. (eds) *Intergovernmental Panel on Climate Change*. IGES, Hayama, Japan.

IPCC (2006) *2006 IPCC Guidelines for National Greenhouse Gas Inventories,* Volume 4: *Agriculture, Forestry and Other Land Use*. http://www.ipcc-nggip.iges.or.jp/public/2006gl/vol4.html (accessed 18 July 2014).

Jenkinson, D.S. (1990) The turnover of organic carbon and nitrogen in soil. *Philosophical Transactions of the Royal Society B* 329, 361–368.

Jenkinson, D.S. and Rayner, J.H. (1977) Turnover of soil organic matter in some of Rothamsted classical experiments. *Soil Science* 23, 298–305.

Jenkinson, D.S., Adams, D.E. and Wild, A. (1991) Model estimates of CO_2 emissions from soil in response to global warming. *Nature* 351, 304–306.

Jenkinson, D.S., Bradbury, N.J. and Coleman, K. (1994) How the Rothamsted classical experiments have been used to develop and test models for the turnover of carbon and nitrogen in soil. In: Leigh, R.A. and Johnston, A.E. (eds) *Long-term Experiments in Agricultural and Ecological Sciences*. CAB International, Wallingford, UK, pp. 117–138.

Jenny, H. (1941) *Factors of Soil Formation*. McGraw Hill, New York.

Kamoni, P.T., Gicheru, P.T., Wokabi, S.M., Easter, M., Milne, E., Coleman, K., Falloon, P. and Paustian, K. (2007) Predicted soil organic carbon stocks and changes in Kenya between 1990 and 2030. *Agriculture, Ecosystems and Environment* 122, 105–113.

Kern, J.S. and Johnson, M.G. (1993) Conservation tillage impacts on national soil and atmosphere carbon levels. *Soil Science Society of America Journal* 57, 200–210.

King, A.W., Post, W.M. and Wullschleger, S.D. (1997) The potential response of terrestrial carbon storage to changes in climate and atmospheric CO_2. *Climate Change* 35, 199–227.

Kotto-Same, J., Woomer, P.L., Appolinaire, M. and Louis, Z. (1997) Carbon dynamics in slash and burn agriculture and land use alternatives of the humid forest zone in Cameroon. *Agriculture Ecosystems and Environment* 65, 227–245.

Leite, L.F.C., Mendonca, E.D., Machado, P., Fenendes, E.I. and Neves, H.C.L. (2004) Simulating trends in soil organic carbon of an Acrisol under no-tillage and disc-plow systems using Century model. *Geoderma* 120(3/4), 283–295.

Li, C.S., Frolking, S. and Harris, R. (1994) Modelling carbon biogeochemistry in agricultural soils. *Global Biogeochemical Cycles* 8(3), 237–254.

Loague, K. and Green, R.E. (1991) Statistical and graphical methods for evaluating solute transport models: overview and application. *Journal of Contaminant Hydrology* 7, 51–73.

McGill, W.B., Hunt, H.W., Woodmansee, R.G. and Reuss, J.O. (1981) PHOENIX, a model of the dynamics of carbon and nitrogen in grassland soils. *Ecological Bulletins* NRF33, 49–115.

Matthews, R.B., Wassmann, R., Knox, J.W. and Buendia, L.V. (2000) Using a crop/soil simulation model and GIS techniques to assess methane emissions from rice fields in Asia. IV. Upscaling to national levels. *Nutrient Cycling in Agroecosystems* 58, 201–217.

Milne, E., Al-Adamat, R., Batjes, N., Bernoux, M., Bhattacharyya, T., Cerri, C.C., Cerri, C.E.P., Coleman, K., Easter, M.J., Falloon, P. *et al.* (2007) National and sub national assessments of soil organic carbon stocks and changes: the GEFSOC modelling system. *Agriculture, Ecosystems and Environment* 122, 3–12.

Milne, E., Paustian, K., Easter, M., Batjes, N.H., Cerri, C.E.P., Kamoni, P., Gicheru, P., Oladipo, E.O., Minxia, M., Stocking, M. *et al*. (2010) Estimating the carbon benefits of sustainable land management projects; The Carbon Benefits Project Component A. In: Gilkes, R.J. and Prakongkep, N. (eds) *Proceedings of the 19th World Congress of Soil Science, Soil Solutions for a Changing World*. 1–6 August, Brisbane, Australia. International Union of Soil Sciences, pp. 73–75. http://www.iuss.org/19th WCSS/Symposium/pdf/1524.pdf (accessed 17 July 2014).

Nye, P.H. and Greenland, D.J. (1960) *Soil Under Shifting Cultivation*. Technical Communication No 51. Commonwealth Agricultural Bureaux, Farnham Royal, UK.

Parton, W.J., Stewart, J.W.B. and Cole, C.V. (1988) Dynamics of C, N, P and S in grassland soils: a model. *Biogeochemistry* 5, 109–131.

Parton, W.J., Scurlock, J.M.O., Ojima, D.S., Gilmanov, T.G., Scholes, R.J., Schimel, D.S., Kirchner, T., Menaut, J.-C., Seastedt, T., Garcia Moya, E., Kamnalrut, A. and Kinyamario, J.I. (1993) Observations and modeling of biomass and soil organic-matter dynamics for the grassland biome worldwide. *Global Biogeochemical Cycles* 7, 785–809.

Paustian, K.P., Parton, W.J. and Persson, J. (1992) Modelling soil organic matter in organic amended and nitrogen fertilised long-term plots. *Soil Science Society of America Journal* 56(2), 476–488.

Paustian, K., Elliot, E.T., Collins, H.P., Vernon Cole, C. and Paul, E. (1995) Using a network of long-term experiments for analysis of soil carbon dynamics and global change: the North American model. *Australian Journal of Experimental Agriculture* 35, 929–939.

Paustian, K., Levine, E., Post, W.M. and Ryzhova, I.M. (1997) The use of models to integrate information and understanding of soil C at the regional scale. *Geoderma* 79, 227–260.

Paustian, K., Elliott, E.T., Killian, K., Cipra, J., Bluhm, G. and Smith, J.L. (2001) Modeling and regional assessment of soil carbon: a case study of the Conservation Reserve Program. In: Lal, R. and McSweeney, K. (eds) *Soil Management for Enhancing Carbon Sequestration*, SSSA Special Publication. Soil Science Society of America, Madison, Wisconsin, pp. 207–225.

Paustian, K., Brenner, J., Killian, K., Cipra, J., Williams, S., Elliott, E.T., Eve, M.D., Kautza, T.G. and Bluhm, G. (2002) State-level analyses of C sequestration in agricultural soils. In: Kimble, J.M., Lal, R. and Follett, R.F. (eds) *Agriculture Practices and Policies for Carbon Sequestration in Soil*. Lewis Publishers/CRC Press, Boca Raton, Florida, pp. 193–204.

Powlson, D.S., Smith, P. and Smith, J.U. (1996) *Evaluation of Soil Organic Matter Models Using Existing Long-term Data Sets*. Springer, Berlin.

Powlson, D.S., Smith, P., Coleman, K., Smith, J.U., Glendining, M.J., Körschens, M. and Franko, U. (1998) A European network of long-term sites for studies on soil organic matter. *Soil and Tillage Research* 47, 263–274.

Sauerbeck, D.R. and Gonzalez, M.A. (1977) Field decomposition of carbon-14-labelled plant residues in various soils of the Federal Republic of Germany and Costa Rica. In: IAEA (ed.) *Soil Organic Matter Studies*, Vol 1. International Atomic Energy Agency, Vienna, pp. 159–170.

Schmidt, M.W.I., Torn, M.S., Abiven, S., Dittmar, T., Guggenberger, G., Janssens, I.A., Kleber, M., Kögel-Knabner, I., Lehmann, J., Manning, D.A.C. *et al*. (2011) Persistence of soil organic matter as an ecosystem property. *Nature* 478, 49–56.

Setia, R., Smith, P., Marschner, P., Gottschalk, P., Baldock, J., Verma, V., Setia, D. and Smith, J.U. (2012) Simulation of salinity effects on past, present, and future soil organic carbon stocks. *Environmental Science and Technology* 46, 1624–1631.

Shirato, Y., Hakamata, T. and Taniyama, I. (2004) Modified Rothamsted carbon model for Andosols and its validation: changing humus decomposition rate constant with pyrophosphate-extractable. *Soil Science and Plant Nutrition* 50, 149–158.

Smith, J., Gottschalk, P., Bellarby, J., Richards, M., Nayak, D., Colemen, K., Hillier, J., Flynn, H., Wattenbach, M., Aitenhead, M. *et al*. (2010) *Model to Estimate Carbon in Organic Soils – Sequestration and Emissions (ECOSSE)*. User Manual. http://www.abdn.ac.uk/staffpages/uploads/soi450/ECOSSE%20User%20 manual%20310810.pdf (accessed 18 July 2014).

Smith, J.U. and Smith, P. (2007) *Environmental Modelling. An Introduction*. Oxford University Press, Oxford, UK, 180 pp.

Smith, J.U., Smith, P. and Addiscott, T.M. (1996a) Quantitative methods to evaluate and compare soil organic matter (SOM) models. In: Powlson, D.S., Smith, P. and Smith, J.U. (eds) *Evaluation of Soil Organic Matter Models Using Existing Long-Term Data Sets*. NATO ASI Series I, Vol 38. Springer, Heidelberg, Germany, pp. 181–200.

Smith, J.U., Smith, P., Wattenbach, M., Zaehle, S., Hiederer, R., Jones, R.J.A., Montanarella, L., Rounsevell, M.D.A., Reginster, I. and Ewert, F. (2005) Projected changes in mineral soil carbon of European croplands and grasslands, 1990–2080. *Global Change Biology* 11, 2141–2152.

Smith, O.L. (1979) An analytical model of the decomposition of soil organic matter. *Soil Biology and Biochemistry* 11, 585–606.

Smith, P., Powlson, D.S., Smith, J.U. and Glendining, M.J. (1996b) The GCTE SOMNET. A global network and database of soil organic matter models and long-term data sets. *Soil Use and Management* 12, 104 pp.

Smith, P., Smith, J.U., Powlson, D.S., McGill, W.B., Arah, J.R.M., Chertov, O.G., Coleman, K., Franko, U., Frolking, S., Jenkinson, D.S. et al. (1997) A comparison of the performance of nine soil organic matter models using seven long-term experimental data sets. In: Smith, P., Powlson, D.S., Smith, J.U. and Elliott, E.T. (eds) *Evaluation and Comparison of Soil Organic Matter Models Using Data Sets from Seven Long-term Experiments*. Geoderma 81, 153–225.

Smith, P., Powlson, D.S., Smith, J.U., Falloon, P. and Coleman, K. (2000) Meeting Europe's climate change commitments: quantitative estimates of the potential for carbon mitigation by agriculture. *Global Change Biology* 6, 525–539.

Smith, P., Smith, J., Flynn, H., Killham, K., Rangel-Castro, I., Foereid, B., Aitkenhead, M., Chapman, S., Towers, W., Bell, J. et al. (2007) *ECOSSE – Estimating Carbon in Organic Soils Sequestration and Emission*. Scottish Executive, Edinburgh, UK, 177 pp.

Tipping, E., Smith, E.J., Bryant, C.L. and Adamson, J.K. (2007) The organic carbon dynamics of a moorland catchment in NW England. *Biogeochemistry* 84, 171–189.

van Bodegom, P.M. and Stams, A.J.M. (1999) Effects of alternative electron acceptors and temperature on methanogenesis in rice paddy soils. *Chemosphere* 39, 167–182.

van Keulen, H. (2001) (Tropical) soil organic matter modelling: problems and prospects. *Nutrient Cycling in Agroecosystems* 61, 33–39.

VCS (Verified Carbon Standard) (2011) *Approved VCS Methodology VM0017: Adoption of Sustainable Agricultural Land Management*. http://www.v-c-s.org/sites/v-c-s.org/files/VM0017%20SALM%20Methodolgy%20v1.0.pdf (accessed 18 July 2014).

Viaud, V., Angers, D. and Walter, C. (2010) Toward landscape-scale modelling of soil organic matter dynamics in agroecosystems. *Soil Science Society of America Journal* 74, 1847–1860.

von Lutzow, M., Kogel Kogel-Knabner, I., Ekschmitt, K., Flessa, H., Guggenberger, G., Matzner, E. and Marschner, B. (2007) SOM fractionation methods: relevance to functional pools and to stabilization mechanisms. *Soil Biology and Biochemistry* 39, 2183–2207.

Whitmore, A.P. (1991) A method for assessing goodness of computer simulations of soil processes. *Journal of Soil Science* 42, 289–299.

Zhang, Y., Li, C., Trettin, C.C., Li, H. and Sun, G. (2002) An integrated model of soil, hydrology and vegetation for carbon dynamics in wetland ecosystems. *Global Biogeochemical Cycles* 16, 9.1–9.17.

Zimmermann, M., Leifeld, J. and Fuhrer, J. (2007a) Quantifying soil organic carbon fractions by infrared-spectroscopy. *Soil Biology and Biochemistry* 39, 224–231.

Zimmermann, M., Leifeld, J., Schmidt, M.W.I., Smith, P. and Fuhrer, J. (2007b) Measured soil organic matter fractions can be related to pools in the RothC model. *European Journal of Soil Science* 58, 658–667.

18 Valuation Approaches for Soil Carbon

David J. Abson*, Unai Pascual and Mette Termansen

Abstract

Valuation of soil carbon can be understood as the process for assigning 'weights' to soil carbon when these are inadequately represented in decision making processes. There are different types of weights or 'values' that can be assigned to soil carbon. One approach is to assign monetary weights to such resources using economic valuation models. The total set of such monetized weights is referred to as total economic value (TEV). The different components of the value of soil carbon differ both conceptually and with respect to how they can be measured or manifested. There are various methods for quantifying soil carbon values that differ with respect to the types of values they are suitable or able to assess. This chapter reviews the various valuation approaches that can be applied to estimate different components of the TEV of soil carbon. In this respect, it discusses how soil carbon values can be estimated through both stated and reveal preferences methods, and places particular emphasis on the production function approach. In addition other approaches are presented, including the preventive or mitigation expenditure (marginal abatement costs) approach and the social cost of carbon approach. Lastly, the chapter addresses the question of how economic values can be included in economic decision making processes. Three main alternatives are explored in terms of their advantages and disadvantages: cost–benefit analysis (CBA), multi-criteria analysis (MCA) and cost-effectiveness analysis (CEA).

Introduction

'The fundamental challenge of valuing ecosystem services lies in providing an explicit description and adequate assessment of the links between the structure and functions of natural systems, the benefits (i.e. goods and services) derived by humanity, and their subsequent values' (Heal et al., 2005, p. 2). Many authors argue that natural ecosystems are assets that produce flows of beneficial goods and services that, in principle, should be valued in a similar manner to other economic goods and services (e.g. Fisher et al., 2008; Gómez-Baggethun et al., 2010; Balmford et al., 2011). However, attempting to measure and value many environmental assets is problematic compared to conventional economic or financial assets (e.g. Limburg et al., 2002; Barbier, 2007; Farley, 2008; Spangenberg and Settele, 2010).

Many of the benefits that humans receive from soil carbon can be categorized as intermediate ecosystem services (Boyd and

*E-mail: abson@leuphana.de

Banzhaf, 2007) – also referred to as indirect ecosystem services (Kaiser and Roumasset, 2002), or supporting ecosystem services (MA, 2005). Intermediate ecosystem services are not directly appropriated by humans in order to increase human well-being; rather, they contribute to directly appropriated (or final) goods and services. Examples of final goods and services supported by soil carbon include the provision of food and fibre via improved nutrient cycling (e.g. Reicosky, 2003; Lal, 2004) or soil erosion control (Kibblewhite et al., 2008) and flood control (Urban et al., 2010; Hopkins and Gregorich, 2013), water quality via improved soil structure (Aitkenhead et al., 1999; Rawls et al., 2003) and air quality (Qian et al., 2003), and climate regulation through the sequestration of atmospheric carbon (Lal, 2004). The often 'hidden' nature of the services provided by soil carbon provides significant challenges in attempts to evaluate their contribution to human well-being.

Valuation of Intermediate Ecosystem Services

Many standard economic methods for valuing ecosystem services are inappropriate or difficult to apply directly to the valuation of soil carbon. Stated preference methods – which seek to elicit values through the direct questioning of individuals' willingness to pay (WTP) or willingness to accept (WTA) some marginal change in the object of choice (Boxall et al., 1996) – cannot be used easily to value changes in soil carbon, as individuals are generally insufficiently informed to ascribe monetary values meaningfully to such hidden services (Abson and Termansen, 2010). Moreover, a general problem with stated preference methods is that they are not based on the observed behaviour of individuals and are therefore open to methodological bias (e.g. Murphy et al., 2005). While it has been suggested that stated preference methods can be applied to value ecosystem functions (Barkmann et al., 2008), these methods have, with a few exceptions (e.g. Colombo et al., 2006; Takatsuka et al., 2009), rarely been applied directly to value the ecosystem services associated with soils.

Similarly, revealed preference methods – which use related, surrogate market values (often land value, or housing prices) as proxies with which to identify the value of a good or service that does not have a market price (Phaneuf et al., 2008) – are difficult to apply to the valuation of soil carbon. The benefits that flow from soil carbon are not always received directly by the individuals making economic choices regarding management of this ecological good. For example, the benefits of flood control or water-quality regulation often accrue to individuals who have made no economic choice in the management of soil carbon. Therefore, proxy market values, such as agricultural land values, provide no information regarding those individual's preferences for flood control or water-quality regulation. However, for some ecosystem services related to soil carbon, insights can be gained from observations of the value of the final services and their association with soil carbon as an input factor. Such valuation approach is often termed the production function approach.

The Production Function Approach to Valuing Soil Carbon

Valuing carbon based on the value of the final marketed goods produced (Chee, 2004; Gómez-Baggethun et al., 2010) may also be challenging because of the intermediate nature of many of the services provided by soil carbon. Soil carbon often represents a single factor, among many in the provision of ecosystem goods and services that are directly valued in the market. For example, agricultural productivity may be increased via careful management of soil carbon (Lal, 2004), but the actual productivity of a given agricultural field depends on many other management decisions such as labour inputs, application of fertilizers, herbicides and pesticides, appropriate choice of cultivars, etc. Using the market value of the final services appropriated by humans (such as the market value of agricultural crops) runs and cost for individual technologies or policy measures (ibid). The MACC approach only considers the mitigation potential and direct costs of a particular technological or policy option. Potential actions may be excluded from a MACC analysis if they involve serious externalities or run counter to other normative goals (such as poverty alleviation). MACC models are also endogenous to the assumed abatement targets. Here, the rationale is that as abatement targets become increasingly difficult to achieve, the costs of reaching those targets for a given society, industrial sector or individual are assumed to increase.

The potential advantage of the MACC in the context of soil carbon is that the marginal abatement cost curves can be based on a particular economic sector (i.e. agriculture or forestry) where the abatement occurs and do not require knowledge of global emissions trajectories. Moreover, the advantage of the MACC approach over SCC is that the abatement costs are based on existing activities and technologies, and can therefore be relatively easily estimated empirically (at least in the present time). It has been suggested that the uncertainty in the empirical estimates of MACC are of two orders of magnitude less than for SCC estimates (Dietz, 2007). Nevertheless, as MACC is projected into the future, it becomes increasingly uncertain as new carbon-reducing activities are required to meet emission targets and new technologies alter the abatement costs.

Joint Production and Ecosystem Service Bundles

Where soil carbon provides an intermediate service to final goods and services for which there is no clearly defined market good or service – e.g. regulation of water quality or flood control – then the production function approach can be used to evaluate the relations between soil carbon and the physical provision of the final service. Subsequently, replacement cost methods, averted behaviour methods or stated preference methods can be used to elicit individuals' preferences for these non-market services. In situations where the benefits of increased soil carbon do not accrue to the producer (for example, as in the case of the regulation of water quality, as mentioned above), then a production function approach should also take into account the cost or benefits incurred by the production of this ecosystem service. This is particularly relevant when the management of soil carbon jointly produces both private goods (such as agricultural commodities) and common pool resources (such as clean water) or public goods and services (such as regulation of floods, climate regulation).

It has recently been suggested that the interdependent, or bundled, nature of the ecosystem services produced in a given location means that they should not be considered in isolation from each other. Rather, different possible bundles of interdependent ecosystem services should be quantified and valued, in order to allow consideration of the trade-offs between the potential provision of different sets of goods and services from a given ecosystem (e.g. Raudsepp-Hearne et al., 2010; Martín-López et al., 2012). Methodologies for identifying ecosystem services bundles are relatively new, with the focus on the spatial mapping of individual services and the trade-offs between those services (e.g. Raudsepp-Hearne et al., 2010; Plieninger et al., 2013). Explicit valuation tools directly related to ecosystem service bundles are rare. One promising approach is the use of stated preference methods to elicit values of, and the societal preferences for, the different bundles of ecosystem services delivered by various types of ecosystems (e.g. Martín-López et al., 2012). In this context, different levels of soil carbon (and the associated land management practices that lead to those soil carbon levels) could be used as factors of production in different bundles of ecosystems services.

Including Economic Values in Decision-making Processes

The methods described above are underpinned by the idea of capturing the intensity of people's preferences as individuals for changes in soil carbon. These economic values are 'relative', as they are associated with how people assign weights to goods and services that affect their well-being. Such approach lends itself to benefit–cost analysis (Wegner and Pascual, 2011). Departing from individual preference-based methods, alternative valuation approaches can be based on participatory valuation. In practice, three main alternative approaches can be explored in terms of their advantages and disadvantages when including economic values in decision-making processes: cost–benefit analysis (CBA), cost-effectiveness analysis (CEA) and participatory multi-criteria analysis (MCA) The first approach puts all of its emphasis on economic efficiency, while the second relates economic efficiency to clearly defined normative goals. The third approach has a broader scope in terms of different socio-economic criteria on which decisions are to be made, such as the distribution of resources. All three approaches can be applied in the context of evaluating soil carbon projects.

There is a burgeoning literature explaining the use of monetary valuation to be integrated in CBA and its merits and drawbacks, both at the technical level as well as from a more philosophical stance (Kelman, 1981; Pearce, 1998). CBA is being used increasingly to evaluate projects and policies that consider ecosystem services to guide the selection of projects that deliver maximum net benefits from the flow of ecosystem services to society (Daily *et al.*, 2009; Wegner and Pascual, 2011).

CBA is an economic evaluation tool that characterizes a person as a self-interested individual who attaches value to entities only if it benefits his or her own welfare, and who is a rational decision maker of consistent choices towards maximizing his or her personal welfare. An individual's welfare is then measured by the satisfaction of personal preferences. In the context of soil carbon projects and policies, CBA focuses on aggregating individuals' willingness to pay (with a 'one person one vote' libertarian view) and monetarily quantifying the welfare impacts on society (Pearce *et al.*, 2006). The ranking of policy alternatives is based on the net present value (NPV) approach, and the one with the highest NPV is selected as it is assumed to bring about the largest increase in social welfare.

Individual preferences are revealed through market-related choices, such as using the production function approach (see above) or, when such markets do not exist, through the recreation of hypothetical or pseudo-markets (through stated preference methods, as explained earlier in the chapter). This approach contains some implicit normative choices when reducing the value dimension to preference utilitarianism, and it categorizes concepts such as freedom by individuals' ability in the marketplace to satisfy individual preferences (Niemeyer and Spash, 2001; Nyborg, 2012).

A key challenge of using CBA is that if different people order preferences according to different rules, the aggregation of all preferences on a single scale may not be feasible (Nyborg, 2000). In view of this limitation, deliberative approaches can be used, especially when the problems of dealing with continuous change, irreversibility, uncertainty and multiple legitimate standpoints in society are thought of as being irreducible (van den Hove, 2000), and when individual preferences may not be held in advance, needing to be discovered or elicited. This implies the need for values to be socially constructed through some form of deliberative communication (Dryzek, 2000). In the realm of environmental valuation, Howarth and Wilson (2006) note that by effective group procedures it is possible to enable participants to access relevant information, share their perspectives and engage in collective thinking in order to make better informed and articulated group choices than by aggregating individual preferences, as done in CBA.

One strand of participatory valuation is based on multi-criteria assessment methods to assess the value of soil carbon in terms of multiple socio-economic objectives. Amidst the multiple approaches in multi-criteria evaluation, social multi-criteria evaluation (SMCE) approaches are being developed to integrate conflicted values in society through the 'orchestration' of non-equivalent representations of diverse scientific disciplines and social views (Munda, 2008). The use of participatory and SMCE processes in environmental policy and natural resource management is gaining momentum. The advantage of SMCE is the possibility of taking into account the diversity of economic and non-economic values that may exist for soil carbon without reducing such diversity of values to a single, all-encompassing unit of measurement such as money. The recognition of the existence of diverse types of values, including economic ones, can be incorporated by means of qualitative and quantitative information, which in turn can be represented through cardinal numbers, ordinal rankings or fuzzy and crisp approaches (Munda, 2008). SMCE is based on: (i) identification of relevant social actors with regard to their confronted interests in soil carbon conservation by means of institutional analysis; (ii) defining the problem of the search for sustainable soil carbon; (iii) development of alternative soil carbon management options and the definition of specific evaluation criteria translated into a multi-criteria impact matrix; (iv) assessment of social actors' preferences and values through in-depth interviews or focus groups; (v) construction of a ranking of the available management alternatives by aggregating criteria scores by means of a mathematical algorithm; and (vi) analysis of the robustness of the analysis by means of a public debate and validation.

CEA considers alternative management options in which both the relative costs and outcomes are taken into account in a systematic way. It is a decision-oriented tool, designed to ascertain which means of attaining a particular normative goal are most efficient (Levin, 1995). With CEA, the costs of different options are expressed in monetary terms, whereas the outcomes are expressed in more 'natural' units (for example, level of conservation goals realized). Different policy options are then compared via the relative ratios of costs of implementation to the outcomes achieved. The CEA approach differs from the more ubiquitous CBA where both the costs and benefits (outcomes) of a given option are expressed in monetary terms. The application of CEA to valuation of soil carbon may be most appropriate when soil carbon contributes to ecosystem services or normative goals that are not expressed easily in monetary terms. For example, the contributions of soil carbon to the maintenance of cultural landscapes, cultural ecosystem services or conservation of biodiversity areas are where CEA might usefully be applied. The advantages of CEA over CBA approaches are that the resulting CEA evaluations of different potential interventions are often understood more intuitively and are less prone to rejection and critiques of the commodification of nature. However, CEA is limited in its application, as (unlike CBA) it can only be used to compare interventions or polices that have commonly defined goals.

Conclusions

Valuing changes in soil carbon through its contribution to maintaining and producing other ecosystem services is becoming increasingly relevant for policy discussions. This is due partially to the realization that the natural carbon cycle and human manipulations of the flow of carbon through both the economic and natural systems is important for understanding the value of multiple interdependent ecosystem services. Some progress has been made, but there are still considerable gaps in the valuation methodologies as well as in the empirical data to support economic assessment of alternative soil carbon conservation/investment paths. We highlight three key areas for further research development. First, the synthesis of the literature suggests that there are considerable advantages in developing further the production function approaches to value the role of carbon through its indirect contribution to a range of ecosystem services, i.e. agricultural food production, climate regulation, to name a few. For this research to be amenable to economic analysis, it is essential that it is taken into account that changing soil carbon levels is only one component in ecosystem service provision. Accounting for other input factors, human control variables, as well as abiotic and biotic factors, will be essential for research to generate meaningful economic insights. Second, in valuing carbon as an input in ecosystem service provision it is important to view soil carbon as a natural capital stock, from which we value continuous flows of services. Ignoring the time dimensions, accumulation and depletion timescales and the mismatch to often relatively shorter-term economic decision making processes will risk ignoring the value of carbon through its maintenance of the long-term productivity of ecosystem service provision. Finally, the chapter highlights that development of an integrated valuation framework is essential, as the future values of soil carbon and the further paths of soil carbon conservation are simultaneously determined. The potential scale of the impacts from alternative soil conservation strategies is such that the economic values of a single unit of soil carbon restored cannot be considered exogenous to the choice of soil conservation strategy.

References

Abson, D.J. and Termansen, M. (2010) Valuing ecosystem services in terms of ecological risks and returns. *Conservation Biology* 25, 250–258.

Aitkenhead, J.A., Hope, D. and Billett, M.F. (1999) The relationship between dissolved organic carbon in stream water and soil organic carbon pools at different spatial scales. *Hydrological Processes* 12, 1289–1302.

Balmford, A., Fisher, B., Green, R.E., Naidoo, R., Strassburg, B., Turner, R.K. and Rodrigues, A.S.L. (2011) Bringing ecosystem services into the real world: an operational framework for assessing the economic consequences of losing wild nature. *Environmental and Resource Economics* 48, 161–175.

Barbier, E.B. (2007) Valuing ecosystem services as productive inputs. *Economic Policy* 22, 179–229.

Barkmann, J., Glenk, K., Keil, A., Leemhuis, C., Dietrich, N., Gerold, G. and Marggraf, R. (2008) Confronting unfamiliarity with ecosystem functions: the case for an ecosystem service approach to environmental valuation with stated preference methods. *Ecological Economics* 65, 48–62.

Bockel, L., Sutter, P., Touchemoulin, O. and Jönsson, M. (2012) Using marginal abatement cost curves to realize the economic appraisal of climate smart agriculture policy options. In: FAO (ed.) *Easypol Resources for Policy Making*. FAO, Rome, pp. 1–36.

Boxall, P.C., Adamowicz, W.L., Swait, J., Williams, M. and Louviere, J. (1996) A comparison of stated preference methods for environmental valuation. *Ecological Economics* 18, 243–253.

Boyd, J. and Banzhaf, S. (2007) What are ecosystem services? The need for standardized environmental accounting units. *Ecological Economics* 63, 616–626.

Chee, Y.E. (2004) An ecological perspective on the valuation of ecosystem services. *Biological Conservation* 120, 549–565.

Clarkson, R. and Deyes, K. (2002) Estimating the social cost of carbon emissions. Government Economic Service Working Paper 140 for DEFRA. HM Treasury, London, pp.1–59.

Colombo, S., Calatrava-Requena, J. and Hanley, N. (2006) Analysing the social benefits of soil conservation measures using stated preference methods. *Ecological Economics* 58, 850–861.

Daily, G.C., Polasky, S., Goldstein, J., Kareiva, P.M., Mooney, H.A., Pejchar, L., Ricketts, T.H., Salzman, J. and Shallenberger, R. (2009) Ecosystem services in decision making: time to deliver. *Frontiers in Ecology and the Environment* 7, 21–28.

DECC (Department of Energy and Climate Change) (2009) *Carbon Valuation in UK Policy Appraisal: A Revised Approach*. Department of Energy and Climate Change, London.

Dietz, S. (2007) *Review of DEFRA Paper: 'The Social Cost of Carbon and the Shadow Rice of Carbon: What They Are, and How to Use Them in Economic Appraisal in the UK'*. Department of Environment, Farming and Rural Affairs, London.

Downing, T.E., Anthoff, D., Butterfield, B., Ceronsky, M., Grubb, M., Guo, J., Hepburn, C., Hope, C., Hunt, A., Li, A. *et al.* (2005) *Scoping Uncertainty in the Social Cost of Carbon*. Department of Environment, Forum for Agricultural Research in Africa, London.

Dryzek, J.S. (2000) *Deliverative Democracy and Beyond. Liberals, Critics, Contestations*. Oxford University Press, Oxford, UK.

Ekins, P. (2007) *Peer Review of Defra's Paper the Social Cost of Carbon and the Shadow Price of Carbon: What They Are, and How to Use Them in Economic Appraisal in the UK*. Department of Environment, Farming and Rural Affairs, London.

Farley, J. (2008) The role of prices in conserving critical natural capital. *Conservation Biology* 22, 1399–1408.

Fisher, B., Turner, K., Zylstra, M., Brouwer, R., de Groot, R., Farber, S., Ferraro, P., Green, R., Hadley, D., Harlow, J. *et al.* (2008) Ecosystem services and economic theory: integration for policy-relevant research. *Ecological Applications: A Publication of the Ecological Society of America* 18, 2050–2067.

Fu, B.J., Su, C.H., Wei, Y.P., Willett, I.R., Lü, Y.H. and Liu, G.H. (2011) Double counting in ecosystem services valuation: causes and countermeasures. *Ecological Research* 26, 1–14.

Gómez-Baggethun, E., de Groot, R., Lomas, P.L. and Montes, C. (2010) The history of ecosystem services in economic theory and practice: from early notions to markets and payment schemes. *Ecological Economics* 69, 1209–1218.

Heal, G.M., Barbier, E.B., Boyle, K.J., Covich, A.P., Gloss, S.P., Hershner, C.H., Hoehn, J.P., Pringle, C.M., Polasky, S., Segerson, K. and Shrader-Frechette, K. (2005) *Valuing Ecosystem Services: Toward Better Environmental Decision Making*. The National Academies Press, Washington, DC.

Hopkins, D.W. and Gregorich, E.G. (2013) Managing the soil–plant system for the delivery of ecosystem services. In: Gregory, P.J. and Nortcliff, S. (eds) *Soil Conditions and Plant Growth*. Blackwell Publishing, Chichester, UK.

Hougner, C., Colding, J. and Söderqvist, T. (2006) Economic valuation of a seed dispersal service in the Stockholm National Urban Park, Sweden. *Ecological Economics* 59, 364–374.

Howarth R.B. and Wilson, M.A. (2006) A theoretical approach to deliberative valuation: aggregation by mutual consent. *Land Economics* 82, 1–16.

Johnston, R.J. and Russell, M. (2011) An operational structure for clarity in ecosystem service values. *Ecological Economics* 70, 2243–2249.

Kaiser, B. and Roumasset, J. (2002) Valuing indirect ecosystem services: the case of tropical watersheds. *Environment and Development Economics* 7, 701–714.

Kelman, S. (1981) Cost–benefit analysis: an ethical critique (with replies), original paper in: *AEI Journal on Government and Society Regulation* 5, 33–40.

Kibblewhite, M.G., Ritz, K. and Swift, M.J. (2008) Soil health in agricultural systems. *Philosophical Transactions of the Royal Society B* 363, 685–701.

Lal, R. (2004) Soil carbon sequestration impacts on global climate change and food security. *Science* 304, 1623–1627.

Levin, H.M. (1995) Cost-effectiveness analysis. In: Carnoy, M. (ed.) *International Encyclopedia of Economics of Education*. Pergamon, Oxford, UK.

Limburg, K.E., O'Neill, R.V., Costanza, R. and Farber, S. (2002) Complex systems and valuation. *Ecological Economics* 41, 409–420.

MA (2005) *Ecosystems and Human Well-Being: Biodiversity Synthesis*. Island Press, Washington, DC.

Martín-López, B., Iniesta-Arandia, I., García-Llorente, M., Palomo, I., Casado-Arzuaga, I., Del Amo, D.G., Gómez-Baggethun, E., Oteros-Rozas, E., Palacios-Agundez, I., Willaarts, B. *et al.* (2012) Uncovering ecosystem service bundles through social preferences. *PLoS ONE* 7, e38970.

Munda, G. (2008) *Social Multi-criteria Evaluation for a Sustainable Economy*. Springer, Heidelberg, Germany.

Murphy, J., Allen, P.G., Stevens, T. and Weatherhead, D. (2005) A meta-analysis of hypothetical bias in stated preference valuation. *Environmental and Resource Economics* 30, 313–325.

Niemeyer, S. and Spash, C.L. (2001) Environmental valuation analysis, public deliberation, and their pragmatic syntheses: a critical appraisal. *Environment and Planning C: Government and Policy* 19, 567–585.

Nyborg, K. (2000) Homo economicus and homo politicus: interpretation and aggregation of environmental values. *Journal of Economic Behaviour and Organization* 42, 305–322.

Nyborg, K. (2012) One reason why environmental valuation is controversial. *European Association of Environmental and Resource Economists, 19th Annual Conference*, 27–30 June 2012, Prague. http://www.webmeets.com/eaere/2012/prog/viewpaper.asp?pid=891 (accessed 24 February 2013).

Pearce, D. (1998) Cost–benefit analysis and environmental policy. *Oxford Review of Economic Policy* 14, 84–100.

Pearce, D. (2003) The social cost of carbon and its policy implications. *Oxford Review of Economic Policy* 19, 362–384.

Pearce, D., Atkinson, G. and Mourato, S. (2006) *Cost–Benefit Analysis and the Environment: Recent Developments*. OECD, Paris.

Pearce, D.W. and Moran, D. (1994) *The Economic Value of Biodiversity*. Earthscan, London.

Phaneuf, D.J., Smith, V.K., Palmquist, R.B. and Pope, J.C. (2008) Integrating property value and local recreation models to value ecosystem services in urban watersheds. *Land Economics* 84, 361–381.

Plieninger, T., Dijks, S., Oteros-Rozas, E. and Bieling, C. (2013) Assessing, mapping, and quantifying cultural ecosystem services at community level. *Land Use Policy* 33, 118–129.

Qian, Y.L., Bandaranayake, W., Parton, W.J., Mecham, B., Harivandi, M.A. and Mosier, A.R. (2003) Long-term effects of clipping and nitrogen management in turfgrass on soil organic carbon and nitrogen dynamics. *Journal of Environmental Quality* 32, 1694–1700.

Raudsepp-Hearne, C., Peterson, G.D. and Bennett, E.M. (2010) Ecosystem service bundles for analyzing tradeoffs in diverse landscapes. *Proceedings of the National Academy of Sciences of the USA* 107, 5242–5247.

Rawls, W.J., Pachepsky, Y.A., Ritchie, J.C., Sobecki, T.M. and Bloodworth, H. (2003) Effect of soil organic carbon on soil water retention. *Geoderma* 116, 61–76.

Reicosky, D.C. (2003) Conservation agriculture: global environmental benefits of soil carbon management. In: García-Torres, I., Benites, J., Martínez-Vilela, A. and Holgado-Cabrera, A. (eds) *Conservation Agriculture: Environment, Farmers Experiences, Innovations, Socio-Economy, Policy*. Kluwer Academic Publishers, Dordrecht, the Netherlands, pp. 3–12.

Richmond, A., Kaufmann, R.K. and Myneni, R.B. (2007) Valuing ecosystem services: a shadow price for net primary production. *Ecological Economics* 64, 454–462.

Spangenberg, J.H. and Settele, J. (2010) Precisely incorrect? Monetising the value of ecosystem services. *Ecological Complexity* 7, 327–337.

Takatsuka, Y., Cullen, R., Wilson, M. and Wratten, S. (2009) Using stated preference techniques to value four key ecosystem services on New Zealand arable land. *International Journal of Agricultural Sustainability* 7, 279–291.

Tol, R.S.J. (2009) Why worry about climate change? *Research Bulletin* 9, 2–7.

Urban, R.A., Bakshi, B.R., Grubb, G.F., Baral, A. and Mitsch, W.J. (2010) Towards sustainability of engineered processes: designing self-reliant networks of technological–ecological systems. *Computers and Chemical Engineering* 34, 1413–1420.

van den Hove, S. (2000) Participatory approaches to environmental policy-making: the European commission climate policy process as a case study. *Ecological Economics* 33, 457–472.

Wegner, G. and Pascual, U. (2011) Cost–benefit analysis in the context of ecosystem services for human well-being: a multidisciplinary critique. *Global Environmental Change* 21, 492–504.

19 Current Soil Carbon Loss and Land Degradation Globally: Where are the Hotspots and Why There?

Hans Joosten*

Abstract

Global soils store in their first metre three times more carbon than all forest biomass of the world combined, and double the CO_2 content of the atmosphere. The natural soil carbon density is controlled by climate, soil properties and vegetation. Land-use intensity, drainage conditions and soil type (organic versus mineral soils) play an important role in controlling soil carbon losses or gains.

Because of its superficial setting, small bulk density and organic constitution, soil organic carbon (SOC) is highly susceptible to water and wind erosion and chemical and physical degradation. The major drivers of SOC loss include demand for fuel, overgrazing, arable agriculture and other overexploitation of vegetation. The resulting depletion of the global SOC pool is estimated at 40–100 Pg.

Three global hotspots can be distinguished where environmental and socio-economic conditions currently lead to large soil carbon losses:

- Peatlands, especially those in the tropics. Drained peatlands on 0.3% of the global land area lose around 0.5 Pg C year^{-1}.
- Drylands (arid, semi-arid and dry subhumid areas). Degraded drylands on 4–8% of the global land area (and home to >2 billion people) lose around 0.3 Pg C year^{-1}.
- Tropical forests, which experience large-scale clearing, with the vast majority of new cropland coming from intact and disturbed rainforests.

The future of SOC will ultimately depend on whether land management will continue to mine soil carbon for short-term gains, but with long-term detriment, or whether it will manage to conserve and enhance soil carbon. Soil carbon management will thus decide whether a legacy of land resources will remain to sustain future generations.

Introduction

Because of its slow formation (~0.1 mm year^{-1}; Pimentel et al., 1995; Stockmann et al., 2014), soil is essentially a finite resource. Soil organic matter (SOM), the mixture of recognizable plant and animal remains and no longer recognizable humus (Oades, 1989), is a vital element of soil. SOM provides and facilitates the biological, chemical and physical capacity of soil to sustain plant growth and is fundamental for soil fertility. SOM strongly determines the formation, stabilization and surface area of soil aggregates, total porosity and pore

*E-mail: joosten@uni-greifswald.de

size distribution, aggregate strength, erodibility and susceptibility to crusting and compaction. It increases plant available water capacity and water infiltration rate, and decreases surface runoff. It improves anion and cation exchange capacity, thereby enhancing nutrient retention and buffering against changes in pH and dissolved elemental concentrations. Furthermore, SOM enhances the activity and diversity of soil organisms, including earthworms, which enrich soil structure, and microbial biomass, which in turn affects C turnover and nitrification/denitrification (Lal, 2012; Fig. 19.1). Managing soils to maintain and enhance SOM stocks is therefore crucial to safeguard and increase food production (see Chapter 10, this volume).

SOM consists of more than half its weight in carbon, and is one of the main pools of carbon on Earth (see Box 19.1). Although the exact amounts of soil carbon are uncertain, it is clear that the top metre of soil holds about three times more carbon than all the forest biomass of the world combined, and around double the CO_2 content of the atmosphere (see Box 19.1; Joosten and Couwenberg, 2008; Lal, 2012). Small losses from this large pool may thus have significant impacts on atmospheric carbon dioxide concentrations. Since the start of agriculture, the soil and biotic carbon pools have been decreasing, leading to soil degradation and increased emissions of CO_2 to the atmosphere.

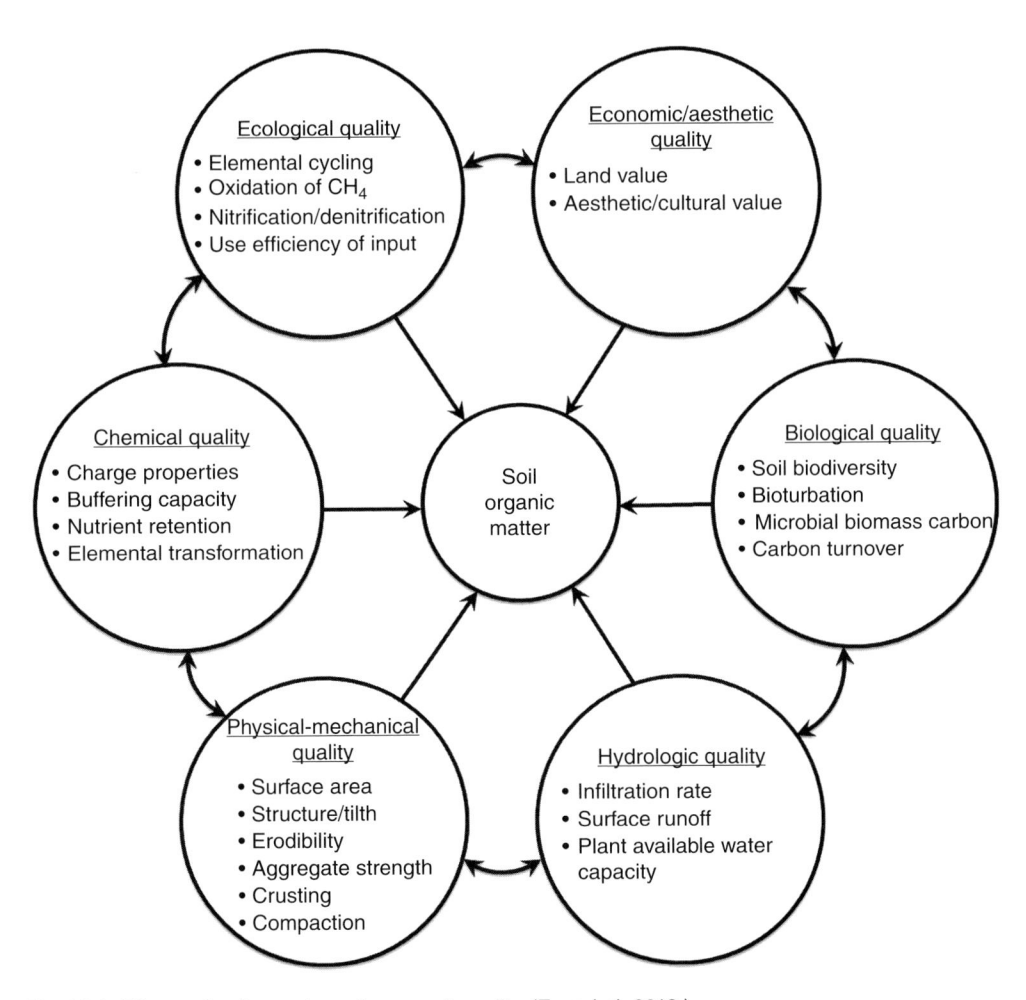

Fig. 19.1. Effects of soil organic matter on soil quality. (From Lal, 2012.)

> **Box 19.1.** Earth's carbon pools
>
> The largest pool of carbon is the ocean, with 38,000 Pg C (1 Pg = 1 petagram = 10^{15} g = 10^9 metric tonnes = 1 Gigatonne). Nearly all ocean carbon exists as dissolved inorganic carbon (DIC), largely as bicarbonate and carbonate ions, whereas some 1000 Pg C are organic (Houghton, 2007). The geologic pool contains 5000–10,000 Pg of organic carbon (as coal, gas and oil) (Lal, 2003; Houghton, 2007). The bedrock carbonates may comprise a similar amount of carbon as the ocean, but are normally disregarded as being largely immobile. Bedrock carbonates are, however, mobilized through metamorphosis in subduction zones or orogenic belts, through weathering at Earth's surface and anthropogenically through mining for lime and cement production.
>
> The soil is the third largest pool of carbon, with an estimated 1550 Pg of soil organic carbon (SOC; Eswaran *et al.*, 1993; Batjes, 1996) and 950 Pg of soil inorganic carbon (SIC; Batjes, 1996; Lal, 2004) in the top metre and 842 Pg of SOC in the next 2 m of depth (Jobbágy and Jackson, 2000). As data on deeper layers are sparse, these estimates are tentative (Lal, 1999). Information is especially incomplete for peat soils, which contain a substantial part of their C pool deeper than 1 m (Jungkunst *et al.*, 2012). Tarnocai *et al.* (2009) report that soils of the northern permafrost region contain 496 Pg of C in the top metre (i.e. double the amount hitherto reported) and 1024 Pg down to a 3 m depth. Globally, the SOC pool to 1 m depth ranges from 30 t ha^{-1} in arid climates to 800 t ha^{-1} in organic soils in cold regions, with a predominant range of 50–150 t ha^{-1} (Lal, 2004; Table 19.1).

The Global Distribution of Soil Organic Carbon

The distribution of soil organic carbon (SOC) is basically controlled by climate (moisture regime, temperature), soil properties (parent material, clay content, cation exchange capacity), vegetation and land use. The worldwide distribution of SOC roughly reflects the zonal distribution of rainfall (McLauchlan, 2006), with more carbon occurring in more humid areas (Plate 2). Most SOC is found in the northern hemisphere, simply because it contains more land mass in humid climates than the southern hemisphere. Also, temperature plays an important role in global SOC distribution (cf. Amundson, 2001), with SOC pools decreasing rapidly with increasing temperatures (Lal, 2002). Cool conditions – in contrast – impede decomposition and limit evapotranspiration, so that soils in cool climates are – with the same amount of rainfall – also wetter than in warm climates. That temperature is not a substantial constraint on SOC formation is illustrated by the occurrence of deep peat deposits in both tropical and polar humid areas. Climate influences both sides of the SOC balance by its effect on both input (primary productivity) and output (decomposition) (McLauchlan, 2006). The resulting effects on SOC dynamics are nicely illustrated by the diversity of peat accumulation in different parts of the Earth:

- low productive, upward-growing mosses (Bryophytes) depositing prevailingly slightly humified peat in subarctic, boreal and Atlantic (i.e. cold and wet-cool) regions;
- downward-growing rhizomes and rootlets of grasses (Poaceae) and sedges (Cyperaceae) producing intermediate humified peat in temperate, more continental and subtropical parts of the world; and
- deep-rooting (lignin-rich) roots of highly productive tall forest trees accumulating strongly humified peat with substantial wood remains in the tropics (Prager *et al.*, 2006).

Within climatic zones, the amount of SOC is determined by soil moisture that is, in addition to climate, determined by parent material, relief and soil texture. Parent material determines the composition and content of clay minerals, which generally stabilize SOM (McLauchlan, 2006). In vertisols, andosols and podzols, complexation and chelation between organic matter and the inorganic matrix occur, whereas soils dominated by kaolinitic clays and rich in iron and aluminium oxides are less prone to C storage.

The C content of soils under different land cover types varies substantially (Table 19.1). The soils of savannahs contain relatively little SOC, but the carbon stocks are globally significant, as this biome covers large areas of land. In contrast, peatlands cover only 3% of the global land area but contain a disproportionately large amount of soil carbon, making them the most spatially intensive carbon store among all terrestrial ecosystems.

Within the biomes, natural soil carbon densities may be modified strongly by land use, which may substantially change the carbon fluxes to and from the soil. Forests generally have the largest (all year-round) input of recalcitrant material, grasslands a large input of often less recalcitrant material, whereas croplands have a small input of rather labile material only when a crop is growing (Smith, 2008). Within the land-use types, land-use intensity, drainage conditions and soil type (cf. organic versus mineral soils) play a further important role in controlling soil carbon content, losses or gains.

Mechanisms and Drivers of Soil Carbon Loss

Three properties make SOM/SOC highly labile and vulnerable (Lal, 2003).

- The superficial setting: whereas ocean and bedrock carbon is largely physically inaccessible, SOC lies close to the surface and is subject to transport and transformation processes.
- The light weight: in comparison to silica-based soil components (gravel, sand, clay), SOM is light. With a bulk density only somewhat larger than that of water, SOM – when not protected by vegetation – is transported easily with water and wind flows. Erosion preferentially removes the light organic fraction.
- The chemical constitution: SOM is – indeed – organic, implying that it oxidizes in the presence of O_2, in some cases completely to CO_2 and H_2O.

Degradation of the SOC pool can take four main forms (Oldeman, 1992; Plate 10; Table 19.2).

Table 19.1. Indicative soil organic carbon pool in various biomes and land cover types. Soil carbon stocks refer to the upper metre of soil, except in the case of peatlands. (From Amundson, 2001; Lal, 2012; Victoria et al., 2012.)

		Soil carbon content	
Biome	Area (10^6 ha)	Mg ha^{-1} (mean)	Mg total
Tundra	880	218	191,840
Boreal desert	200	102	20,400
Cool desert	420	99	41,580
Warm desert	1,400	14	19,600
Tropical desert bush	120	20	2,400
Cool temperate steppe	900	133	119,700
Temperate thorn steppe	390	76	29,640
Tropical woodland and savannah	2,400	54	129,600
Boreal forest, moist	420	116	48,720
Boreal forest, wet	690	193	133,170
Temperate forest, cool	340	127	43,180
Temperate forest, warm	860	71	61,060
Temperate forest, very dry	360	61	21,960
Tropical forest, dry	240	99	23,760
Tropical forest, moist	530	114	60,420
Tropical forest, wet	410	191	78,310
Boreal and subarctic peatland	340	1,340	455,600
Tropical peatland	44	2,000	88,000

Table 19.2. Global extent of severe water and wind erosion. (From Lal, 2003.)

Region	Land area affected by severe erosion (Mha)			Total as a per cent of total land use
	Water erosion	Wind erosion	Total	
Africa	169	98	267	16
Asia	317	90	407	15
South America	77	16	93	6
Central America	45	5	50	25
North America	46	32	78	7
Europe	93	39	132	17
Oceania	4	16	20	3
World	751	296	1047	12

- *Water erosion* occurs when topsoil is removed and rills and gullies are formed; waterborne SOC losses also include leaching and translocation as dissolved (DOC) or particulate organic carbon (POC). Not all effects are, however, negative as water erosion may improve downstream soil fertility, as in river deltas that would not exist without upland erosion.
- *Wind erosion* is caused by a decrease of vegetation cover due to overgrazing or land conversion to arable agriculture, leading to loss of topsoil, deflation hollows and dunes, and overblowing.
- *Chemical degradation* includes the loss of organic matter by oxidation and mineralization following; for example, tillage or drainage.
- *Physical degradation* includes subsidence of organic soils by consolidation, oxidation, compaction and shrinkage (Den Haan *et al.*, 2012).

Since the start of settled agriculture around 10,000 years ago, agricultural activities have led to fundamental changes in the pools and fluxes of carbon (McLauchlan, 2006; McNeill and Winiwarter, 2008). Traditional agricultural systems are based on 'fertility mining'; that is, enhancing the decomposition of SOC, which releases plant nutrients and allows the land to produce more than its natural capacity (Janzen, 2006; Lal, 2012). Agricultural practices that lead to SOC depletion include:

- deforestation, slash and burn traditions and burning of crop remains, which hampers soil carbon replenishment;
- drainage of wetlands; that is, applying 'semi-desert' agriculture to organic soils; worldwide peatland drainage causes carbon-rich peats to disappear at a rate 20 times faster than it took them to accumulate (Joosten, 2010);
- changing the C quality of the crop biomass compared to the native vegetation;
- removal of harvested products and decreasing crop residues by more efficient harvest techniques and increasing the production of silage;
- tillage, levelling and other forms of soil disturbance, which destroys physical structure, exposes SOC to decomposition and enables erosion of the C-rich topsoil by water and wind;
- summer fallowing and clean cultivation;
- soil compaction by heavy machinery, high livestock and agricultural mechanization (repetitive hoeing and ploughing, resulting in a compacted layer at hoe/plough depth);
- increased microbial respiration because of higher fertilization levels; and
- excessive use of pesticides and other chemicals (Harrison *et al.*, 1993; Lal, 2002; McLauchlan, 2006; Smith, 2008; Joosten *et al.*, 2012).

These factors may be self-amplifying; for example, when changing vegetation and soil reflectance alters the temperature regime of the soil, or when decreasing soil productivity reduces the amount of plant residues returned to the soil (Lal, 2003).

The major drivers of SOC losses include:

1. Demand for fuel, for example in form of charcoal, leading to deforestation and removal of other natural vegetation; large-scale deforestation and consequent erosion may, for example, take place in locations in receipt of migrants (Stringer, 2012).
2. Overgrazing, leading to vegetation degradation, soil compaction and wind and water erosion.
3. Arable agriculture for market and subsistence crops.
4. Overexploitation of vegetation for domestic use; for example, for fuel, fencing etc.

Change in land cover and land use generally leads to a decrease in SOC. The largest losses of SOC per area occur where the C stocks are largest; for example, in highly organic soils such as peatlands. SOC tends to be lost when converting grasslands, forest or other native ecosystems to croplands, or by draining, cultivating or liming highly organic soils. SOC tends to increase when restoring grasslands, forests or native vegetation on former croplands, by changing native forest to pasture and by rewetting organic soils (Guo and Gifford, 2002; Smith, 2008).

The decrease in SOC continues in mineral soils until a new, stable level is reached at about 30–50% of the original level (Lal, 2002, 2004), depending on tillage practice and the level of C inputs into the soil as crop residue or animal manure (McLauchlan, 2006). The period to stabilize SOC levels is around 100 years for soils in the temperate region, whereas tropical soils may stabilize more quickly and boreal soils more slowly.

While erosion may have severe consequences for local soil fertility, its effect on carbon emissions to the atmosphere depends on the fate of the SOC exported from the eroded areas. van Oost *et al.* (2007) estimate that as a result of agricultural erosion over the past 50 years globally, ~16–21 Pg C have been moved laterally over Earth's surface and have been buried within agricultural landscapes, constituting a global carbon sink of ~0.12 Pg C year^{-1} (range 0.06–0.27). This figure is, however, admittedly overestimated as it does not account for decomposition losses from the exported SOC, of which the long-term stability is highly uncertain.

Since 1000 AD, an estimated 2 billion ha of once productive land has been irreversibly degraded; that is, more land than is currently under agricultural production (Lal, 2003). The depletion of the global SOC pool as a result of conversion of natural to agroecosystems is estimated at 40–100 Pg (Lal *et al.*, 2004; Smith, 2008; Lal, 2012).

Global Soil Carbon Fluxes

The stock of SOC results from the balance between the input and output of organic matter in the soil. Inputs are primarily from leaf and root detritus, and growth of subsurface plant and associated microbial biomass fed directly by photosynthesis. Outputs consist predominantly of emissions of carbon dioxide (CO_2) to the atmosphere, whereas emissions of volatile organic compounds and methane (CH_4), and transport of DOC and POC may also be important (Lal, 2003; Heimann and Reichstein, 2008).

The rate of input is related to the net primary production (NPP) of the vegetation, which varies with climate, land cover, species composition and soil type. Part of NPP enters the soil as organic matter root growth, exudates and plant litter, where it is partly converted back to CO_2 and CH_4 via (heterotrophic) soil respiration or leaches away as DOC and POC. Also, harvest of belowground biomass and fire (peat fires) may remove belowground carbon. The balance of all these processes determines whether the soil is a source or a sink of C (Smith, 2008).

For the 1990s, the worldwide litter input in the soil was estimated to be 4.0 Pg C, of which 3.9 Pg eventually oxidized to atmospheric CO_2 directly or after leaching and translocation. An estimated 0.1 Pg was added to the long-term SOC pool (Fig. 19.2; Houghton, 2007; cf. van Oost *et al.*, 2007). This assumed small net soil carbon gain,

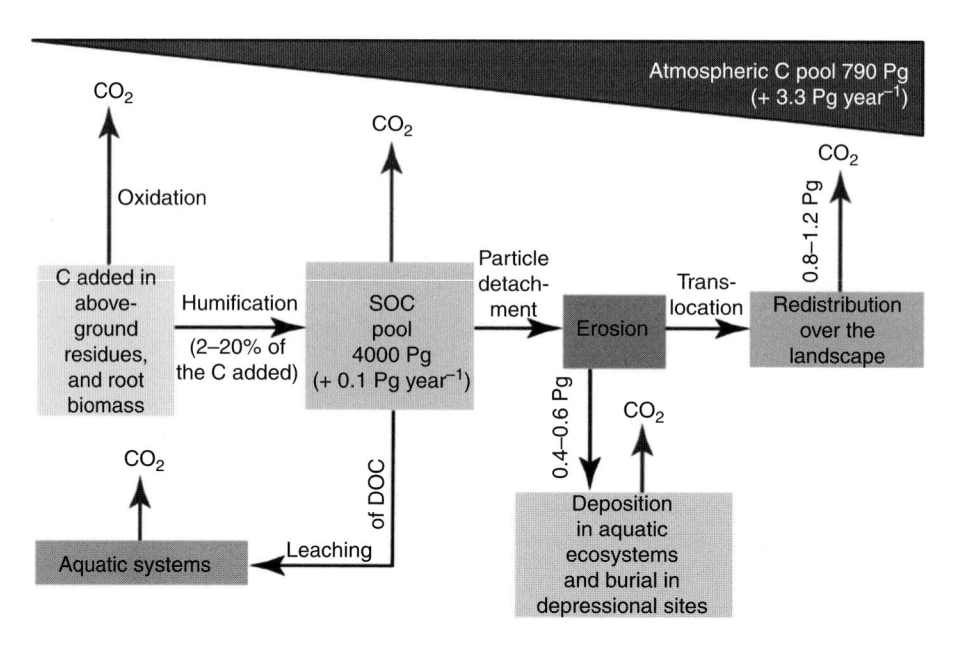

Fig. 19.2. Processes affecting soil organic carbon (SOC) dynamics (peatlands excluded). Arrows pointed upward indicate emissions of CO_2 to the atmosphere. Under anaerobic conditions (peatlands, rice fields) emission of CH_4 also takes place. DOC is dissolved organic carbon. (From Lal, 2004, 2012; Houghton, 2007.)

however, does not include the losses from drained peatland, which have increased from 0.3 Pg C year^{-1} in 1990 to 0.5 Pg C year^{-1} in 2008 (Joosten, 2010), turning the global carbon flux from soil strongly negative.

Hotspots of Soil Carbon Change

In the last two centuries, humans have cleared or converted 70% of the grassland, 50% of the savannah, 45% of the temperate deciduous forest and 27% of the tropical forest biome for agriculture (Foley et al., 2011). Between 1985 and 2005, the world's croplands and pastures expanded by 154 million hectares (Mha) (Foley et al., 2011).

Several global studies have tried to assess the extent of land degradation, with strongly varying results and accuracy (see Stringer, 2012; Plate 11). Oldeman et al. (1991) identified 15% of soil as degraded. The Millennium Ecosystem Assessment (2005) concludes that more than 60% of the ecosystem services have been degraded, but provides limited information on the status of soil. LADA (Land Degradation Assessment in Drylands) finds that 24% of the global land area has been degrading over the past 25 years. The 2011 FAO SOLAW study (Nachtergaele et al., 2011) presents the state of land and water resources for food production and analyses threats to food security and sustainable development.

Rather little focus has been on the loss of SOC because the absolute amount of SOC cannot be used unequivocally for comparing degradation among different types of soil and for different types of ecosystem services (e.g. carbon storage versus food provision; cf. Biancalani et al., 2012; Victoria et al., 2012). Work-in-progress presented at the United Nations Convention to Combat Desertification (UNCCD) Second Scientific Conference 'Economic assessment of desertification, sustainable land management and resilience of arid, semi-arid and dry sub-humid areas' in April 2013 in Bonn, Germany, showed, not surprisingly, that the decline in soil organic matter was related strongly to land use, with India and East Asia as outstanding examples of very intensive agricultural practices.

Some parts of the world are more at risk from land and soil degradation and soil carbon loss than others. Three global hotspot land types can be distinguished where different environmental and socio-economic conditions lead to large soil carbon losses.

Peatlands, especially those in the tropics, have the highest soil carbon density of all land cover categories (Table 19.1). Drained peatlands on 50 Mha worldwide (= 0.3% of the global land area) are currently responsible for carbon emissions of around 0.5 Pg per year, with an increasing trend in emission rate. Half of these emissions stem from rapidly expanding peatland drainage in Southeast Asia (largely for oil palm and pulpwood plantations), with associated peatland fires. Next to these huge emissions, peatland drainage also leads to subsidence, increased flooding, saltwater intrusion and eventually to the loss of large areas of habitable and productive land (Joosten *et al.*, 2012).

Drylands (arid, semi-arid and dry sub-humid areas) occupy 41% of Earth's land area and are home to more than 2 billion people. Some 10–20% of drylands are already degraded, with an estimated 0.3 Pg of soil carbon being lost to the atmosphere from drylands each year. Desertification (land degradation in drylands) has been ranked among the most urgent global environmental challenges (Millennium Ecosystem Assessment, 2005).

Tropical forests are currently experiencing a concentration of land-use change where large-scale forest clearing is taking place (Houghton, 2003; McNeill and Winiwarter, 2008). Between 1980 and 2000, more than half of new cropland came from intact rainforests and another 30% from disturbed forests (Gibbs *et al.*, 2010).

The Future of Soil Organic Carbon

The future of SOC will depend on climate change, land use and land cover, and feedbacks within and between these complex factors.

The response of SOC to climate change depends on the delicate balance between changing primary productivities and changing rates of decomposition as a result of changes in temperature and soil moisture. Rising temperatures could increase biomass production and inputs of organic materials into soils. Warming could, however, also accelerate the microbial decomposition and oxidation of SOM (Houghton, 2007; Lal, 2012), especially in thawing permafrost soils (Heimann and Reichstein, 2008). Central questions with respect to feedback loops are:

1. The temperature sensitivity of soil organic matter decomposition, especially the more recalcitrant pools (cf. Melillo *et al.*, 2002; Knorr *et al.*, 2005).
2. The balance between increased carbon inputs to the soil from increased production and increased losses due to increased rates of decomposition, including the effect of 'microbial priming' (Heimann and Reichstein, 2008).
3. Interactions between global warming and other aspects of global change, including other climatic effects (e.g. changes in precipitation pattern and resulting water balance), changes in atmospheric composition (e.g. increasing atmospheric carbon dioxide concentration and increased atmospheric nitrogen deposition; Heimann and Reichstein, 2008), increase in intensification of pests and diseases (Lal, 2012) and land-use change (cf. McSherry and Ritchie, 2013).

Scenario analyses show a large spatial heterogeneity and regional variation in the response of mineral soil SOC to climate change, which prohibits giving a simple answer on the question of whether SOC will increase or decrease (Gottschalk *et al.*, 2012; see also Chapter 20, this volume). Clearly, the simple picture of gradually increasing CO_2 concentrations and temperatures and non-interactive effects on assimilation and respiration needs to be replaced by a stronger integrative concept of complex interactions between ecosystem processes (Davidson and Janssens, 2006; Heimann and Reichstein, 2008), including land use as a dominant component (cf. van Wesemael *et al.*, 2010; Bell *et al.*, 2011).

Current projections indicate that world population will increase from 6.9 billion people today to over 9.5 billion in 2050. In addition, economic progress, notably in the emerging countries, translates into increased demand for food and other products that the land must deliver (FAO, 2011). Feeding the growing world population may require an additional 2.7–4.9 Mha, biofuel production 1.5–3.9 Mha, urbanization 1.6–3.3 Mha, industrial forestry 1.9–3.6 Mha and protected areas 0.9–2.7 Mha year^{-1}, whereas land degradation will render ≈1–2.9 Mha year^{-1} unsuitable for cultivation. With an additional total land demand of 9.5–26.4 Mha year^{-1}, the current land reserve (i.e. non-forested, non-protected and populated with <25 persons km^{-2}) could be exhausted as early as in the late 2020s, and at the latest by 2050 (Lambin and Meyfroidt, 2011). This means that intensive land uses will continue expanding into areas where SOC stocks are less resilient, soil conditions are marginal for agriculture and lands are prone to degradation. Semi-arid savannahs and grasslands, tropical rainforests and peatlands will all experience tremendous pressure to be further converted to arable land (Victoria et al., 2012).

The future of SOC will ultimately depend on *how* conversion and consequent land management will take place, whether soil carbon will continue to be mined for short-term gains, but with long-term detriment.

The alternative action is to implement land management that will conserve and enhance soil carbon. This will be essential to carry Earth's land-use practice through what could be some of the deepest resource crises of the 21st century and to establish a legacy of land resources that will sustain future generations thereafter.

References

Amundson, R. (2001) The carbon budget in soils. *Annual Review of Earth and Planetary Sciences* 29, 535–562.

Batjes, N.H. (1996) The total C and N in soils of the world. *European Journal Soil Science* 47, 151–163.

Bell, M.J., Worrall, F., Smith, P., Bhogal, A., Black, H., Lilly, A., Barraclough, D. and Merrington, G. (2011) UK land-use change and its impact on SOC: 1925–2007. *Global Biogeochemical Cycles* 25, GB4015.

Biancalani, R., Petri, M., Bunning, S.E., Salvatore, M. and Tubiello, F.N. (2012) The use of soil organic carbon as an indicator of soil degradation. *Energia, Ambiente e Innovazione* 4–5, 73–78.

Davidson, E.A. and Janssens, I.A. (2006) Temperature sensitivity of soil carbon decomposition and feedbacks to climate change. *Nature* 440, 165–173.

Den Haan, E.J., Hooijer, A. and Erkens, G. (2012) Consolidation settlements of tropical peat domes by plantation development. Report, Deltares. Delft, the Netherlands, 44 pp.

Eswaran, H., van den Berg, E. and Reich, P. (1993) Organic carbon in soils of the world. *Soil Science Society of America Journal* 57, 192–194.

FAO (2011) *The State of the World's Land and Water Resources for Food and Agriculture (SOLAW) – Managing Systems at Risk*. Food and Agriculture Organization of the United Nations, Rome and Earthscan, London. http://www.fao.org/docrep/017/i1688e/i1688e.pdf (accessed 20 June 2013).

Foley, J.A., Ramankutty, N., Brauman, K.A., Cassidy, E.S., Gerber, J.S., Johnston, M., Mueller, N.D., O'Connell, C., Ray, D.K., West, P.C. et al. (2011) Solutions for a cultivated planet. *Nature* 478, 337–342.

Gibbs, H.K., Ruesch, A.S., Achard, F., Clayton, M.K., Holmgren, P., Ramankutty, N. and Foley, J.A. (2010) Tropical forests were the primary sources of new agricultural land in the 1980s and 1990s. *Proceedings of the National Academy of Sciences USA* 107, 16732–16737.

Gottschalk, P., Smith, J.U., Wattenbach, M., Bellarby, J., Stehfest, E., Arnell, N., Osborne, T. and Smith, P. (2012) How will organic carbon stocks in mineral soils evolve under future climate? Global projections using RothC for a range of climate change scenarios. *Biogeosciences* 9, 3151–3171.

Guo, L.B. and Gifford, R.M. (2002) Soil carbon stocks and land use change: a meta analysis. *Global Change Biology* 8, 345–360.

Harrison, K.G., Broecker, W.S. and Bonani, G. (1993) The effect of changing land use on soil radiocarbon. *Science* 762, 725–726.

Heimann, M. and Reichstein, M. (2008) Terrestrial ecosystem carbon dynamics and climate feedbacks. *Nature* 451, 289–293.

Houghton, R.A. (2003) Revised estimates of the annual net flux of carbon to the atmosphere from changes in land use and land management 1850–2000. *Tellus* 55B, 378–390.

Houghton, R.A. (2007) Balancing the global carbon budget. *Annual Review of Earth and Planetary Sciences* 35, 313–347.

Janzen, H.H. (2006) The soil carbon dilemma: shall we hoard it or use it? *Soil Biology and Biochemistry* 38, 419–424.

Jobbágy, E.G. and Jackson, R.B. (2000) The vertical distribution of soil organic carbon and its relation to climate and vegetation. *Ecological Applications* 10, 423–436.

Joosten, H. (2010) *The Global Peatland CO_2 Picture. Peatland Status and Drainage Associated Emissions in all Countries of the World*. Wetlands International, Ede, the Netherlands, 10 pp.

Joosten, H. and Couwenberg, J. (2008) Peatlands and carbon. In: Parish, F., Sirin, A., Charman, D., Joosten, H., Minaeva, T. and Silvius, M. (eds) *Assessment on Peatlands, Biodiversity and Climate Change*. Global Environment Centre, Kuala Lumpur and Wetlands International, Wageningen, the Netherlands, pp. 99–117.

Joosten, H., Tapio-Biström, M.-L. and Tol, S. (eds) (2012) *Peatlands – Guidance for Climate Change Mitigation by Conservation, Rehabilitation and Sustainable Use*, 2nd edn. Mitigation of Climate Change in Agriculture Series 5. FAO, Rome, 100 pp.

Jungkunst, H.F., Krüger, J.-P., Heitkamp, F., Erasmi, F., Glatzel, S. and Fiedler, S. (2012) Accounting more precisely for peat and other soil carbon resources. In: Lal, R., Lorenz, K., Hüttl, R.F., Schneider, B.U. and von Braun, J. (eds) *Recarbonization of the Biosphere*. Springer, Amsterdam, pp. 127–157.

Knorr, W., Prentice, I.C., House, J.I. and Holland, E.A. (2005) Long-term sensitivity of soil carbon turnover to warming. *Nature* 433, 298–301.

Lal, R. (1999) Global carbon pools and fluxes and the impact of agricultural intensification and judicious land use. In: Dudal, R. (ed.) *Prevention of Land Degradation, Enhancement of C-Sequestration and Conservation of Biodiversity though Land Use Change and Sustainable Land Management with a Focus on Latin America and the Caribbean*. FAO, Rome, pp. 45–52.

Lal, R. (2002) Soil carbon dynamics in cropland and rangeland. *Environmental Pollution* 116, 353–362.

Lal, R. (2003) Soil erosion and the global carbon budget. *Environment International* 29, 437–450.

Lal, R. (2004) Soil carbon sequestration impacts on global climate change and food security. *Science* 304, 1623–1627.

Lal, R. (2012) World soils and the carbon cycle in relation to climate change and food security. Global Soil Week Issue Paper. http://globalsoilweek.org/wp-content/uploads/2013/05/GSW_IssuePaper_Soils_in_the_Global_Cycle.pdf (accessed 20 June 2013).

Lal, R., Griffin, M., Apt, J., Lave, L. and Morgan, M.G. (2004) Managing soil carbon. *Science* 304, 393.

Lambin, E.F. and Meyfroidt, P. (2011) Global land use change, economic globalization, and the looming land scarcity. *Proceedings of the National Academy of Sciences USA* 108, 3465–3472.

McLauchlan, K. (2006) The nature and longevity of agricultural impacts on soil carbon and nutrients: a review. *Ecosystems* 9, 1364–1382.

McNeill, J.R. and Winiwarter, V. (2008) Breaking the sod: humankind, history, and soil. *Science* 304, 1627–1629.

McSherry, M.E. and Ritchie, M.E. (2013) Effects of grazing on grassland soil carbon: a global review. *Global Change Biology* 19, 1347–1357.

Melillo, J.M., Steudler, P.A., Aber, J.D., Newkirk, K., Lux, H., Bowles, F.P., Catricala, C., Magill, A., Ahrens, T. and Morrisseau, S. (2002) Soil warming and carbon-cycle feedbacks to the climate system. *Science* 298, 2173–2176.

Middleton, N.J. and Thomas, D.S.G. (eds) (1997) *World Atlas of Desertification*, 2nd edn. Edward Arnold, London.

Millennium Ecosystem Assessment (2005) *Ecosystems and Human Well-being: Desertification Synthesis*. World Resources Institute, Washington, DC, 36 pp.

Nachtergaele, F.O., Petri, M., Biancalani, R., van Lynden, G., van Velthuizen, H. and Bloise, M. (2011) Global Land Degradation Information System (GLADIS). An Information database for Land Degradation Assessment at Global Level. Technical Working Paper of the LADA FAO/UNEP Project. http://www.fao.org/nr/lada/index.php?option=com_docman&task=doc_download&gid=773&Itemid=165&lang=en (accessed 15 August 2014).

Oades, J.M. (1989) An introduction to organic matter in mineral soils. In: Dixon, J.B. and Weed, S.B. (eds) *Minerals in Soil Environments*. Soil Science Society of America, Madison, Wisconsin, pp. 89–159.

Oldeman, L.R. (1992) Global extent of soil degradation. ISRIC Biennual Report 1991–1992. International Soil Reference Information Centre (ISRIC), Wageningen, the Netherlands, pp. 19–36.

Oldeman, L.R., Hakkeling, R.T.A. and Sombroek, W.G. (1991) World map of the status of human induced soil degradation. Three maps and an explanatory note (1990, rev. 1991). ISRIC, Wageningen and UNEP, Nairobi, 34 pp.

Pimentel, D., Harvey, C., Resosudarmo, P., Sinclair, D., Kurz, D., McNair, M., Crist, S., Shpritz, L., Fitton, L., Saffouri, R. and Blair R. (1995) Environmental and economic cost of soil erosion and conservation benefits. *Science* 267, 1117–1123.

Prager, A., Barthelmes, A. and Joosten, H. (2006) A touch of tropics in temperate mires: on Alder carrs and carbon cycles. *Peatlands International* 2006/2, 26–31.

Smith, P. (2008) Land use change and soil organic carbon dynamics. *Nutrient Cycling in Agroecosystems* 81, 169–178.

Stockmann, U., Minasny, B. and McBratney, A.B. (2014) How fast does soil grow? *Geoderma* 214, 48–61.

Stringer, L.C. (2012) Global land and soil degradation: challenges to soil. Global Soil Week Issue Paper. http://www.globalsoilweek.org/wp-content/uploads/2013/05/GSW_IssuePaper_Global-Land-and-Soil-Degradation.pdf (accessed 20 June 2013).

Tarnocai, C., Canadell, J.G., Schuur, E.A.G., Kuhry, P., Mazhitova, G. and Zimov, S. (2009) Soil organic carbon pools in the northern circumpolar permafrost region. *Global Biogeochemical Cycles* 23, GB2023, doi:10.1029/2008GB003327.

van Oost, K., Quine, T.A., Govers, G., De Gryze, S., Six, J., Harden, J.W., Ritchie, J.C., McCarty, G.W., Heckrath, G., Kosmas, C. *et al.* (2007) The impact of agricultural soil erosion on the global carbon cycle. *Science* 318, 626–629.

van Wesemael, B., Paustian, K., Meersmans, J., Goidts, E., Barancikova, G. and Easter, M. (2010) Agricultural management explains historic changes in regional soil carbon stocks. *Proceedings of the National Academy of Sciences USA* 107, 14926–14930.

Victoria, R., Banwart, S.A., Black, H., Ingram, H., Joosten, H., Milne, E. and Noellemeyer, E. (2012) The benefits of soil carbon. In: *UNEP Year Book 2012: Emerging Issues in Our Global Environment*. Nairobi, pp. 19–33. http://www.unep.org/yearbook/2012/pdfs/UYB_2012_CH_2.pdf (accessed 20 June 2013).

20 Climate Change and Soil Carbon Impacts

Pete Smith*, Pia Gottschalk and Jo Smith

Abstract

Soils contain vast reserves (~1500 Pg) of carbon (C), about twice that found as carbon dioxide in the atmosphere. Historically, soils in managed ecosystems have lost a portion of this C (40–90 Pg) through land-use change, some of which has remained in the atmosphere. In terms of climate change, most projections suggest soil C changes driven by future climate change will range from small losses to moderate gains, but these global trends show considerable regional variation. The response of soil C in future will be determined by a delicate balance between the impacts of increased temperature and decreased soil moisture on decomposition rates, and the balance between changes in C losses from decomposition and C gains through increased productivity. In terms of using soils to mitigate climate change, soil C sequestration globally has a large, cost-competitive mitigation potential. Nevertheless, limitations of soil C sequestration include time limitation, non-permanence, displacement and difficulties in verification. Despite these limitations, soil C sequestration can be useful to meet short- to medium-term targets, and confers a number of co-benefits on soils, making it a viable option for reducing the short-term atmospheric CO_2 concentration, thus buying time to develop longer-term emission reduction solutions across all sectors of the economy.

Introduction

Soils contain a stock of carbon (C) to a depth of 1 m that is about twice as large as that in the atmosphere and about three times that in vegetation. Small losses from this large pool could have significant impacts upon future atmospheric carbon dioxide concentrations, so the response of soils to global warming is of critical importance when assessing climate C cycle feedbacks (Smith et al., 2008a). Models that have coupled climate and C cycles show a large divergence in the size of the predicted biospheric feedback to the atmosphere. Central questions that still remain when attempting to reduce this uncertainty in the response of soils to global warming are: (i) the temperature sensitivity of soil organic matter, especially of the more recalcitrant pools (Davidson and Janssens, 2006; Smith et al., 2008a); (ii) the balance between increased C inputs to the soil from increased production (from CO_2 fertilization and crop breeding advances) and increased losses due to increased rates of decomposition (Smith and Fang, 2010); and (iii) interactions

*E-mail: pete.smith@abdn.ac.uk

between global warming and other aspects of global change, including other climatic effects (e.g. changes in water balance, fire regimes, pests and diseases), changes in atmospheric composition (e.g. increasing atmospheric carbon dioxide concentration), land-use change (Smith et al., 2005, 2006) and arable land management. In addition to responding to climate change, soils could also play an important role in climate change mitigation; if C can be sequestered in soils, this could be a significant mechanism for reducing atmospheric CO_2 concentrations (Smith et al., 1997, 2000). In this short review, we outline recent evidence of potential responses of soils to climate change, and then present recent evidence on the possible role of soil C sequestration in climate mitigation and discuss some limitations associated with this method of climate mitigation. This review is limited to mineral soils and does not cover peatlands and permafrost soils, since the role of peatlands in climate change has been reviewed recently (Joosten et al., 2014) and space precludes dealing with all soil types in similar depth.

The Impact of Climate Change on Soils

Soils in the global carbon cycle

Globally, soils contain about 1500 Pg (1 Pg = 1 Gt = 10^{15} g) of organic C (Batjes, 1996), about three times the amount of C in vegetation and twice the amount in the atmosphere (IPCC WGI, 2001). The annual fluxes of CO_2 from atmosphere to land (global net primary productivity (NPP)) and land to atmosphere (respiration and fire) are each of the order of 60 Pg C year^{-1} (IPCC WGI, 2001). During the 1990s, fossil fuel combustion and cement production emitted 6.3 ± 1.3 Pg C year^{-1} to the atmosphere, while land-use change emitted 1.6 ± 0.8 Pg C year^{-1} (IPCC WGI, 2001; Schimel et al., 2001). Atmospheric C increased at a rate of 3.2 ± 0.1 Pg C year^{-1}, the oceans absorbed 2.3 ± 0.8 Pg C year^{-1}, with an estimated terrestrial sink of 2.3 ± 1.3 Pg C year^{-1} (IPCC WGI, 2001; Schimel et al., 2001). Soil C pools are smaller now than they were before human intervention. Historically, soils have lost between 40 and 90 Pg C globally through cultivation and disturbance (Schimel, 1995; Houghton, 1999; Houghton et al., 1999; Lal, 1999). The size of the pool of soil organic carbon (SOC) is large compared to gross and net annual fluxes of C to and from the terrestrial biosphere.

Small changes in the SOC pool could have dramatic impacts on the concentration of CO_2 in the atmosphere. The response of SOC to global warming is, therefore, of critical importance. One of the first examples of the potential impact of increased release of terrestrial C on further climate change was given by Cox et al. (2000). Using a climate model with a coupled C cycle, this study showed that the release of terrestrial C under warming would lead to a positive feedback, resulting in increased global warming. Since then, a number of coupled climate C cycle (so-called C4) models have been developed. However, there remains considerable uncertainty concerning the extent of the terrestrial feedback, with the difference between the models amounting to differences in the atmospheric CO_2 concentration of ~250 ppm by 2100 (Friedlingstein et al., 2006). This is of the same order as the difference between fossil fuel C emissions under the IPCC SRES (2000) emissions scenarios. It is clear that better quantifying the response of terrestrial C, a large proportion of which derives from the soil, is essential for understanding the nature and extent of Earth's response to global warming.

The response of soils to future climate change

Despite suggestions during the 1990s that climate change could lead to massive losses of C from the world's soils, more recent studies have suggested that climate change impacts on soil C could be less significant (Smith et al., 2005, 2006; Ciais et al., 2010; Gottschalk et al., 2012). The level of SOC in a particular soil is determined by many factors, including climatic factors (e.g. temperature and moisture regimes) and physical and chemical soil factors (e.g. soil parent material, clay content, cation exchange capacity; Dawson and Smith, 2007). For a given soil type, however, SOC stock can also vary, the stock being determined by the balance of net C inputs to the soil (as organic matter) and net losses of C from the soil (as carbon dioxide, dissolved organic carbon and losses through erosion).

Examining climate impacts on cropland and grassland soils in Europe, Smith et al. (2005) showed that SOC stocks were projected to change little between 1990 and 2080, since increase in productivity, feeding more C into the soil, balanced the increased losses of SOC from faster decomposition under climate change. Further, in some European regions, the future climate is projected to dry so much that the decomposition rate is slowed, despite large increases in soil temperature (Smith et al., 2005). Ciais et al. (2010) reviewed a number of European studies and showed that other modelling studies confirmed this finding, with cropland soil C stocks estimated to change little during 1990–1999, ranging from a small sink of 15 g C m^{-2} $year^{-1}$ to a source of over 1.3–7.6 g C m^{-2} $year^{-1}$ from the ORCHIDEE-STICS, LPJmL and RothC models, respectively. Future changes in cropland SOC were found to be highly dependent on management and land-use change assumptions, but the direct impact of combined climate drivers were not found to be large (Ciais et al., 2010).

Globally, there is some uncertainty about the impact of climate change on mineral soils, related to the complexity of the factors determining the C balance of soils (Smith and Fang, 2010) and uncertainties in the ways models deal with interactions among the climate drivers. Despite this uncertainty, projections are similar. Cramer et al. (2001) used the IS92a anthropogenic emissions scenario (which is comparable to the later IPCC A1b scenario) in conjunction with the HadCM2-SUL version of the Hadley Centre (UK) climate model. Their simulations show a c.10% increase (mean of six Dynamic Global Vegetation Models; DGVMs) between 2000 and 2100. Gottschalk et al. (2012), using the RothC model, driven by a range of climate scenarios and scaled NPP changes from the IMAGE 2.2 model, found a similar impact with a comparable scenario of ~8% increase in SOC stocks. Ito (2005) reported projected global SOC changes for the 21st century using seven climate model realizations of the IPCC A2 scenario, which showed lower climate forcing than the scenarios used by Cramer et al. (2001) and Gottschalk et al. (2012). Unsurprisingly, Ito (2005) projected smaller changes in SOC, in some cases showing a small loss of SOC, where similar scenarios in the studies of Cramer et al. (2001) and Gottschalk et al. (2012) showed small gains. Lucht et al. (2006) used the LPJ model to simulate SOC stock changes from 2000 to 2100, and found similar percentage increases in SOC (~5–6%) to the 3.5% increase projected by Gottschalk et al. (2012) for similar climate-forcing scenarios. The bulk of the evidence from models suggests that, at the global scale, projected changes in SOC in mineral soils are relatively small, and that SOC stocks may well increase under future climate change.

This global finding, however, masks a complex pattern of regional responses (Gottschalk et al., 2012; Arnell et al., 2013, 2014). Whereas SOC stocks increase in most regions, because the increase in NPP offsets the effects of higher temperatures, there is little change, or some loss, in high-latitude parts of Canada and Eastern Europe (Siberia) and parts of East Asia, where the effects of higher temperatures outweigh changes in rainfall and NPP. The complex regional patterns of change in SOC are demonstrated in Plate 12, which shows average trend in SOC stock change from 1971 to 2100 across ten climate scenarios (from Gottschalk et al., 2012). The spatial heterogeneity in the response of SOC to changing climate shows how delicately balanced the competing gain and loss processes are, with subtle changes in temperature, moisture, soil type and land use interacting to determine whether SOC will increase or decrease in the future. Given this delicate balance, we should stop asking the general question of whether soils will increase or decrease in SOC under future climate, as there appears to be no single answer. Instead, we should focus our research efforts on improving our prediction of factors that determine the size and direction of change and the land management practices that can be implemented to protect and enhance SOC stocks, as discussed in Smith et al. (2008b).

The Role of Soils in Mitigating Climate Change

Increasing soil carbon stocks to combat climate change (soil carbon sequestration)

Carbon stocks in the soil can be increased in managed ecosystems by optimizing 'best management practices'. There have been numerous reviews of management to increase soil C stocks (Lal et al., 1998; Lal, 2004; Smith, 2008b), so a full review is not presented here. Increased C stocks in the soil increases soil fertility, workability and water-holding capacity, and reduces erosion risk (Lal, 2004). Increasing soil C stocks can thus reduce the vulnerability of managed soils to future global warming (Smith et al., 2008b; Smith and Olesen, 2010). Management practices effective in increasing SOC stocks include improved plant productivity (through nutrient management, rotations, improved agronomy), reduced/conservation tillage and residue management, more effective use of organic amendments, land-use change (crops to grass/trees), set-aside, agroforestry, optimal livestock densities and legumes/improved species mix (Smith et al., 2008b). While these measures have the technical potential to increase SOC stocks by about 1–1.3 Pg year^{-1} (Lal, 2004; Smith et al., 2008b), the economic potentials for SOC sequestration are estimated to be 0.4, 0.6 and 0.7 Pg C year^{-1} at carbon prices of US$0–20, 0–50 and 0–100 t^{-1} CO_2-equivalent, respectively (Smith et al., 2008b). A small loss of C from permafrost or peatlands could offset this potential sequestration (Joosten et al., 2014), but the increase in SOC engendered by improved management is expected also to reduce the vulnerability of soils to future SOC loss under global warming. As such, soil C sequestration can, in many respects, be regarded as a 'win–win' and a 'no regrets' option (Smith and Powlson, 2003; Smith and Trines, 2007; Smith et al., 2007).

Drawbacks associated with soil carbon sequestration as a climate mitigation measure

While there are many advantages to increasing soil C stocks, and 'win–win' and 'no regrets' options can be identified, there are a number of issues associated with soil C sequestration that make it a risky climate mitigation option (Smith, 2005, 2008a). These issues are: (i) saturation of the C sink (the C is only removed from the atmosphere until the soil reaches a new equilibrium soil C level; Smith, 2005); (ii) non-permanence (C sinks can be reversed at any stage by poor soil management; Smith, 2005); (iii) leakage/displacement (e.g. increasing soil C stocks in one area leads to soil C losses in another; IPCC, 2000); (iv) verification issues (can the sinks be measured and at what depth?; Smith, 2004); and (v) total effectiveness relative to emissions reduction targets (only a fraction of the reduction can be achieved through sinks; IPCC WGI, 2007). These issues are discussed briefly below.

SATURATION OF THE CARBON SINK. The C sink can be defined as the annual removal of C from the atmosphere into soil. When a C sequestration measure is first implemented, the change in soil C is large to begin with, but slows over time as the soil approaches a new equilibrium (see Fig. 20.1) (Smith, 2005). Sink strength therefore decreases over time until the soil reaches a new equilibrium. This phenomenon is termed 'sink saturation'. Compared to reduced emissions of other greenhouse gases, which can continue indefinitely, C sequestration in soils (and indeed in vegetation) is therefore time limited and finite (Smith, 2005). Improved management needs to be maintained indefinitely to maintain the higher soil C stocks, but with no additional sink benefit.

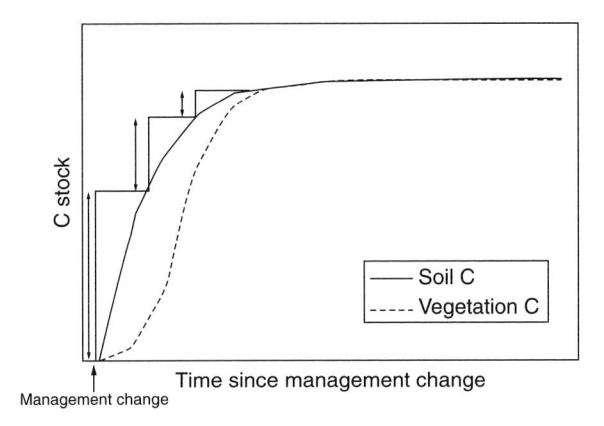

Fig. 20.1. Decline in sink strength over time. Change in soil and vegetation carbon sequestration showing large atmospheric carbon removals (sink strength) soon after management change (large vertical arrow on left-hand side of the figure), but over the subsequent equivalent time removals become smaller as the soils approach a new equilibrium (smaller arrows as soils gain in carbon). (From Smith, 2012.)

NON-PERMANENCE. As well as declining over time, soil C sinks are also reversible. A soil C stock that has been increased by improved soil management will lose C rapidly unless the improved management is maintained. The rate of C loss is more rapid than the rate of gain (Smith, 2005). Compared to reduced emissions of other greenhouse gases, where an emission reduction is permanent, C sequestered in the soil (and in vegetation) is non-permanent, presenting a risk of future release (Smith, 2005).

LEAKAGE/DISPLACEMENT. Increasing soil C stocks does not necessarily lead to a decrease in atmospheric CO_2 concentrations (Powlson et al., 2011). It is possible, for example, to enhance soil C stocks in one area by applying large inputs of organic matter. If, however, the organic matter applied to the area gaining in C would otherwise have been applied in another area, the other area would lose C (i.e. the emissions are displaced; also termed 'leakage' where emissions occur outside the greenhouse gas accounting boundary; IPCC, 2000). In this example, the impact across the two areas would be neutral, leading to no net atmospheric C removal. An increase in soil C stocks in this case, does not constitute a genuine decrease in atmospheric CO_2 concentrations (Smith, 2005). Displacement/leakage also occurs where land-use change to increase C stocks in one area leads to land-use change that causes C release in another area, in a process termed 'indirect land-use change' (Searchinger et al., 2008).

VERIFICATION ISSUES. Changes in soil C are small compared to the large stocks of C present in the soil, meaning that the change in C stock can be difficult to measure, presenting problems for monitoring, reporting and verification (MRV) (Smith, 2004). If the value of the C removed from the atmosphere is less than the cost of measuring the change, MRV costs can make soil C less cost competitive with greenhouse gas reduction measures that are less expensive to demonstrate (Smith, 2004).

TOTAL EFFECTIVENESS RELATIVE TO EMISSION REDUCTION TARGETS. Soil C sequestration is an important climate mitigation strategy, but it is not a panacea for greenhouse gas emissions reduction. Only a fraction of the reduction can be achieved through sinks (IPCC WGI, 2007). Soil C sequestration, therefore, needs to be considered alongside many other greenhouse gas emissions reduction strategies across all sectors.

The problem of attempting to use soil and vegetation to sequester C as a climate mitigation measure has been summarized succinctly by W.H. Schlesinger as 'trying to sequester the geosphere in the biosphere'.

The C that humans are currently releasing through fossil fuel use has been locked up in the geosphere for hundreds of millions of years, and was accumulated over many millions of years. Using the biosphere to capture this geospheric C does not add up – the geospheric C released is too large for the biosphere to store effectively. Given this knowledge, reducing C emissions is obviously more important than attempting to sequester the C after it has been released.

Conclusions

There is still some uncertainty over future responses of soils to climate change, but most projections suggest that, globally, soils either lose only small quantities of soil C or soil C stocks may, in fact, increase. The global picture, however, is underpinned by considerable regional variation in response, with the response determined by a combination of factors, including opposite impacts of increased temperature and decreased soil moisture on decomposition rates and the balance between changes in C losses from decomposition and C gains through increased productivity.

In terms of using soils to mitigate climate change, soil C sequestration globally has a large, cost-competitive mitigation potential. Soil C sequestration can be useful to meet short- to medium-term targets, especially if these targets are large. In addition to the mitigation potential, increasing soil C stocks provides many co-benefits in terms of soil fertility, workability, water-holding capacity, nutrient cycling, reduced emissions risk and a range of other positive soil attributes (Lal, 2004). These arguments for using C sequestration for climate mitigation need to be weighed against the limitations discussed above; for example, time limitation, non-permanence, displacement and difficulties in verification. Despite these limitations, soil C sequestration may have a role in reducing the short-term atmospheric CO_2 concentration, thus buying time to develop longer-term emissions reduction solutions across all sectors of the economy.

Acknowledgements

This project contributes to SmartSoil (EU-FP7-KBBE-2011-5; Grant agreement number: 289694). This paper is an update of a paper published by Smith (2012; Soils and climate change. *Current Opinion in Environmental Sustainability* 4, 539–544).

References

Arnell, N.W., Lowe, J.A., Brown, S., Gosling, S.N., Gottschalk, P., Hinkel, J., Lloyd-Hughes, B., Nicholls, R.J., Osborne, T.M., Smith, P. and Warren, R.A. (2013) The impacts of climate change avoided by climate policy: a global assessment. *Nature Climate Change* 3, 512–519.

Arnell, N.W., Brown, S., Gosling, S.N., Gottschalk, P.A., Huntingford, C.E., Kovats, R.S., Lloyd-Hughes, B., Nicholls, R.J., Osborn, T.J., Osborne, T.M., Smith, P. and Wheeler, T.R. (2014) The impacts of climate change across the globe: a multi-sectoral assessment. *Climatic Change* (in revision).

Batjes, N.H. (1996) Total carbon and nitrogen in the soils of the world. *European Journal of Soil Science* 47, 151–163.

Ciais, P., Wattenbach, M., Vuichard, N., Smith, P., Piao, S.L., Don, A., Luyssaert, S., Janssens, I., Bondeau, A., Dechow, R. *et al.* (2010) The European greenhouse gas balance revisited. Part 2: croplands. *Global Change Biology* 16, 1409–1428.

Cox, P.M., Betts, R.A., Jones, C.D., Spall, S.A. and Totterdell, I.J. (2000) Acceleration of global warming due to carbon-cycle feedbacks in a coupled climate model. *Nature* 408, 184–187.

Cramer, W., Bondeau, A., Woodward, F.I., Prentice, I.C., Betts, R.A., Brovkin, V., Cox, P.M., Fisher, V., Foley, J.A., Friend, A.D. *et al.* (2001) Global response of terrestrial ecosystem structure and function to CO_2 and climate change: results from six dynamic global vegetation models. *Global Change Biology* 7, 357–373.

Davidson, E.A. and Janssens, I.A. (2006) Temperature sensitivity of soil carbon decomposition and feedbacks to climate change. *Nature* 440, 165–173.

Dawson, J.J.C. and Smith, P. (2007) Carbon losses from soil and its consequences for land management. *Science of the Total Environment* 382, 165–190.

Friedlingstein, P., Cox, P., Betts, R., Bopp, L., Von Bloh, W., Brovkin, V., Cadule, P., Doney, S., Eby, M., Fung, I. et al. (2006) Climate–carbon cycle feedback analysis: results from the (CMIP)-M-4 model intercomparison. *Journal of Climate* 19, 3337–3353.

Gottschalk, P., Smith, J.U., Wattenbach, M., Bellarby, J., Stehfest, E., Arnell, N., Osborn, T. and Smith, P. (2012) How will organic carbon stocks in mineral soils evolve under future climate? Global projections using RothC for a range of climate change scenarios. *Biogeosciences* 9, 3151–3171, doi:10.5194/bg-9-3151-2012.

Houghton, R.A. (1999) The annual net flux of carbon to the atmosphere from changes in land use 1850 to 1990. *Tellus* 50B, 298–313.

Houghton, R.A., Hackler, J.L. and Lawrence, K.T. (1999) The US carbon budget: contributions form land-use change. *Science* 285, 574–578.

IPCC (Intergovernmental Panel on Climate Change) (2000) *Special Report on Land Use, Land Use Change, and Forestry*. Cambridge University Press, Cambridge, UK.

IPCC SRES (2000) *Special Report on Emissions Scenarios*. Cambridge University Press, Cambridge, UK.

IPCC WGI (2001) *Climate Change: The Scientific Basis*. Cambridge University Press, Cambridge, UK.

IPCC WGI (2007) *Climate Change. The Physical Science Basis*. Cambridge University Press, Cambridge, UK.

Ito, A. (2005) Climate-related uncertainties in projections of the twenty-first century terrestrial carbon budget: off-line model experiments using IPCC greenhouse-gas scenarios and AOGCM climate projections. *Climate Dynamics* 24, 435–448.

Joosten, H., Sirin, A., Couwenberg, J., Laine, J. and Smith, P. (2014) The role of peatlands in climate regulation. In: Bonn, A., Allott, T., Evans, M., Joosten, H. and Stoneman, R. (eds) *Peatland Restoration and Ecosystem Services*. Cambridge University Press, Cambridge, UK (in press).

Lal, R. (1999) Soil management and restoration for C sequestration to mitigate the accelerated greenhouse effect. *Progress in Environmental Science* 1, 307–326.

Lal, R. (2004) Soil carbon sequestration impacts on global climate change and food security. *Science* 304, 1623–1627.

Lal, R., Kimble, J.M., Follet, R.F. and Cole, C.V. (eds) (1998) *The Potential of U.S. Cropland to Sequester Carbon and Mitigate the Greenhouse Effect*. Ann Arbor Press, Chelsea, Michigan.

Lucht, W., Schaphoff, S., Erbrecht, T., Heyder, U. and Cramer, W. (2006) Terrestrial vegetation redistribution and carbon balance under climate change. *Carbon Balance and Management* 1, 6, doi:10.1186/1750-0680-1-6.

Powlson, D.S., Whitmore, A.P. and Goulding, K.W.T. (2011) Soil carbon sequestration to mitigate climate change: a critical re-examination to identify the true and the false. *European Journal of Soil Science* 62, 42–55.

Schimel, D.S. (1995) Terrestrial ecosystems and the carbon-cycle. *Global Change Biology* 1, 77–91.

Schimel, D.S., House, J.I., Hibbard, K.A., Bousquet, P., Ciais, P., Peylin, P., Braswell, B.H., Apps, M.J., Baker, D., Bondeau, A. et al. (2001) Recent patterns and mechanisms of carbon exchange by terrestrial ecosystems. *Nature* 414, 169–172.

Searchinger, T., Heimlich, R., Houghton, R.A., Dong, F., Elobeid, A. and Fabiosa, J. (2008) Use of U.S. croplands for biofuels increases greenhouse gases through emissions from land use change. *Science* 319, 1238–1240.

Smith, J.U., Smith, P., Wattenbach, M., Zaehle, S., Hiederer, R., Jones, R.J.A., Montanarella, L., Rounsevell, M.D.A., Reginster, I. and Ewert, F. (2005) Projected changes in mineral soil carbon of European croplands and grasslands, 1990–2080. *Global Change Biology* 11, 2141–2152.

Smith, P. (2004) Monitoring and verification of soil carbon changes under Article 3.4 of the Kyoto protocol. *Soil Use and Management* 20, 264–270.

Smith, P. (2005) An overview of the permanence of soil organic carbon stocks: influence of direct human-induced, indirect and natural effects. *European Journal of Soil Science* 56, 673–680.

Smith, P. (2008a) Do agricultural and forestry carbon offset schemes encourage sustainable climate solutions? *International Journal of Agricultural Sustainability* 6, 169–170.

Smith, P. (2008b) Land use change and soil organic carbon dynamics. *Nutrient Cycling in Agroecosystems* 81, 169–178.

Smith, P. (2012) Soils and climate change. *Current Opinion in Environmental Sustainability* 4, 539–544.

Smith, P. and Fang, C. (2010) A warm response by soils. *Nature* 464, 499–500.

Smith, P. and Olesen, J.E. (2010) Synergies between mitigation of, and adaptation to, climate change in agriculture. *Journal of Agricultural Science* 148, 543–552.

Smith, P. and Powlson, D.S. (2003) Sustainability of soil management practices – a global perspective. In: Abbott, L.K. and Murphy, D.V. (eds) *Soil Biological Fertility – A Key to Sustainable Land Use in Agriculture*. Kluwer Academic Publishers, Dordrecht, the Netherlands, pp. 241–254.

Smith, P. and Trines, E. (2007) Agricultural measures for mitigating climate change: will the barriers prevent any benefits to developing countries? *International Journal of Agricultural Sustainability* 4, 173–175.

Smith, P., Powlson, D.S., Glendining, M.J. and Smith, J.U. (1997) Potential for carbon sequestration in European soils: preliminary estimates for five scenarios using results from long-term experiments. *Global Change Biology* 3, 67–79.

Smith, P., Powlson, D.S., Smith, J.U., Falloon, P.D. and Coleman, K. (2000) Meeting Europe's climate change commitments: quantitative estimates of the potential for carbon mitigation by agriculture. *Global Change Biology* 6, 525–539.

Smith, P., Smith, J.U., Wattenbach, M., Meyer, J., Lindner, M., Zaehle, S., Hiederer, R., Jones, R., Montanarella, L., Rounsevell, M., Reginster, I. and Kankaanpää, S. (2006) Projected changes in mineral soil carbon of European forests, 1990–2100. *Canadian Journal of Soil Science* 86, 159–169.

Smith, P., Martino, D., Cai, Z., Gwary, D., Janzen, H.H., Kumar, P., McCarl, B., Ogle, S., O'Mara, F., Rice, C. et al. (2007) Policy and technological constraints to implementation of greenhouse gas mitigation options in agriculture. *Agriculture, Ecosystems and Environment* 118, 6–28.

Smith, P., Fang, C., Dawson, J.J.C. and Moncreiff, J.B. (2008a) Impact of global warming on soil organic carbon. *Advances in Agronomy* 97, 1–43.

Smith, P., Martino, D., Cai, Z., Gwary, D., Janzen, H.H., Kumar, P., McCarl, B., Ogle, S., O'Mara, F., Rice, C. et al. (2008b) Greenhouse gas mitigation in agriculture. *Philosophical Transactions of the Royal Society, B* 363, 789–813.

21 Impacts of Land-use Change on Carbon Stocks and Dynamics in Central-southern South American Biomes: Cerrado, Atlantic Forest and Southern Grasslands

Heitor L.C. Coutinho*, Elke Noellemeyer, Fabiano de Carvalho Balieiro, Gervasio Piñeiro, Elaine C.C. Fidalgo, Christopher Martius and Cristiane Figueira da Silva

Abstract

Land-use changes (LUC) are one of most significant global change processes of the current era, with noticeable consequences on habitat loss, due mainly to agricultural expansion and urbanization. The carbon cycle dynamics can be affected significantly by LUC, with impacts on carbon sequestration and emission rates. Considering the direct effect of carbon gases enrichment of the atmosphere on climate change, it is of utmost importance to improve the knowledge base on the impacts of agricultural-based LUC on carbon sinks, such as soils. This chapter reviews the available data on the effects of LUC on soil carbon stocks in three major biomes of the southern portion of the South American continent (the Cerrado, the Southern Grasslands and the Atlantic Forest). The area of soybean crops has expanded almost four times in the La Plata Basin Grasslands of Argentina over the past decade, and near ten times in the Brazilian Cerrado since the mid-1980s. The area under sugarcane crops in Brazil has almost doubled since the mid-1990s, occupying approximately 8.5 million ha (Mha) in 2009. In 2011, forestry plantations occupied 28% more land in Brazil than in 2005, with a total area of 6.5 Mha (75% with *Eucalyptus* and 25% with *Pinus*). In general, all conversions of natural vegetation to agricultural land-use systems in the different biomes have resulted in significant losses of soil carbon stocks. The conversion of pastures and grasslands to annual croplands in the Rolling Pampas grasslands has decreased C stocks by 50% over the last century. This represents a much faster loss rate than the loss triggered by the introduction of domestic herbivores over the course of the previous nearly four centuries (22%). These results imply that soil degradation caused by annual crops is very rapid and results in a strong decrease in carbon stocks. However, adopting soil and water conservation management strategies and increasing the complexity of the cropping systems – through adoption of no-tillage (NT) agriculture, well-managed pasture systems, integrated crop–livestock–forestry systems, multiple cropping and crop rotation with legume cover species, for example, can improve soil carbon sequestration rates by up to nearly 2.0 Mg C ha^{-1} year^{-1}. The elimination of preharvest burning practices in sugarcane crops alone can result in gains of up to 0.93 t C year^{-1} ha^{-1}. Improving soil and crop management to boost carbon sequestration in agricultural systems, while at the same time increasing resilience by improving soil quality, is a potential climate change mitigation option for farmers in South America.

*E-mail: heitor.coutinho@embrapa.br

Introduction

The central and south-eastern part of the South American continent is characterized by heterogeneous biomes and ecosystems, with diverse soil, vegetation, climatic, geologic, hydrologic and social configurations (Plate 13). The varying topography adds natural complexity to the region, resulting in a combination of extensive savannah-like plateaus (Cerrado in Brazil), grasslands (Pampas in Argentina and Uruguay), the largest tropical wetland area in the world (Pantanal in Brazil) and both dry and humid forest biomes (Espinal, Chaco and Atlantic Forests, respectively). Together with this environmental heterogeneity, the diverse human societies that compose the region have led to complex patterns of human-induced land-use change. This has accelerated since the mid-1970s, as high demands for commodities such as soybeans, maize, wheat and sunflowers have driven major land-use change processes in the different biomes, mainly in the Cerrado and the Southern Grasslands. These changes resulted in substitution of native ecosystems and traditional extensive cattle ranching. More recently, considerable portions of both grain crops and pastureland have been converted to sugarcane and forestry, to help meet the growing demands of the biofuel, timber and cellulose industries.

Common effects of agricultural expansion on tropical soil resources described in the literature include depletion of soil organic matter, with consequent reduction of carbon stocks and fertility; soil compaction and reduction in water-holding capacity; erosion and desertification; and biological degradation. These changes in the soil, coupled with other natural resource impairments such as biodiversity degradation, can result in severe disruptions in ecosystem function, affecting services such as hydrologic regulation, water quality and erosion control, as well as increasing CO_2 emissions to the atmosphere.

Considering the role of agricultural soils as important modulators of carbon sequestration and greenhouse gas emissions, we need to understand how land-use and management changes affect the rates and dynamics of carbon sequestration by soils in regions prone to intense land-use change processes in order to be able to improve climate change mitigation and adaptation in land use.

In this chapter, we describe the main processes of past and current agricultural expansion in the Cerrado, Southern Grasslands and the Atlantic Forest, representative biomes of the central and south-eastern portions of South America. We review data derived from published research on the impact assessments of land-use and management changes in the region with a focus on carbon stocks and dynamics. We end with suggestions of land-use and management-related policies, as well as soil and crop management options, to mitigate some of the reported environmental impacts of LUC in the region.

Land-use Change

Four agricultural sectors have been responsible for most of the land-use change (LUC) in the region since the mid-1970s: international commodities (mainly soybeans, maize, sunflowers and wheat), wood (eucalypt and pine), meat (cattle) and biofuels (sugarcane). Soybeans and other oilseeds (used for food, feed and biofuel production), grain crops and cultivated pastures (*Brachiaria* spp., lucerne–grass mixtures) are widespread and have replaced portions of all biomes considered in this review. Biofuel crops, especially sugarcane, are expanding to meet the demand of growing international markets and national policies (Cerqueira Leite *et al.*, 2009; Lago, 2012). Sugarcane cultivation is concentrated in the Brazilian State of São Paulo (approximately 60% of the Brazilian sugarcane production) and has expanded most rapidly in the states of Mato Grosso do Sul and Goiás (Goes *et al.*, 2008; Uriarte *et al.*, 2009; Monteiro *et al.*, 2010; Rudorff *et al.*, 2010). In the Southern Grasslands, traditional cattle breeding, a more low-impact, environmentally sustainable land-use system, has preserved most of the biome's integrity (Rótolo *et al.*, 2007), but agriculture and forestry expansions have been decimating the natural ecosystems at an alarming rate since the late 20th century (Baldi and Paruelo, 2008).

Cerrado

The Cerrado biome covers approximately 2 million km², which represents approximately 25% of Brazil (Sano et al., 2008). In size, the Cerrado biome is the second largest ecosystem in Brazil after the Amazon. With 6500 plant species, the Cerrado has only 23% fewer species than the Amazon; it is one of the most diverse tropical biomes and the most species-rich tropical savannah in the world (UNEP, 2010). A vast portion of this biome contains the headwaters of the Parana and Paraguay River sub-basins, components of the La Plata River Basin. The original Cerrado landscape suffered widespread transformation after agricultural innovations provided technologies and crop varieties adapted to the Cerrado soils and climate (Miyasaka and Medina, 1981; Spehar, 1995). Several authors have suggested that the Cerrado is being modified due to land-use expansion, particularly for grain crops and pastures (Fearnside, 2001; Klink and Moreira, 2002). This effect is due partly to the fact that the more open, savannah-type structure invites human settlements and developments. The southern portion of the Cerrado biome, located in the La Plata River Basin, is where most of the land-use conversion to agriculture and pasture has taken place (Plate 13). The Brazilian states of São Paulo, Goiás and Mato Grosso do Sul were reported to have retained only 15–32% of their native Cerrado vegetation (2002 land-use data) (Sano et al., 2008).

LUC in the Cerrado initially aimed at opening the frontier for cattle grazing. Typically, the occupation began with the removal of native woody Cerrado vegetation for charcoal production, mostly used as an energy source for steel mills located elsewhere in Brazil (Uhlig et al., 2008).

The second major LUC process was the expansion of cash crops, especially soybeans (Spehar, 1995; Alves et al., 2003) (Fig. 21.1). Currently, more than 60% of Brazil's soybean production derives from the Cerrado (Souza et al., 2007), where it occupies an estimated area of approximately 13 Mha.

Land is steadily converted to forest plantations in the Brazilian Cerrado at an average annual rate of 3.5%. From 2005 to 2010, the forestry area grew by 23.0% (ABRAF, 2012). The Brazilian states with the highest *Eucalyptus* spp. plantation expansion rates during this period were Mato Grosso do Sul (30%), Maranhão (10.2%), Tocantins (7.3%) and Minas Gerais (7.7%), all under characteristic Cerrado vegetation and climate.

Southern Grasslands

Agricultural expansion into the marginal semi-arid regions of the La Plata River Basin, such as the provinces of La Pampa and San Luis, increased deforestation of the native savannah vegetation significantly in favour of sunflower, soybean and maize crops. Despite the relatively adverse climatic conditions for soybean cropping, the area planted to this crop in La Pampa expanded by 370% during the past decade, and grain production followed a similar trend with a 360% increase (Fig. 21.2).

Traditionally, the cash crops in this semi-arid, marginal agricultural region have been wheat since the early 20th century and sunflower since the 1970s. The soybean expansion has reduced the planted areas of both these crops, especially that of wheat (Fig. 21.3). The total cropped area, however, has decreased by about 30% during the last decade, from 942,000 ha in 2000 to 659,000 ha in 2008, due to drought events in 2004 and 2007–2009 (Gobierno de La Pampa, 2009).

Apart from expanding, agriculture has also intensified and industrialized throughout the region. While traditional agriculture was based on diversified mixed systems that implied rotations of cash crops with permanent pastures for livestock production, at present, especially in the Argentinean Grasslands, monocultures of cash crops predominate.

At the same time, afforestation of marginal lands has become an important feature of LUC in the Southern Grasslands biome due to increasing demands from the paper industry. Results from ongoing studies show that forest cover on soils originally formed under grassland in the Argentinean Pampas has significant effects on their chemical and physical properties, resulting in significant impacts on soil fertility, carbon stocks and hydrology.

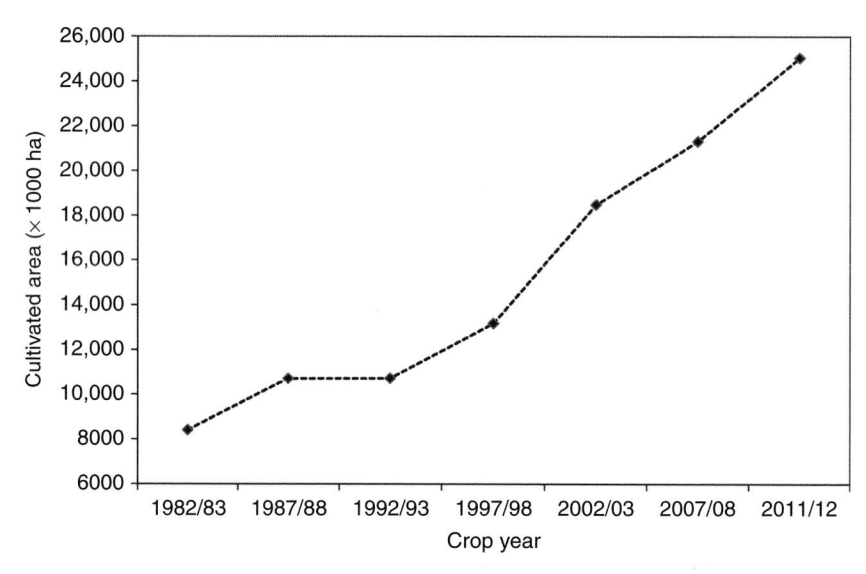

Fig. 21.1. Soybean expansion rates in the Cerrados (1982–2012). (From Conab, 2013.)

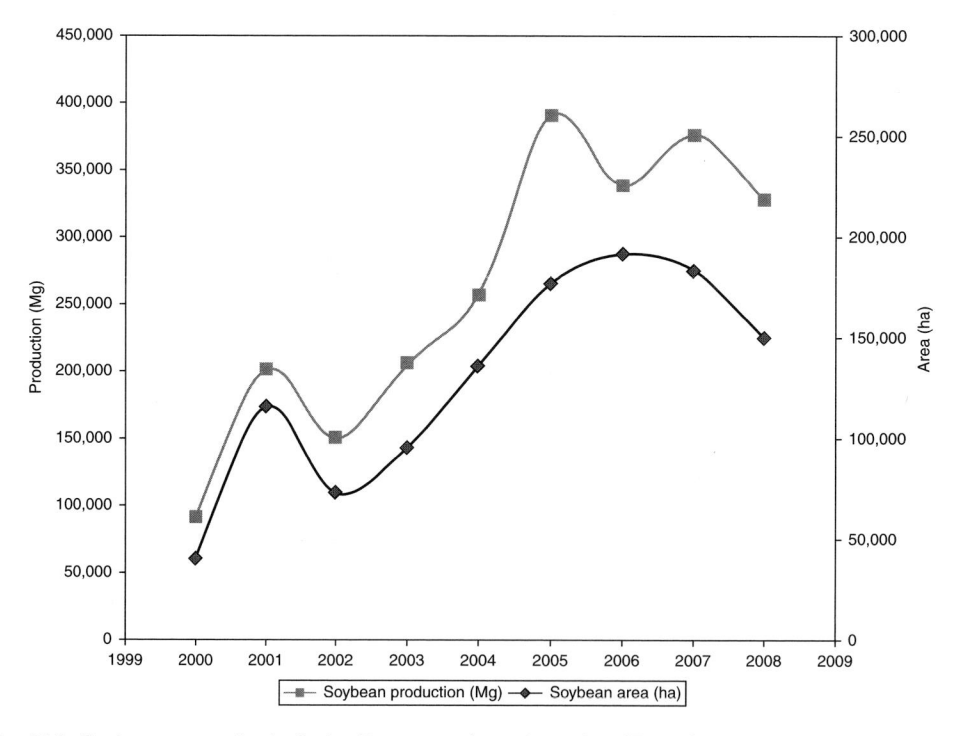

Fig. 21.2. Soybean expansion in the La Pampa province, Argentina. (From Gobierno de La Pampa, 2009, p. 135.)

Atlantic Forest

The main agricultural, urban and industrial development in the 19th and early 20th centuries was concentrated in the Atlantic Rainforest biome, in addition to the Southern Grasslands, with the Cerrado coming into play in the 1970s for cattle grazing

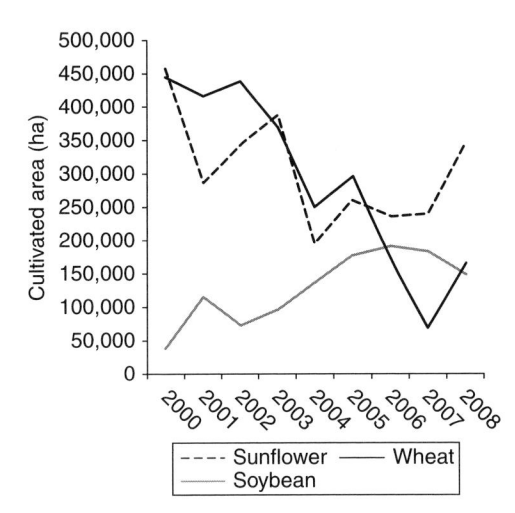

Fig. 21.3. Evolution of areas cultivated with wheat, sunflowers and soybeans in the La Pampa province, Argentina, during the past decade. (From Gobierno de La Pampa, 2009.)

(Neto *et al.*, 2010). The Atlantic Forest is the most threatened biome in Brazil (Schaffer and Prochnow, 2002). Originally covering about 1,315,000 km² (Plate 13), it was reduced to a discontinuous area of approximately 100,000 km²; that is, approximately 8% of its original extent (Morellato and Haddad, 2000; Atlântica, 2012). This biome contains a species diversity higher than most of the Amazon forests, and is characterized by high levels of endemism (Brown and Brown, 1992).

Land-use dynamics of the Atlantic Forest biome could be summarized by the onset of the sugarcane (*Saccharum officinarum*) cycle, Brazil's first major export crop, which started being planted by the Portuguese settlers in the 16th century, followed by coffee, introduced in the 18th century. Timber extraction and charcoal production has been a constant activity in this biome (Dean, 1995; Morellato and Haddad, 2000; Boddey *et al.*, 2006; Salemi *et al.*, 2012). The impact of forest clearing for agriculture, followed by pastures, created degraded landscapes, with significant impacts on the provisioning of ecosystem services (biodiversity conservation, soil water retention, soil stabilization and carbon sequestration) (Boddey *et al.*, 2006; Macedo *et al.*, 2008; Machado *et al.*, 2010; Tabarelli *et al.*, 2010).

Annual Crops and Pasture: Impacts on Carbon Stocks and Dynamics

Carbon stocks in Cerrado soils are close to those found under Brazilian rainforests, with values varying from 133 to 236 Mg C ha^{-1} in the 0–100 cm layer (Lopes-Assad, 1997; Corazza *et al.*, 1999; Bustamante *et al.*, 2006). Analysing data from the literature, we estimate carbon stocks in the surface soil layer (0–20 cm) of different Cerrado phytophysionomies ('cerradão', 'cerrado *stricto sensu*', 'campo limpo', 'campo sujo' and 'forest') to be 46 ± 15 t C ha^{-1}, being an important carbon reservoir (Corazza *et al.*, 1999; Freitas *et al.*, 2000; D'Andréa *et al.*, 2004; Neves *et al.*, 2004; Silva *et al.*, 2004; Bayer *et al.*, 2006; Corbeels *et al.*, 2006; Siqueira Neto, 2006; Frazão, 2007; Machado *et al.*, 2007; Rangel and Silva, 2007; Dieckow *et al.*, 2009; Matias *et al.*, 2009; Salton *et al.*, 2011).

In general, the conversion of native vegetation to agricultural systems, especially under conventional tillage, results in significant reductions of these stocks (Corazza *et al.*, 1999; D'Andréa *et al.*, 2004; Oliveira *et al.*, 2004; Silva *et al.*, 2004; Bayer *et al.*, 2006; Bustamante *et al.*, 2006; Jantalia *et al.*, 2007).

Soil carbon stocks (SCSs) (0–30 cm) under different land uses and soil classes in Brazil were estimated by processing data from the Embrapa Soil Information System's database (http://www.bdsolos.cnptia.embrapa.br) and land-use maps available at the Brazilian Ministry of Environment website (Fidalgo *et al.*, 2007). Latosols (Oxisols, USDA Soil Taxonomy) are the dominant soil order (49% of total area) in the Cerrado biome, and the discrimination between land uses is noticeable, with 43.8, 40.6 and 34.2 Mg C ha^{-1} on average for native vegetation, pasture and agriculture, respectively, in a total of 113 soil profiles (Elaine C.C. Fidalgo, personal communication).

In the rolling Pampas grassland (northern part of the biome), where soil organic matter is the main C pool, converting pastures and grasslands to annual croplands decreased C stocks by 50% in one century

(Alvarez, 2001). Considering the rolling Pampas natural ecosystems (herbivore exclusion zones) harboured a total SCS of 68 t C ha^{-1} (Andriulo et al., 1999), a century after annual crops were established in the biome, these stocks were reduced to 34 Mg C ha^{-1} (Alvarez, 2001). On the other hand, Piñeiro et al. (2006) used the Century model to estimate SCS losses associated with herbivory and concluded that 370 years after livestock was introduced in the Pampas grasslands, there was a 22% reduction in SCSs in the 0–20 cm depth. This is a dramatically different response to LUC.

LUC in the semi-arid Pampas caused rapid soil C losses, with C half-lives just above 10 years and no evidence of long-term stabilized C in any of the soil fractions (Zach et al., 2006). In this study, soils originally under long-term pasture or natural vegetation, with initial C contents between 24 and 33 mg C g^{-1} bulk soil, lost 33–57% of their original C within 12–18 years of continuous cultivation. In degraded soils that had been restored with pasture, C accretion was rapid, but levelled off well below the original C levels (Zach et al., 2006).

The conversion of natural pastures to cash cropping brings about similar carbon stock losses, as does the deforestation of native vegetation (Noellemeyer et al., 2008). A 16% reduction was observed only 2 years after converting pasture to agriculture, and 14 years later, soil C stocks were reduced by 32% after livestock was introduced.

Conservation or no-tillage agriculture (NT) can improve the carbon sequestration in soils under agricultural use in the semi-arid and subhumid regions of the Pampas. Long-term field experiments have shown that under NT significantly higher amounts of carbon were sequestered by the soil than under conventional tillage (CT) (16.6 versus 13.2 Mg C ha^{-1} under NT and CT, respectively, 9 years after establishment of crops). This finding indicates that soils under NT can act as a carbon sink, while those under CT as a carbon source (Quiroga et al., 2009). However, in NT production systems without crop rotation this might not be the case (Díaz Zorita et al., 2002; Steinbach and Alvarez, 2006; Álvarez et al., 2009).

LUC and the intensification of agricultural systems also alter the biological activity in soils, with potential effects on soil carbon stabilization and storage. Respiration rates of different soil aggregate-size classes are directly related to their carbon contents. Both were substantially lower in agricultural than in pasture soils, indicating higher biological activity in the latter, which increases the cycling rates and stabilization of organic carbon in the soil as organo-mineral complexes (Denef et al., 2009). This was corroborated by the higher levels of silt- and clay-associated carbon in these soils as compared to their agricultural counterparts (Noellemeyer et al., 2008). As indicated above, soils under NT accrue more carbon than those under CT, especially in the topsoil. This provides more substrate for microbial metabolism, and, subsequently, higher respiration rates have been found under NT when compared to CT in incubation experiments, indicating an active soil biota and conditions that might favour stabilization of residue-derived carbon. Although one could conclude that higher respiration would lead to carbon loss, the enhanced microbial activity that processes the organic matter inputs and the physical protection of soils under no-tillage systems increase soil aggregation, which leads to stabilization of soil organic carbon decomposition products (Fernández et al., 2010).

Modern industrialized agricultural production systems generally focus on very few crops that only occupy the land surface during short periods of the year. The average growing season in the Southern Grasslands is 5 months for soybean and many maize hybrids and 4 months for sunflowers and modern wheat hybrids. Thus, in most situations, the soil is fallow during more than half of the year, exposing its surface to wind and water erosion and causing important soil water evaporation losses. Double cropping, such as growing wheat during winter and soybeans during summer, could effectively counteract these effects by keeping the soil protected. In Brazil, a lot of effort is being placed on the development of integrated crop–livestock–forestry systems, shown to be economically and technically feasible in the

Cerrados by ensuring long-term ground cover, enhanced carbon fixation, increases in soil organic matter content and reduction in the emission of greenhouse gases, when compared to conventional systems (Pacheco et al., 2012).

Long fallow periods and lower root residue input to the soil imply that biological activity is concentrated at the soil surface, which is more evident under NT, and that deeper soil layers receive less organic matter inputs to maintain biological functions. The consequences are well documented as an increase in bulk density, a decrease in infiltration rates and compaction of soil at depths varying from 30 to 50 cm, leading to increased runoff, erosion and waterlogging. Maintaining good soil structure throughout the soil profile requires a sustained biological activity responsible for aggregate formation and pore stabilization. An approach to solve this problem for industrialized monocultures is using cover crops, which are non-harvest crops grown during the fallow interval to provide soil cover and protection against wind and water erosion, and providing an active rhizosphere to promote aggregate formation and carbon storage. Preliminary studies in the semi-arid and subhumid region of the Pampas have shown that cover crops can provide significant amounts of residue cover on the soil surface, which in turn immobilizes nitrogen and prevents its leaching into the groundwater and that cover crops do not decrease water availability to subsequent cash crops (Fernández et al., 2010).

Sugarcane Production and Impacts on Carbon Stocks and Dynamics

Knowledge on the impacts of sugarcane occupation on soil attributes derives mainly from research in the Atlantic Forest biome, where most Brazilian sugarcane is cultivated (Campos et al., 2001; Resende et al., 2006; Balieiro et al., 2008; Sant'Anna et al., 2009; Pinheiro et al., 2010), but some research has been undertaken in the hotspot sugarcane expansion areas of the Cerrados, in the states of Mato Grosso do Sul and Goiás (Galdos et al., 2009; Angelini et al., 2010; Rachid, 2010; Rachid et al., 2012).

In the last decade, sugarcane plantations for bioethanol and sugar production have notably expanded in the Cerrado in response to the rapidly growing demand for biofuels and favoured by the suitable topography, soil characteristics, water availability and climate, and by national policies (Manzatto et al., 2009; Oosterveer and Mol, 2010). Most of the sugarcane plantations established in Brazil in recent years replaced pastureland (approximately 65% during the 2007/08 season), but also soybeans, orange and maize crops (together 27%) – food staples of global importance (Gauder et al., 2011). The result has been a fast and steady expansion of sugarcane for bioethanol and sugar production in the Cerrados, mainly in the Brazilian states of Mato Grosso do Sul and Goiás (INPE, 2011) (Fig. 21.4).

Most authors report a significant reduction in soil carbon and nitrogen stocks in sugarcane plantations when compared with native forests (Campos et al., 2004; Resende et al., 2006; Pinheiro et al., 2010; INPE, 2011). Campos et al. (2004), for example, showed that 22 years after forest conversion to pasture in the Atlantic Forest, 29% of the SCSs were lost. On a neighbouring field where a 10-year-old sugarcane field succeeded 12 years under pasture, SCS losses increased to 43%, compared to that under the native forest (as inferred by soils collected from a neighbouring forest) (Fig. 21.5).

Some labile organic matter fractions, such as microbial biomass and particulate organic matter, have been shown to be sensitive even to the slightest soil-quality change, being good indicators of the ecosystem services that are necessary for a sustainable agricultural system, as demonstrated by several authors (Graham and Haynes, 2006; Silva et al., 2007; Galdos et al., 2009; Sant'Anna et al., 2009; Souza et al., 2012). Souza et al. (2012) showed that native vegetation and unburned sugarcane exhibited qMic values (microbial quotient = carbon on microbial biomass/total organic carbon) of 2.64 and 2.84%, respectively,

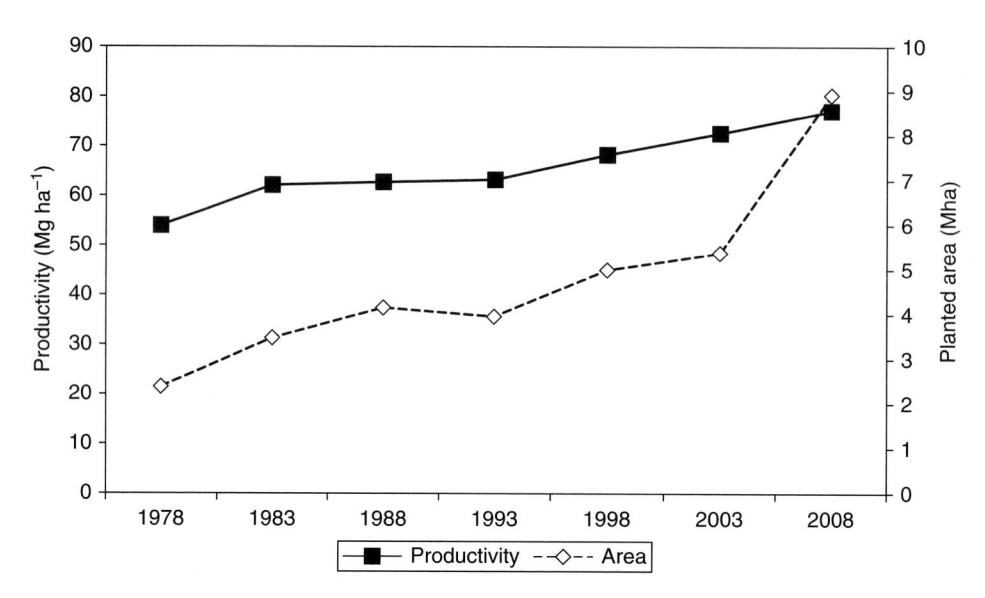

Fig. 21.4. Total sugarcane crop area (Mha) and yield (Mg ha^{-1}) in Brazil, from 1978 to 2008. (From Brasil/Ministério da Agricultura Pecuária e Abastecimento (2009), data source: IBGE.)

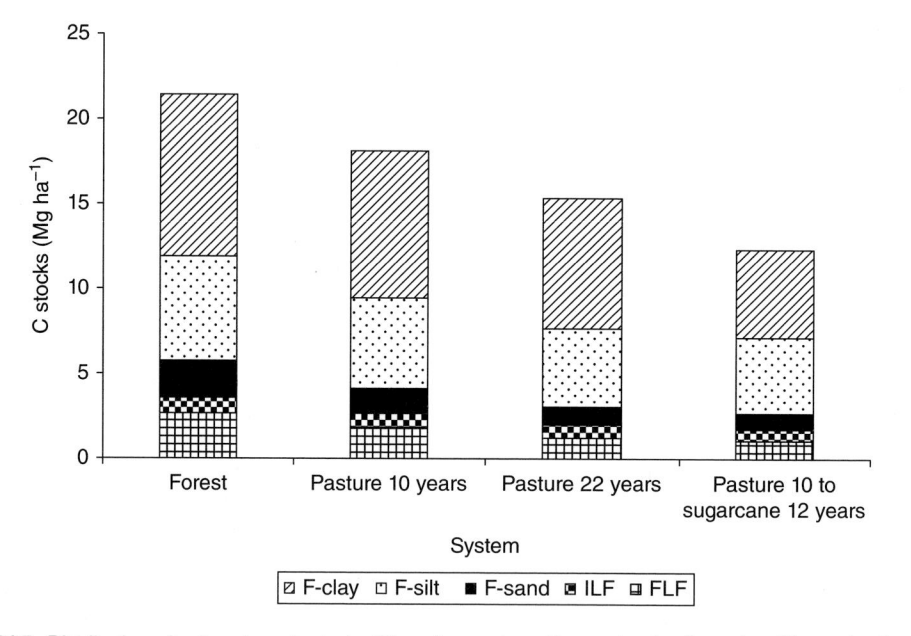

Fig. 21.5. Distribution of soil carbon stocks in different organic matter pools of soils under different land uses in the Atlantic Forest biome (FLF, free light fraction; ILF, intra-aggregate light fraction); F-sand, F-silt and F-clay: organic matter associated to sand, silt and clay particles, respectively. (From Campos et al., 2004).

significantly higher than burnt sugarcane (2.01%). This indicates a higher availability of substrates to microorganisms, and a positive influence on microbial biomass. Rachid et al. (2012) have observed the effects of sugarcane crop management on soil bacterial community structure and nitrogen cycle functional gene diversity (ammonia oxidation and denitrification) in the Cerrado and related those to physical and chemical soil attributes. A significantly higher impact was caused by preharvest burning, and a high correlation was found between the microbial metagenomic data and attributes such as soil bulk density and water-filled pore spaces.

Forest Plantations and Impacts on Carbon Stocks and Dynamics

Eucalyptus and pine forests are the most important sources of wood, cellulose and charcoal for industry worldwide; thus, afforestation with these species is expanding significantly all over the world. Planted forests account for an estimated 7% of the total forest area of the world (FAO, 2010). Between 1980 and 2005, global annual industrial wood production increased from 1450 to 1710 million m^3 year^{-1}, while that for energy production increased from 1530 to 1840 million m^3 year^{-1} (FAO, 2008). In Brazil, *Eucalyptus* and *Pinus* spp. forests occupied about 6.51 Mha in 2011, with 75% covered by eucalyptus alone (ABRAF, 2012).

In the Cerrados, especially in degraded pastures, several authors observed a positive effect of *Eucalyptus* spp. afforestation on soil organic carbon stocks. For example, during the aggrading period, Lima et al. (2006) observed carbon fixation rates of up to 0.57 Mg C ha^{-1} year^{-1}. These values are similar to those found in Cerrado oxisol soils (0–20 cm) under annual no-till crops (Sá et al., 2001; Bayer et al., 2006; Cerri et al., 2007). Higher rates of carbon sequestration were found by Corazza et al. (1999) in soils under eucalyptus plantations (1.22 Mg C ha^{-1} year^{-1}), as well as under no-till systems (1.43 Mg C ha^{-1} year^{-1}) in Cerrado sites of the Distrito Federal, in Brazil.

Forest stand management can influence SCSs significantly. Studying the impacts of different land uses (savannah, pasture and *Eucalyptus* spp. plantations) and management (60 years under short rotation versus 60 years under continuous growth) on soil carbon and nitrogen stocks, Maquère et al. (2008) found that significant SCSs increased with *Eucalyptus* spp. under short-rotation management (approximately 25%), while SCSs in the continuous forest plantation increased by 15%, both compared with soils under native vegetation.

The impacts of *Eucalyptus* spp. on SCSs varies according to land-use history, previous crop management, local climate and spatial variability of soil attributes. The available data show that good silvicultural management practices (reduced tillage, residue maintenance over the soil and especially consortia with nitrogen-fixing tree species) improve the C sequestration rates under *Eucalyptus* spp. afforestation (Balieiro et al., 2002, 2008; Forrester et al., 2006; Chaer and Totola, 2007; Coelho et al., 2007; Laclau et al., 2008). The positive effect of including legume trees in the eucalyptus management system was demonstrated by Balieiro et al. (2008), who showed that 5 years after the afforestation of pasture fields in a sandy soil of the Atlantic Forest biome, a mixed plantation of *Eucalyptus grandis* and *Pseudosamanea guachapele* (a legume tree native to Central America) resulted in an increase of 6.6 Mg C ha^{-1} in the soil C stocks (0–40 cm), in comparison with single plantations of *E. grandis*. However, no differences in C stocks were observed between soils under pasture (24.24 Mg C ha^{-1}) and mixed plantation (23.83 Mg C ha^{-1}). On the other hand, a study developed by Voigtlaender et al. (2012) with mixed and pure plantations of *Acacia mangium* and *E. grandis* in the Cerrado environment demonstrated that mixed plantations largely increased the turnover rate of nitrogen (N) in the topsoil (124 kg N ha^{-1} year^{-1} for mixed versus 64 kg N ha^{-1} year^{-1} for pure eucalyptus stands), despite a 44% reduction in soil C stocks. They conclude that introducing *A. mangium* trees might improve mineral N availability in soils where commercial eucalyptus plantations have been managed for a long time.

The introduction of nitrogen-fixing trees has also been used as a strategy of land reclamation in Brazil. Significant levels (0–60 cm: 88.1 Mg C ha^{-1}) and rates (1.73 Mg C ha^{-1} year^{-1}) of soil C storage have been shown to occur 13 years after reforestation using N_2-fixing legumes (*A. mangium, Acacia auriculiformis, Enterolobium contortisiliquum, Gliciridia sepium, Leucena leucocephala, Mimosa caesalpiniifolia* and *Paraseroanthes falcataria*) to reclaim an area degraded by gully erosion processes in the Atlantic Forest biome (Macedo *et al.*, 2008).

Results from ongoing studies in the La Plata Basin Grasslands show that the forest cover on soils originally formed under grassland strongly affects their chemical and physical properties, significantly impacting soil fertility, carbon stocks and hydrology (Jobbagy and Jackson, 2003; Jobbágy *et al.*, 2006; Farley *et al.*, 2008; Berthrong *et al.*, 2009a). Tree plantations are often regarded as a global C sequestration option. However, in the Southern Grasslands, the plantations set off part of their biomass C gains with soil C losses, particularly in grassland environments, under humid climates and in first rotations (Berthrong *et al.*, 2009a,b, 2012). The extent of SCS changes due to afforestation will depend mainly on the age of the plantation and the mean annual precipitation (Balieiro *et al.*, 2004, 2008; Resende *et al.*, 2006; Berthrong *et al.*, 2009a,b, 2012; Eclesia, 2011) (Fig. 21.6).

Afforestation usually increases SCSs in dry areas and decreases them in humid environments, depending on the stand age. In specific cases, where afforestation is used to recover degraded agricultural soils, increases in soil C stocks have been observed, depending on the forest species, the presence of mycorrhiza, whether they are grown in consortia with legumes and the soil type (Riestra *et al.*, 2012).

LUC is affecting the albedo (surface reflectance) of large areas in the region significantly (Lee and Berbery, 2012). Albedo changes after afforestation are a direct climate effect that offsets that of C storage on climate. The impact of albedo changes becomes more relevant towards temperate and drier zones. Other unwanted side effects of afforestation in the Southern Grasslands are high water consumption, soil cation depletion and biodiversity losses (Farley *et al.*, 2008; Jackson *et al.*, 2008).

Land Use, Crop Management and Soil Organic Carbon Sequestration

Several researchers have argued that SCSs in the Cerrado biome could be improved if good agricultural practices were adopted (Corazza *et al.*, 1999; Bayer *et al.*, 2006; Bustamante *et al.*, 2006; Jantalia *et al.*, 2007; Carvalho *et al.*, 2010a). Systems under NT, crop rotation or consociation with nitrogen-fixing legume species show the highest carbon sequestration potentials, with a special emphasis on crop–livestock, or crop–livestock–forestry integrated systems (Sisti *et al.*, 2004; Bayer *et al.*, 2006; Bustamante *et al.*, 2006; Corbeels *et al.*, 2006; Jantalia *et al.*, 2007; Batlle-Bayer *et al.*, 2010; Carvalho *et al.*, 2010b). The positive effects of NT on SCSs can be observed after a period of organic matter

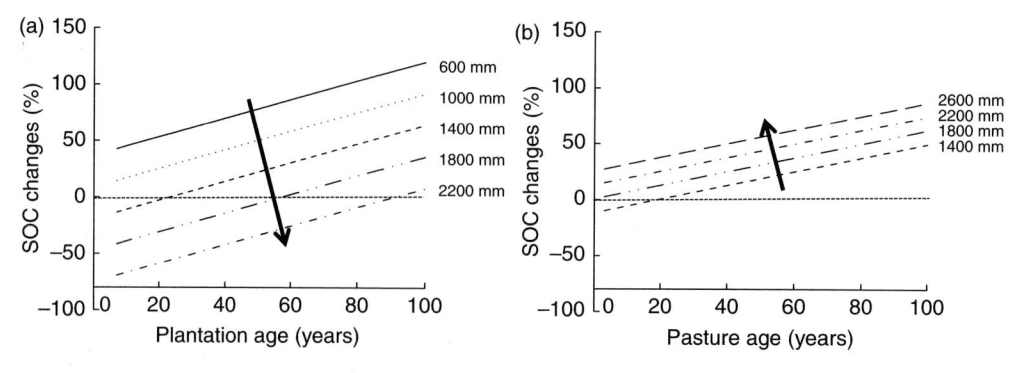

Fig. 21.6. Multiple regression model between the effect of age since conversion to (a) tree plantations and to (b) pasture, and changes in SOC contents at different levels of mean annual precipitation (MAP). The arrows indicate the direction of increasing MAP isohyets. (From Eclesia, 2011.)

accumulation and stabilization, as shown by Jantalia et al. (2007) (Fig. 21.7). In a soil under NT for one year, the SCSs fell by 33% and 17% when compared with native vegetation and conventional tillage (CT), respectively. However, 12 years after the crop management change, SCSs were similar to those under the native vegetation and 26% higher than under CT. Corazza et al. (1999) found high rates of carbon sequestration in soils under no-till systems (1.43 Mg C ha^{-1} year^{-1}) in sites of the Federal District of Brazil, located in the Cerrado biome. This dependency on time was also shown by Siqueira Neto and co-workers (Siqueira Neto, 2006; Siqueira Neto et al., 2010).

The literature shows a wide range of carbon sequestration rates in Cerrado biome soils according to land use, crop and soil management (0.3–1.91 Mg ha^{-1} year^{-1}).

A literature survey, including 40 scientific articles and several papers presented at Brazilian soil scientific meetings reporting SCS measurements under different soil types, land uses and management in the Cerrado biome shows that, on average, NT systems consistently promote the highest levels of SCSs, an average of 76.73 Mg C ha^{-1} year^{-1}, at a depth of 0–40 cm (Table 21.1).

In general, the conversion of native Cerrado and Atlantic Forest vegetation into annual or perennial crops (including sugarcane and eucalyptus, with the exception of pasture) reduces SCSs drastically (Guo and Gifford, 2002; Carvalho et al., 2010b; Melillo et al., 2011). These results were corroborated by Kaschuk et al. (2011), who used data integration and meta-analysis approaches to analyse soil biological and chemical attributes from different Brazilian biomes and land uses. The authors demonstrate that, in general, the conversion of native vegetation to annual crops has a strong negative impact on all variables studied (carbon in the soil microbial biomass, total organic soil carbon and metabolic coefficient), this being more pronounced in the Cerrado biome.

The literature data show a wide range of carbon sequestration rate values in Cerrado biome soils according to land use, crop, soil and postharvest management. Systems under NT, crop rotation or consociation with nitrogen-fixing legume species show the highest carbon sequestration potentials, with

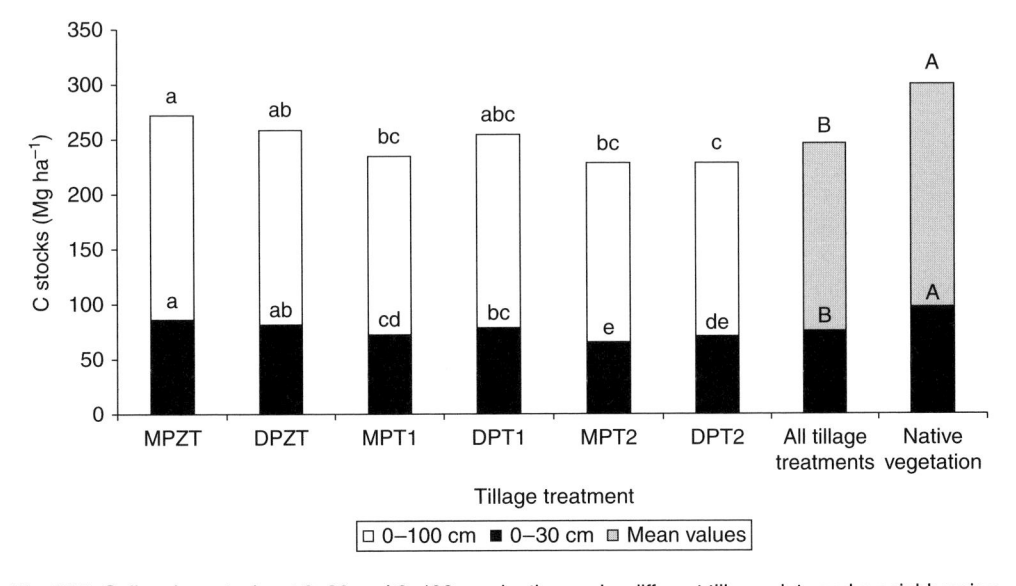

Fig. 21.7. Soil carbon stocks at 0–30 and 0–100 cm depths, under different tillage plots and a neighbouring area with native Cerrado vegetation, at a field experiment in Embrapa Cerrados (Planaltina, DF, Brazil). DP and MP represent plots tilled with a disc and mouldboard plough, respectively. ZT, T1 and T2 correspond to no-tillage, tilling annually before sowing and tilling annually before sowing and after harvest, respectively. Means in the same column followed by the same letter are not significant at $P < 0.05$ (t-test). (From Jantalia et al., 2007.)

Table 21.1. Carbon stocks (average ± standard error) and texture (average ± standard error) of Cerrado biome soils under different land use sampled at depths of 0–20, 0–30 and 0–40 cm. (From Corazza et al., 1999; Freitas et al., 2000; Chapuis Lardy et al., 2002; Ruggiero et al., 2002; Machado et al., 2003; Roscoe and Buurman, 2003; Bayer et al., 2004, 2006; D'Andréa et al., 2004; Neves et al., 2004; Silva et al., 2004; Corbeels et al., 2006; Siqueira Neto, 2006; Araújo et al., 2007; Frazão, 2007; Jantalia et al., 2007; Machado et al., 2007; Metay et al., 2007; Paiva and Faria, 2007; Rangel and Silva, 2007; Luca et al., 2008; Maquère et al., 2008; Beutler et al., 2009; Carvalho et al., 2009; Czycza et al., 2009; Dieckow et al., 2009; Faria et al., 2009; Leite et al., 2009; Maia et al., 2009; Matias et al., 2009; Moreira et al., 2009; Pulrolnik et al., 2009; Rossi et al., 2009; Salton et al., 2011.)

	C stock (Mg C ha^{-1})		Sand (g kg^{-1})		Clay (g kg^{-1})		Clay + silt (g kg^{-1})	
0–20 cm								
NV	44.09	± 2.72	371.35	± 9968	493.00	± 54.55	661.84	± 91.78
P	41.6	± 2.37	336.11	± 99.08	532.77	± 43.83	597.30	± 99.01
CT	32.02	± 1.78	431.57	± 88.83	453.88	± 52.60	579.47	± 95.13
NT	45.98	± 1.89	244.48	± 76.90	488.32	± 46.81	804.52	± 55.15
PF	41.36	± 4.72	205.17	± 17.98	571.00	± 84.94	928.25	± 18.86
0–30 cm								
NV	49.91	± 2.82	387.79	± 186.65	470.00	± 71.89	609.77	± 144.28
P	46.35	± 2.76	455.58	± 181.28	513.64	± 55.90	541.75	± 179.39
CT	44.38	± 2.26	462.55	± 121.28	469.00	± 67.06	559.30	± 85.66
NT	56.48	± 3.42	453.33	± 230.96	430.00	± 108.64	572.53	± 127.56
PF	53.88	± 5.54	202.00	± 18.02	325.00	± 112.70	759.00	± 25.67
0–40 cm								
NV	65.87	± 4.97	472.28	± 97.06	202.71	± 65.55	794.54	± 64.04
P	67.49	± 2.66	498.43	± 56.18	337.58	± 71.50	706.92	± 112.68
CT	53.09	± 3.67	386.07	± 85.02	370.15	± 82.26	616.09	± 92.57
NT	76.73	± 1.97	523.25	± 82.73	225.00	± 90.00	775.00	± 90.00
PF	72.09	± 7.99	466.00	± 188.98	291.00	± 68.58	740.00	± 77.57

NV, native vegetation; P, pasture; CT, conventional tillage; NT, no tillage; PF, planted forest.

a special emphasis on crop–livestock and crop–livestock–forestry integrated systems (Sisti et al., 2004; Bayer et al., 2006; Bustamante et al., 2006; Corbeels et al., 2006; Jantalia et al., 2007; Batlle-Bayer et al., 2010; Carvalho et al., 2010b) (Table 21.2).

Conservation or NT agriculture can also improve carbon sequestration in soils under agricultural use in the semi-arid and subhumid regions of the Pampas. Long-term field experiments have shown that under NT, significantly higher amounts of carbon were sequestered in stabilized C fractions than under conventional tillage (CT) (9.2 versus 11.3 Mg C ha^{-1} in the <50 µm fraction under CT and NT, respectively). This finding indicates that soils under NT can act as a carbon sink, while soils under CT as a source (Quiroga et al., 2009). However, this might not be the case in NT production systems without crop rotation (Díaz Zorita et al., 2002; Steinbach and Alvarez, 2006; Álvarez et al., 2009).

The response of SCSs to different agricultural systems varies significantly according to soil type, particularly regarding its textural properties, which controls vegetation structure, biomass production and soil carbon dynamics (Hassink, 1997; Zinn et al., 2002; Noellemeyer et al., 2006; Luca et al., 2008; Gili et al., 2010). Sandy soils are highly susceptible to degradation under LUC and will lose carbon faster than soils with finer textures (Noellemeyer et al., 2006).

Enhancement of soil fertility has also been associated with soil carbon build-up and retention. Native grassland areas in the Cerrado biome, converted to pastures, have been shown to store significantly more carbon in the soil profile when fertilized. Pastures (*Brachiaria brizantha*) on soils enriched with mineral (50 kg N ha^{-1}, as ammonium sulfate)

Table 21.2. Carbon sequestration rates resulting from land-use change and soil management in the Cerrado, Atlantic Forest and Southern Grasslands biomes.

Biome	Agriculture system/ land-use change	Practice/age	Soil	SOC sequestration rate (Mg ha^{-1} year^{-1})	Reference
Cerrado	NV (Cerradao) => agriculture	NT (13 years)	Typic Hapludox	−1.44 (0–30 cm)	Carvalho et al. (2010b)
Cerrado	NV (Cerrado stricto sensu) => agriculture	NT; CR (21 years)	Typic Hapludox	−0.69 (0–30 cm)	Carvalho et al. (2010b)
Cerrado	Agriculture: CR => ICLS (low fertility)	1–2/9–10 years	Typic Hapludox	1.03–1.35 (0–30 cm)	Carvalho et al. (2010b)
Cerrado	NV => pasture (low fertility)	13 years	Typic Hapludox	−1.53 (0–30 cm)	Carvalho et al. (2010b)
Cerrado	NV (savannah) => agriculture	DMC (30 years)[a]	Various	1.00	Bustamante et al. (2006)
Cerrado	NV => pasture	(Various)	Various	0.87–3.00 (0–20 cm)	Bustamante et al. (2006)
Cerrado/Atlantic Forest	Agriculture: CT => NT	NT (various)[b]	Various	0.50 (0–10 cm)	Cerri et al. (2007)
Cerrado/Atlantic Forest	Agriculture: CT => NT	NT (various)[b]	Various	0.41 (0–20 cm)	La Scala et al. (2012)
Cerrado	Agriculture: CT => NT	DMC (0–12 years)[a]	Ferralsol	0.70–1.15 (0–40 cm)	Corbeels et al. (2006)
Atlantic Forest	NV => agriculture	CT (56 years)[c]	Typic Hapludults	−1.05 (0–20 cm)	Leite et al. (2009)
		NT (56 years)[c]	Typic Hapludults	−0.36 (0–20 cm)	Leite et al. (2009)
Cerrado	NV => agriculture	CT (46 years)[c]	Typic Hapludox	−0.82 (0–20 cm)	Leite et al. (2009)
		NT (46 years)[c]	Typic Hapludox	−0.30 (0–20 cm)	Leite et al. (2009)
Atlantic Forest	NV => agriculture	CT (15 years)	Typic Hapludults	−0.86 (0–20 cm)	Leite et al. (2009)
		NT (15 years)	Typic Hapludults	−0.17 (0–20 cm)	Leite et al. (2009)
Cerrado	NV => agriculture	CT (12 years)	Typic Hapludox	−0.84 (0–20 cm)	Leite et al. (2009)
		NT (12 years)	Typic Hapludox	−0.14 (0–20 cm)	Leite et al. (2009)
Cerrado	Agriculture: CT => NT	NT (9 years; 11 years)	Typic Hapludox	−0.17–1.00 (0–20 cm)	Salton et al. (2011)
	Agriculture: CT => ICLS	NT; CR (9 years; 11 years)	Typic Hapludox	0.14–0.44 (0–20 cm)	Salton et al. (2011)
	Agriculture (CT) => pasture	PP (9 years)	Typic Hapludox	0.65 (0–20 cm)	Salton et al. (2011)
	CT => pasture (more clayed)	PP (11 years)	Typic Hapludox	0.94 (0–20 cm)	Salton et al. (2011)
	CT => pasture (more sandy)	Pp (11 years)	Typic Hapludox	0.65 (0–20 cm)	Salton et al. (2011)
	Agriculture: CT => NT	NT (15 years)		0.3 (0–20 cm)	Bayer et al. (2006)
	Agriculture: CT => NT	NT (20 years)		0.6 (0–20 cm)	Bayer et al. (2006)
Atlantic Forest/Southern Grasslands	Agriculture: CT=>NT	NT[b]	Various	0.19–0.81 (0–20 cm)	Bayer et al. (2006)
Atlantic Forest/Southern Grasslands	Agriculture: CT => NT	NT; CR	Typic Haplorthox	0.04–0.88 (0–30 cm)	Boddey et al. (2010)
Atlantic Forest/Southern Grasslands	Agriculture: CT => NT	NT; CR	Typic Haplorthox	0.48–1.53 (0–30 cm)	Boddey et al. (2010)

[a]Simulation using G'DAY model; [b]from literature review; [c]CQESTR model.
CR, crop rotation; CT, conventional tillage; ICLS, integrated crop-livestock system; NV, native vegetation; NT/DMC, no tillage or direct seeding mulch-based system; PP, permanent pasture with grass only; PP+L, permanent pasture with grass + legumes.

or biologically fixed nitrogen (consociation with *Stylosanthes guianiensis*) showed a gain of 10.6–13.3 Mg C ha⁻¹ (0–100 cm), 3–4 years after establishment. In contrast, soils without nitrogen amendment, under *Brachiaria decumbens*, showed early signs of degradation, and loss of 2.6 Mg C ha⁻¹ (0–100 cm) (Silva et al., 2004), in the same period. Cerrado biome topsoils (0–5 cm), with higher clay contents, lost lower amounts of soil organic carbon (19% in clay oxisols versus 48% in sandy entisols) (Zinn et al., 2002). In deeper soil strata (0–60 cm), carbon losses were also higher in sandy entisols under *Eucalyptus* spp., while no net losses occurred in oxisols.

The influence of soil texture on the response of carbon sequestration to sugarcane harvest management is very significant, especially in traditional burning harvest systems, when impacts on soil carbon are much higher. Burning sugarcane prior to harvesting results in higher SCS losses than unburned sugarcane areas, especially for sandy soils (Resende et al., 2006; Balieiro et al., 2008; Luca et al., 2008; Galdos et al., 2009). Eight years after replanting sugarcane at a site in the Atlantic Forest biome (São Paulo), the unburned plots showed an increase in total soil carbon at the 0–10 cm layer of 41%, compared to the burned fields (23.30 × 16.57 g C kg⁻¹) (Galdos et al., 2009).

Conclusions

To meet the increasing global demand for food and bioenergy, the sharp increases demanded for agricultural production in rural landscapes must be balanced with greater environmental and social sustainability. This review focused on the understanding of the LUCs in the Cerrado, Atlantic Forest and Southern Grasslands, major biomes in central-southern South America, during the past 40 years. Conversion of the native vegetation to agricultural systems was shown to cause a significant reduction in SCSs, for all biomes analysed. We suggest soil carbon losses are derived mainly from lower C inputs to the soil under crop production, and to a lesser extent to higher C outputs (respiration) from the soil. The global need for a reduction of greenhouse gases in the atmosphere calls for the implementation of agricultural systems with lower carbon footprints. This review has shown that improved soil and crop management in agricultural systems can counteract most of the soil carbon losses derived from LUC in the region.

Soil and crop integrated management systems can partially reverse agricultural soil carbon losses. The magnitude of these losses and gains varies greatly with land-use history, soil management, climate, soil type and soil cover. In the Argentinian Southern Grasslands, conventional agriculture with annual crops led to an overall loss of 34 Mg C ha⁻¹ (50% loss), considering the last 100 years (Alvarez, 2001). In the Cerrado biome, conventional tillage was shown to lose 33% of the SCS in one year (Jantalia et al., 2007). On the contrary, SCS could be increased significantly in degraded Cerrado biome pastureland converted to conservation agriculture systems, such as no-tillage including nitrogen-fixing legumes in the crop rotation scheme. Forestry can be beneficial in the northern parts of the region, where afforestation in the Cerrado and Atlantic Forest biomes can result in soil carbon sequestration rates of 1.22 (Corazza et al., 1999) and 1.73 Mg C ha⁻¹ year⁻¹ (Macedo et al., 2008), respectively, the latter with N_2-fixing trees. However, in humid environments of the southern portion of the basin, afforestation with eucalyptus or pine monocultures has been shown to cause a decrease in SCS.

The data presented show the importance of no-tillage systems for soil carbon sequestration and storage. It also reinforces the need to maintain high levels of plant residue inputs to the soil, so as to boost carbon sequestration and stabilization in the soil, reducing the agricultural carbon footprint in all biomes studied. In the Cerrado biome, soil carbon sequestrations up to 1.43 Mg C ha⁻¹ year⁻¹ have been reported (Corazza et al., 1999) and in the Southern Grasslands, a nearly 26% increase in SCSs were obtained by no-tillage systems compared to conventional systems, 9 years after their establishment (16.6 × 13.2 t C ha⁻¹ in NT × CT systems, respectively) (Quiroga et al., 2009).

Conversion of burned to unburned sugarcane production systems results in undisputable gains in SCSs in the surface layers. A study in a São Paulo plantation showed a 41% gain in the surface layer 8 years after replanting (Galdos et al., 2009). However, more research should be done to reveal the SCS levels at greater depths.

Soil degradation brought about by agricultural expansion can be very rapid and implies a strong decrease of SCSs, especially in sandy soils. Soil texture was found to be one of the most important factors that controlled land-use impacts on vegetation structure (Gili et al., 2010), biomass production and carbon sequestration or losses.

Most of the results shown were obtained from experimental sites, where agricultural management was optimized and the variability of soil, topography and other biophysical and agronomical features were minimized, to ensure scientific precision. Ideally, full carbon accounting of the different agricultural systems in the region should be carried out, considering indirect impacts, such as increased use of fossil fuel-derived fertilizers and other inputs with high carbon footprints in the management systems.

Although the global need for a reduction of greenhouse gases in the atmosphere calls for the implementation of agricultural systems with lower carbon footprints, the local role of soil carbon storage in soil fertility and agricultural production emerges as an important regional issue. Therefore, intensification of current agricultural and livestock production systems, using existing technology and developing innovative solutions, is the most effective way to ensure the sustainable development of rural areas.

The data presented support the importance of no-tillage systems for soil carbon capture and storage, hydrologic regulation, soil erosion control and resilience to extreme events, but suggest that new agricultural practices aimed to increase carbon uptake by the vegetation and into the soil are needed. Such objectives can be achieved by developing site-specific management of multi-cropping systems that capture more efficiently the available water, carbon and nutrients, rendering farmers with more options of marketable products, while increasing farm agrobiodiversity and the delivery of ecosystem services. Incentive policies aimed at the full adoption of no-tillage systems with crop rotations with legume cover species, commercial forestry consociated with nitrogen-fixing legume trees, acceleration of the implementation of sugarcane anti-burning legislation and improvement of multi-cropping systems should be implemented to stimulate a low carbon agriculture throughout the continent.

Acknowledgements

This work received funding from the Inter-American Institute for Global Change Research (IAI), through the CRN II Program (Project 2031), which is supported by the US National Science Foundation (NSF Grant GEO-0452325) and the IAI Project 'Land-use change, biofuels and rural development in the Rio de La Plata Basin', supported by the International Development Research Center (IDRC, grant 104783-001). The authors would like to thank Edson Sano for supplying land-use maps with the delimitation of the Cerrado biome and Taíssa Santos for helping with graphs and formatting.

References

ABRAF (2012) *ABRAF Statistical Yearbook 2012* base year 2011. Associação Brasileira de Produtores de Florestas Plantadas, Brasília, DF, Brazil.

Álvarez, C.R., Duggan, M.T., Chamorro, E.R., D´Ambrosio, D. and Taboada, M.A. (2009) Descompactación de suelos franco limosos en siembra directa: efectos sobre las propriedades edáficas y los cultivos. *Ciencia del Suelo* 27, 159–169.

Alvarez, R. (2001) Estimation of carbon losses by cultivation from soils of the Argentine Pampa using the Century model. *Soil Use and Management* 17, 62–66.

Alves, B.J.R., Boddey, R.M. and Urquiaga, S. (2003) The success of BNF in soybean in Brazil. *Plant and Soil* 252, 1–9.

Andriulo, A., Guérif, J. and Mary, B. (1999) Evolution of soil carbon with various cropping sequences on the rolling pampas. Determination of carbon origin using variations in natural 13C abundance. *Agronomie* 19, 349–364.

Angelini, G.A.R., Balieiro, F.C., Júnior, O.J.S., Zanatta, J.A., Coutinho, H.L.C., Salton, J.C. and Franco, A.A. (2010) Impacto da Retirada da Palhada de Áreas com Cana-de-Açúcar sobre os Fungos Micorrízicos Arbusculares em Solo de Cerrado, em Dourados – MS. XIII Reunião Brasileira sobre Micorrizas. Brazilian Soil Science Society, Viçosa, Minas Gerais.

Araújo, F.S., Leite, L.F.C., Júnior, J.O.L., Sagrilo, E., Araújo, A.R. and Lopes, A.N.C. (2007) Estoques totais de carbono e nitrogênio em Latossolo Amarelo sob diferentes sistemas de culturas e floresta nativa de cerrados. V Fertbio, 2006 Bonito – MS. Brazilian Soil Science Society, Viçosa, Minas Gerais.

Atlântica, S.M. (2012) *A Mata Atlântica* [Online]. SOS Mata Atlântica (http://www.sosma.org.br/nossa-causa/a-mata-atlantica, accessed 19 February 2013).

Baldi, G. and Paruelo, J.M. (2008) Land-use and land cover dynamics in South American temperate grasslands. *Ecology and Society* 13, article 6.

Balieiro, F.C., Fontes, R.L.F., Dias, L.E., Franco, A.A., Campello, E.F.C. and Faria, S.M. (2002) Accumulation and distribution of aboveground biomass and nutrients under pure and mixed stands of Guachapele e Eucalyptus. *Journal of Plant Nutrition* 25, 2639–2654.

Balieiro, F.C., Franco, A.A., Pereira, M.G., Campello, E.F.C., Dias, L.E., Faria, S.M. and Alves, B.J.R. (2004) Dinâmica da serapilheira e transferência de nitrogênio ao solo, em plantios de *Pseudosamanea guachapele* e *Eucalyptus grandis*. *Pesquisa Agropecuária Brasileira* 39, 597–601.

Balieiro, F.C., Pereira, M.G., Alves, B.J.R., Resende, A.S. and Franco, A.A. (2008) Soil carbon and nitrogen in pasture soil reforested with eucalyptus and guachapele. *Revista Brasileira de Ciência do Solo* 32, 1253–1260.

Batlle-Bayer, L., Batjes, N.H. and Bindraban, P.S. (2010) Changes in organic carbon stocks upon land use conversion in the Brazilian Cerrado: a review. *Agriculture, Ecosystems and Environment* 137, 47–58.

Bayer, C., Martin-Neto, L., Mielniczuk, J. and Pavinato, A. (2004) Armazenamento de carbono em frações lábeis da matéria orgânica de um Latossolo Vermelho sob plantio direto. *Pesquisa Agropecuária Brasileira* 39, 677–683.

Bayer, C., Martin-Neto, L., Mielniczuk, J., Dieckow, J. and Amado, T.J.C. (2006) C and N stocks and the role of molecular recalcitrance and organomineral interaction in stabilizing soil organic matter in a subtropical Acrisol managed under no-tillage. *Geoderma* 133, 258–268.

Berthrong, S.T., Jobbágy, E.G. and Jackson, R.B. (2009a) A global meta-analysis of soil exchangeable cations, pH, carbon, and nitrogen with afforestation. *Ecological Applications* 19, 2228–2241.

Berthrong, S.T., Schadt, C.W., Pineiro, G. and Jackson, R.B. (2009b) Afforestation alters the composition of functional genes in soil and biogeochemical processes in South American grasslands. *Applied and Environmental Microbiology* 75, 6240–6248.

Berthrong, S.T., Piñeiro, G., Jobbagy, E. and Jackson, R. (2012) Soil C and N changes with afforestation of grasslands across gradients of precipitation and plantation age. *Ecological Applications* 22, 76–86.

Beutler, S.J., Loss, A., Pereira, M.G., Perin, A. and Anjos, L.H.C. (2009) Estoques totais de carbono das frações químicas e físicas da matéria orgânica do solo em sistema plantio direto no Cerrado. In: *XXXII Brazilian Congress of Soil Science, 2009 Recife, PE*. Brazilian Society of Soil Science, Viçosa, Minas Gerais, abstract 1968.

Boddey, R.M., Alves, B.J.R. and Urquiaga, S. (2006) Leguminous biological nitrogen fixation in sustainable tropical agroecosystems. In: Uphoff, N. (ed.) *Biological Approaches to Sustainable Soil Systems*. CRC Press, Boca Raton, Florida, pp. 401–408.

Boddey, R.M., Jantalia, C.P., Conceição, P.C., Zanatta, J.A., Bayer, C., Mielniczuk, J., Dieckow, J., dos Santos, H.P., Denardin, J.E., Aita, C. *et al.* (2010) Carbon accumulation at depth in Ferralsols under zero-till subtropical agriculture. *Global Change Biology* 16, 784–795.

Brasil/Ministério Da Agricultura Pecuária E Abastecimento (2009) *Anuário estatístico da agroenergia*. Mapa/ACS, Brasília.

Brown, K.S.J. and Brown, G.G. (1992) Habitat alteration and species loss in Brazilian forests. In: Whitmore, T.C. and Slayer, J.A. (eds) *Tropical Deforestation and Species Extinction*. Chapman and Hall, London, pp. 119–142.

Bustamante, M.M.C., Corbeels, M., Scopel, E. and Roscoe, R. (2006) Soil carbon storage and sequestration potential in the Cerrado region of Brazil. In: Lal, R., Cerri, C.C., Bernoux, M., Etchevers, J. and Cerri, E. (eds) *Carbon Sequestration in Soils of Latin America*. Food Products Press, Binghamton, New York, pp. 285–304.

Campos, D.V.B., Machado, P.L.O., Pinheiro, E.F.M., Silva, C.A., Tarré, R., Macedo, R., Alves, B.J.R., Urquiaga, S., Boddey, R.M. and Santos, G.A. (2001) Use of 13C to trace the substitution of forest derived C in organic matter fractions of soil under an elephant grass *Pennisetum purpureum*) pasture. In: Embrapa Solos (ed.) *International Congress of Land Degradation. rtbi2001 Rio de Janeiro, RJ, Brasil*. Embrapa Solos, Rio de Janeiro.

Campos, D.V.B., Machado, P.L.O.A., Braz, S.P., Santos, G.A., Lima, E., Alves, B.J.R., Urquiaga, S. and Boddey, R.M. (2004) Decomposition of soil carbon derived from forest in an ultisol under sugar cane or *Brachiaria* sp. in the atlantic forest region of Brazil. In: *XII International Meeting of International Humic Substances Society*. IHSS, São Pedro – São Paulo, pp. 647–649.

Carvalho, J.L.N., Cerri, C.E.P., Feigl, B.J., Píccolo, M.C., Godinho, V.P. and Cerri, C.C. (2009) Carbon sequestration in agricultural soils in the Cerrado region of the Brazilian Amazon. *Soil and Tillage Research* 103, 342–349.

Carvalho, J.L., Avanzi, J.C., Silva, M.L.N., Mello, C.R. and Cerri, C.E.P. (2010a) Potencial de sequestro de carbono em diferentes biomas do Brasil. *Revista Brasileira de Ciência do Solo* 34, 277–289.

Carvalho, J.L.N., Raucci, G.S., Cerri, C.E.P., Bernoux, M., Feigl, B.J., Wruck, F.J. and Cerri, C.C. (2010b) Impact of pasture, agriculture and crop-livestock systems on soil C stocks in Brazil. *Soil and Tillage Research* 110, 175–186.

Cerqueira Leite, R.C., Verde Leal, M.R.L., Barbosa Cortez, L.A., Griffin, W.M. and Gaya Scandiffio, M.I. (2009) Can Brazil replace 5% of the 2025 gasoline world demand with ethanol? *Energy* 34, 655–661.

Cerri, C.E.P., Sparovek, G., Bernoux, M., Easterling, W.E., Melillo, J.M. and Cerri, C.C. (2007) Tropical agriculture and global warming: impacts and mitigation options. *Scientia Agricola* 64, 83–99.

Chaer, G.M. and Totola, M.R. (2007) Impacto do manejo de resíduos orgânicos durante a reforma de plantios de eucalipto sobre indicadores de qualidade do solo. *Revista Brasileira de Ciencia do Solo* 31, 1381–1396.

Chapuis Lardy, L., Brossard, M., Lopes Assad, M.L. and Laurent, J.Y. (2002) Carbon and phosphorus stocks of clayey Ferralsols in Cerrado native and agroecosystems, Brazil. *Agriculture, Ecosystems and Environment* 92, 147–158.

Coelho, S.R.F., Gonçalves, J.L.M., Mello, S.L.M., Moreira, R.M., Silva, E.V. and Laclau, J.P. (2007) Crescimento, nutrição e fixação biológica de nitrogênio em plantios mistos de eucalipto e leguminosas arbóreas. *Pesquisa Agropecuária Abrasileira* 42, 759–768.

Conab (2013) Soja – Brasil – Série Histórica de Produtividade. *Séries Históricas*. Companhia Nacional de Abastecimento, Brasilia. (http://www.conab.gov.br/conteudos.php?a=1252&&Pagina_objcmsconteudos =3#A_objcmsconteudos, accessed 21 October 2014).

Corazza, E.J., Silva, J.E., Resck, D.V.S. and Gomes, A.C. (1999) Comportamento de diferentes sistemas de manejo como fonte ou depósito de carbono em relação à vegetação de Cerrado. *Revista Brasileira de Ciência do Solo* 23, 425–432.

Corbeels, M., Scopel, E., Cardoso, A., Bernoux, M., Douzet, J.-M. and Neto, M.S. (2006) Soil carbon storage potential of direct seeding mulch-based cropping systems in the Cerrados of Brazil. *Global Change Biology* 12, 1773–1787.

Czycza, R.V., Signor, D. and Cerri, C.E.P. (2009) Estoques de carbono e nitrogênio do solo e biomassa microbiana em sistema de colheita com e sem queima da cana-de-açúcar. XXXII Brazilian Congress of Soil Science, 2009 Recife, Pernambuco. Brazilian Society of Soil Science, Viçosa, Minas Gerais.

D'Andréa, A.F., Silva, M.L.N., Curi, N. and Guilherme, L.R.G. (2004) Estoque de carbono e formas de nitrogênio mineral em solo submetido a diferentes sistemas de manejo. *Pesquisa Agropecuária Brasileira* 39, 179–186.

Dean, W. (1995) *With Broadax and Firebrand: The Destruction of the Brazilian Atlantic Forest*. University of California Press, Berkeley, California.

Denef, K., Plante, A.F. and Six, J. (2009) Characterization of soil organic matter. In: Kutsch, W.L., Bahn, M. and Heinemeyer, A. (eds) *Soil Carbon Dynamics – An Integrated Methodology*. Cambridge University Press, Cambridge, UK, pp. 91–126.

Díaz Zorita, M., Duarte, G.A. and Grove, J.H. (2002) A review of no till systems and soil management for sustainable crop production in the subhumid and semiarid Pampas of Argentina. *Soil Tillage Research* 65, 1–18.

Dieckow, J., Bayer, C., Conceição, P.C., Zanatta, J.A., Martin-Neto, L., Milori, D.B.M., Salton, J.C., Macedo, M.M., Mielniczuk, J. and Hernani, L.C. (2009) Land use, tillage, texture and organic matter stock and composition in tropical and subtropical Brazilian soils. *European Journal of Soil Science* 60, 240–249.

Eclesia, P. (2011) *Consecuencias del remplazo de ecosistemas naturales sudamericanos por forestaciones y pasturas megatérmicas: efectos sobre el carbono orgánico edáfico*. MSc thesis, Universidad de Buenos Aires.

FAO (2008) Contribution of the forestry sector to national economies, 1990–2006, by A. Lebedys. *Forest Finance Working Paper*. ftp://ftp.fao.org/docrep/fao/011/k4588e/k4588e00.pdf (accessed 19 August 2014).

FAO (2010) *Global Forest Resources Assessment 2010*. Main Report. FAO Forestry Paper. FAO, Rome.

Faria, G.E., Araújo, B.L., Sabino, D., Dantas, J.S., Wendling, B. and Arnhold, E. (2009) Alterações nos estoques de carbono no solo sob campo sujo e Eucalipto. In: *XXXII Brazilian Congress of Soil Sciences, 2009 Recife, PE*. Brazilian Society of Soil Sciences, Viçosa, Minas Gerais.

Farley, K.A., Piñeiro, G., Palmer, S.M., Jobbágy, E.G. and Jackson, R.B. (2008) Stream acidification and base cation losses with grassland afforestation. *Water Resources Research* 44, W00A03.

Fearnside, P.M. (2001) Status of South American natural ecosystems. In: Simon, A.L. (ed.) *Encyclopedia of Biodiversity*. Elsevier, New York, pp. 345–359.

Fernández, R., Quiroga, A., Zorati, C. and Noellemeyer, E. (2010) Carbon contents and respiration rates of aggregate size fractions under no-till and conventional tillage. *Soil and Tillage Research* 109, 103–109.

Fidalgo, E.C.C., Benites, V.M., Machado, P.L.O.A., Madari, B.E., Coelho, M.R., Moura, I.B. and Lima, C.X. (2007) Estoque de Carbono nos Solos do Brasil. *Boletim de Pesquisa e Desenvolvimento* 121, 1–27.

Forrester, D.I., Bauhus, J., Cowie, A.L. and Vanclay, J.K. (2006) Mixed-species plantations of Eucalyptus with nitrogen-fixing trees: a review. *Forest Ecology and Management* 233, 211–230.

Frazão, L.A. (2007) Conversão do Cerrado em pastagem e sistemas agrícolas: efeitos na dinâmica da matéria orgânica do solo. MSc thesis, University of São Paulo.

Freitas, P.L., Blancaneaux, P., Gavinelli, E., Larré-Larrouy, M.C. and Feller, C. (2000) Nível e natureza do estoque orgânico de latossolos sob diferentes sistemas de uso e manejo. *Pesquisa Agropecuária Brasileira* 35, 157–170.

Galdos, M.V., Cerri, C.C. and Cerri, C.E.P. (2009) Soil carbon stocks under burned and unburned sugarcane in Brazil. *Geoderma* 153, 347–352.

Gauder, M., Graeff-Hönninger, S. and Claupein, W. (2011) The impact of a growing bioethanol industry on food production in Brazil. *Applied Energy* 88, 672–679.

Gili, A., Trucco, R., Niveyro, S., Balzarini, M.D.E., Quiroga, A. and Noellemeyer, E. (2010) Soil texture and carbon dynamics in savannah vegetation patches of Central Argentina. *Soil Science Society American Journal* 74, 647–657.

Gobierno de La Pampa (2009) *Statistical Yearbook La Pampa 2009*. Direccion de Estadisticas y Censo, Gobierno de La Pampa.

Goes, T., Marra, R. and Silva, G.S.E. (2008) Setor sucroalcooleiro no Brasil: Situação atual e perspectivas. *Revista de Política Agrícola* 17, 39–52.

Graham, M.H. and Haynes, R.J. (2006) Organic matter status and the size, activity and metabolic diversity of the soil microbial community in the row and inter-row of sugarcane under burning and trash retention. *Soil Biology and Biochemistry* 38, 21–31.

Guo, L.B. and Gifford, R.M. (2002) Soil carbon stocks and land use change: a meta analysis. *Global Change Biology* 8, 345–360.

Hassink, J. (1997) The capacity of soils to preserve organic and N by their association with clay and silt particle. *Plant and Soil* 191, 77–87.

INPE (2011) CANASAT – Mapeamento da cana via imagens de satélite de observação da Terra. INPE. http://www.dsr.inpe.br/laf/canasat/tabelas.html (accessed 19 August 2014).

Jackson, R.B., Randerson, J.T., Canadell, J.G., Anderson, R.G., Avissar, R., Baldocchi, D.D., Bonan, G.B., Caldeira, K., Diffenbaugh, N.S., Field, C.B. et al. (2008) Protecting climate with forests. *Environmental Research Letters* 3, 044006.

Jantalia, C.P., Resck, D.V.S., Alves, B.J.R., Zotarelli, L., Urquiaga, S. and Boddey, R.M. (2007) Tillage effect on C stocks of a clayey Oxisol under a soybean-based crop rotation in the Brazilian Cerrado region. *Soil and Tillage Research*, 95, 97–109.

Jobbágy, E., Nosetto, M., Paruelo, J.M. and Piñeiro, G. (2006) Las forestaciones rioplatenses y el agua. *Ciencia Hoy* 16, 12–21.

Jobbagy, E.G. and Jackson, R.B. (2003) Patterns and mechanisms of soil acidification in the conversion of grasslands to forests. *Biogeochemistry* 64, 205–229.

Kaschuk, G., Alberton, O. and Hungria, M. (2011) Quantifying effects of different agricultural land uses on soil microbial biomass and activity in Brazilian biomes: inferences to improve soil quality. *Plant and Soil* 338, 467–481.

Klink, C.A. and Moreira, A.G. (2002) Past and current human occupation and land-use. In: Marquis, P.S.O.R.J. (ed.) *The Cerrado of Brazil. Ecology and Natural History of a Neotropical Savanna*. Columbia University Press, New York, pp. 69–88.

La Scala, N., Figueiredo, E.B. and Panosso, A.R. (2012) A review on soil carbon accumulation due to the management change of major Brazilian agricultural activities. *Brazilian Journal of Biology* 72, 775–785.

Laclau, J.P., Bouillet, J.P., Gonçalves, J.L.M., Silva, E.V., Jourdan, C., Cunha, M.C.S., Moreira, M.R., Saint-André, L., Maquère, V., Nouvellon, Y. and Ranger, J. (2008) Mixed-species plantations of *Acacia mangium* and *Eucalyptus grandis* in Brazil: 1. Growth dynamics and aboveground net primary production. *Forest Ecology and Management* 255, 3905–3917.

Lago, A.A.C. (2012) International negotiations on bioenergy sustainability. *Sustainability of Sugarcane Bioenergy*. Center for Strategic studies and management (CGEE), Brasília, Distrito Federal.

Lee, S. and Berbery, E.H. (2012) Land cover change effects on the climate of the la plata Basin. *Journal of Hydrometeorology* 13, 84–102.

Leite, L.F.C., Doraiswamy, P.C., Causarano, H.J., Gollany, H.T., Milak, S. and Mendonca, E.S. (2009) Modeling organic carbon dynamics under no-tillage and plowed systems in tropical soils of Brazil using CQESTR. *Soil and Tillage Research* 102, 118–125.

Lima, A.M.N., Silva, I.R., Neves, J.C.L., Novais, R.F., Barros, N.F., Mendonça, E.S., Smyth, T.J., Moreira, M.S. and Leite, F.P. (2006) Soil organic carbon dynamics following afforestation of degraded pastures with eucalyptus in southeastern Brazil. *Forest Ecology and Management* 235, 219–231.

Lopes-Assad, M.L. (1997) Fauna do Solo. In: Vargas, M. and Hungria, M. (eds) *Biologia dos Solos dos Cerrados*. Embrapa Cerrados, Planaltina, Brazil, pp. 363–444.

Luca, E.F.D., Feller, C., Cerri, C.C., Barthès, B., Chaplot, V., Campos, D.C. and Manechini, C. (2008) Avaliação de atributos físicos e estoques de carbono e nitrogênio em solos com queima e sem queima de canavial. *Revista Brasileira de Ciência do Solo* 32, 789–800.

Macedo, M.O., Resende, A.S., Garcia, P.C., Boddey, R.M., Jantalia, C.P., Urquiaga, S., Campello, E.F.C. and Franco, A.A. (2008) Changes in soil C and N stocks and nutrient dynamics 13 years after recovery of degraded land using leguminous nitrogen-fixing trees. *Forest Ecology and Management* 255, 1516–1524.

Machado, L.D.O., Lana, Â.M.Q., Lana, R.M.Q., Guimarães, E.C. and Ferreira, C.V. (2007) Variabilidade espacial de atributos químicos do solo em áreas sob sistema plantio convencional. *Revista Brasileira de Ciência do Solo* 31, 591–599.

Machado, P.L.O.A., Sohi, S.P. and Gaunt, J.L. (2003) Effect of no-tillage on turnover of organic matter in a Rhodic Ferralsol. *Soil Use and Management* 19, 250–256.

Machado, R.L., Resende, A.S.D., Campello, E.F.C., Oliveira, J.A. and Franco, A.A. (2010) Soil and nutrient losses in erosion gullies at different degrees of restoration. *Revista Brasileira de Ciência do Solo* 34, 945–954.

Maia, S.M.F., Ogle, S.M., Cerri, C.E.P. and Cerri, C.C. (2009) Effect of grassland management on soil carbon sequestration in Rondônia and Mato Grosso states, Brazil. *Geoderma* 149, 84–91.

Manzatto, C.V., Assad, E.D., Bacca, J.F.M., Zaroni, M.J. and Pereira, S.E.M. (2009) Zoneamento Agroecológico da Cana-de-Açúcar: Expandir a produção, preservar a vida, garantir o futuro. Embrapa Solos/Documentos. (https://www.embrapa.br/solos/busca-de-publicacoes/-/publicacao/579169/zoneamento-agroecologico-da-cana-de-acucar-expandir-a-producao-preservar-a-vida-garantir-o-futuro, accessed 9 September 2014).

Maquère, V., Laclau, J.P., Bernoux, M., Saint-André, L., Gonçalves, J.L.M., Cerri, C.C., Piccolo, M.C. and Ranger, J. (2008) Influence of land use (savanna, pasture, eucalyptus plantations) on soil carbon and nitrogen stocks in Brazil. *European Journal of Soil Science* 59, 863–877.

Matias, M.C.B., Salviano, A.A.C., Leite, L.F.C. and Araújo, A.S.F. (2009) Biomassa microbiana e estoques de C e N do solo em diferentes sistemas de manejo, no cerrado do Estado do Piauí. *Acta Scientiarum. Agronomy* 1, 517–521.

Melillo, J.M., Butler, S., Johnson, J., Mohan, J., Steudler, P., Lux, H., Burrows, E., Bowles, F., Smith, R., Scott, L. *et al*. (2011) Soil warming, carbon–nitrogen interactions, and forest carbon budgets. *Proceedings of the National Academy of Sciences* 108, 9508–9512.

Metay, A., Moreira, J.A.A., Bernoux, M., Boyer, T., Douzet, J.-M., Feigl, B., Feller, C., Maraux, F., Oliver, R. and Scopel, E. (2007) Storage and forms of organic carbon in a no-tillage under cover crops system on clayey Oxisol in dryland rice production (Cerrados, Brazil). *Soil and Tillage Research* 94, 122–132.

Miyasaka, S. and Medina, J.C.A. (eds) (1981) *A Soja no Brasil*. Instituto Agronômico de Campinas, Campinas, São Paulo.

Monteiro, J.M.G., Elabras-Veiga, L.B. and Coutinho, H.L.C. (2010) Projeto SENSOR: políticas públicas relacionadas à expansão da cana-de-açúcar para a produção de biocombustíveis. *Documentos* 124. (http://www.infoteca.cnptia.embrapa.br/bitstream/doc/882167/4/documentos124.pdf, accessed 23 October 2014).

Moreira, J.A.A., Madari, B.E., Meneguci, J.L.P., Carvalho, M.T.M., Matos, G.R. and Costa, A.R. (2009) Matéria orgânica e estoque de carbono em solo sob produção integrada de citrus. In: *XXXII Brazilian Congress of Soil Science, 2009 Recife, PR*. Brazilian Society of Soil Science, Viçosa, Minas Gerais, abstract 2472.

Morellato, L.P.C. and Haddad, C.F.B. (2000) Introduction: the Brazilian Atlantic Forest. *Biotropica* 32, 786–792.

Neto, M.S., Scopel, E., Corbeels, M., Cardoso, A.N., Douzet, J.-M., Feller, C., Piccolo, M.D.C., Cerri, C.C. and Bernoux, M. (2010) Soil carbon stocks under no-tillage mulch-based cropping systems in the Brazilian Cerrado: an on-farm synchronic assessment. *Soil and Tillage Research* 110, 187–195.

Neves, C.M.N., Silva, M.L.N., Curi, N., Macedo, R.L.G. and Tokura, A.M. (2004) Estoque de carbono em sistemas agrossilvopastoril, pastagem e eucalipto sob cultivo convencional na Região Noroeste do Estado de Minas Geraisx. *Ciência e Agrotecnologia* 28, 1038–1046.

Noellemeyer, E., Quiroga, A.R. and Estelrich, D. (2006) Soil quality in three range soils of the semi-arid Pampa of Argentina. *Journal of Arid Environments* 65, 142–155.

Noellemeyer, E., Frank, F., Alvarez, C., Morazzo, G. and Quiroga, A. (2008) Carbon contents and aggregation related to soil physical and biological properties under a land-use sequence in the semiarid region of central Argentina. *Soil and Tillage Research* 99, 179–190.

Oliveira, O.C., Oliveira, I.P., Alves, B.J.R., Urquiaga, S. and Boddey, R.M. (2004) Chemical and biological indicators of decline/degradation of Brachiaria pastures in the Brazilian Cerrado. *Agriculture, Ecosystems and Environment* 103, 289–300.

Oosterveer, P. and Mol, A.P.J. (2010) Biofuels, trade and sustainability: a review of perspectives for developing countries. *Biofuels, Bioproducts and Biorefining* 4, 66–76.

Pacheco, A.R., Chaves, R.Q. and Nicoli, C.M.L. (2012) Integration of crops, livestock, and forestry: a system of production for the Brazilian Cerrados. In: Hershey, C.H. (ed.) *Eco-Efficiency: From Vision to Reality*. Centro Internacional de Agricultura Tropical (CIAT), Cali, Colombia, pp. 51–61.

Paiva, A.O. and Faria, G.E. (2007) Estoque de carbon do solo sob cerrado sensu strict no Distrito Federal, Brasil. *Revista Trópica-Ciências Agrárias e Biológicas* 1, 59–65.

Piñeiro, G., Paruelo, J.M. and Oesterheld, M. (2006) Potential long-term impacts of livestock introduction on carbon and nitrogen cycling in grasslands of Southern South America. *Global Change Biology* 12, 1267–1284.

Pinheiro, É.F.M., Lima, E., Ceddia, M.B., Urquiaga, S. and Alves, B.J.R. (2010) Impact of pre-harvest burning versus trash conservation on soil carbon and nitrogen stocks on a sugarcane plantation in the Brazilian Atlantic forest region. *Plant and Soil* 333, 71–80.

Pulrolnik, K., Barros, N.F., Silva, I.R., Novais, R.F. and Brandani, C.B. (2009) Estoques de carbono de frações lábeis e estáveis da matéria orgânica em solos sob diferentes tempos de cultivo de Eucalipto no vale do Jequitinhonha-MG. In: *XXXII Brazilian Congress of Soil Sciences, 2009 Recife, PE*. Brazilian Society of Soil Sciences, Viçosa, Minas Gerais, abstract 648.

Quiroga, A., Fernández, R. and Noellemeyer, E. (2009) Grazing effect on soil properties in conventional and no-till systems. *Soil and Tillage Research* 105, 164–170.

Rachid, C.T., Piccolo, M., Leite, D.C., Balieiro, F., Coutinho, H.L., Van Elsas, J., Peixoto, R. and Rosado, A. (2012) Physical-chemical and microbiological changes in Cerrado Soil under differing sugarcane harvest management systems. *BMC Microbiology* 12, 170. (http://www.biomedcentral.com/1471-2180/12/170, accessed 21 October 2014).

Rachid, C.T.R.C. (2010) Comunidade bacteriana, atributos do solo e fluxo de gases em solo sob Cerrado e cana-de-açúcar. MSc Dissertation, Universidade de São Paulo.

Rangel, O.J.P. and Silva, C.A. (2007) Estoques de carbono e nitrogênio e frações orgânicas de Latossolo submetido a diferentes sistemas de uso e manejo. *Revista Brasileira de Ciência do Solo* 31, 1609–1623.

Resende, A.S., Xavier, R.P., Oliveira, O.C., Urquiaga, S. and Alves, B.J.R. (2006) Long-term effects of pre-harvest burning and nitrogen and vinasse applications on yield of sugar cane and soil carbon and nitrogen stocks on a plantation in Pernambuco, N.E. Brazil. *Plant and Soil* 343, 339–351.

Riestra, D., Noellemeyer, E. and Quiroga, A. (2012) Soil texture and forest species condition the effect of afforestation on soil quality parameters. *Soil Science* 177, 279–287.

Roscoe, R. and Buurman, P. (2003) Tillage effects on soil organic matter in density fractions of a Cerrado Oxisol. *Soil and Tillage Research* 70, 107–119.

Rossi, J., Govaerts, A., De Vos, B., Verbist, B., Vervoort, A., Poesen, J., Muys, B. and Deckers, J. (2009) Spatial structures of soil organic carbon in tropical forests—a case study of Southeastern Tanzania. *Catena* 77, 19–27.

Rótolo, G.C., Rydberg, T., Lieblein, G. and Francis, C. (2007) Emergy evaluation of grazing cattle in Argentina's Pampas. *Agriculture, Ecosystems and Environment* 119, 383–395.

Rudorff, B.F.T., Aguiar, D.A., Silva, W.F., Sugawara, L.M., Adami, M. and Moreira, M.A. (2010) Studies on the rapid expansion of sugarcane for ethanol production in São Paulo State (Brazil) using landsat data. *Remote Sensing* 2, 1057–1076.

Ruggiero, P.G.C., Batalha, M.A., Pivello, V.R. and Meirelles, S.T. (2002) Soil-vegetation relationships in cerrado (Brazilian savanna) and semideciduous forest, Southeastern Brazil. *Plant Ecology* 160, 1–16.

Sá, J.C.M., Cerri, C.C., Dick, W.A., Lal, R., Vesnke-Filho, S.P., Piccolo, M.C. and Feigl, B.E. (2001) Organic matter dynamics and carbon sequestration rates for a tillage chronosequence in a Brazilian Oxisol. *Soil Science Society of America Journal* 65, 1486–1499.

Salemi, L.F., Groppo, J.D., Trevisan, R., Seghesi, G.B., Moraes, J.M.D., Ferraz, S.F.D.B. and Martinelli, L.A. (2012) Hydrological consequences of land-use change from forest to pasture in the Atlantic rain forest region. *Ambiente & Água – An Interdisciplinary Journal of Applied Science* 7, 127–140 (in Portuguese). (http://www.ambi-agua.net/seer/index.php/ambi-agua/article/view/927/pdf_728, accessed 19 August 2014.)

Salton, J.C., Tomazi, M., Comunello, E., Zanatta, J.A. and Rabello, L.M. (2011) Condutividade elétrica e atributos físicos e químicos de um Latossolo após 15 anos sob sistemas de manejo em Mato Grosso do Sul. In: Inamasu, R.Y., Naime, J.M., Resende, A.V., Bassoi, L.H. and Bernardi, A.C.C. (eds) *Agricultura de precisão: um novo olhar* (in Portuguese). Embrapa Instrunmentação, São Carlos, São Paulo, pp. 254–260. (http://www.macroprograma1.cnptia.embrapa.br/redeap2/laboratorio-nacional-de-agricultura-de-precisao/livro-agricultura-de-precisao-um-novo-olhar/4.12, accessed 19 August 2014.)

Sano, E.E., Rosa, R., Brito, J.L.S. and Ferreira, L.G. (2008) Mapeamento semidetalhado do uso da terra do Bioma Cerrado. *Pesquisa Agropecuária Brasileira* 43, 153–156.

Sant'Anna, S.A.C., Fernandes, M.F., Ivo, W.M.P.M. and Costa, J.L.S. (2009) Evaluation of soil quality indicators in sugarcane management in sandy loam soil. *Pedosphere* 19, 312–322.

Schaffer, W.B. and Prochnow, M. (2002) *A Mata Atlântica e Você: Como Preservar, Recuperar e se Beneficiar da Mais Ameaçada Floresta Brasileira*. APREMAVI, Brasília.

Silva, A.J.N., Ribeiro, M.R., Carvalho, F.G., Silva, V.N. and Silva, L.E.S.F. (2007) Impact of sugarcane cultivation on soil carbon fractions, consistence limits and aggregate stability of a Yellow Latosol in Northeast Brazil. *Soil and Tillage Research* 94, 420–424.

Silva, J.E., Resck, D.V.S., Corazza, E.J. and Vivaldi, L. (2004) Carbon storage in clayey Oxisol cultivated pastures in the 'Cerrado' region, Brazil. *Agriculture, Ecosystems and Environment* 103, 357–363.

Siqueira Neto, M. (2006) Estoque de Carbono e Nitrogênio do Solo com Diferentes usos no Cerrado em Rio Verde. DSc, Universidade de São Paulo.

Siqueira Neto, M., Scopel, E., Corbeels, M., Cardoso, A.N., Douzet, J.-M., Feller, C., Piccolo, M.D.C., Cerri, C.C. and Bernoux, M. (2010) Soil carbon stocks under no-tillage mulch-based cropping systems in the Brazilian Cerrado: an on-farm synchronic assessment. *Soil and Tillage Research* 110, 187–195.

Sisti, C.P.J., Dos Santos, H.P., Kohhann, R., Alves, B.J.R., Urquiaga, S. and Boddey, R.M. (2004) Change in carbon and nitrogen stocks in soil under 13 years of conventional or zero tillage in southern Brazil. *Soil and Tillage Research* 76, 39–58.

Souza, P.I.M., Moreira, C.T., Neto, A.L.F., Silva, S.A., Silva, N.S., Almeida, L.A. and Kihl, R.A.S. (2007) A conquista do cerrado pela soja. In: Faleiro, F.G. and Sousa, E.S. (eds) *Pesquisa, desenvolvimento e inovação para o Cerrado*. Embrapa Cerrados, Planaltina, Distrito Federal, pp.129–138.

Souza, R.A., Telles, T.S., Machado, W., Hungria, M., Filho, J.T. and Guimarães, M.D.F. (2012) Effects of sugarcane harvesting with burning on the chemical and microbiological properties of the soil. *Agriculture, Ecosystems and Environment* 155, 1–6.

Spehar, C.R. (1995) Impact of strategic genes in soybean on agricultural development in the Brazilian tropical savannahs. *Field Crops Research* 41, 141–146.

Steinbach, H.S. and Alvarez, R. (2006) Changes in soil organic carbon contents and nitrous oxide emissions after introduction of no-till in pampean agroecosystems. *Journal of Environmental Quality* 35, 3–13.

Tabarelli, M., Aguiar, A.V., Ribeiro, M.C., Metzger, J.P. and Peres, C.A. (2010) Prospects for biodiversity conservation in the Atlantic Forest: lessons from aging human-modified landscapes. *Biological Conservation* 143, 2328–2340.

Uhlig, A., Goldemberg, J. and Coelho, S.T. (2008) O Uso de carvão vegetal na indústria siderúrgica brasileira e o impacto sobre as mudanças climáticas. *Revista Brasileira de Energia* 14, 67–85.

UNEP (2010) *Global Environment Outlook: Latin America and the Caribbean (GEO LAC 3)*. United Nations Environment Programme (UNEP), Panama City, Panama.

Uriarte, M., Yackulic, C.B., Cooper, T., Flynn, D., Cortes, M., Crk, T., Cullman, G., McGinty, M. and Sircely, J. (2009) Expansion of sugarcane production in São Paulo, Brazil: implications for fire occurrence and respiratory health. *Agriculture, Ecosystems and Environment* 132, 48–56.

Voigtlaender, M., Laclau, J.-P., Gonçalves, J.D., Piccolo, M.D., Moreira, M., Nouvellon, Y., Ranger, J. and Bouillet, J.-P. (2012) Introducing *Acacia mangium* trees in *Eucalyptus grandis* plantations: consequences for soil organic matter stocks and nitrogen mineralization. *Plant and Soil* 352, 99–111.

Zach, A., Tiessen, H. and Noellemeyer, E. (2006) Carbon turnover and carbon-13 natural abundance under land use change in semiarid savanna soils of La Pampa, Argentina. *Soil Science Society American Journal* 70, 1541–1546.

Zinn, Y.L., Resck, D.V.S. and Da Silva, J.E. (2002) Soil organic carbon as affected by afforestation with Eucalyptus and Pinus in the Cerrado region of Brazil. *Forest Ecology and Management* 166, 285–294.

22 Basic Principles of Soil Carbon Management for Multiple Ecosystem Benefits

Elke Noellemeyer* and Johan Six

Abstract

Management of soil organic matter (SOM) has traditionally focused on improving crop productivity and hence been considered mainly as a source of plant nutrients. Recently, there has been an increasing focus on SOM as a reservoir for carbon (C) and a mechanism of C sequestration, but far less emphasis has been placed on managing the regulating, cultural and supporting ecosystem services.

Soils are living bodies, and their multiple ecosystem functions are intimately related to SOM transformations and dynamics, which are mediated by soil biotic activity and soil structural dynamics. Hence, soil management for multiple ecosystem services needs to focus on the link between SOM, soil structure and soil biota, and the regulating factors for this link. Stable or even increased C stocks can potentially be achieved by using zero, reduced or conservation tillage, which diminishes the frequency and aggressiveness of ploughing and harrowing, thereby maintaining soil structure and soil biota. On the other hand, organic residue input to soil must be increased in order to stabilize or enhance C stocks. This implies that crop stubbles should not be burnt and/or only minimally grazed, and that pastures should not be overgrazed, leaving the maximum amount of above- and especially below-ground plant material to be stabilized as SOM. These practices not only affect soil C stocks but also will prevent soil losses caused by wind and water erosion and will improve soil water infiltration, potentially avoiding flooding and runoff. In order to maintain these vital soil ecosystem functions, it is necessary to restore soil structure and the associated soil biodiversity.

Multiple ecosystem services can be preserved by soil management that favours C sequestration and biological activity. Most of the vital soil functions that sustain these ecosystem services are related to soil structure, and C plays a central role in aggregate formation. Soil management recommendations will vary according to the environmental, social and economic conditions of any region, but will also have to be accompanied by political and economic actions that favour their implementation.

Introduction

Carbon (C) is the most abundant element in soil organic matter (SOM) and is the one whose presence or absence drives most biological, physical and chemical processes that occur in the soil.

Management of SOM has traditionally focused on improving crop productivity. Hence, SOM has been considered mainly as a source of plant nutrients, and agricultural practices were developed with the premise to cycle, and hence more easily extract, nutrients

*E-mail: noellemeyer@agro.unlpam.edu.ar/enoellemeyer@gmail.com

during the crop phase of the rotation and to replenish nutrient reserves during the non-cropping phase (Whitbread et al., 2000). The rotation of field crops and grass fallows conferred stability to the soil system and was a sustainable agricultural production system until the advent of inorganic fertilizers, herbicides, genetic improvement for high-yielding crops and technological innovation of mechanized tillage and other field operations, which all together constituted the so-called 'green revolution'. These global changes in agriculture brought about a substantial improvement in food production, but also produced a decoupling of the biological processes in the soil and essential nutrient concentrations (Tonitto et al., 2006). The availability of inexpensive synthetic nitrogen sources and efficient herbicides promoted this trend and enabled vast areas of the world's most productive land to be cultivated to one type of crop for prolonged periods (Tilman et al., 2002). The inherent problem associated with this type of land management is a drastic reduction of aboveground plant diversity, which also results in a decline of soil microbial activity and diversity (Milcu et al., 2010), and thus in a loss of vital soil functions (Nielsen et al., 2011). Biomass production, protection of humans and the environment, gene reservoir, physical basis of human activities, source of raw materials and geogenic and cultural heritage have been identified as key soil functions (Blum, 2005). Soils are living bodies and their multiple ecosystem functions are intimately related to SOM transformations and dynamics, which are mediated by soil biotic activity and soil structural dynamics (Six et al., 2002b). Hence, soil management for multiple ecosystem services needs to focus on the link between SOM, soil structure and soil biota, and the regulating factors of this link (Six et al., 2004; Wardle et al., 2004).

Recently, there has been a strong focus on SOM as a reservoir for C and a mechanism of C sequestration and climate change protection (Lal, 2004; Powlson et al., 2011), but far less emphasis has been placed on managing the regulating, cultural and supporting ecosystem services. The importance of soil carbon in relation to addressing pressing global issues through the provisioning of various ecosystem services has only recently been recognized by policy makers (Victoria et al., 2012).

The Soil's Natural Capital and Related Ecosystem Services

The term 'natural capital' was brought to prominence by Robert Costanza (Costanza et al., 1997), who defined natural capital as 'the extension of the economic notion of capital (manufactured means of production) to environmental goods and services. A functional definition of capital in general is: a stock that yields a flow of valuable goods or services into the future. Natural capital is thus the stock of natural ecosystems that yields a flow of valuable ecosystem goods or services into the future. For example, a stock of trees or fish provides a flow of new trees or fish, a flow that can be sustained indefinitely. Natural capital may also provide services like recycling wastes or water capture and erosion control. Since the flow of services from ecosystems requires that they function as whole systems, the structure and diversity of the system are important components of natural capital' (http://www.eoearth.org/article/Natural_capital, accessed 22 January 2013).

Only recently, these concepts have been applied to soils (Dominati et al., 2010; Robinson et al., 2012), despite the obvious relevance of natural capital and ecosystem services to soil science. The lack of consistent typology or terminology for ecosystem services means that properties, processes, functions and services become used interchangeably, leading to confusion (Robinson et al., 2012), and often the focus on final goods and services ignores the importance of soils in their delivery. In an attempt to clarify the concepts and develop a natural capital–ecosystem services framework for soils, Dominati and colleagues (Dominati et al., 2010) defined the soils' natural capital through its inherent physical properties such as depth, clay contents and type, along with manageable properties such as nutrient availability, SOM, pH, etc. (Fig. 22.1). These properties can change under the influence of natural processes, such as climate and geological processes, or due to anthropogenic factors, such as land use and farming technology. Through the natural capital of soils, ecosystem services can be defined that fulfil human needs (Fig. 22.1).

Most of the manageable soil properties can be directly related to soil C. For instance, SOM contains more than 50% of total soil C; nitrogen and sulfur availability are directly linked to the C cycle (Cadisch *et al.*, 1996); macroporosity, aggregate size, bulk density and other soil physical properties have been shown to be directly related to soil C (Plante and McGill, 2002; Scott *et al.*, 2002; Six *et al.*, 2002a, 2004; Holeplass *et al.*, 2004; Dexter *et al.*, 2008; Noellemeyer *et al.*, 2008; Urbanek *et al.*, 2011).

Ecosystem Services and Soil-based Processes

Regardless of the type of ecosystem service, soil processes provide key functioning to satisfy human needs. These soil-based processes can provide agricultural goods in the form of food and fibre, as well as non-agricultural services (Fig. 22.2).

The non-agricultural services provided by soil-based processes are vital to human needs, and all of the processes that deliver these services are directly related to the soil's C stocks and cycle. Soil management for sustaining water supply and quality will also improve erosion control, atmospheric composition and climate regulation. All these ecosystem services are very intimately dependent on the soil's structure and soil organic carbon (SOC) dynamics. The basic principles for soil management that improves SOC and soil structure consist of

Soil natural capital
- Inherent properties
  - slope
  - orientation
  - depth
  - clay types
  - texture
  - subsoil aggregates
  - stoniness
  - subsoil strength
  - wetness
- Manageable properties
  - soluble phosphate
  - mineral nitrogen
  - soil organic matter
  - carbon content
  - temperature
  - pH
  - land cover
  - macroporosity
  - bulk density
  - topsoil strength
  - topsoil aggregates

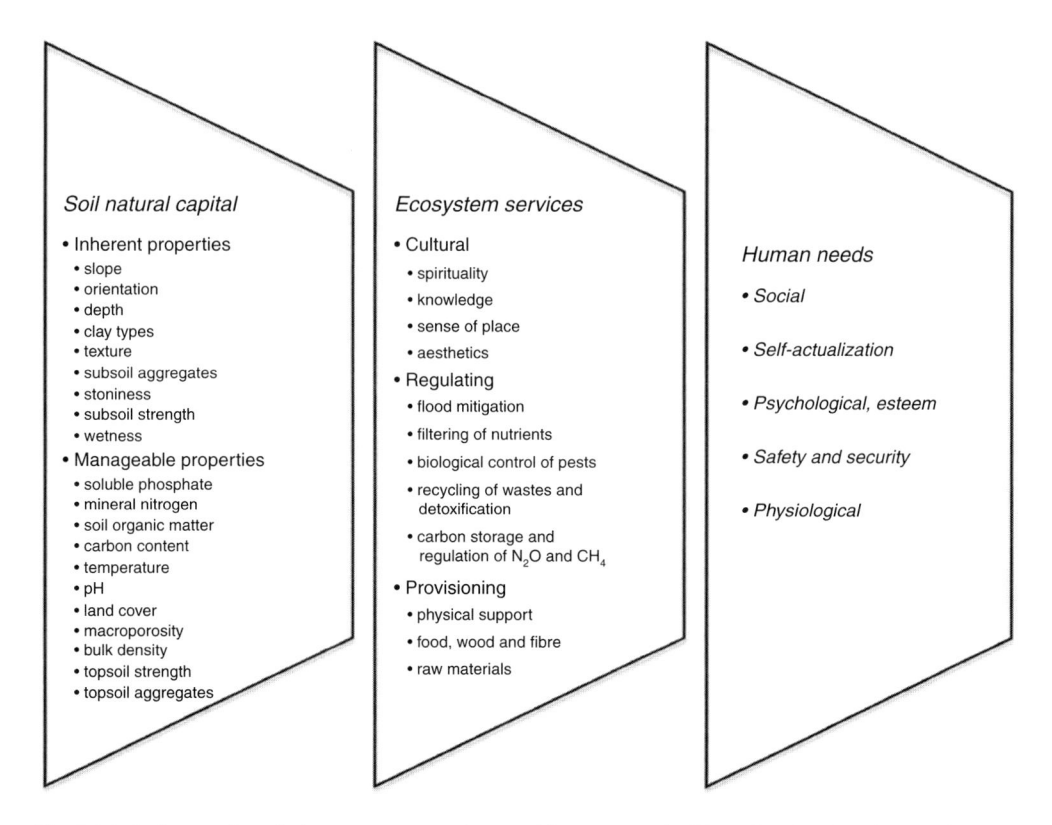

Ecosystem services
- Cultural
  - spirituality
  - knowledge
  - sense of place
  - aesthetics
- Regulating
  - flood mitigation
  - filtering of nutrients
  - biological control of pests
  - recycling of wastes and detoxification
  - carbon storage and regulation of N_2O and CH_4
- Provisioning
  - physical support
  - food, wood and fibre
  - raw materials

Human needs
- Social
- Self-actualization
- Psychological, esteem
- Safety and security
- Physiological

Fig. 22.1. Soil natural capital, ecosystem services and human needs. (From Dominati *et al.*, 2010.)

Agricultural goods	Soil-based delivery processes
Food and fibre	Nutrient capture and cycling
	OM input decomposition
	SOM dynamics
	Soil structure maintenance
	Biological population regulation
Non-agricultural services	Soil-based delivery processes
Water quality and supply	Soil structure maintenance
	Nutrient cycling
Erosion control	Soil structure maintenance
Atmospheric composition and climate regulation	SOM dynamics
Pollutant attenuation and degradation	Decomposition
	Nutrient cycling
Pest and disease control	Biological population regulation
Biodiversity conservation	Habitat provision
	Biological population regulation

Aggregate ecosystem functions
1. C transformations
2. Nutrient cycling
3. Soil structure maintenance
4. Biological population regulation

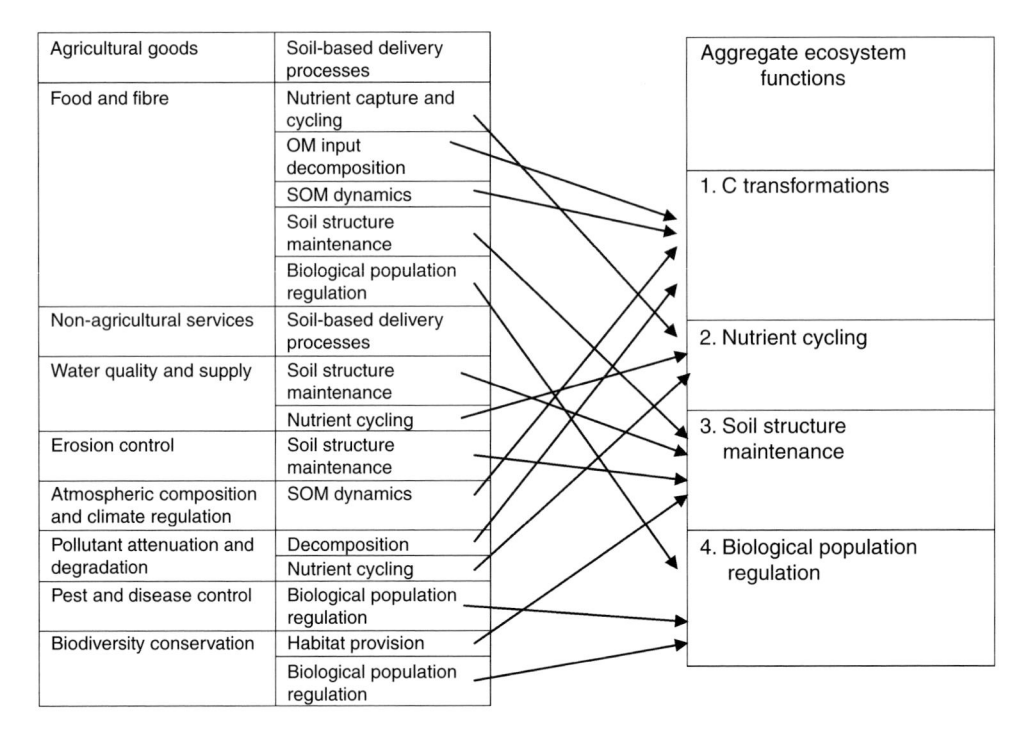

Fig. 22.2. Ecosystem services and the related soil-based processes. (From Pulleman et al., 2012.)

reducing carbon losses through excessive respiration rates that are associated to conventional ploughing operations. Zero tillage is a widely used technology that produces only minimal disturbance of the soil and therefore induces less microbial respiration than the traditional mouldboard plough. This technique also confers permanent residue cover to the soil, preventing bare soil from being exposed to eroding winds or rainfall (specific examples of regional agricultural practices that improve the soil's carbon budget and enhance ecosystem services are given in Chapters 23–28 of this volume.) Crop rotations that include perennial crops, such as grass leys, also contribute to reduce C emissions, especially during the ley phase. High-yielding crops provide the basis for high residue returns to the soil. Residue management that removes large proportions of postharvest crop remnants, as for instance for biofuel production or straw bales, has a negative impact on the SOC content of soils (Lal, 2009). The biological activity of the soil is crucial for building and maintaining good soil structure (Six, 2004; Ayuke et al., 2011). Apart from aboveground plant materials, root biomass and exudates constitute an important food source for microorganisms, earthworms and other higher organisms as well (Kong and Six, 2010; Kong et al., 2011).

Soil Carbon Management for Specific and Multiple Ecosystem Services

Soil structure maintenance and improvement

Soil structure is the result of the interaction between soil mineral, inorganic and organic constituents that form stable aggregates (Niewczas, 2003; Elmholt et al., 2008). In most soils, organic matter is the most important binding agent contributing to aggregate stability (Chivenge et al., 2011), although in many tropical soils metal and alkali ions are very important in maintaining structure

(Six et al., 2000; Barthes et al., 2008). The arrangement of stable aggregates defines the soil pore volume and size distribution, which in turn affects most soil properties related to water and gas transport (Horn and Smucker, 2005). Many studies have shown that soils under permanent pastures have better structure than agricultural soils that are tilled frequently (Pulleman and Marinissen, 2004; Zach et al., 2006; Noellemeyer et al., 2008; Berhongaray et al., 2013). This is one way in which crop rotations that include grass fallows contribute to soil structure maintenance. No-till (NT) agriculture (see Chapter 23, this volume) also has been shown to improve soil structure when compared to traditional ploughed cultivation systems through reducing soil disturbance (Six et al., 2002b; Smith and Bolton, 2003; Hollinger et al., 2005; Zotarelli et al., 2005; Lorenz et al., 2006; Lal et al., 2007; Quiroga et al., 2009; Fernández et al., 2010; López et al., 2012). However, C input into a soil is also a major factor in stabilizing soil structure and soil C (Kong and Six, 2010). Therefore NT combined with other agricultural practices that enhance C input such as mulching (Rockström et al., 2009) and planting of cover crops will further stabilize soil structure and increase soil C (Ding et al., 2006; Alletto et al., 2011; dos Santos et al., 2011; Restovich et al., 2012; Zhu et al., 2012).

Soil carbon management for erosion control

The loss of mineral and organic soil particles through water and wind erosion cause an irreversible loss of resources that sustain soil-based ecosystem services. In some cases, erosional processes transport both organic matter and clay particles short distances, and result in a spatial redistribution of these elements (Polyakov and Lal, 2004; Li et al., 2008a,b): in many cases, however, the eroded particles are transported much greater distances and lost from productive lands (Ballantine et al., 2005). The prevention of soil erosion depends largely on the stability of soil structure (Fattet et al., 2011) and therefore is affected strongly by SOM dynamics. SOM content is an important factor determining aggregate stability (Cerda, 2000; Eynard et al., 2004), and the dry aggregate size fraction (<0.84 mm) can be used as an indicator of the susceptibility of soils to wind erosion (Zobeck et al., 2003; López et al., 2007). A shift from conventional cultivation to conservation tillage systems can reduce water (Schuller et al., 2007) and wind erosion drastically (Hevia et al., 2007). This beneficial effect of conservation tillage has been related to higher SOM contents and greater surface porosity under conservation till (Kirkby et al., 2000; Rhoton and Shipitalo, 2002). Another factor contributing to erosion control is the permanent soil cover provided by living or dead plants under conservation tillage (López et al., 2003; Soane et al., 2011) or natural vegetation systems (Adema et al., 2004). Repeated ploughing to a fixed depth can lead to subsurface soil compaction and the formation of plough layers (Hamza and Anderson, 2005), which reduce water infiltration and hence increase erosion (Zink et al., 2011). Rotational and integrated cropping systems that promote subsurface root development can mitigate the effect of repeated ploughing (Keller et al., 2012). In rangelands or pastures, grazing intensity is an important driver of erosion; overgrazing promotes erosion (Oztas, 2003) by compacting the soil's surface layer (Franzluebbers and Stuedemann, 2008; Steffens et al., 2008; Barto et al., 2010) and reducing above- and belowground C inputs.

Soil management for climate regulation

Soil C stocks are determined by inherent soil properties such as clay content and type, as well as climate, vegetation and land use (Swift, 2001; Noellemeyer et al., 2006; Gili et al., 2010). Soils have a finite quantity of C they can store, and the potential for C sequestration in a given area will be greatest for degraded soils with C contents well below their saturation level (Six et al.,

2002a; Stewart et al., 2007). This could be the case for revegetation of degraded agricultural lands in marginal zones (Zach et al., 2006; Powlson et al., 2011). The UN Commission on Sustainable Development agrees on the need to find viable mitigation and adaptation strategies to address climate change (Bierbaum et al., 2007) that include reduction of CO_2 emissions and C sequestration. The protection of current soil C stocks has an important impact on the global C budget (Ingram and Fernandes, 2001), and it is imperative to prevent deforestation for cultivation and to encourage simultaneously the adoption of conservation tillage, cover crops and higher crop production systems across the globe (Follett, 2001). Strategies to mitigate climate change should also include soil management that increases the residence time of C in soils (Jastrow et al., 2006) by using perennial crops and less frequent tillage that favour a slower turnover rate of SOC. These practices may also reduce soil N_2O emissions (Drury et al., 2004; Petersen et al., 2011). Excessive applications of nitrogen (N) fertilizer should be avoided, since high N availability increases C turnover and N_2O emissions, favouring C emissions (Gärdenäs et al., 2011). Furthermore, Powlson et al. (2011) argue that reduced tillage coupled with agricultural intensification through enhanced plant productivity with synthetic fertilizers and irrigation water can offset gains in soil C through increased emissions related to fertilizer and herbicide production and irrigation energy requirements, and therefore cannot be considered net C gains. In semi-arid regions, inorganic carbon in the form of calcium carbonates also can play an important role as C sinks (Laudicina et al., 2013), and soil acidification can produce excessive solubilization and C losses in these environments.

Soil management for improved pollutant attenuation and degradation

Properties that account for pollutant attenuation and degradation functions are mostly related to biological activity and SOC dynamics in soils, but also to inorganic and mineral constituents that determine ion exchange and sorption capacity. Soil drainage and aeration, both strongly related to soil structure, also have a marked effect on the speciation of metals, thus regulating the bioavailability and biotoxicity of metal ions. Soil biological activity and SOC dynamics determine the cycling and storage of elements that can become pollutants when leached to the groundwater, such as nitrogen and phosphorus. The implementation of catch and cover crops can, however, be a very effective option to retain these nutrients in the soil and promote their slower release (Tonitto et al., 2006; Piotrowska and Wilczewski, 2012; Zhu et al., 2012), thereby preventing pollution and eutrophication of water bodies.

Soil management for pest and disease control

Before the advance of synthetic pesticides, crop rotations were the only means to control agricultural crop pests and diseases. The principle consists of diminishing the amount of harmful organisms that can affect crops by changing the type of plants cultivated and thus preventing the same host plants each season for pests and disease propagation. For example, the rotation of crops has been shown to reduce root damage by Fusarium, a common soil fungal pathogen (Nayyar et al., 2009). In modern agriculture, however, agrochemicals are applied to control pests and diseases, and how these substances affect the soil's biosphere is not fully understood (Zhang et al., 2010). Nevertheless, it is well known that higher biological activity improves crop health (Pulleman et al., 2012; Thiele-Bruhn et al., 2012) and suppresses plant diseases (Brussaard et al., 2007). Thus, soil management that includes crop rotations and maintains high levels of C input will favour the soil's biological activity and contribute to the control of pests and crop diseases.

Soil management for biodiversity conservation

Biodiversity is crucial for all soil functions (Altieri, 1999), specifically those that are related to SOC cycling, and the loss of diversity in modern agricultural monocultures implies a social cost (Hietala-Koivu et al., 2004). Yet, simply adding more species to the cropped fields might not result in enhanced soil-based ecosystem functions (Jackson et al., 2007); instead, well-conceived analogues to natural systems must be envisioned (Kirschenmann, 2007); for example, the traditional Mayan home gardens (Flores-Delgadillo et al., 2011). Land use, management intensity and fertilization have important effects on soil biota (Jangid et al., 2008); particularly, the presence of roots stimulates soil biological activity and diversity (Kong et al., 2011). Reduced soil disturbance such as with conservation tillage, residue retention and the use of cover crops promote biological activity and favour microbial biomass (Kushwaha et al., 2001; Dinesh, 2004; Helgason et al., 2010; Nielsen et al., 2011).

Incentives and Actions Needed for Implementing Sustainable Management Practices

New incentives and policies for ensuring the sustainability of agriculture and ecosystem services will be crucial if we are to meet the demands of improving food security without compromising environmental integrity or public health (Tilman et al., 2002). There are several successful examples of policies and incentives that have promoted the conservation of environmental goods and services. For instance, Bawa et al. (2007) report that, at several sites in India's biodiversity hotspots, interventions of a non-governmental organization (NGO) have improved the livelihoods of several rural communities by providing increased income from non-timber forest products (NTFPs), diversification of livelihoods and enhanced agricultural production. These interventions have improved the prospects for sustainable land use in forest–agriculture ecotones. Simultaneously, these interventions have strengthened a range of village level and regional institutions that play a critical role in the rural economy and the conservation of biodiversity. Jackson et al. (2007) considered that farmers might avoid intensification by investing in the agrobiodiversity and conservation of heterogeneous environments, as shown for land races of maize that are maintained in Mexico (Flores-Delgadillo et al., 2011). In contrast, in landscapes with less environmental and cultural heterogeneity, farmers usually disinvest in agrobiodiversity as an asset due to the lack of incentives offered by markets and other institutions at both local and larger scales, especially when synthetic inputs are available at low cost (Pascual and Perrings, 2007). Strategies that intend to reverse these trends must take into account stakeholder and local knowledge to define the benefits and management options of increased diversity (Barrios and Trejo, 2003; Payraudeau and Vanderwerf, 2005). In many countries, there are stakeholder organizations that have explicit objectives of soil conservation through collaboration with research; for example, in Australia they have actively promoted precision agriculture (Jochinke et al., 2007), whereas in Argentina they have focused on no-till farming (Caride et al., 2012). Nevertheless, the importance of economics in the implementation of sustainable land management (Osinski, 2003) implies that policies for good agricultural practices have to be developed at a regional scale (Piorr, 2003). Local land-use strategies may be highly diverse and, depending on the socioeconomic and biophysical context, range from subsistence farming to industrialized cash crops (Mertz et al., 2005). We will have to solve these problems by integrating biophysical, socio-economical and cultural constraints (Spiertz, 2012) for optimal land management to achieve carbon sequestration and other identified positive ecosystem services that are driven by soil C.

Conclusions

Multiple ecosystem services can be preserved by soil management that favours SOC stabilization and soil biological activity.

Most of the vital soil functions that sustain these ecosystem services are related to soil structure, and C plays a central role in the formation of a stable soil structure. Soil management recommendations will vary according to the environmental, social and economic conditions of any region, but will also have to be accompanied by political and economic actions that favour their implementation.

References

Adema, E.O., Buschiazzo, D.E., Babinec, F.J., Rucci, T.E. and Hermida, V.F.G. (2004) Mechanical control of shrubs in a semiarid region of Argentina and its effect on soil water content and grassland productivity. *Agricultural Water Management* 68, 185–194.

Alletto, L., Coquet, Y. and Justes, E. (2011) Effects of tillage and fallow period management on soil physical behaviour and maize development. *Agricultural Water Management* 102, 74–85.

Altieri, M.A. (1999) The ecological role of biodiversity in agroecosystems. *Agriculture, Ecosystems and Environment* 74, 19–31.

Ayuke, F.O., Brussaard, L., Vanlauwe, B., Six, J., Lelei, D.K., Kibunja, C.N. and Pulleman, M.M. (2011) Soil fertility management: impacts on soil macrofauna, soil aggregation and soil organic matter allocation. *Applied Soil Ecology* 48, 53–62.

Ballantine, J.-A.C., Okin, G.S., Prentiss, D.E. and Roberts, D.A. (2005) Mapping North African landforms using continental scale unmixing of MODIS imagery. *Remote Sensing of Environment* 97, 470–483.

Barrios, E. and Trejo, M.T. (2003) Implications of local soil knowledge for integrated soil management in Latin America. *Geoderma* 111, 217–231.

Barthes, B., Kouakoua, E., Larrelarrouy, M., Razafimbelo, T., Deluca, E., Azontonde, A., Neves, C., Defreitas, P. and Feller, C. (2008) Texture and sesquioxide effects on water-stable aggregates and organic matter in some tropical soils. *Geoderma* 143, 14–25.

Barto, E.K., Alt, F., Oelmann, Y., Wilcke, W. and Rillig, M.C. (2010) Contributions of biotic and abiotic factors to soil aggregation across a land use gradient. *Soil Biology and Biochemistry* 42, 2316–2324.

Bawa, K.S., Joseph, G. and Setty, S. (2007) Poverty, biodiversity and institutions in forest-agriculture ecotones in the Western Ghats and Eastern Himalaya ranges of India. *Agriculture, Ecosystems and Environment* 121, 287–295.

Berhongaray, G., Alvarez, R., De Paepe, J., Caride, C. and Cantet, R. (2013) Land use effects on soil carbon in the Argentine Pampas. *Geoderma* 192, 97–110.

Bierbaum, R., Holdren, J.P., McCracken, M., Moss, R.A. and Raven, P.H. (2007) Confronting climate change: avoiding the unmanageable and managing the unavoidable. Scientific Expert Group Report on Climate Change and Sustainable Development. Report prepared for the United Nations Commission on Sustainable Development. Sigma Xi, Research Triangle Park, North Carolina, and the United Nations Foundation, Washington, DC, 144 pp.

Blum, W.E.H. (2005) Functions of soil for society and the environment. *Reviews in Environmental Science and Bio/Technology* 4, 75–79.

Brussaard, L., De Ruiter, P.C. and Brown, G.G. (2007) Soil biodiversity for agricultural sustainability. *Agriculture, Ecosystems and Environment* 121, 233–244.

Cadisch, G., Imhof, H., Urquiaga, S., Boddey, R. and Giller, K. (1996) Carbon turnover ([delta] 13C) and nitrogen mineralization potential of particulate light soil organic matter after rainforest clearing. *Soil Biology and Biochemistry* 28, 1555–1567.

Caride, C., Piñeiro, G. and Paruelo, J.M. (2012) How does agricultural management modify ecosystem services in the Argentine Pampas? The effects on soil C dynamics. *Agriculture, Ecosystems and Environment* 154, 23–33.

Cerda, A. (2000) Aggregate stability against water forces under different climates on agriculture land and scrubland in southern Bolivia. *Plant and Soil* 57, 159–166.

Chivenge, P., Vanlauwe, B., Gentile, R. and Six, J. (2011) Comparison of organic versus mineral resource effects on short-term aggregate carbon and nitrogen dynamics in a sandy soil versus a fine textured soil. *Agriculture, Ecosystems and Environment* 140, 361–371.

Costanza, R., Arge, R., de Groot, R., Farber, S., Grasso, M., Hannon, B., Limburg, K., Naeem, S., O'Neill, R.V., Paruelo, J. *et al.* (1997) The value of the world's ecosystem services and natural capital. *Nature* 387, 253–260.

Dexter, A., Czyz, E., Richard, G. and Reszkowska, A. (2008) A user-friendly water retention function that takes account of the textural and structural pore spaces in soil. *Geoderma* 143, 243–253.

Dinesh, R. (2004) Long-term influence of leguminous cover crops on the biochemical properties of a sandy clay loam Fluventic Sulfaquent in a humid tropical region of India. *Soil and Tillage Research* 77, 69–77.

Ding, G., Liu, X., Herbert, S., Novak, J., Amarasiriwardena, D. and Xing, B. (2006) Effect of cover crop management on soil organic matter. *Geoderma* 130, 229–239.

Dominati, E., Patterson, M. and Mackay, A. (2010) A framework for classifying and quantifying the natural capital and ecosystem services of soils. *Ecological Economics* 69, 1858–1868.

dos Santos, N.Z., Dieckow, J., Bayer, C., Molin, R., Favaretto, N., Pauletti, V. and Piva, J.T. (2011) Forages, cover crops and related shoot and root additions in no-till rotations to C sequestration in a subtropical Ferralsol. *Soil and Tillage Research* 111, 208–218.

Drury, C., Yang, X., Reynolds, W. and Tan, C. (2004) Influence of crop rotation and aggregate size on carbon dioxide production and denitrification. *Soil and Tillage Research* 79, 87–100.

Elmholt, S., Schjønning, P., Munkholm, L.J. and Debosz, K. (2008) Soil management effects on aggregate stability and biological binding. *Geoderma* 144, 455–467.

Eynard, A., Schumacher, T., Lindstrom, M. and Malo, D. (2004) Aggregate sizes and stability in cultivated South Dakota prairie Ustolls and Usterts. *Soil Science Society of America Journal* 68, 1360–1365.

Fattet, M., Fu, Y., Ghestem, M., Ma, W., Foulonneau, M., Nespoulous, J., Le Bissonnais, Y. and Stokes, A. (2011) Effects of vegetation type on soil resistance to erosion: relationship between aggregate stability and shear strength. *Catena* 87, 60–69.

Fernández, R., Quiroga, A., Zorati, C. and Noellemeyer, E. (2010) Carbon contents and respiration rates of aggregate size fractions under no-till and conventional tillage. *Soil and Tillage Research* 109, 103–109.

Flores-Delgadillo, L., Fedick, S.L., Solleiro-Rebolledo, E., Palacios-Mayorga, S., Ortega-Larrocea, P., Sedov, S. and Osuna-Ceja, E. (2011) A sustainable system of a traditional precision agriculture in a Maya homegarden: soil quality aspects. *Soil and Tillage Research* 113, 112–120.

Follett, R.F. (2001) Soil management concepts and carbon sequestration in cropland soils. *Soil and Tillage Research* 61, 77–92.

Franzluebbers, A. and Stuedemann, J. (2008) Soil physical responses to cattle grazing cover crops under conventional and no tillage in the Southern Piedmont USA. *Soil and Tillage Research* 100, 141–153.

Gärdenäs, A.I., Ågren, G.I., Bird, J.A., Clarholm, M., Hallin, S., Ineson, P., Kätterer, T., Knicker, H., Nilsson, S.I., Näsholm, T. *et al.* (2011) Knowledge gaps in soil carbon and nitrogen interactions – from molecular to global scale. *Soil Biology and Biochemistry* 43, 702–717.

Gili, A., Trucco, R., Niveyro, S., Balzarini, M., Estelrich, D., Quiroga, A. and Noellemeyer, E. (2010) Soil texture and carbon dynamics in savannah vegetation patches of central Argentina. *Soil Science Society of America Journal* 74, 647–657.

Hamza, M. and Anderson, W. (2005) Soil compaction in cropping systems: a review of the nature, causes and possible solutions. *Soil and Tillage Research* 82, 121–145.

Helgason, B.L., Walley, F.L. and Germida, J.J. (2010) No-till soil management increases microbial biomass and alters community profiles in soil aggregates. *Applied Soil Ecology* 46, 390–397.

Hevia, G.G., Mendez, M. and Buschiazzo, D.E. (2007) Tillage affects soil aggregation parameters linked with wind erosion. *Geoderma* 140, 90–96.

Hietala-Koivu, R., Lankoski, J. and Tarmi, S. (2004) Loss of biodiversity and its social cost in an agricultural landscape. *Agriculture, Ecosystems and Environment* 103, 75–83.

Holeplass, H., Singh, B.R. and Lal, R. (2004) Carbon sequestration in soil aggregates under different crop rotations and nitrogen fertilization in an inceptisol in southeastern Norway. *Nutrient Cycling in Agroecosystems* 70, 167–177.

Hollinger, S.E., Bernacchi, C.J. and Meyers, T.P. (2005) Carbon budget of mature no-till ecosystem in North Central Region of the United States. *Agricultural and Forest Meteorology* 130, 59–69.

Horn, R. and Smucker, A. (2005) Structure formation and its consequences for gas and water transport in unsaturated arable and forest soils. *Soil and Tillage Research* 82, 5–14.

Ingram, J. and Fernandes, E. (2001) Managing carbon sequestration in soils: concepts and terminology. *Agriculture, Ecosystems and Environment* 87, 111–117.

Jackson, L., Pascual, U., Brussaard, L., Deruiter, P. and Bawa, K. (2007) Biodiversity in agricultural landscapes: investing without losing interest. *Agriculture, Ecosystems and Environment* 121, 193–195.

Jangid, K., Williams, M., Franzluebbers, A., Sanderlin, J., Reeves, J., Jenkins, M., Endale, D., Coleman, D. and Whitman, W. (2008) Relative impacts of land-use, management intensity and fertilization upon soil microbial community structure in agricultural systems. *Soil Biology and Biochemistry* 40, 2843–2853.

Jastrow, J.D., Amonette, J.E. and Bailey, V.L. (2006) Mechanisms controlling soil carbon turnover and their potential application for enhancing carbon sequestration. *Climatic Change* 80, 5–23.

Jochinke, D., Noonon, B., Wachsmann, N. and Norton, R. (2007) The adoption of precision agriculture in an Australian broadacre cropping system – challenges and opportunities. *Field Crops Research* 104, 68–76.

Keller, T., Sutter, J.A., Nissen, K. and Rydberg, T. (2012) Using field measurement of saturated soil hydraulic conductivity to detect low-yielding zones in three Swedish fields. *Soil and Tillage Research* 124, 68–77.

Kirkby, M.J., Le Bissonais, Y., Coulthard, T.J., Daroussin, J. and McMahon, M.D. (2000) The development of land quality indicators for soil degradation by water erosion. *Agriculture, Ecosystems and Environment* 81, 125–135.

Kirschenmann, F.L. (2007) Potential for a new generation of biodiversity in agroecosystems of the future. *Agronomy Journal* 99, 373–376.

Kong, A.Y.Y. and Six, J. (2010) Tracing root vs. residue carbon into soils from conventional and alternative cropping systems. *Soil Science Society of America Journal* 74, 1201–1210.

Kong, A.Y.Y., Scow, K.M., Córdova-Kreylos, A.L., Holmes, W.E. and Six, J. (2011) Microbial community composition and carbon cycling within soil microenvironments of conventional, low-input, and organic cropping systems. *Soil Biology and Biochemistry* 43, 20–30.

Kushwaha, C.P., Tripathi, S.K. and Singh, K.P. (2001) Soil organic matter and water-stable aggregates under different tillage and residue conditions in a tropical dryland agroecosystem. *Applied Soil Ecology* 16, 229–241.

Lal, R. (2004) Soil carbon sequestration to mitigate climate change. *Geoderma* 123, 1–22.

Lal, R. (2009) Soil quality impacts of residue removal for bioethanol production. *Soil and Tillage Research* 102, 233–241.

Lal, R., Reicosky, D.C. and Hanson, J.D. (2007) Evolution of the plow over 10,000 years and the rationale for no-till farming. *Soil and Tillage Research* 93, 1–12.

Laudicina, V.A., Scalenghe, R., Pisciotta, A., Parello, F. and Dazzi, C. (2013) Pedogenic carbonates and carbon pools in gypsiferous soils of a semiarid Mediterranean environment in south Italy. *Geoderma* 192, 31–38.

Li, J., Okin, G.S., Alvarez, L. and Epstein, H. (2008a) Effects of wind erosion on the spatial heterogeneity of soil nutrients in two desert grassland communities. *Biogeochemistry* 88, 73–88.

Li, J., Okin, G.S., Alvarez, L.J. and Epstein, H.E. (2008b) Sediment deposition and soil nutrient heterogeneity in two desert grassland ecosystems, southern New Mexico. *Plant and Soil* 319, 67–84.

López, M.V., Moret, D., Gracia, R. and Arrúe, J.L. (2003) Tillage effects on barley residue cover during fallow in semiarid Aragon. *Soil and Tillage Research* 72, 53–64.

López, M.V., De Dios Herrero, J.M., Hevia, G.G., Gracia, R. and Buschiazzo, D.E. (2007) Determination of the wind-erodible fraction of soils using different methodologies. *Geoderma* 139, 407–411.

López, M.V., Blanco-Moure, N., Limón, M.Á. and Gracia, R. (2012) No tillage in rainfed Aragon (NE Spain): effect on organic carbon in the soil surface horizon. *Soil and Tillage Research* 118, 61–65.

Lorenz, K., Lal, R. and Shipitalo, M.J. (2006) Stabilization of organic carbon in chemically separated pools in no-till and meadow soils in Northern Appalachia. *Geoderma* 137, 205–211.

Mertz, O., Wadley, R. and Christensen, A. (2005) Local land use strategies in a globalizing world: subsistence farming, cash crops and income diversification. *Agricultural Systems* 85, 209–215.

Milcu, A., Thebault, E., Scheu, S. and Eisenhauer, N. (2010) Plant diversity enhances the reliability of belowground processes. *Soil Biology and Biochemistry* 42, 2102–2110.

Nayyar, A., Hamel, C., Lafond, G., Gossen, B.D., Hanson, K. and Germida, J. (2009) Soil microbial quality associated with yield reduction in continuous-pea. *Applied Soil Ecology* 43, 115–121.

Nielsen, U.N., Ayres, E., Wall, D.H. and Bardgett, R.D. (2011) Soil biodiversity and carbon cycling: a review and synthesis of studies examining diversity-function relationships. *European Journal of Soil Science* 62, 105–116.

Niewczas, J. (2003) Index of soil aggregates stability as linear function value of transition matrix elements. *Soil and Tillage Research* 70, 121–130.

Noellemeyer, E., Quiroga, A. and Estelrich, D. (2006) Soil quality in three range soils of the semi-arid Pampa of Argentina. *Journal of Arid Environments* 65, 142–155.

Noellemeyer, E., Frank, F., Alvarez, C., Morazzo, G. and Quiroga, A. (2008) Carbon contents and aggregation related to soil physical and biological properties under a land-use sequence in the semiarid region of central Argentina. *Soil and Tillage Research* 99, 179–190.

Osinski, E. (2003) Economic perspectives of using indicators. *Agriculture, Ecosystems and Environment* 98, 477–482.

Oztas, T. (2003) Changes in vegetation and soil properties along a slope on overgrazed and eroded rangelands. *Journal of Arid Environments* 55, 93–100.

Pascual, U. and Perrings, C. (2007) Developing incentives and economic mechanisms for in situ biodiversity conservation in agricultural landscapes. *Agriculture, Ecosystems and Environment* 121, 256–268.

Payraudeau, S. and Vanderwerf, H. (2005) Environmental impact assessment for a farming region: a review of methods. *Agriculture, Ecosystems and Environment* 107, 1–19.

Petersen, S.O., Mutegi, J.K., Hansen, E.M. and Munkholm, L.J. (2011) Tillage effects on N_2O emissions as influenced by a winter cover crop. *Soil Biology and Biochemistry* 43, 1509–1517.

Piorr, H. (2003) Environmental policy, agri-environmental indicators and landscape indicators. *Agriculture, Ecosystems and Environment* 98, 17–33.

Piotrowska, A. and Wilczewski, E. (2012) Effects of catch crops cultivated for green manure and mineral nitrogen fertilization on soil enzyme activities and chemical properties. *Geoderma* 189–190, 72–80.

Plante, A.F. and McGill, W.B. (2002) Soil aggregate dynamics and the retention of organic matter in laboratory-incubated soil with differing simulated tillage frequencies. *Sciences-New York* 66, 79–92.

Polyakov, V. and Lal, R. (2004) Modeling soil organic matter dynamics as affected by soil water erosion. *Environment International* 30, 547–556.

Powlson, D.S., Whitmore, A.P. and Goulding, K.W.T. (2011) Soil carbon sequestration to mitigate climate change: a critical re-examination to identify the true and the false. *European Journal of Soil Science* 62, 42–55.

Pulleman, M., Creamer, R., Hamer, U., Helder, J., Pelosi, C., Pérès, G. and Rutgers, M. (2012) Soil biodiversity, biological indicators and soil ecosystem services – an overview of European approaches. *Current Opinion in Environmental Sustainability* 4, 529–538.

Pulleman, M.M. and Marinissen, J.C.Y. (2004) Physical protection of mineralizable C in aggregates from long-term pasture and arable soil. *Geoderma* 120, 273–282.

Quiroga, A., Fernández, R. and Noellemeyer, E. (2009) Grazing effect on soil properties in conventional and no-till systems. *Soil and Tillage Research* 105, 164–170.

Restovich, S.B., Andriulo, A.E. and Portela, S.I. (2012) Introduction of cover crops in a maize–soybean rotation of the Humid Pampas: effect on nitrogen and water dynamics. *Field Crops Research* 128, 62–70.

Rhoton, F. and Shipitalo, M. (2002) Runoff and soil loss from midwestern and southeastern US silt loam soils as affected by tillage practice and soil organic matter content. *Soil and Tillage Research* 66, 1–11.

Robinson, D.A., Hockley, N., Cooper, D., Emmett, B.A., Keith, A.M., Lebron, I., Reynolds, B., Tipping, E., Tye, A.M., Watts, C.W. *et al.* (2012) Natural capital and ecosystem services, developing an appropriate soils framework as a basis for valuation. *Soil Biology and Biochemistry* 57, 1023–1033.

Rockström, J., Kaumbutho, P., Mwalley, J., Nzabi, A.W., Temesgen, M., Mawenya, L., Barron, J., Mutua, J. and Damgaard-Larsen, S. (2009) Conservation farming strategies in East and Southern Africa: yields and rain water productivity from on-farm action research. *Soil and Tillage Research* 103, 23–32.

Schuller, P., Walling, D.E., Sepúlveda, A., Castillo, A. and Pino, I. (2007) Changes in soil erosion associated with the shift from conventional tillage to a no-tillage system, documented using 137Cs measurements. *Soil and Tillage Research* 94, 183–192.

Scott, N.A., Tate, K.R., Giltrap, D.J., Tattersall Smith, C., Wilde, R.H., Newsome, P.F.J. and Davis, M.R. (2002) Monitoring land-use change effects on soil carbon in New Zealand: quantifying baseline soil carbon stocks. *Environmental Pollution (Barking, Essex: 1987)* 116(Suppl), S167–S186.

Six, J. (2004) A history of research on the link between (micro)aggregates, soil biota, and soil organic matter dynamics. *Soil and Tillage Research* 79, 7–31.

Six, J., Elliott, E.T. and Paustian, K. (2000) Soil macroaggregate turnover and microaggregate formation: a mechanism for C sequestration under no-tillage agriculture. *Journal of Soil Science* 32, 2099–2103.

Six, J., Conant, R.T., Paul, E.A. and Paustian, K. (2002a) Stabilization mechanisms of soil organic matter: implications for C-saturation of soils. *Plant and Soil* 241, 155–176.

Six, J., Feller, C., Denef, K., Ogle, S.M., de Moraes Sa, J.C. and Albrecht, A. (2002b) Soil organic matter, biota and aggregation in temperate and tropical soils – effects of no-tillage. *Agronomie* 22, 755–775.

Six, J., Bossuyt, H., Degryze, S. and Denef, K. (2004) A history of research on the link between (micro) aggregates, soil biota, and soil organic matter dynamics. *Soil and Tillage Research* 79, 7–31.

Smith, J.M.B.J.L. and Bolton, V.L.B.H. (2003) Priming effect and C storage in semi-arid no-till spring crop rotations. *Canadian Journal of Soil Science* 37, 237–244.

Soane, B.D., Ball, B.C., Arvidsson, J., Basch, G., Moreno, F. and Roger-Estrade, J. (2011) No-till in northern, western and south-western Europe: a review of problems and opportunities for crop production and the environment. *Soil and Tillage Research* 118, 66–87.

Spiertz, H. (2012) Avenues to meet food security. The role of agronomy on solving complexity in food production and resource use. *European Journal of Agronomy* 43, 1–8.

Steffens, M., Kolbl, A., Totsche, K. and Kogelknabner, I. (2008) Grazing effects on soil chemical and physical properties in a semiarid steppe of Inner Mongolia (P.R. China). *Geoderma* 143, 63–72.

Stewart, C.E., Paustian, K., Conant, R.T., Plante, A.F. and Six, J. (2007) Soil carbon saturation: concept, evidence and evaluation. *Biogeochemistry* 86, 19–31.

Swift, R. (2001) Sequestration of carbon by soil. *Soil Science* 166, 858–871.

Thiele-Bruhn, S., Bloem, J., De Vries, F.T., Kalbitz, K. and Wagg, C. (2012) Linking soil biodiversity and agricultural soil management. *Current Opinion in Environmental Sustainability* 4, 523–528.

Tilman, D., Cassman, K.G., Matson, P.A., Naylor, R. and Polasky, S. (2002) Agricultural sustainability and intensive production practices. *Nature* 418, 671–677.

Tonitto, C., David, M. and Drinkwater, L. (2006) Replacing bare fallows with cover crops in fertilizer-intensive cropping systems: a meta-analysis of crop yield and N dynamics. *Agriculture, Ecosystems and Environment* 112, 58–72.

Urbanek, E., Smucker, A.J.M. and Horn, R. (2011) Total and fresh organic carbon distribution in aggregate size classes and single aggregate regions using natural 13C/12C tracer. *Geoderma* 164, 164–171.

Victoria, R., Noellemeyer, E., Banwart, S., Black, E., Ingram, J., Joosten, P. and Milne, E. (2012) On the benefits of soil carbon. In: *UNEP Yearbook 2012*. UNEP, Nairobi, pp. 19–32.

Wardle, D.A., Bardgett, R.D., Klironomos, J.N., Setälä, H., van der Putten, W.H. and Wall, D.H. (2004) Ecological linkages between aboveground and belowground biota. *Science* 304, 1629–1633.

Whitbread, A.M., Blair, G.J. and Lefroy, R.D.B. (2000) Managing legume leys, residues and fertilisers to enhance the sustainability of wheat cropping systems in Australia: 1. The effects on wheat yields and nutrient balances. *Soil and Tillage Research* 54, 63–75.

Zach, A., Tiessen, H. and Noellemeyer, E. (2006) Carbon turnover and Carbon-13 natural abundance under land use change in semiarid savanna soils of La Pampa, Argentina. *Soil Science Society of America Journal* 70, 1541–1546.

Zhang, C., Xu, J., Liu, X., Dong, F., Kong, Z., Sheng, Y. and Zheng, Y. (2010) Impact of imazethapyr on the microbial community structure in agricultural soils. *Chemosphere* 81, 800–806.

Zhu, B., Yi, L., Guo, L., Chen, G., Hu, Y., Tang, H., Xiao, C., Xiao, X., Yang, G., Acharya, S.N. and Zeng, Z. (2012) Performance of two winter cover crops and their impacts on soil properties and two subsequent rice crops in Dongting Lake Plain, Hunan, China. *Soil and Tillage Research* 124, 95–101.

Zink, A., Fleige, H. and Horn, R. (2011) Verification of harmful subsoil compaction in loess soils. *Soil and Tillage Research* 114, 127–134.

Zobeck, T.M., Popham, T.W., Skidmore, E.L., Lamb, J.A., Merrill, S.D., Lindstrom, M.J., Mokma, D.L. and Yoder, R.E. (2003) Aggregate-mean diameter and wind-erodible soil predictions using dry aggregate-size distributions. *Soil Science Society of America Journal* 67, 425–436.

Zotarelli, L., Alves, B., Urquiaga, S., Torres, E., Dos Santos, H., Paustian, K., Boddey, R. and Six, J. (2005) Impact of tillage and crop rotation on aggregate-associated carbon in two Oxisols. *Soil Science Society of America Journal* 69, 482–491.

23 Managing Soil Carbon for Multiple Ecosystem Benefits – Positive Exemplars: Latin America (Brazil and Argentina)

Carlos Eduardo P. Cerri*, Newton La Scala Jr, Reynaldo Luiz Victoria, Alberto Quiroga and Elke Noellemeyer

Abstract

Agriculture provides food, fibre and energy, which have been the foundation for the development of all societies. Soil carbon plays an important role in providing essential ecosystem services. Historically, these have been viewed in terms of plant nutrient availability only, with agricultural management being driven to obtain maximum benefits of this soil function. However, recently, agricultural systems have been envisioned to provide a more complete set of ecosystem services, in a win–win situation, in addition to the products normally associated with agriculture. The expansion and growth of agricultural production in Brazil and Argentina brought about a significant loss of soil carbon stocks, and consequently the associated ecosystem services, such as flooding and erosion control, water filtration and storage. There are several examples of soil carbon management for multiple benefits in Brazil and Argentina, with new soil management techniques attempting to reverse this trend by increasing soil carbon (C) stocks. One example is zero tillage, which has the advantage of reducing CO_2 emissions from the soil and thus preserving or augmenting C stocks. Crop rotations that include cover crops have been shown to sequester significant amounts of C, both in Brazilian subtropical regions as well as in the Argentinean Pampas. Associated benefits of zero tillage and cover crop rotations include flood and erosion control and improved water filtration and storage. Another positive example is the adoption of no-burning harvest in the vast sugarcane area in Brazil, which also contributes to reduced CO_2 emissions, leaving crop residues on the soil surface and thus helping the conservation of essential plant nutrients and improving water storage.

Introduction

The concept of ecosystem services has been cited as a major issue, and it has been possible to quantify services related to air quality, hydrological cycles, watershed protection and biodiversity, among others. New evaluation techniques, as well as the adoption of models that allow us to extrapolate those services on spatial and temporal scales, are the fundamentals for accounting the externalities associated with ecosystems. Once one ecosystem is replaced by another, the disservices can also be quantified in economical terms and according to their market price; this could make an important

*E-mail: cepcerri@usp.br

contribution towards the conservation and protection of ecosystems. Of the many services provided by ecosystems, soil carbon sequestration and biomass carbon potential are examples of ones that have been more ambitious to date, as carbon markets have been already been established.

Among the properties that most influence soil quality is the carbon content, which is affected strongly by agricultural management. Hence, appropriate soil management and the induced changes in soil carbon stocks are critical aspects when an agricultural productive scenario is considered. It is well recognized that under appropriate production scenarios, agriculture could impact much less in neighbouring ecosystems, and the possible disservices caused by its externalities could be minimized. Recently, agricultural systems have been envisioned as providing similar services as ecosystems in addition to the products normally associated with agriculture, which results in a win–win situation.

The challenge of modern agriculture is that it should provide goods as well as ecosystem services such as climate regulation, hydrological services, greenhouse gas (GHG) mitigation and the improvement of biodiversity (GBEP, 2011). As many authors have indicated, all these aspects are directly related to soil organic carbon (SOC). Here, we present some results pointing to the benefits of maintaining or building up soil carbon levels in Latin America.

Burned Versus Green Harvested Sugarcane

Brazil is the main sugarcane producer in the world; with nearly twice the harvested area and almost 2.5 times more production than India, which ranks second (FAOSTAT, 2012). In 2010, the sugarcane harvested area in Brazil was more than 8 million hectares (Mha) and production close to 625 million tonnes (Mt), an increase of 3.4% compared with the previous year. Around 54% of sugarcane produced in Brazil is for ethanol production, generating approximately 27,670 million litres (Ml) (CONAB, 2011).

Residue deposition in no-burning harvest areas is equivalent to 15% of the dry matter productivity of sugarcane, with a mean value of 13.9 Mg of dry matter ha^{-1} $year^{-1}$ (Campos, 2003). No-burning harvest systems have several benefits: for instance, higher crop longevity and lower costs for renewing areas; recycling and gradual release of nutrients by straw decomposition; decrease in gas emissions; and less nutrient losses (Canellas et al., 2003). There is also improvement of physical soil conditions, such as moisture retention, which is especially important during drought periods (Resende et al., 2006), increase in soil aggregate stability (Szakács, 2007; de Luca et al., 2008) and improvement in soil structure, mainly in sandy soils with an original low level of soil carbon (C) (de Luca et al., 2008).

Higher organic matter levels in soils improve chemical and physical soil properties, as discussed above, and also contribute to the mitigation of global warming by increasing soil C stocks. Sugarcane crops without burning accumulate more organic C in the soil than those harvested with fire (20% more C in 0–5 cm and 15% more in 0–10 cm soil depth). The main difference in organic carbon levels between the two systems occurs in the 0–2 μm fraction, where there is 35% more C under no-burning management (Razafimbelo et al., 2006), indicating the stabilization of the accumulated C. This was also shown by higher humification indexes of soil organic matter (SOM) (Panosso et al., 2011) in no-burn sugarcane. SOM under this management has up to four times more C of aromatic compounds and less C of carboxylic groups (Canellas et al., 2003). However, the effect of the harvest management system on the quality of soil C seems to occur slowly. The results reported by Canellas et al. (2003) occurred 55 years after the adoption of green harvest, while Czycza (2009), considering a period of 12 years without burning, did not observe differences in carboxylic and phenolic group concentrations of humic acids due to sugarcane harvest management. Czycza (2009) also compared 12- and 19-year-old areas and verified higher aromatic group concentration of humic acids in the superficial layer (0–10 cm depth) of older areas, while in the subsuperficial layer (10–20 cm depth), there was no difference due to time of system adoption.

About 50% of total sugarcane in Brazil (approximately 4 Mha) is still burnt prior to harvest (Canasat, 2011). Once those areas are converted from burning to mechanized harvest, a huge amount of crop residue is left on the soil surface, in some places close to 15 t ha^{-1}, which is equivalent to 6 t of C. Several studies have been conducted on sugarcane areas converted from burned to green harvest and have shown an important enhancement of soil carbon stocks due to this conversion (Cerri *et al.*, 2011). The increases in soil carbon stocks reported in these field studies, at least in the first years of green harvest adoption, would be enough to compensate for all emissions associated with other agricultural practices. Recent estimations of the amount of GHGs emitted to the atmosphere associated with all sources in the agricultural management of sugarcane fields in southern Brazil mention about 3 t CO_2 equivalents ha^{-1} year^{-1} (De Figueiredo and La Scala, 2011). A soil carbon accumulation rate of 1 t ha^{-1} year^{-1}, which has been observed in many field studies (Cerri *et al.*, 2011; La Scala *et al.*, 2012), would be enough to compensate for the emissions associated with crop production, and the ethanol derived from this agricultural management would have close to a 'zero emission' footprint.

Renovation operations with intensive soil tillage promote mineralization of SOM (Silva-Olaya *et al.*, 2013) and attenuate differences between burning and no-burning harvest systems (Resende *et al.*, 2006). To understand better the carbon balance and the system potential to increase C stocks in no-burning sugarcane areas, it is important to take into account the tillage system during the renovation period (De Figueiredo and La Scala, 2011). La Scala *et al.* (2006) evaluated the effects of conventional tillage (mouldboard ploughing followed by two passes of offset disk harrows), reduced tillage (chisel ploughing) and no-till on CO_2 emissions from sugarcane soils. The CO_2 emissions during 1 month after soil tillage were increased by 160%, and by 71% when soils were prepared with conventional and reduced tillage as compared to no-till, respectively. The results suggest that in a 1-month period after tillage, 30% of soil carbon input in sugarcane crop residues could be lost after ploughing tropical soils, when compared to the no-till plot emissions.

The same set of studies has also pointed to another important aspect: once the sugarcane fields are reformed and tillage is applied, a large amount of CO_2 is emitted from soil, and soil carbon stocks are depleted dramatically (Cerri *et al.*, 2011). Hence, the adoption of green harvest in sugarcane fields, with the input of large amounts of residues on soil surface should, desirably, be combined with a reduced or even no-till practice. This would be an ideal production scenario, a win–win situation where less fossil fuel and synthetic fertilizer use would result in higher soil carbon stocks.

In addition to reducing the dependence on fossil fuels, another objective of ethanol use is to mitigate GHG emissions (Cerri *et al.*, 2007; Goldemberg *et al.*, 2008). Brazilian sugarcane ethanol presents a mean decrease of 85% in GHG emissions compared to fossil fuels, while American maize ethanol presents a reduction of only 25% (Börjesson, 2009). Galdos *et al.* (2010) presented data for Brazilian ethanol production showing that most ethanol GHG emissions occurred in the field during sugarcane production. De Figueiredo *et al.* (2010) quantified the carbon footprint of sugar production in two Brazilian mills and observed that 241 kg of CO_2 equivalent were emitted to produce 1 t of sugar, 44% of this from burning, 20% due to mineral fertilizer use and about 18% derived from fossil fuel combustion, confirming the information reported by Galdos *et al.* (2010). Brazilian ethanol has another advantage: a lower production cost per litre in relation to fossil fuel extraction and refinement (Luo *et al.*, 2009).

Soil Organic Carbon Stocks in Conventional Versus No-tillage Systems in Latin America (Brazil and Argentina)

No-tillage (NT) is presumed to be the oldest soil management system in agriculture and, in some parts of the tropics, NT is still practised in slash-and-burn agriculture where, after forest clearing by controlled burning, seeds are placed directly into the soil without any tillage operation. As mankind developed more systematic agricultural systems, cultivation of the soil became an accepted practice as a means of preparing a more suitable seedbed and environment for plant growth. Indeed, tillage as symbolized by the mouldboard plough became almost synonymous with agriculture (Dick and Durkalski, 1997). No-tillage can be defined as a crop production system where soil is left continuously undisturbed, except in a narrow strip where seed and fertilizer are placed.

Conversion of native vegetation to cultivated cropland under a conventional tillage (CT) system has resulted in a significant decline in SOM content (Paustian et al., 2000; Lal, 2002; Zach et al., 2006). Farming methods that use mechanical tillage, such as the mouldboard plough for seedbed preparation or disking for weed control, can promote soil C loss by several mechanisms: (i) they disrupt soil aggregates that protect SOM from decomposition (Six et al., 2002; Soares et al., 2005); (ii) they stimulate short-term microbial activity through enhanced aeration, resulting in increased levels of CO_2 and other gases released to the atmosphere (Bayer et al., 2000a,b; Kladivko, 2001); and (iii) they mix fresh residues into the soil where conditions for decomposition are often more favourable than on the surface (Plataforma Plantio Direto, 2009). Furthermore, tillage can leave soils more prone to erosion, resulting in further loss of soil C (Bertol et al., 2005; Lal, 2006).

However, no-tillage farming, due to less soil disturbance, often results in significant accumulation of SOC (Bayer et al., 2000b; Sá et al., 2001; Schuman et al., 2002) and in consequent reductions of GHG emissions to the atmosphere, especially CO_2 (Lal, 1998; Paustian et al., 2000), compared to CT. There is considerable evidence that the main effect on SOC is in the topsoil layers (Six et al., 2002; Abril et al., 2005; Noellemeyer et al., 2008), but significant increments in SOC have also been reported for layers below 30 cm in depth in NT soils with high-input cropping systems (Sisti et al., 2004; Dieckow et al., 2005a,b).

Worldwide, approximately 63 Mha are currently being managed under NT farming, with the USA having the largest area (Lal, 2006), followed by Brazil and Argentina. In Brazil, NT farming began in the southern states in the 1970s as an alternative to the misuse of land that was leading to unacceptable levels of soil losses by water erosion (Denardin and Kochhann, 1993). Similarly, in Argentina, NT began to be used in the central rolling Pampas, where water erosion also had become a major problem when soybean–wheat double cropping was introduced (Alvarez and Steinbach, 2009), and NT was shown to reduce runoff velocity and sediment load effectively (Castigilioni et al., 2006). The underlying land management principles that led to the development of NT systems were prevention of surface sealing, caused by rainfall impact on the soil surface, achievement and maintenance of an open soil structure and reduction of the volume and velocity of surface runoff. Consequently, NT was based on two essential farm practices: (i) not tilling and (ii) keeping soil covered all the time. This alternative strategy expanded quickly and the cropped area under NT has since then increased exponentially.

In the early 1990s, the NT area in Brazil was about 1 Mha, increased ten times by 1997, and currently is approximately 24 Mha. This expansion includes the conversion from CT in the southern region (72%) and expansion of the agricultural frontier, clearing the natural savannah in the central-western area (28%). Recently, due to the high profits, ranchers in the Amazon region are converting old pastures to soybean/millet under NT.

Changes in soil C stocks under NT have been estimated in earlier studies for temperate and tropical regions. Reicosky et al. (1995) reviewed various publications and found that organic matter increased under conservation management systems, with rates ranging from 0 to 1.15 t C ha^{-1} year^{-1}, with the highest accumulation rates generally occurring in temperate conditions. In the tropics, specifically in Brazil, the rate of C accumulation has been estimated in the two main regions under NT systems (the south and central-western regions). In the southern region, Sá (2001) and Sá et al. (2001) estimated sequestration rates of 0.8 t C ha^{-1} year^{-1} in the 0–20 cm layer and 1.0 t C ha^{-1} year^{-1} at 0–40 cm soil depth after 22 years under NT compared to the same period under CT. The authors mentioned that the accumulated C was generally greater in the coarse fraction (>20 µm), indicating that most of this additional C was relatively labile.

Bayer et al. (2000a,b) found a C accumulation rate of 1.6 t ha^{-1} year^{-1} for a 9-year NT system compared with 0.10 t ha^{-1} year^{-1} for the CT system in the first 30 cm layer of an Acrisol in the southern part of Brazil. Corazza et al. (1999) reported an additional accumulation of approximately 0.75 t C ha^{-1} year^{-1} in the 0–40 cm soil layer due to NT in the savannah region located in the centre-west. Estimates by Amado et al. (1998, 1999) indicated an accumulation rate of 2.2 t ha^{-1} year^{-1} of soil organic C in the first 10 cm layer. Other studies considering the NT system carried out in the centre-west region of Brazil (Lima et al., 1994; Castro Filho et al., 1998; Riezebos and Loerts, 1998; Vasconcellos et al., 1998; Peixoto et al., 1999; Spagnollo et al., 1999; Resck et al., 2000) reported soil C sequestration rates due to NT varying from 0 up to 1.2 t C ha^{-1} year^{-1} for the 0–10 cm layer. Bernoux et al. (2006) reported that most studies of Brazilian soils gave annual rates of carbon storage in the top 40 cm of the soil varying from 0.4 to 1.7 t C ha^{-1}, with the highest rates in the Cerrado region. However, the authors stressed that caution must be taken when analysing NT systems in terms of carbon sequestration. Comparisons should include changes in trace gas fluxes and should not be limited to a consideration of carbon storage in the soil alone if the full implications of global warming are to be assessed. The adoption of NT management in subtropical Brazilian soils has led to SOC accumulation rates of 0.19–0.81 Mg ha^{-1} year^{-1} in the 0–20 cm layer (Bayer et al., 2006), due to the less oxidative environment and the physical protection mechanism imparted by the stable aggregates in NT soils (Eiza et al., 2005).

The Importance of Cover Crops for Sustainable Carbon Management in Tropical and Subtropical Agroecosystems

Soils in tropical and subtropical environments are often exposed to strong rain and long-term drought events during the year. Hence, either in sugarcane or annual crops, any management that leaves more crop residues covering the soil surface is beneficial for soil protection and C sequestration.

The importance of residue cover to avoid soil erosion or food web respiration, and for maintaining SOM matter in annual crops through grass–legume rotations has been discussed by many authors (Lal et al., 1998; Magdoff and Weil, 2004). In southern Brazil, many experiments showed SOC accumulation due to the conversion of systems based on intensive tillage to NT with crop rotation, with topsoil SOC gains of up to 91% (Zanatta et al., 2007).

Dieckow et al. (2005a,b) evaluated soil organic C and N losses during a period of conventional cultivation (1969–1983) that followed native grassland and the potential of four long-term (17 years) no-till cereal- and legume-based cropping systems with different N fertilization levels to increase the C and N stocks of a southern Brazilian Acrisol. The C content in the 0–17.5 cm soil layer decreased by 22% (8.6 Mg C ha^{-1}) and N decreased by 14% (0.44 Mg N ha^{-1}) during the period of conventional cultivation. Legume-based cropping systems increased C and N stocks due to the higher residue input. Although the major soil management effects were found in the 0–100 cm layer, up to 24% of the overall C losses and 63%

of the gains of the whole 0–17.5 cm soil profile occurred below the 17.5 cm depth, reinforcing the importance of subsoil as a C source or sink. The average C sequestration rate of legume-based cropping systems (with N) were 0.83 Mg C ha^{-1} year^{-1} in the top 0–17.5 cm layer and 1.42 Mg C ha^{-1} year^{-1} in the profile, indicating the remarkable potential of legume cover crops and N fertilization under no-tillage to improve soil carbon stocks and thus soil and environmental quality in humid subtropical regions.

The adoption of NT and soil cover also brings about favourable soil physical conditions that improve water infiltration and storage (Fernández et al., 2008) and prevent water and wind erosion (Hevia et al., 2007), which is very important in semi-arid regions. For temperate environments in eastern Argentina, NT combined with pasture rotations was shown to be a sustainable agricultural practice that combined high yields with carbon storage in soils (Studdert et al., 1997; Studdert and Echeverría, 2000). Similar results were also obtained in Uruguay when NT was incorporated into crop–pasture rotations (Garciaprechac, 2004). Other benefits of NT include higher biological activity of the soil ecosystem (Quiroga et al., 2009; Fernández et al., 2010a), which also promotes the diversity of soil organisms, as confirmed by studies on Chilean and Argentinean NT soils (Abril et al., 1995; Borie et al., 2006). Crops cultivated under NT are usually more efficient in water use and produce higher yields (Noellemeyer et al., 2013). All of the mentioned benefits were enhanced, without negative impacts on crop yields, when cover crops were used to improve soil cover and residue biomass input to the soil (Fernández et al., 2010b; Mohammadi, 2010; dos Santos et al., 2011; Restovich et al., 2012). Cover crops or double cropping can also help to retain more water in the soils of a region compared to single crops under conventional tillage (Nosetto et al., 2012), by reducing losses through deep drainage and surface runoff. Cover crops and NT technology has been widely adopted both in Brazil and Argentina, and is also applied in important areas of Chile, Paraguay, Bolivia and Uruguay, resulting in manifold benefits. Generally, crop yields are higher under NT, resulting in improved provision of goods, but also many ecosystem services such as water filtration and storage, erosion prevention, soil formation and biodiversity conservation are enhanced.

In conclusion, crop residue management is a key point of NT systems, and includes selecting crops that produce sufficient quantities of residues (e.g. maize, sorghum, etc.) and the introduction of cover crops in rotation schemes that provide an effective ground cover. Rather than turning under plant materials or crop residues following harvest, the residues are left on the soil surface to protect the soil against the erosive forces of rainfall and wind. Crop residue management is also a key point for enhancing SOC under energy crops such as sugarcane. Here, there is room to improve management simply by changing the tillage system during the reform of the sugarcane fields.

References

Abril, A., Caucas, V. and Nuñez Vazquez, F. (1995) Sistemas de labranza y dinámica microbiana del suelo en la región central de la Provincia de Córdoba (Argentina). *Ciencia del Suelo* 13, 104–106.

Abril, A., Salas, P., Lovera, E., Kopp, S. and Casado-Murillo, N. (2005) Efecto acumulativo de la siembra directa sobre algunas características del suelo en la región semiárida central de la Argentina. *Ciencia del Suelo* 23, 179–188.

Alvarez, R. and Steinbach, H.S. (2009) A review of the effects of tillage systems on some soil physical properties, water content, nitrate availability and crops yield in the Argentine Pampas. *Soil and Tillage Research* 104, 1–15.

Amado, T.J.C., Fernandez, S.B. and Mielniczuk, J. (1998) Nitrogen availability as affected by ten years of cover crop and tillage systems in Southern Brazil. *Journal of Soil and Water Conservation* 53, 268–271.

Amado, T.J., Pontelli, C.B., Júnior, G.G., Brum, A.C.R., Eltz, F.L.F. and Pedruzzi, C. (1999) Seqüestro de carbono em sistemas conservacionistas na Depressão Central do Rio Grande do Sul. In: Wildner, L.P. (ed.) *Reunión Bienal de la Red Latinoamericana de Agricultura Conservacionista*. Universidade Federal de Santa Catarina, Florianópolis, Brazil, pp. 42–43.

Bayer, C., Martin-Neto, L., Mielniczuk, J. and Ceretta, C.A. (2000a) Effect of no-till cropping systems on soil organic matter in a sandy clay loam Acrisol from southern Brazil monitored by electron spin resonance and nuclear magnetic resonance. *Soil and Tillage Research* 53, 95–104.

Bayer, C., Mielniczuk, J., Amado, T.J.C., Martin-Neto, L. and Fernandes, S.V. (2000b) Organic matter storage in a sandy clay loam Acrisol affected by tillage and cropping systems in southern Brazil. *Soil and Tillage Research* 54, 101–109.

Bayer, C., Martin-Neto, L., Mielniczuk, J., Pavinato, A. and Dieckow, J. (2006) Carbon sequestration in two Brazilian Cerrado soils under no-till. *Soil and Tillage Research* 86, 237–245.

Bernoux, M., Cerri, C.C., Cerri, C.E.P., Siqueira Neto, M., Metay, A., Perrin, A.S., Scopel, E., Blavet, D., Piccolo, M.C., Pavei, M. and Milne, E. (2006) Cropping systems, carbon sequestration and erosion in Brazil, a review. *Agronomy for Sustainable Development* 26, 1–8.

Bertol, I., Guadagnin, J.C., Gonzalez, A.P., Amaral, A.J. and Brignoni, L.F. (2005) Soil tillage, water erosion, and calcium, magnesium and organic carbon losses. *Scientia Agricola* 62, 578–584.

Borie, F., Rubio, R., Rouanet, J.L., Morales, A., Borie, G. and Rojas, C. (2006) Effects of tillage systems on soil characteristics, glomalin and mycorrhizal propagules in a Chilean Ultisol. *Soil and Tillage Research* 88, 253–261.

Börjesson, P. (2009) Good or bad bioethanol from a greenhouse gas perspective – what determines this? *Applied Energy* 86, 589–594.

Campos, D.C. (2003) Potencialidade do sistema de colheita sem queima da cana de açúcar par ao seqüestro de carbono. PhD thesis, ESALQ/USP, 103 pp.

Canasat (2011) *Sugarcane Crop Mapping in Brazil by Earth Observing Satellite Images*. Maps and Graphs. http://www.dsr.inpe.br/laf/canasat/en/map.html (accessed in 2012).

Canellas, L.P., Velloso, A.C.X., Marciano, C.R., Ramalho, J.F.G.P., Rumjanek, V.M., Rezende, C.E. and Santos, G.A. (2003) Propriedades químicas de um cambissolo cultivado com cana-de-açúcar, com preservação do palhiço e adição de vinhaça por longo tempo. *Revista Brasileira de Ciência do Solo* 27, 935–944.

Castigilioni, M., Chagas, C., Massobrio, M., Santanatoglia, O. and Bujan, A. (2006) Análisis de los escurrimientos de una microcuenca de Pampa ondulada bajo diferentes sistemas de labranza. *Revista de la Asociación Argentina de la Ciencia del Suelo* 24, 169–176.

Castro Filho, C., Muzilli, O. and Podanoschi, A.L. (1998) Estabilidade dos agregados e sua relação com o teor e carbono orgânico num latossolo distrófico, em função de sistemas de plantio, rotações de culturas e métodos de preparo das amostras. *Revista Brasileira de Ciencia do Solo* 22, 527–538.

Cerri, C.E.P., Sparovek, G., Bernoux, M., Easterling, W.E., Melillo, J.M. and Cerri, C.C. (2007) Tropical agriculture and global warming: impacts and mitigation options. *Scientia Agricola* 64, 83–99.

Cerri, C.C., Galdos, M.V., Maia, S.M.F., Bernoux, M., Feigl, B.J., Powlson, D. and Cerri, C.E.P. (2011) Effect of sugarcane harvesting systems on soil carbon stocks in Brazil: an examination of existing data. *European Journal of Soil Science* 62, 23–28.

CONAB – Companhia Nacional de Abastecimento (2011) Ministério da Agricultura, Pecuária e Abastecimento. Acompanhamento da safra brasileira: cana-de-açúcar, safra 2009/2010, primeiro levantamento. (http://www.conab.gov.br/conteudos.php?a=1253&, accessed 21 October 2014).

Corazza, E.J., Silva, J.E., Resck, D.V.S. and Gomes, A.C. (1999) Comportamento de diferentes sistemas de manejo como fonte ou depósito de carbono em relação a vegetação de Cerrado. *Revista Brasileira de Ciência do Solo* 23, 425–432.

Czycza, R. (2009) Quantidade e qualidade da matéria orgânica do solo em sistemas de colheita com e sem queima da cana-de-açúcar. Master dissertation, ESALQ/USP, 92 pp.

De Figueiredo, E.B. and La Scala, N. (2011) Greenhouse gas balance due to the conversion of sugarcane areas from burned to green harvest in Brazil. *Agriculture, Ecosystems and Environment* 141, 77–85.

De Figueiredo, E.B., Panosso, A., Romão, R. and La Scala, N. (2010) Greenhouse gas emission associated with sugar production in southern Brazil. *Carbon Balance and Management* 5, 3–10.

de Luca, E.F., Feller, C., Cerri, C.C., Barthès, B., Chaplot, V., Campos, D.C. and Manechini, C. (2008) Avaliação de atributos físicos e estoques de carbono e nitrogênio em solos com queima e sem queima de canavial. *Revista Brasileira de Ciência do Solo* 32, 789–800.

Denardin, J.E. and Kochhann, R.A. (1993) Requisitos para a implementação e a manutenção do plantio direto. In: EMBRAPA-CNPT, FUNDACEP-FETRIGO and Fundação ABC (eds) *Plantio direto no Brasil, EMBRAPA*. Editora Aldeia Norte, Passo Fundo, Brazil, pp. 19–27.

Dick, W.A. and Durkalski, J.T. (1997) No-tillage production agriculture and carbon sequestration in a Typic Fragiudalf soil of Northeastern Ohio. In: Lal, R., Kimble, J., Follett, R.F. and Stewart, B.A. (eds) *Management of Carbon Sequestration in Soil*. CRC Lewis Publishers, Boca Raton, Florida, pp. 59–71.

Dieckow, J., Mielniczuk, J., Knicker, H., Bayer, C., Dick, D.P. and Kögel-Knabner, I. (2005a) Soil C and N stocks as affected by cropping systems and nitrogen fertilisation in a southern Brazil Acrisol managed under no-tillage for 17 years. *Soil and Tillage Research* 81, 87–95.

Dieckow, J., Mielniczuk, J., Knicker, H., Bayer, C., Dick, D.P. and Knabner, I.K. (2005b) Carbon and nitrogen stocks in physical fractions of a subtropical Acrisol as influenced by long-term no-till cropping systems and N fertilisation. *Plant and Soil* 268, 319–328.

dos Santos, N.Z., Dieckow, J., Bayer, C., Molin, R., Favaretto, N., Pauletti, V. and Piva, J.T. (2011) Forages, cover crops and related shoot and root additions in no-till rotations to C sequestration in a subtropical Ferralsol. *Soil and Tillage Research* 111, 208–218.

Eiza, M.J., Fioriti, N., Studdert, G.A and Echeverría, H.E. (2005) Fracciones de carbono orgánico en la capa arable: efecto de los sistemas de cultivo y de la fertilización nitrogenada. *Ciencia del Suelo* 23, 59–67.

FAOSTAT (Food and Agriculture Organization of the United Nations) (2012) FAO Statistics Division (http://faostat.fao.org/site/339/default.aspx, accessed 28 November 2012).

Fernández, R., Quiroga, A., Noellemeyer, E., Funaro, D., Montoya, J., Hitzmann, B. and Peinemann, N. (2008) A study of the effect of the interaction between site-specific conditions, residue cover and weed control on water storage during fallow. *Agricultural Water Management* 95, 1028–1040.

Fernández, R., Quiroga, A., Zorati, C. and Noellemeyer, E. (2010a) Carbon contents and respiration rates of aggregate size fractions under no-till and conventional tillage. *Soil and Tillage Research* 109, 103–109.

Fernández, R., Saks, J., Arguello, J., Quiroga, A. and Noellemeyer, E. (2010b) *Cultivo de cobertura, Una alternativa viable para la region semiarida pampeana?* Reunión Técnica SUCS -ISTRO, Colonia, Uruguay, pp. 1–6.

Galdos, M.V., Cerri, C.C., Lal, R., Bernoux, M., Feigl, B.J. and Cerri, C.E.P. (2010) Net greenhouse gas fluxes in Brazilian ethanol production systems. *GCB Bioenergy* 2, 37–44.

Garciaprechac, F. (2004) Integrating no-till into crop-pasture rotations in Uruguay. *Soil and Tillage Research* 77, 1–13.

GBEP (Global Bioenergy Partnership) (2011) *Sustainability Indicators for Bioenergy*. FAO. http://www.csrees.usda.gov/nea/plants/pdfs/gbep_indicat_list.pdf (accessed 21 August 2014).

Goldemberg, J., Coelho, S.T. and Guardabassi, P.M. (2008) The sustainability of ethanol production from sugarcane. *Energy Policy* 36, 2086–2097.

Hevia, G.G., Mendez, M. and Buschiazzo, D.E. (2007) Tillage affects soil aggregation parameters linked with wind erosion. *Geoderma* 140, 90–96.

Kladivko, E. (2001) Tillage systems and soil ecology. *Soil and Tillage Research* 61, 61–76.

Lal, R. (1998) Long-term tillage and maize monoculture effects on a tropical Alfisol in Western Nigeria. II. Soil chemical properties. *Soil and Tillage Research* 42, 161–174.

Lal, R. (2002) Soil carbon dynamic in cropland and rangeland. *Environmental Pollution* 116, 353–362.

Lal, R. (2006) Enhancing crop yields in the developing countries through restoration of the soil organic carbon pool in agricultural lands. *Land Degradation and Development* 17, 197–209.

Lal, R., Kimble, J., Follett, R.F. and Cole, C.V. (1998) *The Potential of U.S. Cropland to Sequester Carbon and Mitigate the Greenhouse Effect*. Ann Arbor Press, Ann Arbor, Michigan, 123 pp.

La Scala, N., Bolonhezi, D. and Pereira, G.T. (2006) Short-term soil CO_2 emission after conventional and reduced tillage of a no-till sugar cane area in southern Brazil. *Soil and Tillage Research* 91, 244–248.

La Scala, N., De Figueiredo, E.B. and Panosso, A.R. (2012) A review on soil carbon accumulation due to the management change of major Brazilian agricultural activities. *Brazilian Journal of Biology* 72, 775–785.

Lima, V.C., Lima, J.M.C., Eduardo, B.J.P. and Cerri, C.C. (1994) Conteúdo de carbono e biomassa microbiana em agrosistemas: comparação entre métodos de preparo do solo. *Revista do Setor de Ciências Agrárias* 13, 297–302.

Luo, L., Voet, E. and Huppes, G. (2009) Life cycle assessment and life cycle costing of bioethanol from sugarcane in Brazil. *Renewable and Sustainable Energy Reviews* 13, 1613–1619.

Magdoff, F. and Weil, R.R. (2004) *Soil Organic Matter in Sustainable Agriculture*. CRC Press, Boca Raton, Florida, 398 pp.

Mohammadi, G.R. (2010) The effects of different autumn-seeded cover crops on subsequent irrigated corn response to nitrogen fertilizer. *Agricultural Sciences* 1, 148–153.

Noellemeyer, E., Frank, F., Alvarez, C., Morazzo, G. and Quiroga, A. (2008) Carbon contents and aggregation related to soil physical and biological properties under a land-use sequence in the semiarid region of central Argentina. *Soil and Tillage Research* 99, 179–190.

Noellemeyer, E., Fernández, R. and Quiroga, A. (2013) Crop and tillage effects on water productivity of dryland agriculture in Argentina. *Agriculture* 3, 1–11.

Nosetto, M.D., Jobbágy, E.G., Brizuela, A.B. and Jackson, R.B. (2012) The hydrologic consequences of land cover change in central Argentina. *Agriculture, Ecosystems and Environment* 154, 2–11.

Panosso, A.R., Marques, J., Milori, D.M.B.P., Ferraudo, A.S., Barbieri, D.M., Pereira, G.T. and La Scala, N. (2011) Soil CO_2 emission and its relation to soil properties in sugarcane areas under slash-and-burn and green harvest. *Soil and Tillage Research* 111, 190–196.

Paustian, K., Six, J., Elliott, E.T. and Hunt, H.W. (2000) Management options for reducing CO_2 emissions from agricultural soils. *Biogeochemistry* 48, 147–163.

Peixoto, R.T., Stella, L.M., Machulek Jr, A., Mehl, H.U. and Batista, E.A. (1999) Distibução das frações granulométricas da matéria orgânica em função do manejo do sols. In: Martin-Neto, L., Milori, D.M.B.P. and Silva, W.T.L. (eds) *Encontro brasileiro sobre substâncias húmicas*. Iconos, Santa Maria, Brazil, pp. 346–348.

Plataforma Plantio Direto (2009) Sistema Plantio Direto. (http://www22.sede.embrapa.br/plantiodireto, accessed 23 October 2014.)

Quiroga, A., Fernández, R. and Noellemeyer, E. (2009) Grazing effect on soil properties in conventional and no-till systems. *Soil and Tillage Research* 105, 164–170.

Razafimbelo, T., Barthès, B., Larré-Larrouy, M.C., Luca, E.F., Laurent, J.Y., Cerri, C.C. and Feller, C. (2006) Effect of sugarcane residue management (mulching versus burning) on organic matter in a clayey Oxisol from southern Brazil. *Agriculture, Ecosystems and Environment* 115, 285–289.

Reicosky, D.C., Kemper, W.D., Langdale, G.W., Douglas, C.L. and Rasmunssen, P.E. (1995) Soil organic matter changes resulting from tillage and biomass production. *Journal of Soil and Water Conservation* 50, 253–261.

Resck, D.V.S., Vasconcellos, C.A., Vilela, L. and Macedo, M.C.M. (2000) Impact of conversion of Brazilian Cerrados to cropland and pastureland on soil carbon pool and dynamics. In: Lal, R., Kimble, J.M. and Stewart, B.A. (eds) *Global Climate Change and Tropical Ecosystems*. CRC Press, Boca Raton, Florida, pp. 169–196.

Restovich, S.B., Andriulo, A.E. and Portela, S.I. (2012) Introduction of cover crops in a maize–soybean rotation of the Humid Pampas: effect on nitrogen and water dynamics. *Field Crops Research* 128, 62–70.

Resende, A.S., Xavier, R.P., Oliveira, O.C., Urquiaga, S., Alves, B.J.R. and Boddey, R.M. (2006) Long-term effects of pre-harvest burning and nitrogen and vinasse applications on yield of sugar cane and soil carbon and nitrogen stocks on a plantation in Pernambuco, N.E. Brazil. *Plant and Soil* 281, 339–351.

Riezebos, H.T.H. and Loerts, A.C. (1998) Influence of land use change and tillage practice on soil organic matter in southern Brazil and eastern Paraguay. *Soil and Tillage Research* 49, 271–275.

Sá, J.C.M. (2001) Dinâmica da matéria orgânica do solo em sistemas de manejo convencional e plantio direto no estado do Paraná. Piracicaba. Thesis (PhD) – Universidade de São Paulo, 114 pp.

Sá, J.C.M., Cerri, C.C., Lal, R., Dick, W.A., Venzke Filho, S., Piccolo, M.C. and Feigl, B. (2001) Organic matter dynamics and carbon sequestration rates for a tillage chronosequence in a Brazilian Oxisol. *Soil Science Society of America Journal* 65, 1486–1499.

Schuman, G.E., Janzen, H.H. and Herrick, J.E. (2002) Soil carbon dynamics and potential carbon sequestration by rangelands. *Environmental Pollution* 116, 391–396.

Silva-Olaya, A.M., Cerri, C.C., La Scala, N. Jr, Cerri, C.E.P. and Dias, C.T.S. (2013) Carbon dioxide emissions under different soil tillage systems in the cultivation of mechanically harvested sugarcane. *Environmental Research Letters* 8. (http://iopscience.iop.org/1748-9326/8/1/015014, accessed 21 August 2014.)

Sisti, C.P.J., Santos, H.P., Kohhann, R., Alves, B.J.R., Urquiaga, S. and Boddey, R.M. (2004) Change in carbon and nitrogen stocks in soil under 13 years of conventional or zero tillage in Southern Brazil. *Soil and Tillage Research* 76, 39–58.

Six, J., Feller, C., Denef, K., Ogle, S.M., Sa, J.C.M. and Albrecht, A. (2002) Soil organic matter, biota and aggregation in temperate and tropical soils – effects of no-tillage. *Agronomie* 22, 755–775.

Soares, J.L.N., Espindola, C.R. and Pereira, W.L.M. (2005) Physical properties of soils under intensive agricultural management. *Scientia Agricola* 62, 165–172.

Spagnollo, E., Bayer, C., Prado Wildner, L., Ernani, P.R., Albuquerque, J.A. and Proença, M.M. (1999) Influência de plantas intercalare ao milho no rendimento de grãos e propriedades químicas do sols em differentes sistemas de cultivo. In: Martin-Neto, L., Milori, D.M.B.P. and Silva, W.T.L. (eds) *Encontro Brasileiro sobre Substâncias Húmicas*. Antares, Santa Maria, Brazil, pp. 229–231.

Studdert, G. and Echeverría, H.E. (2000) Crop rotations and nitrogen fertilization to manage soil organic carbon dynamics. *Soil Science Society of America Journal* 64, 1496–1503.

Studdert, G.A., Echeverría, H.E. and Casanovas, E.M. (1997) Crop pasture rotation for sustaining the quality and productivity of a Typic Argiudoll. *Soil Science Society of America Journal* 61, 1466–1472.

Szakács, G.G.L. (2007) *Estoques de carbono e agregados do solo cultivado com cana-de-açúcar: efeito da palhada e do clima no centro-sul do Brasil*. PhD thesis CENA/USP, 106 pp.

Vasconcellos, C.A., Figueiredo, A.P.M., França, G.E., Coelho, A.M. and Bressan, W. (1998) Manejo do solo e a atividade microbiana em latossolo vermelho-escuro da região de Sete Lagoas, MG. *Pesquisa Agropecuaria Brasileira* 33, 1897–1905.

Zach, A., Tiessen, H. and Noellemeyer, E. (2006) Carbon turnover and Carbon-13 natural abundance under land use change in semiarid savanna soils of La Pampa, Argentina. *Soil Science Society of America Journal* 70, 1541–1546.

Zanatta, J.A., Bayer, C., Dieckow, J., Vieira, F.C.B. and Mielniczuk, J. (2007) Soil organic carbon accumulation and carbon costs related to tillage, cropping systems and nitrogen fertilization in a subtropical Acrisol. *Soil and Tillage Research* 94, 510–519.

24 Managing Soil Carbon for Multiple Benefits – Positive Exemplars: North America

Rich Conant*

Abstract

Implementing land-use practices that maximize the soil carbon (C) stocks can simultaneously lead to additional economic and social benefits while maintaining or enhancing the ecological support functions of the land resources. This enhances the ability to meet our near-term needs while ensuring the ability of future generations to meet theirs. Within North America, numerous opportunities exist to increase soil fertility, enhance soil water balance, increase production efficiency and reduce reliance on external inputs, which will enhance the resilience of production and yields in the face of climate variability. Practices that build resilience in the face of current climate variability are also expected to ameliorate some of the effects of the forecast increase in future extreme events; thus, building C stocks can foster adaptation to a changing climate. This chapter reviews North American agricultural and grazing land management practices that can sequester C in soils, their potential to mitigation greenhouse gas emissions and the additional benefits that arise from these practices.

Introduction

The main processes governing the carbon (C) balance of agricultural and grazing lands are the same as for other ecosystems: the photosynthetic uptake and assimilation of CO_2 into organic compounds and the release of gaseous carbon through respiration (primarily CO_2 but also CH_4) and fire. In agricultural lands, carbon assimilation is directed towards the production of food, fibre and forage by manipulating species composition and growing conditions (soil fertility, irrigation, etc.). Biomass in agricultural and grassland systems (being predominantly herbaceous, i.e. non-woody), is a small, transient carbon pool (compared to forests) and hence soils constitute the dominant carbon stock. Cropland systems can be among the most productive ecosystems, but in some cases restricted growing season length, fallow periods and grazing-induced shifts in species composition or production can reduce carbon uptake relative to that in other ecosystems. These factors, along with tillage-induced soil disturbances and the removal of plant carbon through harvest, have depleted soil carbon stocks by 20–40% or more in relation to pre-cultivated conditions (Davidson and Ackerman, 1993; Houghton and Goodale, 2004). Soil organic carbon (SOC) stocks in grazing lands in North America

---

*E-mail: conant@nrel.colostate.edu

have been depleted to a lesser degree than in croplands (Ogle et al., 2004), and in some regions biomass has increased due to the suppression of disturbance and subsequent woody encroachment. Woody encroachment is potentially a significant sink for atmospheric CO_2, but the magnitude of the sink is not well quantified (Houghton et al., 1999; Pacala et al., 2001). Disturbance-induced increases in decomposition rates of aboveground litter and harvest removal of some (30–50% of forage in grazing systems, 40–50% in grain crops) or almost all (e.g. maize for silage) of the aboveground biomass have altered carbon cycling drastically within agricultural lands, and thus the sources and sinks of CO_2 to the atmosphere.

Much of the carbon lost from agricultural soil and biomass pools can be recovered with changes in management practices that increase carbon inputs, stabilize carbon within the system or reduce carbon losses, while still maintaining outputs of food, fibre and forage (Eagle and Olander, 2012). Increased production, increased residue C inputs to the soil and increased organic matter additions have reversed historic soil C losses in long-term experimental plots (e.g. Buyanovsky and Wagner, 1998). However, the management practices that promote soil carbon sequestration would need to be maintained over time to avoid subsequent losses of sequestered carbon. Across Canada and the USA, mineral soils have been sequestering 2.5 and 17.0 ± 0.45 million t of carbon (Mt C) per year (Smith et al., 1997, 2001; Ogle et al., 2003), respectively, largely through increased production and improved management practices on annual cropland. Conversion of agricultural land to grassland, like under the Conservation Reserve Program in the USA (7.6–11.5 Mt C year^{-1} on 12.5 million ha (Mha) of land), and afforestation have also sequestered carbon in agricultural and grazing lands (Follett et al., 2001).

Increasing carbon stocks is accompanied by increased soil fertility, enhanced soil water balance, increased production efficiency and reduced reliance on external inputs. These factors tend to enhance the resilience of production and yields in the face of climate variability (Post et al., 2012). Practices that build resilience to current climate variability are also expected to ameliorate some of the effects of the forecasted increase in future extreme weather events (Easterling et al., 2007; Goklany, 2007), thus enhancing carbon stocks and the rehabilitation of degraded lands. In addition, the implementation of sustainable land management practices can foster adaptation to a changing climate (Lal, 2009).

This chapter documents agronomic practices in North America capable of enhancing production, yield and income, reducing greenhouse gas emissions and mitigating greenhouse gas emissions from other sectors, and enhancing resilience in the face of inter-annual climate variability, and thus resilience to climate change (Table 24.1). It is possible to advance agricultural development agendas by focusing on a single aspect of agriculture (e.g. yield). However, expanding the role of land management for greenhouse gas mitigation presents us with an opportunity to increase investments in land management practices that will rebuild resilience – increasing current yields while enhancing the ability to meet future needs.

Positive Exemplars: Soil Carbon Storage in North America

Conservation tillage and no-tillage

Tillage has been used ubiquitously in agriculture to prepare the seedbed, to incorporate fertilizer, manure and residues into the soil, to relieve compaction and to control weeds (Phillips et al., 1980; Leij et al., 2002). However, tilling the soil is disruptive and can promote soil erosion, high rates of soil moisture loss, degradation of soil structure and depletion of soil nutrients and C stocks. Following long-term tillage, soil C stocks can be reduced by as much as 20–50% (Haas et al., 1957; Davidson and Ackerman, 1993; Murty et al., 2002; Ogle et al., 2003). Conservation tillage reduces the negative

Table 24.1. Assessment of socio-economic and environmental benefits, greenhouse gas mitigation and adaptation to climate change for different cropping and grazing management systems. Conventional and degrading land management practices are indicated in grey.

Activity	Practice	Yield/productivity	Socio-economic benefits	Environmental impacts	Greenhouse gas mitigation	Adaptation to climate change
Tillage	Conventional tillage	Variable		Weed control; reduces compaction	−	Reduced weed pressure
	Conservation tillage No-tillage	Variable	Reduced machinery, energy, labour requirements	Soil fertility; increased ground cover; reduced erosion; $+H_2O$ balance	+	Drought resilience; dependable soil fertility; reduced external inputs
Cropping system	Continuous cropping	+	Food security	Increased disease susceptibility	−	
	Cover crops/green manuring/catch crops	+	Food security; forage crop; increased land-use efficiency; higher returns; low-tech, low-cost nitrogen inputs, reduced mineral nitrogen requirement; weed control	Nitrogen fixation; increased soil fertility; reduced erosion; reduced crop disease incidence; enhanced H_2O balance; pest/disease control	+	Drought resilience; more dependable soil fertility
	Frequent fallow	−	Enhanced reliability of soil moisture	Increased erosion; decreased H_2O efficiency	−	Some drought resilience
	Reduced bare fallow frequency Mulching/residue management	+	Food security; crop diversification Food security; low-tech, low-cost soil enhancement; enhanced net income	Soil fertility; reduced erosion; enhanced H_2O use efficiency	+	Enhanced income, product diversification Drought resilience
Fertilizer/soil amendment	Conventional fertilizers	+	Food security; enhances fertility of poor land/soils	Nitrogen leaching, runoff; energy use	−	
	Efficient fertilizer use	+	Food security; reduced fertilizer expenses	Reduced nitrogen runoff	+	Dependable fertility; reduced dependence on external inputs; drought resilience

impacts of tillage, preserves soil resources and can lead to the accrual of much of the soil carbon lost during tillage (Lal, 1997; Paul et al., 1997; Paustian et al., 1997) and to net reduction in greenhouse gas emissions (e.g. Dendooven et al., 2012). A change from conventional tillage to no-till can sequester carbon, although results vary as a function of soil type, climate and land-use history (West and Post, 2002). The accumulation of SOC will continue (provided the soil is not tilled), and SOC levels can be expected to peak after 5–10 years, with SOC reaching a new equilibrium in 15–20 years. Conservation tillage is one of the largest potential sources of greenhouse gas mitigation within the agricultural sector (Smith et al., 2008) and, coupled with associated declines in fuel use, could make an immediate, substantial contribution to offsetting and reducing greenhouse gas emissions (Caldeira et al., 2004; CAST, 2004).

No-tillage and minimum-tillage systems provide opportunities for increasing soil water by trapping snow and reducing evaporative losses during the early part of the growing season; crop yields are significantly higher for field peas, flax and spring wheat (in stubble) in conservation tillage systems (most likely due to increased soil water) (Lafond et al., 1992). Based on these results and previous studies (e.g. Gupta and Larson, 1979), conservation tillage could increase soil water to the point where fallow cropping could be eliminated without increasing production (Lafond et al., 1992). Tillage experiments in several semi-arid regions demonstrate consistently higher soil water contents under no-tillage (Scopel et al., 2005; De Vita et al., 2007) and increased rain-use efficiency (Kronen, 1994; Rockström et al., 2009).

No-tillage management increases residue cover and decreases evaporative losses from the soil profile, thus enhancing moisture availability and reducing water stress for plant production (Holland, 2004). Tillage management can also reduce energy use associated with irrigation. However, the adoption of no-till can increase emissions of N_2O, particularly in the short term (Six et al., 2004). Reducing tillage intensity has been found to minimize CO_2 losses from decomposition in drained histosols due to less aeration and drier conditions at the soil surface in the absence of intensive tillage (i.e. switch ploughing).

Conservation tillage systems were originally developed to address problems of water quality, soil erosion and agricultural sustainability (Woodfine, 2009). Conservation tillage reduces evaporation from the surface, increases infiltration and shades the soil, decreasing soil temperature – all of which alter the water balance. In semi-arid environments, no-tillage can lead to enhanced water balance, enabling increased cropping intensity and greater return per unit land area (Peterson and Westfall, 2004). The impacts of no-tillage on yields tend to be greatest for the driest sites (Cantero-Martinez et al., 2007). By making more efficient use of existing precipitation, conservation tillage builds carbon stocks, pays dividends in terms of increased yields and ensures greater yields during dry years by conserving soil moisture (Wang et al., 2006).

Cover crops, green manuring, catch crops

Cover cropping, green manuring and catch crops (crops that are not planted for harvest but to take up and retain nitrogen) all increase carbon inputs to the soil by extending the time over which plants are fixing atmospheric CO_2. Green manures and catch crops have the benefit that they enhance system nitrogen balance, which further increases productivity. Growing cover crops enhances soil protection and groundwater quality, controls pests and increases C stocks by enhancing carbon inputs when the ground would otherwise lay fallow (Bowman et al., 1999; Govaerts et al., 2009). Using green manures can simultaneously build soil C and nitrogen stocks (Vanden-Bygaart et al., 2003), enhancing soil fertility and sequestering carbon in the soil, but likely increasing N_2O emissions. Use of catch crops also tends to sequester carbon in the soil (Friebauer et al., 2004; Christensen et al., 2009).

Enhancing inputs of nitrogen and nitrogen retention and minimizing soil erosion are keys to maximizing production and ensuring long-term sustainability of agricultural systems (Greenland, 1975). Use of nitrogen-fixing crops – green manures: nitrogen-fixing plants that are not harvested – and non-crop nitrogen fixers is an important component of soil fertility. Sowing non-leguminous cover crops between harvested crops can enhance nitrogen inputs through fixation and plough-under, while having a minimal impact on crop growing season length (Fageria, 2007). By enhancing soil fertility and nitrogen supply to crops, green manuring can have immediate impacts on the growth of subsequent crops. Similarly, catch crops sown following a crop take up residual nutrients in the soil; subsequently ploughing those catch crops into the soil before the next crop increases the availability of nutrients to subsequent crops (Hansen et al., 2007). Cover cropping can employ nitrogen-fixing crops to bolster soil fertility, but cover cropping with non-leguminous crops also provides surface cover that minimizes soil erosion between crop growing seasons.

Reduced bare fallow frequency

Reduced carbon inputs associated with more frequent bare summer fallow in semi-arid regions reduces the level of soil organic matter in dryland agricultural systems (Rasmussen et al., 1998). Bare fallows are increasingly recognized in most agroecosystems as generally not beneficial, and indeed may be harmful. Reducing the frequency of bare fallow leads to carbon sequestration by increasing the time over which carbon is taken up by plants in input to the soil (VandenBygaart et al., 2003). In semi-arid regions, alternate-year fallow is used to collect soil moisture; conservation tillage is often required to enable the reduction of fallow frequency (Peterson et al., 1996). Conservation tillage coupled with reduced fallow frequencies have been used successfully in a variety of semi-arid environments to sequester carbon in soils (Peterson et al., 1996; VandenBygaart et al., 2003; Erenstein and Laxmi, 2008; Govaerts et al., 2009). If production and carbon inputs decrease with decreasing fallow frequency, soil carbon stocks can decrease (Olsson and Ardo, 2002).

Bare fallows can be beneficial for crops growing in certain dryland situations, where a period of bare fallow can increase soil moisture storage, subsequently increasing crop yields, enabling a wider variety of crops, and reducing inter-annual yield variability. However, bare fallow periods reduce the period over which crops can be grown; gross income is also reduced. New tillage, crop rotation, fertilization and residue management practices can be integrated to conserve and use precipitation efficiently, broadening the portion of semi-arid areas acceptable for a given crop. Reducing bare fallow can result in substantial increases in income and food security in semi-arid ecosystems (Peterson and Westfall, 2004). Because bare ground is directly exposed to intense heat, wind and rainfall, erosion is greater for bare fallow than cropped land; reducing bare fallow will reduce soil erosion.

Mulching/residue management

Within annual croplands, much of the carbon in plant biomass is removed during harvest. Although most carbon storage in agricultural systems is in the soil, plant and residue management are directly related to soil carbon inputs and are important for erosion prevention. Plants and plant residues can also be used as a renewable energy source, offsetting the use of fossil fuels. Understanding how much crop residue must remain on the soil surface to prevent erosion is an important practical question. The answer to the question will dictate management strategies to prevent erosion and maintain soils. An equivalent question for soil carbon stocks is how much crop residue carbon must be input into the soil to maintain soil carbon stocks? Quantitative answers to this question are very important for understanding the impacts of harvest and biomass removal practices, but very few quantitative studies have been carried out to date. One study did evaluate the impacts of maize harvest residues on soil carbon stocks, and found that under mulch tillage 50% of the aboveground residues could be removed without impacting soil carbon stocks negatively (top 20 cm) (Sheehan et al., 2004). Another (Corbeels et al., 2006) suggested that most of the carbon sequestration benefits of direct-seeding mulch-based cropping systems in Brazil were realized due to increased carbon inputs from a mulch crop. Under no-tillage, the proportion of surface residues that could be removed increased to nearly three-quarters. Maintaining vegetative cover increases water infiltration, O_2 diffusion into the soil (limiting denitrification) and CH_4 diffusion into the soil (promoting CH_4 consumption); it also promotes the uptake of mineral nitrogen, slows nitrogen loss and maintains soil organic nitrogen stocks, all of which limit denitrification.

In the US Southern Great Plains, surface residues slow the movement of water across the surface, providing more time for infiltration, evaporation is reduced with residues due to reduced wind speeds and temperatures at the surface, and soil water storage at many semi-arid locations increased with increasing amounts of crop residue maintained on the surface (Unger et al., 1991). In semi-arid areas, conservation tillage (especially no-tillage) can enhance production, but may be limited on severely degraded soils because of low organic matter contents, low soil fertility, poor physical condition, low water infiltration rates and poor plant productivity. Because crop residues are often a valuable resource for other purposes (fodder, bedding, fuel, etc.), use of residues for mulching may require a trade-off with these other uses (Giller et al., 2009). Also, in some cases immobilization of mineral nitrogen in decomposing litter may detract from soil fertility.

Retaining mulch and other crop residues enhances water storage in soil, enabling adaptation to extreme events (drought, heat stress), which are expected to become more frequent as the climate changes. Surface residues absorb radiation, keeping the soil temperature cooler (Steiner, 1989; Moreno et al., 1997). They also increase surface roughness, decrease wind-driven drying of the soil surface and insulate the soil (Lampurlanes and Cantero-Martinez, 2006). In total, surface mulch reduces crop water requirements by 30% (FAO, 2007). Surface mulches also absorb energy from falling raindrops, reducing soil erosion. In situations where precipitation is lower/more erratic, the physical presence of crop residues on the soil surface protects the upper soil layer, reducing soil temperatures and hence water loss, both important factors for the adaptation of plant growth as the climate warms (Maruthi et al., 2008).

Positive Exemplars: Reducing N_2O Emissions in North America

Plant growth is enhanced by adding fertilizers – mineral or organic. However, while several studies have documented soil carbon sequestration in response to added fertilizers (Alvarez, 2005), even if C inputs to the soil increase in response to fertilization, added fertilizers can accelerate soil C outputs (decomposition) offsetting any C sequestration (Russell et al., 2009). Perhaps most importantly, increased N_2O emissions are likely to offset most C sequestration in almost all cases (Schlesinger, 1999).

Synchronization of nitrogen supply and nitrogen demand limits the build-up of mineral nitrogen (either NH_4^+ or NO_3^-) in the soil, thus limiting N_2O production via either nitrification or denitrification. Practices that promote the synchronization of nitrogen supply and demand – increasing the efficiency of nitrogen use – are likely to reduce N_2O fluxes (Mosier et al., 2004). For example, autumn fertilizer applications lead to greater emission than spring applications, because new crops or re-emerging forage grass seedlings in the spring remove mineral nitrogen from the soil, limiting the availability of nitrogen for microbially mediated nitrification and denitrification. Plants typically demand the greatest amount of nitrogen early in the season, and thus application at this time should reap the most benefit for reducing N_2O emissions, while still providing the necessary nutrients for high productivity. Hence, the best management strategy is to reduce mineral nitrogen additions, if possible, or at least fertilize during periods of the year when the crops or forage grasses are most actively growing (Drinkwater, 2004).

N_2O emissions can increase shortly after the adoption of no-tillage due to greater denitrification; this is probably due to increased water-filled pore space with improved soil aggregation. However, high N_2O emission rates decline with long-term (10–20 years) adoption of no-tillage (Six *et al.*, 2004). Applying minimal amounts of nitrogen fertilizer and ensuring that the timing of applications coincides with the active growth of crops or forages will minimize N_2O emissions from soils.

systems, offsetting greenhouse gas emissions while simultaneously leading to economic and environmental co-benefits. These ancillary benefits make adoption of those practices more compelling. Co-benefits are not distributed evenly across all regions of North America; in particular, the soil moisture benefits for adaptation to climate change variability and climate change seem most important in regions susceptible to periodic droughts. The ancillary benefits of C-sequestering management practices carry some risk: they raise additionality barriers to participation in carbon markets (Conant, 2011) and may be susceptible to reversals if they lead to the creation of a resource that could be harvested, such as soil nitrogen (Janzen, 2006). Nevertheless, the value of co-benefits must be considered in projects to encourage management practices that sequester carbon in North American croplands and grazing lands.

Conclusion

Many agricultural management practices can lead to increases in soil carbon stocks in North American cropland and grazing land

Acknowledgement

This work was supported by a grant from the UN Food and Agriculture Organization.

References

Alvarez, R. (2005) A review of nitrogen fertilizer and conservation tillage effects on soil organic storage. *Soil Use and Management* 21, 38–52.
Bowman, R.A., Vigil, M.F., Mielsen, D.C. and Anderson, R.L. (1999) Soil organic matter changes in intensively cropped dryland systems. *Soil Science Society of America Journal* 63, 186–191.
Buyanovsky, G.A. and Wagner, G.H. (1998) Carbon cycling in cultivated land and its global significance. *Global Change Biology* 4, 31–141.
Caldeira, K., Granger Morgan, M., Baldocchi, D., Brewer, P.G., Chen, C.-T.A., Nabbuurs, G.-J., Nakicenovic, N. and Robertson, G.P. (2004) A portfolio of carbon management options. In: Field, C.B. and Raupach, M.R. (eds) *The Global Carbon Cycle: Integrating Humans, Climate, and the Natural World*. Island Press, Washington, DC, pp. 103–130.
Cantero-Martinez, C., Angas, P. and Lampurlanes, J. (2007) Long-term yield and water use efficiency under various tillage systems in Mediterranean rainfed conditions. *Annals of Applied Biology* 150, 293–305.
CAST (Council for Agricultural Science and Technology) (2004) *Climate Change and Greenhouse Gas Mitigation: Challenges and Opportunities for Agriculture*. CAST, Ames, Iowa.
Christensen, B.T., Rasmussen, J., Eriksen, J. and Hansen, E.M. (2009) Soil carbon storage and yields of spring barley following grass leys of different age. *European Journal of Agronomy* 31, 29–35.
Conant, R.T. (2011) Sequestration through forestry and agriculture. *Wiley Interdisciplinary Reviews, Climate Change* 2, 238–254.
Corbeels, M., Scopel, E., Cardoso, A., Bernoux, M., Douzet, J.M. and Neto, M.S. (2006) Soil carbon storage potential of direct seeding mulch-based cropping systems in the Cerrados of Brazil. *Global Change Biology* 12, 1773–1787.

Davidson, E.A. and Ackerman, I.L. (1993) Change in soil carbon inventories following cultivation of previously untilled soils. *Biogeochemistry* 20, 161–193.

De Vita, P., Di Paolo, E., Fecondo, G., Di Fonzo, N. and Pisante, M. (2007) No-tillage and conventional tillage effects on durum wheat yield, grain quality and soil moisture content in southern Italy. *Soil and Tillage Research* 92, 69–78.

Dendooven, L., Gutierrez-Oliva, V.F., Patino-Zuniga, L., Ramirez-Villanueva, D.A., Verhulst, N., Luna-Guido, M., Marsch, R., Montes-Molina, J., Gutierrez-Miceli, F.A., Vasquez-Murrieta, S. and Govaerts, B. (2012) Greenhouse gas emissions under conservation agriculture compared to traditional cultivation of maize in the central highlands of Mexico. *Science of the Total Environment* 431, 237–244.

Drinkwater, L.E. (2004) Improving fertilizer nitrogen use efficiency through an ecosystem-based approach. In: Mosier, A.R., Syers, J.K. and Freney, J.R. (eds) *Agriculture and the Nitrogen Cycle*. Island Press, Washington, DC, pp. 93–102.

Eagle, A.J. and Olander, L.P. (2012) Greenhouse gas mitigation with agricultural land management activities in the United States-a side-by-side comparison of biophysical potential. *Advances in Agronomy* 115, 79–179.

Easterling, W.E., Aggarwal, P.K., Batima, P., Brander, K.M., Erda, L., Howden, S.M., Kirilenko, A., Morton, J., Soussana, J.-F., Schmidhube, J. and Tubiello, F.N. (2007) Food, fibre and forest products. In: Parry, M.L., Canziani, O.F., Palutikof, J.P., van der Linden, P.J. and Hanson, C.E. (eds) *Climate Change 2007: Impacts, Adaptation and Vulnerability*. Contribution of Working Group II to the Fourth Assessment Report of the Intergovernmental Panel on Climate Change. Cambridge University Press, Cambridge UK, pp. 273–313.

Erenstein, O. and Laxmi, V. (2008) Zero tillage impacts in India's rice-wheat systems: a review. *Soil and Tillage Research* 100, 1–14.

Fageria, N.K. (2007) Green manuring in crop production. *Journal of Plant Nutrition* 30, 691–719.

FAO (Food and Agriculture Organization) (2007) Adaptation to climate change in agriculture, forestry and fisheries: perspective, framework and priorities. FAO, Rome.

Follett, R.F., Kimble, J.M. and Lal, R. (2001) The potential of US grazing lands to sequester soil carbon. In: Follett, R.F., Kimble, J.M. and Lal, R. (eds) *The Potential of US Grazing Lands to Sequester Soil Carbon*. CRC Press, Chelsea, Michigan, pp. 401–430.

Friebauer, A., Rounsevell, M., Smith, P. and Verhagen, A. (2004). Carbon sequestration in agricultural soils in Europe. *Geoderma* 122, 1–23.

Giller, K.E., Witter, E., Corbeels, M. and Tittonell, P. (2009) Conservation agriculture and smallholder farming in Africa: the heretics' view. *Field Crops Research* 114, 23–34.

Goklany, I. (2007) Integrated strategies to reduce vulnerability and advance adaptation, mitigation, and sustainable development. *Mitigation and Adaptation Strategies for Global Change* 12, 755–786.

Govaerts, B., Verhulst, N., Castellanos-Navarrete, A., Sayre, K.D., Dixon, J. and Dendooven, L. (2009) Conservation agriculture and soil carbon sequestration: between myth and farmer reality. *Critical Reviews in Plant Sciences* 28, 97–122.

Greenland, D.J. (1975) Bringing the green revolution to the shifting cultivator. *Science* 190, 841–844.

Gupta, S.C. and Larson, W.E. (1979) Estimating soil-water retention characteristics from particle-size distribution, organic-matter percent, and bulk-density. *Water Resources Research* 15, 1633–1635.

Haas, H.J., Evans, C.E. and Miles, E.F. (1957) Nitrogen and carbon changes in Great Plains soils as influenced by cropping and soil treatments. *Technical Bulletin* 1164, USDA, Washington, DC.

Hansen, E.M., Eriksen, J. and Vinther, F.P. (2007) Catch crop strategy and nitrate leaching following grazed grass-clover. *Soil Use and Management* 23, 348–358.

Holland, J.M. (2004) The environmental consequences of adopting conservation tillage in Europe: reviewing the evidence. *Agriculture, Ecosystems and Environment* 103, 1–25.

Houghton, R.A. and Goodale, C.L. (2004) Effects of land-use change on the carbon balance of terrestrial ecosystems. Ecosystem and land use change. *Geophysical Monograph Series* 53, 85–96.

Houghton, R.A., Hackler, J.L. and Lawrence, K.T. (1999) The U.S. carbon budget: contributions from land-use change. *Science* 285, 574–578.

Janzen, H.H. (2006) The soil carbon dilemma: shall we hoard it or use it? *Soil Biology and Biochemistry* 38, 419–424.

Kronen, M. (1994) Water harvesting and conservation techniques for smallholder crop production systems. *Soil and Tillage Research* 32, 71–86.

Lafond, G.P., Loeppky, H. and Derksen, D.A. (1992) The effects of tillage systems and crop rotations on soil-water conservation, seedling establishment and crop yield. *Canadian Journal of Plant Science* 72, 103–115.

Lal, R. (1997) Residue management, conservation tillage and soil restoration for mitigating greenhouse effect by CO_2 enrichment. *Soil and Tillage Reseach* 43, 81–107.

Lal, R. (2009) Soils and food sufficiency: a review. *Agronomy for Sustainable Development* 29, 113–133.

Lampurlanes, J. and Cantero-Martinez, C. (2006) Hydraulic conductivity, residue cover and soil surface roughness under different tillage systems in semiarid conditions. *Soil and Tillage Research* 85, 13–26.

Leij, F.J., Ghezzehei, T.A. and Or, D. (2002) Modeling the dynamics of the soil pore-size distribution. *Soil and Tillage Research* 64, 61–78.

Maruthi, V., Srinivas, K., Reddy, G.S., Reddy, B.S., Reddy, K.S., Reddy, P.R., Sudhakar, R. and Ramakrishna, Y.S. (2008) Value addition to crop residues: an indigenous resource conserving and soil fertility enhancing technology of India. *Journal of Sustainable Agriculture* 31, 5–27.

Moreno, F., Pelegrin, F., Fernandez, J.E. and Murillo, J.M. (1997) Soil physical properties, water depletion and crop development under traditional and conservation tillage in southern Spain. *Soil and Tillage Research* 41, 25–42.

Mosier, A.R., Syers, J.K. and Freney, J.R. (2004) *Agriculture and the Nitrogen Cycle*. Island Press, Washington, DC.

Murty, D., Kirschbaum, M.U.F., McMurtrie, R.E. and McGilvray, H. (2002) Does conversion of forest to agricultural land changes soil carbon and nitrogen? A review of the literature. *Global Change Biology* 8, 105–123.

Ogle, S.M., Breidt, F.J., Eve, M.D. and Paustian, K. (2003) Uncertainty in estimating land use and management impacts on soil organic carbon storage for US agricultural lands between 1982 and 1997. *Global Change Biology* 9, 1521–1542.

Ogle, S.M., Conant, R.T. and Paustian, K. (2004) Deriving grassland management factors for a carbon accounting method developed by the intergovernmental panel on climate change. *Environmental Management* 33, 474–484.

Olsson, L. and Ardo, J. (2002). Soil carbon sequestration in degraded semiarid agro-ecosystems – perils and potentials. *Ambio* 31, 471–477.

Pacala, S.W., Hurtt, G.C., Baker, D., Peylin, P., Houghton, R.A., Birdsey, R.A., Heath, L., Sundquist, E.T., Stallard, R.F., Ciais, P. et al. (2001) Consistent land- and atmosphere-based U.S. carbon sink estimates. *Science* 292, 2316–2320.

Paul, E.A., Paustian, K., Elliott, E.T. and Cole, C.V. (eds) (1997) *Soil Organic Matter in Temperate Agroecosystems*. CRC Press, Boca Raton, Florida.

Paustian, K., Collins, H.P. and Paul, E.A. (1997) Management controls on soil carbon. In: Paul, E.A., Paustian, K., Elliot, E.T. and Cole, C.V. (eds) *Soil Organic Matter in Temperate Agroecosystems*. CRC Press, Boca Raton, Florida, pp. 15–49.

Peterson, G.A. and Westfall, D.G. (2004) Managing precipitation use in sustainable dryland agroecosystems. *Annals of Applied Biology* 144, 127–138.

Peterson, G.A., Schlegel, A.J., Tanaka, D.L. and Jones, O.R. (1996) Precipitation use efficiency as affected by cropping and tillage systems. *Journal of Production Agriculture* 9, 180–186.

Phillips, R.E., Blevins, R.L., Thomas, G.W., Frye, W.W. and Phillips, S.H. (1980) No-tillage agriculture. *Science* 208, 1108–1113.

Post, W.M., Izaurralde, R.C., West, T.O., Liebig, M.A. and King, A.W. (2012) Management opportunities for enhancing terrestrial carbon dioxide sinks. *Frontiers in Ecology and the Environment* 10, 554–561.

Rasmussen, P.E., Albrecht, S.L. and Smiley, R.W. (1998) Soil C and N changes under tillage and cropping systems in semi-arid Pacific Northwest agriculture. *Soil and Tillage Research* 47, 197–205.

Rockström, J., Kaurnbutho, P., Mwalley, J., Nzabi, A.W., Temesgen, M., Mawenya, L., Barron, J., Mutua, J. and Damgaard-Larsen, S. (2009) Conservation farming strategies in East and Southern Africa: yields and rain water productivity from on-farm action research. *Soil and Tillage Research* 103, 23–32.

Russell, A.E., Cambardella, C.A., Laird, D.A., Jaynes, D.B. and Meek, D.W. (2009) Nitrogen fertilizer effects on soil carbon balances in Midwestern US agricultural systems. *Ecological Applications* 19, 1102–1113.

Schlesinger, W.H. (1999) Carbon and agriculture: carbon sequestration in soils. *Science* 284, 2095.

Scopel, E., Findeling, A., Guerra, E.C. and Corbeels, M. (2005) Impact of direct sowing mulch-based cropping systems on soil carbon, soil erosion and maize yield. *Agronomy for Sustainable Development* 25, 425–432.

Sheehan, J., Aden, A., Paustian, K., Killian, K., Brenner, J., Walsh, M. and Nelson, R. (2004) Energy and environmental aspects of using corn stover for fuel ethanol. *Journal of Industrial Ecology* 7, 117–146.

Six, J., Ogle, S.M., Briedt, F.J., Conant, R.T., Mosier, A.R. and Paustian, K. (2004) The potential to mitigate global warming with no-tillage management is only realized when practiced in the long term. *Global Change Biology* 10, 155–160.

Smith, P., Martino, D., Cai, Z., Gwary, D., Janzen, H., Kumar, P., McCarl, B., Ogle, S., O'Mara, F., Rice, C. et al. (2008) Greenhouse gas mitigation in agriculture. *Philosophical Transactions of the Royal Society B–Biological Sciences* 363, 789–813.

Smith, W.N., Rochette, P., Monreal, C., Desjardins, R.L., Pattey, E. and Jaques, A. (1997) The rate of carbon change in agricultural soils in Canada at the landscape level. *Canadian Journal of Soil Science* 77, 219–229.

Smith, W.N., Desjardins, R.L. and Grant, B. (2001) Estimated changes in soil carbon associated with agricultural practices in Canada. *Canadian Journal of Soil Science* 81, 221–227.

Steiner, J.L. (1989) Tillage and surface residue effects on evaporation from soils. *Soil Science Society of America Journal* 53, 911–916.

Unger, P.W., Stewart, B.A., Parr, J.F. and Singh, R.P. (1991) Crop residue management and tillage methods for conserving soil and water in semiarid regions. *Soil and Tillage Research* 20, 219–240.

VandenBygaart, A.J., Gregorich, E.G. and Angers, D.A. (2003) Influence of agricultural management on soil organic carbon: a compendium and assessment of Canadian studies. *Canadian Journal of Soil Science* 83, 363–380.

Wang, X.B., Cai, D.X., Hoogmoed, W.B., Oenema, O. and Perdok, U.D. (2006) Potential effect of conservation tillage on sustainable land use: a review of global long-term studies. *Pedosphere* 16, 587–595.

West, T.O. and Post, W.M. (2002) Soil organic carbon sequestration rates by tillage and crop rotation: a global data analysis. *Soil Science Society of America Journal*, 66, 1930–1946.

Woodfine, A. (2009) *The Potential of Sustainable Land Management Practices for Climate Change Mitigation and Adaptation in Sub-Saharan Africa.* Food and Agricultural Organization of the United Nations, Rome.

25 Managing Soil Carbon in Europe: Paludicultures as a New Perspective for Peatlands

Hans Joosten*, Greta Gaudig, René Krawczynski, Franziska Tanneberger, Sabine Wichmann and Wendelin Wichtmann

Abstract

Conventional peatland agriculture and forestry is based on drainage, which enhances peat oxidation, causes massive greenhouse gas emissions and eventually destroys the peatland subsistence base. In contrast, paludicultures use biomass from wet and rewetted peatlands under conditions that maintain the peat body, facilitate peat accumulation and provide the associated natural peatland ecosystem services. In the temperate, subtropical and tropical zones, i.e. those zones of the world where plant productivity is high, peat is generally formed by roots and rhizomes, and peatlands by nature hold vegetation of which aboveground parts can be harvested without substantially harming peat conservation and formation.

Besides traditional yields of food, feed, fibre and fuel, the biomass can be used as a raw material for industrial biochemistry, for producing high-quality liquid or gaseous biofuels and for further purposes like extracting and synthesizing pharmaceuticals and cosmetics. Some outstanding examples are introduced, including low-intensity grazing with water buffalos, biofuels from fens, common reed as industrial raw material and sphagnum farming for horticultural growing media.

Paludicultures may support substantial co-benefits, including the preservation and sequestration of carbon, regulation of water dynamics (flood control) and quality, and conservation and restoration of typical peatland flora and fauna. They can provide sustainable income from sites that have been abandoned or degraded.

In many cases, paludicultures can compete effectively with drainage-based peatland agriculture and forestry, certainly when external costs are adequately considered. Various technical and political constraints, however, still hamper large-scale implementation of this promising type of land use.

Introduction

Drainage for agriculture and forestry is the main cause of carbon losses from peatland. Drained peatlands are found primarily in the temperate zone and the (sub)tropics, i.e. in those areas that are densely populated and climatically favourable for agriculture.

Drained peatland soils are subject to inherent degradation, which burdens the environment and continuously lowers their productive value (Joosten et al., 2012).

Since 1990, east and central Europe have witnessed the abandonment of millions of hectares of agriculturally used peatlands through the combination of progressive soil

*E-mail: joosten@uni-greifswald.de

degradation, increasing costs of drainage and changed political and economic conditions. This abandonment did not decrease the environmental problems. On the contrary: peatland fires arose as a new phenomenon in the drained and deserted peatlands, especially in Russia and Ukraine (Abel *et al.*, 2011), and importantly, the abandonment caused large-scale rural unemployment and social disintegration.

Vast areas of drained and deeply subsided peatlands have meanwhile been flooded, because maintenance of the drainage infrastructure was no longer cost-effective. This rewetting has to some extent re-established the ecosystem services of wet peatlands, including carbon storage, flood control, water purification and the consolidation of biodiversity and wilderness conditions (Succow and Joosten, 2001; Theuerkauf *et al.*, 2006; Trepel, 2010a; Tanneberger and Wichtmann, 2011), but while the delivery of these regulating services was improved, that of provisioning services usually ceased.

The introduction of sustainable productive land-use options on wet and rewetted peatlands is an innovative development to (re-)install economic carriers and to strengthen rural livelihoods. This chapter presents an overview of such 'paludicultures' in Europe.

Environmental Drawbacks of Conventional Peatland Utilization

Conventional peatland utilization requires a lowering of the water table. As peat consists largely of water, peatland drainage leads to subsidence and compaction of the peat. Drainage, furthermore, leads to oxidation of the peat that is no longer water-saturated, resulting in huge emissions of greenhouse gases (CO_2 and N_2O) to the atmosphere and nitrate to adjacent surface waters.

Drained peatland loses – depending on the climate – some millimetres up to several centimetres of peat per year (Couwenberg *et al.*, 2010, 2011). These losses are accelerated by ploughing and by the addition of lime, fertilizers and clastic materials (which increase peat oxidation; Clymo, 1983), by wind and water erosion (by which bare peat is blown or washed away; Evans and Warburton, 2007), and by (subsurface) peat fires. The resulting lowering of the surface necessitates – in the case of continued exploitation – a continuous deepening of the drainage ditches, which again enhances peat oxidation, surface lowering, ditch deepening, etc., a phenomenon known as 'the vicious circle of peatland utilization' (Kuntze, 1982). The continuously lowering surface makes gravity drainage increasingly difficult, and eventually necessitates the establishment of expensive polder systems with dykes and pumps. Furthermore, subsidence increases the risk of floods and saltwater intrusion (Joosten *et al.*, 2012).

In drier, more continental climates, continuous peat shrinkage and swelling as a result of water-level fluctuations cause the formation of fissures in the drained peat, which impede capillary water flow and lead to more frequent and deeper drying out of the soil. Through the activity of soil organisms, drained peat soils become loosened and fine-grained and may eventually become totally hydrophobic (Succow and Joosten, 2001), so that after a few decades, agriculture becomes impossible on the remaining black deserts (Plate 14).

It is clear that peatland exploitation by peatland drainage is a dead-end street. Urgently new production techniques have to be developed that combine the (re-)instalment of productive use with the restoration/maintenance of the ecosystem services of wet peatlands.

The Principles of Paludiculture

Conventional peatland agriculture is based on drainage, which enhances peat oxidation and eventually destroys the peatland subsistence base. In contrast, paludicultures (Latin '*palus*' = swamp) use biomass from wet and rewetted peatlands under conditions that maintain the peat body, facilitate peat accumulation and provide the associated natural peatland ecosystem services.

Paludicultures use that part of net primary production that is dispensable for peat formation. In the temperate, subtropical and tropical zones, i.e. those zones of the world where plant productivity is high, peat is generally formed by roots and rhizomes, and peatlands by nature hold vegetation of which aboveground parts can be harvested without substantially harming peat formation (Wichtmann and Joosten, 2007).

Paludiculture comprises any biomass use from wet and rewetted peatlands, from harvesting spontaneous vegetation on natural sites to artificially established crops on rewetted sites. Besides traditional yields of food, feed, fibre and fuel, the biomass can be used as a raw material for industrial biochemistry, for producing high-quality liquid or gaseous biofuels, and for further purposes like extracting and synthesizing pharmaceuticals and cosmetics.

Land-use Options for Rewetted Peatlands in Temperate Europe

Since early times, the utilization of biomass is part of the ambiguous relationship of humans and wet peatlands (Moore, 1987; Joosten, 2009). Already in 17th century in England, the value of fen peatlands in providing fodder for horses, cattle and sheep, as a store of 'osier, reed and sedge' and as 'nurseries and seminaries' of fish and fowl was recognized (Wheeler, 1896).

Land-use options of wet peatlands may entail the collection of plants and hunting of animals for direct consumption without much management intervention. In the boreal zone of Eurasia, a wide variety of wild edible berries (*Vaccinium, Empetrum, Rubus, Ribes*) and mushrooms are gathered for food and vitamins (Joosten and Clarke, 2002), and these services have been major justifications in Russia and Belarus to protect and restore mires. In other parts of the world, a variety of plants for human nutrition or medical use are collected from wet peatlands, such as wild rice, *Zizania aquatica* (North America), or *Menyanthes trifoliata*, *Acorus calamus* and *Hierochloe odorata* (Europe) (Joosten and Clarke, 2002). Other traditional, low-intensity uses include hunting and fishing.

More intensive land-use options include the site-adapted cultivation of crops and livestock grazing. Some of them revitalize traditional forms of land use through new utilization schemes. Others, such as the cultivation of plants for biofuels, provide newly developed products for new market demands (Table 25.1). Some outstanding examples are introduced in the following sections. All concepts have in common that the peatland mean water level is near the soil surface.

Low-intensity grazing with water buffalos

Peatlands have over centuries served as wild pasture for (semi-)domestic animals, but this kind of low-impact use has decreased drastically in recent decades (Tanneberger and Wichtmann, 2011). Cattle grazing for dairy production generally require peatland drainage. However, on drained sites, the quantity and quality of the available fodder cannot keep up with the increasing quality needs of high-production dairy cows. As a consequence, the drained peatland grasslands are often abandoned or (in the European Union, EU) only managed symbolically for securing EU subsidies.

Grazing with water buffalos, *Bubalus bubalis*, after rewetting may provide an alternative. Water buffalos forage on biomass with low energy content and are well adapted to permanent wet conditions. They suffer much less from parasites and hoof diseases, and can move much better on wet soils than normal cattle (Krawczynski *et al.*, 2008). Wet grassland management with buffalos seems to have important benefits for biodiversity, as diverse reed beds have a variety of microhabitats. As buffalos hardly need medication, in contrast to conventional dairy cattle, their dung is not contaminated by anthelmintics and supports rich microflora and fauna, which forms the basis

Table 25.1. Biomass utilization from wet peatlands in temperate Europe.

Category	Products	Producing	Exploitation method[a]	Harvesting time	Quality demands	Peat accumulation rate
Food	Berries and other fruits	Shrubs, dwarf shrubs	S, A	Autumn		Medium
	Mushrooms	Mushrooms	S	Summer/autumn	High	Medium
	Meat	Game, fowl, fish	S	Hunting season		Medium
Fodder	Ex situ (hay, silage)	Wet meadows, reeds	S	Summer	High	Possible
	In situ (grazing)	Wet meadows, reeds	S	Entire year	High	Possible
Fibre	Roofing materials	Reeds	S, A	Winter	High	Medium
	Building panels/boards	Reeds	S, A	Winter	High/no	Medium
	Insulation materials	Reed, cattail	S, A	Winter	High	Medium
	Timber/veneer	Alder, birch, pine	S, A	Winter-frost	High	Medium
	Wattle and basketware	Willow, bulrush	S, A	Autumn	High	Possible
	Form bodies	Wet meadows, reeds	S, A	Autumn/winter	Medium	Medium
	Growing media	Peat moss	S, A	Entire year	High	Possible
	Litter	Sedge meadows, reeds	S, A	Summer/autumn	No	Possible
	Compost	Wet meadows, reeds	S, A	Late summer	No	Possible
Fuel	Direct combustion	Alder/birch/willow, reeds	S, A	Autumn/winter	No	Medium
	Pellets, briquettes	Wet meadows, reeds	S, A	Winter/early spring	No	Medium
	Biogas	Wet meadows, reeds	S, A	Early summer	Possible	Possible
	Liquid biofuels	Wet meadows, reeds	S, A	Entire year	No	Medium
	Biochar	Wet meadows, reeds	S, A	Winter	No	Medium
Further	Pharmaceutics	Many herb and forb species	S, A	Summer	High	Possible
	Flavours	Various herb, forbs, grasses	S, A	Summer	High	Possible
	Cosmetics	Various herb species	S, A	Summer	High	Possible

[a]S, spontaneous occurrence; A, artificially established.

for a rich food web. Buffalos can be kept outside year-round so that dung, and its microflora and -fauna, is available year-round for birds and bats. In Germany, 13 years of buffalo breeding has shown that buffalos have no difficulty with temperatures below −20°C if they can shelter from wind and have a type of straw bed to lie on.

Biofuels from fens

Biofuel paludicultures maintain and reinstall ecosystem services of undrained peatlands, in sharp contrast to 'biofuels' from drained peatland. Biogas from maize (*Zea mays*) grown on drained peatlands, for example, may cause over 800 t CO_2-eq (equivalent) of greenhouse gas emissions per terajoule of energy produced because of the carbon losses from peat oxidation, whereas burning fossil coal produces only 100 t per terajoule (Couwenberg, 2007). From a climate point of view, it is much better to burn the peat directly than to cultivate 'biofuels' on drained peatland. The absurd practice of biofuel production on drained peatland is supported, perversely, by EU agricultural subsidies and the Kyoto Protocol, which considers biofuels as climate neutral in the energy sector but fails to account for the huge peat carbon losses under the land-use sector (Joosten, 2011). Maize cultivation on drained peat soils has in recent years expanded massively in Germany, stimulated by subsidies under the German Renewable Energy Sources Act.

On wet peatlands, winter-mown common reed, *Phragmites australis*, with average yields of 8 t dry weight ha^{-1} $year^{-1}$, can compete economically with biofuels from mineral soils (cereal straw, miscanthus), even when adapted machinery for harvesting under wet conditions is required (Wichmann and Wichtmann, 2009). The type of utilization (liquid biofuels, biogas, direct heat supply to single consumers, combined heat power for district heating, and others) eventually determines the economic revenues. With a yield of 8 t dry weight ha^{-1} $year^{-1}$ and a heating value of 17.5 MJ kg^{-1} dry weight, common reed from 1 ha can replace fossil fuels in a combined heat power (CHP) plant that would otherwise emit 10 t CO_2-eq. With emissions from handling (mowing, transport, storage, delivery and operation of the cogeneration plant) amounting to 2 t CO_2-eq ha^{-1} (Wichtmann et al., 2009), the emissions reduction from the rewetting (15 t CO_2-eq ha^{-1} $year^{-1}$) and replacement of fossil fuel (10 t CO_2-eq ha^{-1} $year^{-1}$) adds up to about 23 t CO_2-eq ha^{-1} $year^{-1}$ (Wichtmann and Wichmann, 2011b).

Common reed as industrial raw material

From a climatic perspective, the use of biomass for construction or handicraft is to be favoured over its use as a fuel, as the carbon sequestered in the biomass remains stored long-term. With respect to using fibre from peatlands, common reed for thatch has the longest tradition in Europe (Rodewald-Rudescu, 1974; Moir and Letts, 1999; Häkkinen, 2007). In Germany, the Netherlands, Denmark and the UK, the demand for high-quality reed is so large that it cannot be satisfied by inland harvest, and reed has to be imported from southern and eastern Europe, Turkey and China (Schäfer, 1999; Haslam, 2010). Weaving of reed is a long-established way of manufacturing construction and insulation materials.

The economics of reed as a construction material depend strongly on processing and use (Schäfer, 1999). Thatch produced from wet peatlands showed negative revenues when high costs for planting and processing met low biomass yields and prices, but calculated proceeds turned positive in more realistic settings (Schäfer, 2004). Further mechanization of planting, improved harvesting and the development of new products may improve the results. In comparison, cattail, *Typha* spp., cultivation, with an average dry matter yield of 15 t ha^{-1} $year^{-1}$, proved to be profitable in calculations based on pilot trials in the Donaumoos (Germany) (Wild et al., 2001) when used for the production of high-quality insulation

Sphagnum farming for horticultural growing media

The cultivation of vegetables, fruits and flowers is increasingly soil-less, i.e. takes place in pre-prepared 'growing media' that allow uniform, high-quality plants to be grown at very high productivity levels. Sphagnum peat has, in the last decades, emerged as the foremost constituent of these growing media (Joosten, 1995; Alexander et al., 2008) because of its structural stability, low bulk density, high porosity and low pH, nutrient and nitrogen immobilization levels (Schmilewski, 2008). Annually, some 30 million m³ of peat in the EU are used for the production of growing media that support a modern horticulture, accounting for a turnover of €1.3 billion and 11,000 jobs (Altmann, 2008).

The highest-quality peat is slightly humified sphagnum peat ('white' peat), which over the past 3000 years has formed from sphagnum mosses. In most countries of western and central Europe, including Germany, the stocks of white peat are nearly depleted (Joosten, 2012), and raised bogs are protected as a priority habitat under the EU Habitat Directive (92/43/EEG). To satisfy horticultural demands, white peat is, in increasing volumes, imported from Scandinavia, Canada and, especially, the Baltic States (Joosten, 1995). Peat extraction thus progressively destroys raised bogs with their typical biodiversity, carbon storage capacity, water regulation function and palaeo-environmental archive. The white peat used in Germany, for example, leads to annual emissions of 20 Mt of CO_2, that is, a volume similar to that from aviation (http://unfccc.int/national_reports/annex_i_ghg_inventories/national_inventories_submissions/items/6598.php).

A reduction of these negative environmental effects of horticultural production needs urgent attention (Verhagen et al., 2009; LLUR, 2012). The UK and Switzerland have already decided to reduce and eventually phase out the use of peat entirely in their countries (Secretary of State for Environment, Food and Rural Affairs, 2011; http://www.news.admin.ch/message/index.html?lang=de&msg-id=47174). Until now, however, these decisions have had little effect on the volume of peat consumed, as environmentally friendly, economically competitive and high-quality alternatives have been lacking.

The most promising alternative for peat as a constituent of growing media is sphagnum biomass, which has similar physical and chemical properties as white peat. Since 2004, peat moss cultivation ('sphagnum farming') has been studied in greenhouse and field experiments by the University of Greifswald (Germany) and associated research and industrial partners. A commercial scale pilot site of 5 ha was installed successfully on former bog grassland in spring 2011 (Joosten et al., 2013). Plant cultivation experiments by the Horticulture Research Station, Hanover and the von Humboldt University, Berlin have shown that growing media of sphagnum biomass – even up to a proportion of 100% – enable professional plant cultivation without loss of quality compared to peat. Next to providing a renewable alternative to fossil peat, sphagnum farming may enable a climate-friendly, sustainable after-use option for abandoned cut-over bogs and degraded bog grasslands (http://www.sphagnumfarming.com).

Benefits of Paludicultures

Compared to land use on drained peat soils, paludicultures have important environmental benefits (Wichtmann and Wichmann, 2011a). Rewetting as a precondition for paludiculture may restore drained peatlands to peat-forming ecosystems, enhancing important regulating services such as climate regulation through protecting carbon stores and new carbon sequestration, water-quality regulation by providing sinks for nutrients and water filtration (Trepel,

2010b), and wildfire control by decreasing fire incidence and its associated damage.

Rewetting substantially reduces greenhouse gas emissions from peat oxidation (Couwenberg et al., 2011). Even more emissions are avoided when the produced biomass is used to replace fossil raw materials and fossil fuels. Combining bioenergy generation and rewetting of drained peatlands thus makes paludiculture an extraordinarily cost-effective climate change mitigation option that can generate income both from carbon credits and from biomass production (Tanneberger and Wichtmann, 2011).

Rewetting drained peatlands can also support typical peatland flora and fauna, which is often severely threatened due to habitat loss. Within a few years after installation, many typical bog species had spontaneously established in our first 1200 m^2 large sphagnum farming pilot plot in north-west Germany, including several red list plant species, an extremely rare bog myxomycete (*Badhamia lilacina*, second observation in Germany) and a similarly rare bog spider (*Bathyphantes setiger*, fourth observation in Germany since 1950).

Rewetted peatlands that were used for intensive, drainage-based agriculture are often nutrient overloaded, highly productive and hardly ever harbour rare species. By harvesting biomass in summer, significant amounts of nutrients may be removed, which improves habitat quality for characteristic mire species of more nutrient-poor conditions. In contrast, by collecting dead biomass in winter, most nutrients are left on site, and stable yields can be achieved over long periods.

In sites designated for conservation, paludiculture must be considered as a cost-effective management option, instrumental but ancillary to conservation (Wichtmann et al., 2010a,b).

Last but not least, paludicultures improve rural livelihoods. They can provide sustainable income from sites that have been abandoned or where use took place that degraded the land. Autumn and winter harvest leads to more consistent employment throughout the year, whereas biomass processing may create net added value and generate additional jobs.

Perspectives

The area of drained peatlands in the world amounts to some 500,000 km^2 (Joosten, 2010), with problems associated with degradation occurring everywhere (Joosten et al., 2012). Practical experiences and model calculations show that, in many cases, paludicultures can compete effectively with drainage-based peatland agriculture and forestry, certainly when external costs are adequately considered. Various conditions, however, still hamper large-scale implementation.

Rewetting of peat soils often requires investments over the entire hydrological unit (polder, catchment area), as it is unfeasible to rewet single plots surrounded by fields that continue to be drained. Such rewetting will demand hydrological restructuring, land reallotment and consolidation of a similar scale as the huge projects that drained the peatlands in the first place. Furthermore, similar to drainage schemes, rewetting must be considered to be virtually irreversible on the level of the individual enterprise. Because of the important ecosystem services generated for wider beneficiaries, it is reasonable that peatland rewetting projects are supported by central planning and public financing. In the EU, however, agricultural subventions still cause a substantial market distortion by subsidizing drainage-based peatland agriculture; for example, for biofuel production of maize on peatlands, but not similarly supporting paludicultures. There, the subsidy system of the EU Common Agricultural Policy should be modified to target mainstream support to paludicultures (direct payments), to sharpen cross-compliance requirements towards protecting carbon-rich organic soils, to introduce agri-environmental schemes and agroclimate programmes for raising water levels (in average to 20 cm below surface or higher) and to enable long-term agreements (cf. irreversibility) to secure peatland rewetting.

At the regional level, paludicultures can only be implemented effectively if the entire life cycle from production, harvesting and processing is in place. This can be encouraged through investments in market analysis, site-adapted machinery, processing facilities and production lines, product placement and agricultural consultation for site-adapted peatland use.

Most paludicultures are still in their infancy. Further research is required into the identification of possible crops, optimization of cultivation techniques, selection and propagation of suitable varieties, development of site-adapted machinery and long-term environmental effects (peat hydraulics, peat formation, emissions, biodiversity). Research and pilot implementation are not only necessary for the temperate zone, where paludicultures are being developed on rewetted peatlands but also especially in the tropics, where paludicultures must be established as an alternative for rapidly expanding drainage-based peatland agriculture and forestry (Dommain et al., 2012).

References

Abel, S., Haberl, A. and Joosten, H. (2011) *A Decision Support System for Degraded Abandoned Peatlands Illustrated by Reference to Peatlands of the Russian Federation*. Michael Succow Foundation for the Protection of Nature, Greifswald, Germany (in Russian and English).

Alexander, P.D., Bragg, N.C., Meade, R., Padelopoulos, G. and Watts, O. (2008) Peat in horticulture and conservation: the UK response to a changing world. *Mires and Peat* 3, Article 08, 1–10.

Altmann, M. (2008) Socio-economic impact of the peat and growing media industry on horticulture in the EU. *Study for EPAGMA* by CO CONCEPT. CO CONCEPT Marketingberatung, Luxemburg.

Clymo, R.S. (1983) Peat. Mires: swamp, bog, fen and moor. In: Gore, A.J.P. (ed.) *General Studies*. Elsevier, Amsterdam, pp. 159–224.

Couwenberg, J. (2007) Biomass energy crops on peatlands: on emissions and perversions. *IMCG Newsletter* 2007(3), 12–14.

Couwenberg, J., Dommain, R. and Joosten, H. (2010) Greenhouse gas fluxes from tropical peatlands in south-east Asia. *Global Change Biology* 16, 1715–1732.

Couwenberg, J., Thiele, A., Tanneberger, F., Augustin, J., Bärisch, S., Dubovik, D., Liashchynskaya, N., Michaelis, D., Minke, M., Skuratovich, A. and Joosten, H. (2011) Assessing greenhouse gas emissions from peatlands using vegetation as a proxy. *Hydrobiologia* 674, 67–89.

Dommain, R., Barthelmes, A., Tanneberger, F., Bonn, A., Bain, C. and Joosten, H. (2012) 5. Country-wise opportunities. In: Joosten, H., Tapio-Biström, M.-L. and Tol, S. (eds) *Peatlands – Guidance for Climate Change Mitigation by Conservation, Rehabilitation and Sustainable Use. Mitigation of Climate Change in Agriculture Series 5*. FAO, Rome, pp. 45–82.

Evans, M. and Warburton, J. (2007) *The Geomorphology of Upland Peat: Pattern, Process, Form*. Blackwell, Oxford, UK.

Häkkinen, J. (2007) Traditional use of reed. Read up on reed! In: Ikonen, I. and Hagelberg, E. (eds) *Part IV, Touch and Thatch*. Southwest Finland Regional Environment Centre, Turku, Finland, pp. 62–72.

Haslam, S.M. (2010) *A Book of Reed: (Phragmites australis (Cav.) Trin. ex Steudel, Phragmites communis Trin.)*. Forrest, Tresaith, UK.

Joosten, H. (2009) Human impacts: farming, fire, forestry and fuel. In: Maltby, E. and Barker, T. (eds) *The Wetlands Handbook*. Wiley-Blackwell, Oxford, UK, pp. 689–718.

Joosten, H. (2010) *The Global Peatland CO_2 Picture. Peatland Status and Drainage Associated Emissions in All Countries of the World*. Wetlands International, Ede, the Netherlands, 10 pp. + tables.

Joosten, H. (2011) Selling peatland rewetting on the compliance carbon market. In: Tanneberger, F. and Wichtmann, W. (eds) *Carbon Credits from Peatland Rewetting. Climate – Biodiversity – Land Use*. Schweizerbart, Stuttgart, Germany, pp. 99–105.

Joosten, H. (2012) Zustand und Perspektiven der Moore weltweit. *Natur und Landschaft* 87, 50–55.

Joosten, H. and Clarke, D. (2002) *Wise Use of Mires and Peatlands – Background and Principles Including a Framework for Decision-Making*. International Mire Conservation Group/ International Peat Society, Saarijärvi, Finland.

Joosten, H., Tapio-Biström, M.-L. and Tol, S. (eds) (2012) *Peatlands – Guidance for Climate Change Mitigation by Conservation, Rehabilitation and Sustainable Use*, 2nd edn. Mitigation of Climate Change in Agriculture Series 5. FAO, Rome, 100 pp.

Joosten, H., Gaudig, G. and Krebs, M. (2013) Peat-free growing media: sphagnum biomass. *Peatlands International* 1, 28–31.

Joosten, J.H.J. (1995) The golden flow: the changing world of international peat trade. *Gunneria* 70, 269–292.

Krawczynski, R., Biel, P. and Zeigert, H. (2008) Wasserbüffel als Landschaftspfleger. Erfahrungen zum Einsatz in Feuchtgebieten. *Naturschutz und Landschaftsplanung* 40, 133–139.

Kuntze, H. (1982) Die Anthropogenese nordwestdeutscher Grünlandböden. *Abhandlungen Naturwissenschaftlicher Verein zu Bremen* 39, 379–395.

LLUR (Landesamt für Landwirtschaft, Umwelt und ländliche Räume des Landes Schleswig-Holstein; Endredaktion) (2012) *Eine Vision für Moore in Deutschland. Potentiale und Ziele zum Moor- und Klimaschutz*. Gemeinsame Erklärung der Naturschutzbehörden. Landesamt für Landwirtschaft, Umwelt und ländliche Räume des Landes Schleswig-Holstein, Flintbek, Germany, 38 pp.

Moir, J. and Letts, J. (1999) Thatch, Thatching in England 1790–1940. *English Heritage Research Transactions*, Volume 5. James and James, London.

Moore, P.D. (1987) Man and mire: a long and wet relationship. *Transactions of the Botanical Society of Edinburgh* 45, 77–95.

Rodewald-Rudescu, L. (1974) *Das Schilfrohr*. Schweizerbart, Stuttgart, Germany.

Schäfer, A. (1999) Schilfrohrkultur auf Niedermoor – Rentabilität des Anbaus und der Ernte von Phragmites australis. *Archiv für Naturschutz und Landschaftsforschung* 38, 193–216.

Schäfer, A. (2004) Umwelt als knappes Gut – Ökonomische Aspekte von Niedermoorrenaturierung und Gewässerschutz. *Archiv für Naturschutz und Landschaftsforschung* 43, 87–105.

Schätzl, R., Schmitt, F., Wild, U. and Hoffmann, H. (2006) Gewässerschutz und Landnutzung durch Rohrkolbenbestände. *WasserWirtschaft* 11, 24–27.

Schmilewski, G. (2008) The role of peat in assuring the quality of growing media. *Mires and Peat* 3, Article 2, 8 pp. http://mires-and-peat.net/pages/volumes/map03/map0302.php (accessed 18 August 2014).

Secretary of State for Environment, Food and Rural Affairs (2011) *The Natural Choice: Securing the Value of Nature*. The Stationery Office, 77 pp. (https://www.gov.uk/government/uploads/system/uploads/attachment_data/file/228842/8082.pdf, accessed 23 October 2014).

Succow, M. and Joosten, H. (eds) (2001) *Landschaftsökologische Moorkunde*, 2nd edn. Schweizerbart, Stuttgart, Germany.

Tanneberger, F. and Wichtmann, W. (2011) *Carbon Credits from Peatland Rewetting. Climate – Biodiversity – Land Use*. Schweizerbart, Stuttgart, Germany.

Theuerkauf, M., Couwenberg, J., Joosten, H., Kreyer, D. and Tanneberger, F. (eds) (2006) *New Nature in North-Eastern Germany. A Field Guide*. Institute of Botany and Landscape Ecology, Greifswald, Germany.

Trepel, M. (2010a) Assessing the cost-effectiveness of the water purification function of wetlands for environmental planning. *Ecological Complexity* 7, 320–326.

Trepel, M. (2010b) Nährstoffrückhaltung in Feuchtgebieten – Prozesse, Risiken, Kosten und Potenziale. *Norddeutsche Tagung für Abwasserwirtschaft und Gewässerentwicklung, Tagungsband* 22, 19–27.

Verhagen, A., van den Akker, J.J.H., Blok, C., Diemont, W.H., Joosten, J.H.J., Schouten, M.A., Schrijver, R.A.M., den Uyl, R.M., Verweij, P.A. and Wösten, J.H.M. (2009) Peatlands and carbon flows. Outlook and Importance for the Netherlands. Report WAB 500102 027. Netherlands Environmental Assessment Agency PBL, Bilthoven, the Netherlands, 50 pp.

Wheeler, W.H. (1896) *A History of the Fens of South Lincolnshire*, 2nd edn. Newcomb, Boston, UK.

Wichmann, S. and Wichtmann, W. (2009) *Bericht zum Forschungs- und Entwicklungsprojekt Energiebiomasse aus Niedermooren (ENIM)*. Institut für Botanik und Landschaftsökologie, Greifswald, Germany. (http://www.duene-greifswald.de/doc/enim_endbericht-2009.pdf, accessed 23 October 2014).

Wichtmann, W. and Joosten, H. (2007) Paludiculture: peat formation and renewable resources from rewetted peatlands. *IMCG-Newsletter* 3, 24–28.

Wichtmann, W. and Wichmann, S. (2011a) Environmental, social and economic aspects of a sustainable biomass production. *Journal of Sustainable Energy and Environment* Special Issue 2011, 77–83.

Wichtmann, W. and Wichmann, S. (2011b) Paludikultur: standortgerechte Bewirtschaftung wiedervernässter Moore (Paludiculture – site adapted management of re-wetted peatlands). *Telma Beiheft* 4, 215–234.

Wichtmann, W., Couwenberg, J. and Kowatsch, A. (2009) Standortgerechte Landnutzung auf wiedervernässten Niedermooren – Klimaschutz durch Schilfanbau. *Ökologisches Wirtschaften* 1, 25–27.

Wichtmann, W., Wichmann, S. and Tanneberger, F. (2010a) Paludikultur – Nutzung nasser Moore: perspektiven der energetischen Verwertung von Niedermoorbiomasse. *Naturschutz und Landschaftspflege in Brandenburg* 19, 211–218.

Wichtmann, W., Tanneberger, F., Wichmann, S. and Joosten, H. (2010b) Paludiculture is paludifuture: climate, biodiversity and economic benefits from agriculture and forestry on rewetted peatland. *Peatlands International* 1, 48–51.

Wild, U., Kamp, T., Lenz, A., Heinz, S. and Pfadenhauer, J. (2001) Cultivation of *Typha* spp. in constructed wetlands for peatland restoration. *Ecological Engineering* 17, 49–54.

26 Managing Soil Organic Carbon for Multiple Benefits: The Case of Africa

Peter T. Kamoni* and Patrick T. Gicheru

Abstract

Organic matter is of great importance in soil, because it impacts on the physical, chemical and biological properties of soils. Physically, it promotes aggregate stability and therefore water infiltration, percolation and retention. Biologically, it stimulates the activity and diversity of organisms in soil. Decomposing organic matter releases nutrients, such as nitrogen (N), phosphorus (P), sulfur (S) and potassium (K), essential for plant and microbial growth. Sustainable land management practices enhance carbon sequestration and sustain agricultural productivity, thus mitigating against climate change. In Western Africa, there is a rapid decline of soil organic carbon (SOC) levels with continuous cultivation. For the sandy soils, average annual losses may be as high as 4.7%, whereas with sandy loam soils, losses are lower, with an average of 2%. In the equatorial forest zone with higher rainfall, abundant moisture favours high biomass production, which in turn brings about higher SOC (~24.5 g kg^{-1} organic C) and nitrogen contents. In the Sudan savannah, organic carbon (~3.3–6.8 g kg^{-1}) and total nitrogen are very low, because of low biomass production and high rates of decomposition. Estimates of SOC stocks and changes made for Kenya using the Global Environmental Facility Soil Organic Carbon (GEFSOC) Modelling System indicated soil C stocks of 1.4–2.0 Pg (0–20 cm), which compared well with a soil and terrain (SOTER)-based approach that estimated ~1.8–2.0 Pg (0–30 cm) of soil C between 2000 and 2030 in Kenya. Direct field sampling and laboratory measurements of soil carbon in Kenya has been going on for over half a century, and the data exist in the form of numerous technical and research reports, theses, journal papers and workshop proceedings, Kenya Agricultural Research Institute annual reports and geographic information system (GIS) databases. A combination of biomass measurements and empirical equations has also been employed in Kenya to measure organic carbon stocks. Stratified random sampling of herbaceous standing crop has been carried out at Nairobi National Park to estimate primary production of the grassland savannah. The Carbon Benefits Project, developed between 2009 and 2012 by Colorado State University (USA) in collaboration with Kenya, Nigeria, Niger and China, is able to estimate carbon stocks and greenhouse gas (GHG) emissions. Increasing soil organic matter content can both improve soil fertility and reduce the impact of drought, improving adaptive capacity, making agriculture less vulnerable to climate change, while also sequestering carbon. Agronomic practices in western Kenya include using improved crop varieties, extending crop rotations, notably those with perennial crops that allocate more C below ground, and avoiding or reducing the use of bare unplanted fallow among others. Increasing the soil carbon in farms improves soil fertility, hence improves food security, increases economic returns from carbon revenues and creates business development opportunities for farmers to diversify income-generating activities.

*E-mail: pkamoni@gmail.com

Population pressures, declining plot sizes and resource constraints in Africa have led to agricultural intensification and continuous cropping with insufficient inputs, leading to rapid decline in SOC stocks. Sustainable management of organic resources will require interventions by regional governments to help farmers access inputs cheaply, as well as educating farmers on sustainable land management practices.

Introduction

The distribution of organic matter resources varies with agroecological zones, land use in the ecosystems and rainfall within the continent from the Sahel to the Congo basin. Due to this variability, the mapping of soil organic carbon (SOC) is essential for appropriate land uses and climate mitigation and adaptation strategies. In the Sahel and the horn of Africa, soil organic matter (SOM) has proven to be an indicator of soil quality and has been severely degraded due to human activities, which leads to soil degradation through erosion and loss of soil fertility, but this trend can be alleviated through appropriate management. In forest ecosystems, notably the Congo basin, the parameter is an indicator of soil health and ecosystems equilibrium. SOC has been used for centuries by rural communities to yield multiple benefits, such as: (i) improve food security and nutrition with the cropping of mushrooms and other meso fauna species; (ii) increase yield and diversity of on-farm and out-farm crops; and (iii) ensure the resistance of housing using dark rich soils, reducing rainfall penetration.

At subnational, national and regional levels, increased attention is being given to SOC in the context of climate change as a result and indicator of the relevance of adaptation and mitigation technologies. Managed appropriately, carbon stocks will ensure the conservation of multiple benefits such as the diversification of crops, wildlife, timber and non-timber, livestock and fisheries products. To maintain and improve soil productivity and its derived benefits, organic carbon should be managed according to a specific context. Among the biotic factors, biodiversity conservation in cropping, pastures, water and forest systems, along with carbon conservation, may provide a pathway to sustainable development.

Land-use changes such as conversion of forest or grassland to agricultural use have continued to be of great concern to environmentalists (Lal et al., 1995). Land-use conversion from forest to annual crop cultivation and grazing land has been shown to influence change in soil properties (Kironchi and Mbuvi, 1996; Gal et al., 2006). Gal et al. (2006) have shown that there is considerable decline of total organic carbon, SOC and dissolved organic carbon in the world carbon stocks. Land-use changes, especially the cultivation of deforested land, may diminish soil quality rapidly, leading to severe land degradation (Kang and Juo, 1986; Nardi et al., 1996; Islam et al., 1999). The conversion of forest to cropland has been associated with the reduction in the organic matter content of the topsoil (Ross, 1993; Singh and Singh, 1996) and the subsequent decline in productivity (Sanchez et al., 1997; Palm et al., 2001).

Population pressure, decreasing plot sizes and resource constraints in Kenya and sub-Saharan Africa have resulted in continuous cultivation that has led to the depletion of SOM. The manure used in Kenya is generally of poor quality, due to poor storage and management. Continuous cropping, removal of field crop residues for feeding ruminants and overgrazing between cropping seasons with little or no external inputs have reduced the productive capacity of arable lands throughout sub-Saharan Africa (Onyango et al., 2000). High input and transport costs for agrochemicals make the use of inorganic fertilizers on staple food crops uneconomical for most smallholder farmers. Lekasi et al. (2005) report that most farmers in the Sasumua catchment, a humid area in Kenya, hardly use fertilizer to replenish the soil nutrients harvested through crop and animal produce. Most farmers in this catchment have observed a decline in yields of food and vegetable crops. This long-term decline is related to cropping intensity on shrinking smallholder farms (less than 1 ha) and to the limited use of organic and inorganic fertilizers (Lekasi et al., 2005). In Kenya's semi-arid regions like the Kiboko Makindu area in eastern Kenya, farmers do not apply fertilizers and manure to maize because of the unreliable rainfall and also the assumption that their soils are fertile enough. Nutrient depletion through crop cultivation and soil erosion in sloping areas results in substantial decline in SOM content.

Role of Organic Carbon in Sustainable Land Management

Organic matter is of great importance in soil, because it impacts on the physical, chemical and biological properties of soils (Bationo et al., 2013). Physically, it promotes aggregate stability, water infiltration, percolation and retention. It impacts on soil chemistry by increasing cation exchange capacity, soil buffer capacity and nutrient supply. Biologically, it stimulates the activity and diversity of organisms in the soil. The global carbon cycle involves the exchange of CO_2 between the atmosphere and the biosphere, apart from oceans. Plants fix CO_2 from the atmosphere during photosynthesis to produce organic matter, which is stored above and below ground. The bulk of the biomass in above- and belowground plant parts is eventually transferred to the dead organic matter pool, or is oxidized or burnt. Dead organic matter, which consists of deadwood (standing as well as fallen) and litter, is either decomposed or oxidized, or stored for longer periods above or below the ground as detritus. CO_2 fixed by plants ends up in soil as organic matter or finer forms as humus through the decomposition process. Thus, CO_2 removed from the atmosphere is stored as dead and living biomass or soil carbon in the biosphere. Sustainable land management practices enhance carbon sequestration and sustain agricultural productivity, thus mitigating against climate change. SOM has a stabilizing effect on soil moisture, improves the moisture retention and release characteristics of soil and protects the soil against erosion (Batjes and Sombroek, 1997). Decomposing organic matter releases nutrients, such as nitrogen (N), phosphorus (P), sulfur (S) and potassium (K), essential for plant and microbial growth.

Status of Soil Carbon

Soils in West Africa are poorly endowed when it comes to soil fertility. Unlike, for example, the Rift Valley area of East Africa, West African soils never enjoyed volcanic rejuvenation (Bationo et al., 1998). In the equatorial forest zone with higher rainfall, abundant moisture favours high biomass production, which in turn brings about higher SOC (~24.5 g kg^{-1} organic C) and nitrogen contents. In the Sudan savannah, organic carbon (~3.3–6.8 g kg^{-1}) and total nitrogen are very low, because of low biomass production and high rates of decomposition. With kaolinite being the main clay type, the cation exchange capacity of the soils in this region are often less than 1 cmol kg^{-1}, depending on the level of SOC (Bationo et al., 2005). There is a rapid decline of SOC levels with continuous cultivation. For sandy soils, the average annual losses may be as high as 4.7%, whereas with sandy loam soils, losses are lower, with an average of 2%.

Estimates of SOC stocks and changes made for Kenya using the Global Environmental Facility Soil Organic Carbon (GEFSOC) Modelling System indicated soil C stocks of 1.4–2.0 Pg (0–20 cm) (Kamoni et al., 2007), which compared well with a soil and terrain (SOTER)-based approach that estimated ~1.8–2.0 Pg (0–30 cm) (Batjes, 2004). In 1990, 48% of the country had SOC stocks of <18 t C ha^{-1} and 20% had SOC stocks of 18–30 t C ha^{-1}, whereas in 2000, 56% of the country had SOC stocks of <18 t C ha^{-1} and 31% had SOC stocks of 18–30 t C ha^{-1}. Conversion of natural vegetation to annual crops led to the greatest soil C losses. All three methods involved in the GEFSOC system estimated that there would be a net loss of soil C between 2000 and 2030 in Kenya.

Monitoring of Soil Carbon

Soil databases that only hold data on total organic carbon content and limited information on land use (history) can provide only limited information on the dynamics of carbon during land use or climate-induced changes in different agroecosystems (Batjes and Sombroek, 1997). None the less, they remain critical in estimating the size of the global soil C and pools. Direct field sampling and laboratory measurements of soil carbon in Kenya have been going on for over half a century and the data exist in the form of numerous technical and research reports, theses, journal papers and workshop proceedings, annual reports and geographic information system (GIS) databases. A combination of biomass measurements and empirical equations has also been employed in Kenya (Woomer, 2003; Kamoni and Macharia, 2011). For trees, this normally involves measurements of the diameter at breast height (1.3 m from the ground) of a number of trees within each farm, and the total number of trees in the farm estimated visually. The aboveground biomass (AGB) is then estimated using allometric equations (Woomer, 2003), originally from FAO (1997):

in dry zones (<1500 mm year^{-1})

$Y = \exp^{(-1.996 + 2.32 \ln D)}$

and in moist zones (1500–4000 mm year^{-1})

$Y = \exp^{(-2.134 + 2.53 \ln D)}$

where Y is the aboveground tree biomass in kg tree^{-1}, exp = 2.71828 ... or 22/7 and D is the measured tree diameter at breast height (DBH) in cm (calculated from the circumference–diameter relationship, i.e. $C = \pi D$).

The equations below (Hairiah *et al.*, 2001) were used to calculate the aboveground biomass for bananas and coffee:

$Y = 0.303 D^{2.1345}$

$Y = 0.281 D^{2.0635}$

Grass and shrubs

Whole grass and shrub cover in 1×1 m plots are cut and weighed in the field and subsamples dried in the oven. The average dry weights per 1×1 m space are then calculated and used to determine the aboveground biomass per hectare.

Root biomass

The total root biomass for trees, fruit trees, bananas, coffee and tea are calculated using the estimates below (Woomer, 2003).

Total root biomass = root biomass (0.35 AGB) + leaf drop (0.15 AGB) + fine root turnover (0.15 AGB) but for maize, beans, cowpeas, watermelon, grass and shrubs, only the 0.35 AGB factor was used.

Maize and legumes

The relationship between yield, total biomass and harvest index (HI), that is total biomass = yield × HI, was used to estimate the aboveground biomass for maize, beans and cowpeas (Kamoni and Macharia, 2011). Information on crop yields and area planted per crop was provided by the farmers. Area measurements for small, planted plots were taken during interviews.

Converting Biomass to Carbon Stocks

Total biomass (kg ha^{-1}) is multiplied with a factor of 0.47 to convert to C (kg ha^{-1}) (Woomer, 2003).

Kamau *et al.* (2008) used destructive methods (uprooting) to measure tea biomass in Kericho, Kenya. Batjes (2004, 2011) used information contained in the International Soil Reference and Information Centre (ISRIC) soil profile pits database for Kenya (bulk density, per cent of carbon, thickness of soil layer and volume of fraction >2 mm) to calculate the soil carbon stocks for Kenya and Upper Tana, respectively.

Kamoni et al. (2007) used a modelling approach using Century and Roth C models and the Intergovernmental Panel on Climate Change (IPCC) system to predict soil organic carbon stocks and changes in Kenya between 1990 and 2030. Stratified random sampling of herbaceous standing crops have been carried out in Nairobi National Park to estimate the primary production of the grassland savannah (Desmukh, 1986; Kinyamario and Imbamba, 1986), although their results were not converted to carbon stocks. The Carbon Benefits Project, developed between 2009 and 2012 by a consortium of partners including Colorado State University in collaboration with Kenya, Brazil, Nigeria, Niger and China, provides tools to estimate carbon stocks and greenhouse gas (GHG) emissions.

Management of Organic Carbon

Land management practices that increase net primary productivity, reduce the rate of heterotrophic respiration, or both, lead to an increase in ecosystem C storage. Examples include the planting of trees, reducing the intensity of tillage on cropland or restoring grasslands on degraded (SCC-VI Agroforestry East Africa, 2008) land. Soil fertility improvement increases plant biomass, hence increasing carbon sequestration (the storage of carbon dioxide usually captured from the atmosphere), and controls climate change. Decline in soil fertility causes substantial net losses of soil carbon, resulting in increased carbon flux to the atmosphere. Increasing SOM content can both improve soil fertility and reduce the impact of drought, improving adaptive capacity and making agriculture less vulnerable to climate change, while also sequestering carbon. Burning of fossil fuels (coal, oil and gas) and land-use conversion (agriculture, deforestation) releases GHGs that influence atmospheric cycling, or a net positive radiative forcing that triggers an increase in temperature, which eventually affects climatic patterns like rainfall, pressure or cyclones, and later interacts with whole spheres of the earth (SCC-VI Agroforestry East Africa, 2008). GHGs are the gases released by human activity that are responsible for climate change and global warming. Agriculture (through improved management practices) and forestry provide, in principle, a significant potential for GHG mitigation. Improved and sustainable crop husbandry practices increase productivity, leading to increased SOC storage (SCC-VI Agroforestry East Africa, 2008). Examples of agronomic practices in western Kenya include using improved crop varieties, extending crop rotations, notably those with perennial crops that allocate more C belowground, and avoiding or reducing the use of bare unplanted fallow among others. This allows for better vegetation cover, protection of the soil and spread of the harvest within the farm. Forests can store 20–100 times more carbon than other vegetation types on the same land area, or around 30–60 t C ha^{-1}.

Conclusions

Population pressures, declining plot sizes and resource constraints in Africa have led to agricultural intensification and continuous cropping with insufficient inputs, leading to rapid decline in SOC stocks. Sustainable management of organic resources will require interventions by regional governments to help farmers access inputs cheaply, as well as educating farmers on sustainable land management practices. Specific strategies to increase the soil carbon pool have been identified and include degraded land restoration and vegetative regeneration, no-till farming, cover crops, organic residues, composting, nutrient management, manuring, improved grazing, water conservation and harvesting, efficient irrigation, agroforestry practices and growing energy crops on fallow land.

References

Bationo, A., Lompo, F. and Koala, S. (1998) Research on nutrient flows and balances in West Africa: state-of-the-art. *Agriculture, Ecosystems and Environment* 71, 19–35.

Bationo, A., Kihara, J., Vanlauwe, B., Waswa, B. and Kimetu, J. (2005) Soil organic carbon dynamics, functions and management in West African agro-ecosystems. *Agricultural Systems* 94, 13–25.

Bationo, A., Waswa, B.S. and Kihara, J. (2013) *Soil Carbon and Agricultural Productivity: Perspectives from Sub Saharan Africa*. Alliance for a Green Revolution in Africa, Accra, Ghana. International Center for Tropical Agriculture (CIAT), Nairobi.

Batjes, N. (2004) Soil carbon stocks and projected changes according to land use management: a case study of Kenya. *Soil Use and Management* 20, 350–356.

Batjes, N. (2011) Projected changes in soil organic carbon stocks upon adoption of recommended soil and water conservation practices in the Upper Tana River Catchment, Kenya. *Land Degradation and Development* 25, 278–287.

Batjes, N. and Sombroek, W. (1997) Possibilities for carbon sequestration in tropical and subtropical soils. *Global Change Biology* 3, 161–173.

Desmukh, I. (1986) Primary production of grassland in Nairobi National Park, Kenya. *Journal of Applied Ecology* 23(1), 115–123.

Food and Agriculture Organization of the United Nations (FAO) (1997) *Estimating Biomass and Biomass Change of Tropical Forests: A Primer*. FAO Forestry Paper 134. FAO, Rome, 55 pp.

Gal, A., Szegi, T., Simon, B., Szeder, B., Michael, E., Tombacz, E., Zsolnay, A. and Akagi, J. (2006) Indicators of soil degradation processes on a chernozem field in Hungary. In: *Proceedings of the 18th World Congress of Soil Science July 9–15, Philadelphia, Pennsylvania*. Soil Science Society of America, Madison, Wisconsin.

Hairiah, K., Sitompul, S.M., van Noordwijk, M.V. and Palm, C. (2001) *Methods for Sampling Carbon Stocks Above and Below Ground*. International Centre for Research in Agroforestry (ICRAF) Report. ICRAF, Nairobi.

Islam, K.R., Kamaluddin, M., Bhuiyam, M.K. and Badruddin, A. (1999) Comparative performance of exotic and indigenous forest species for tropical semi-green degraded forestland reforestation in Bangladesh. *Land Degradation and Development* 10, 241–249.

Kamau, D.M., Spiertz, J.H. and Oenema, O. (2008) Carbon and nutrient stocks of tea plantations differing in age, genotype and plant population density. *Plant Soil* 307, 29–39.

Kamoni, P.T. and Macharia, P.N. (2011) *Assessment of Above and Below-ground Carbon Stocks in Upper Tana Catchment*. Miscellaneous Report No M141. Kenya Soil Survey, Nairobi.

Kamoni, P.T., Gicheru, P.T., Wokabi, S.M., Easter, M., Milne, E., Coleman, C., Falloon, P. and Paustian, K. (2007) Predicted soil organic carbon stocks and changes in Kenya between 1990 and 2030. *Agriculture, Ecosystems and Environment* 122, 105–113.

Kang, B.T. and Juo, A.S.R. (1986) Effects of forest clearing on soil chemical properties and crop performance. In: Lal, R., Sanchez, P.A. and Cumming, R.W. (eds) *Land Clearing and Development in the Tropics*. Belkerma, Rotterdam, the Netherlands, pp. 383–394.

Kinyamario, J.I. and Imbamba, S.K. (1986) *Savannah at the Nairobi National Park, Nairobi*. University of Nairobi, Kenya.

Kironchi, G. and Mbuvi, J. (1996) Effects of deforestation on soil fertility on the North Western slopes of Mt. Kenya. *ITC Journal* 3/4, 260–263.

Lal, R., Kimble, J., Levine, E. and Whitman, C. (1995) World soils and greenhouse effect. An overview. In: Lal R. (ed.) *Soils and Global Change*. Lewis Publishing, Boca Raton, Florida, pp. 1–7.

Lekasi, J.K., Gicheru, P.T., Gachimbi, L.N., Sijali, I.V. and Nyagwa'ra, M.K. (2005) *Small Farming, Rural Livelihoods, Biodiversity in Kinale/Kikuyu Catchment: Baseline Survey Report*. Kinale Technical Report, SLM, Report No 6. Kenya Agricultural Productivity and Sustainable Management Project, KARI, Nairobi.

Nardi, S., Cucheri, G. and Dell'Angolla, G. (1996) Biological activity of humus. In: Piccolo, A.A. (ed.) *Humic Substances in Terrestial Ecosystems*. Elsevier, Amsterdam, pp. 361–406.

Onyango, R.M.A., Mwangi, A.T.J., Kiiya, W.W., Kamidi, M.K. and Wanyonyi, M.W. (2000) Evaluation of organic and inorganic fertilisers for small holder maize production in North Rift Kenya. In: Mureithi, J.G., Gachene, C.K.K., Muyekho, F.N., Onyango, M., Mose, L. and Magenya, O. (eds) *Participatory Technology Development for Soil Management by Small Holders in Kenya*. Kenya Agricultural Research Institute, Nairobi, pp. 3–12.

Palm, C.A., Gachengo, C.N., Delve, R.S., Cadish, G. and Giller, K.E. (2001) Organic inputs for soil fertility management in tropical agro ecosystems: application of an organic resource database. *Agriculture and Environment* 83, 27–42.

Ross, S.M. (1993) Organic matter in tropical soils; current conditions, concerns and prospects for conservation. *Progress in Physical Geography* 17, 265–305.

Sanchez, P.A., Sherherd, K.D., Soule, M.J., Place, F.M., Mokwunye, A.U., Buresh, R.J., Kwesiga, F.R., Izac, A.M., Nderitu, C.G. and Woomer, P.L. (1997) Soil fertility replenishment in Africa: an investment in natural resource capital. In: Buresh, R.J., Sanchez, P.A. and Calhoun, F. (eds) *Replenishing Soil Fertility in Africa*. Special Publication, 51. Soil Science Society of America and ICRAF, Madison, Wisconsin.

SCC-VI Agroforestry East Africa (2008) *Sustainable Agriculture Land Management Guidelines Towards Agricultural Greenhouse Gas Mitigation and Adaptation to Climate Change*. Lake Victoria Regional Environmental and Sustainable Agricultural Productivity Programme (RESAPP). SCC-VI Agroforestry Eastern Africa, Kisumu, Kenya.

Singh, S. and Singh, J.S. (1996) Water-stable aggregates and associated organic matter in forest, savannah and cropland soils of a seasonally dry tropical region. *India Biology and Fertility of Soils* 22, 76–82.

Woomer, P.L. (2003) *Monitoring Plan, Carbon Sequestration Projections and Verification Protocols*. Western Kenya Integrated Environmental Management Project. SACRED Africa and World Agroforestry Centre, Nairobi, 49 pp.

27 Benefits of SOM in Agroecosystems: The Case of China

Genxing Pan*, Lianqing Li, Jufeng Zheng, Kun Cheng, Xuhui Zhang, Jinwei Zheng and Zichuan Li

Abstract

While soil organic matter (SOM) content can be directly correlated with crop yield at the country/province level, abundance in SOM content may also be linked to economical output such as gross domestic production per capita at the county or province scale. Benefits of SOM include increasing nitrogen (N) efficiency, enhancing soil biodiversity and the health of soil food web systems, as well as aiding the degradation of toxic pollutants. Among these, the enhancement of the soil microbial community, and hence microbiochemical functions, is of key importance for productivity in croplands. Many studies have shown that the metabolic quotient of the soil microbial community (the specific quotient of soil respiration to SOM content) is unchanged or lowered as SOM accumulates. In order to characterize these benefits, several parameters are needed: (i) microbial abundance on the base of SOM content (microbial quotient, %); (ii) soil basal respiration to microbial biomass carbon (C); and (iii) normalized enzyme activity on the bases of soil microbial biomass carbon and soil organic carbon. As yet, there is no evidence that there is an SOM limit for Asian agricultural soils. A conceptual model of the role of SOM and benefits from the interaction with mineral particles and the formation of aggregates is hypothesized to gain an understanding of the benefits of SOM in croplands.

The Importance of Soil Organic Matter in China's Agriculture

Organic matter is considered as vital to the soil fertility of croplands, and has been proposed as a key soil parameter for the characterization of soil quality, productivity and ecosystem functioning (Tiessen et al., 1994). Moreover, topsoil soil organic matter (SOM) content has been suggested as the most important key soil quality parameter for the European Union's (EU) agricultural and forestry sectors (EC, 2002). A case study of tropical farming systems (Dawe et al., 2003) demonstrated that SOM had a significant control on crop productivity and functioning. This was later highlighted by Manlay et al. (2007). Recently, when reviewing the

---

*E-mail: pangenxing@aliyun.com

historical development of world agriculture, scientists have put increasing emphasis on the degradation of agroecosystems due to the depletion of SOM under intensified cropping and increasing chemical fertilization (Feller et al., 2012).

Based on statistics of cropland topsoil organic matter contents and the crop yield changes across a time span from 1949 to 1999, Pan et al. (2009a) were able to draw a linear relationship between mean cereal productivity and average cropland SOM contents in provinces of China. However, the incremental response depends on climatic conditions and socio-economical and technological factors. The dependency of crop yield on SOM appears to have decreased in recent times, probably due to the enhanced technology input with economical development. This is also seen in an analysis of crop productivity in relation to SOM and the variation between the regions of China, where mean provincial cereal yield in the major crop producing provinces is strongly related to mean topsoil SOM contents in north China as well as in Jiangsu and Shanghai during 1949–1999 (Pan et al., 2013). Taking the example of Jiangsu, one of most developed provinces in China, a good correlation between both agricultural output and gross domestic product (GDP) could be found with cropland SOM levels for the time before 1985 and in 2004 (Pan et al., 2013). These findings clearly show that the SOM level is not only very important for agricultural production but also for regional economic development, presumably due to enhanced production and ecosystem functioning with SOM accumulation in soils.

Dynamics of SOM in China's Croplands over the Past Two Decades

Depletion of SOM in croplands in the history of agricultural development in China

In the history of agricultural development, the depletion of SOM from cultivated lands has been a cause for concern from the perspective of both land degradation and greenhouse gas (GHG) emissions (Feller et al., 2012). As estimated by Lal (1999), up to 5% of the original SOM has been lost from croplands worldwide due to land use changing from natural soils to croplands. This has given rise to an accumulative global CO_2 emission of 55 Pg C (IPCC, 1995). In a literature review of agricultural soil research in China before 1960, Lindert et al. (1996) voiced serious concerns over the historical loss of SOM from a wide range of China's ecosystems due to agricultural land use since the 1950s. Utilizing the Denitrification–Decomposition (DNDC) model, Li (2000) found there had been a significant decline of organic carbon (C) storage in China's croplands of up to 70 Tg since the 1970s. Lal (2004b), however, estimated a total loss of 3.5 Pg since agricultural development in China, including approximately 2 Pg from land desertification caused by irrational land use and management, as previously estimated (Lal, 2002),whereas a statistical analysis by Wu et al. (2003) of soil organic carbon storage from the archived second national soil survey data of cultivated soils showed a loss of whole soil organic carbon stock from China's croplands of 7–8 Pg C compared with natural soils. In their work, a dramatic decline in organic carbon stock occurred mainly in the north and other arid and semi-arid regions of China. Nevertheless, it has been argued that there have been large areas and soil–land-use associations where changes in SOM contents were either minimal or even positive, especially in irrigated areas (Pan et al., 2003). Using a similar method to Wu et al. (2003), but using topsoil data from the national soil survey conducted during 1982–1985, Song et al. (2005) compared soil organic carbon (SOC) levels of cultivated soils to uncultivated soils in a comparative analysis, and revealed that cultivation induced a loss of topsoil SOC stock of up to 14.8 ± 15.1 Mg ha^{-1}. This gave a total SOM decline of 2 Pg C due to the cultivation of natural soils for the whole of mainland China. In their work, over 60% of this loss was observed to occur in soils of north-east China, north-west China and south-west China. Therefore, significant loss of SOM occurred under intensive cultivation and in degraded ecosystems due to land desertification and/or climate change impacts. Of course, such a great loss of SOM could have led to historical emissions of CO_2 to the atmosphere, leading to land-use change-induced climate forcing. Such a large decline in SOM would also lead to low soil fertility for croplands and degraded ecosystem services.

The trend of increasing SOM in China's croplands over the past two decades

There has been much work done to characterize SOM changes with agricultural development since the 1980s, motivated by increasing concerns for soil resilience and sustainable production, as well as climate change mitigation through SOC sequestration. Since the second national soil survey, which recognized soil fertility and nutrient status as constraints on China's agriculture, great efforts have been made to increase SOM levels and enhance soil resilience throughout China. Efforts have included policies to encourage straw return and conservation tillage, as well as combined use of organic and inorganic fertilizers. A series of long-term experiments has been established across mainland China for monitoring soil fertility and fertilizer use efficiency changes (Pan et al., 2009b), in particular in the major crop production regions. Among these is a national long-term fertilizer experimental network, which includes 70 sites across the main agricultural production area of China. It was initiated in the early 1980s, and in 1990 nine sites, representing the major crop regions, were updated to a national long-term monitoring network for soil fertility and fertilizer efficiency. There was also a network of long-term ecosystem experiments managed by the Chinese Academy of Sciences, which included 16 sites situated across ecological gradients in China. In addition, the Chinese Ministry of Agriculture has been managing cropland fertility and productivity in 299 sites across the main crop production areas of China since the mid-1980s.

As shown in Table 27.1, over the period 1985–2006, the monitoring sites and studies considered here have shown a trend for the overall increase of SOM. However, a continuous decline may also be observed in some regions, such as north-east and south-west China where there is soil erosion due to intensified agriculture and/or climate change (Huang and Sun, 2006; Cheng et al., 2009). SOM levels have been increased by about 10% over the past few decades, with increases being greater in rice paddies than in dry croplands. For the last decade, there have been an increasing number of studies addressing the significance of this increase in soil SOC and the contribution this makes to the mitigation of CO_2 emissions from agriculture. Of the croplands in China, rice paddies had been shown to deliver greater levels of carbon sequestration due multiple mechanisms e.g. physical protection, chemical binding with oxyhydrates, as well as molecular stabilization (Pan et al., 2007, 2009b). Using the data relating to changes in SOC, one can infer that a potential saturation of SOM could be as high as 18 g kg^{-1} and 32 g kg^{-1}, respectively, for Chinese dry croplands and rice paddies (Cheng et al., 2009). Clearly, most of China's croplands are still a long way from this saturation level, with present levels being far below the general levels found for EU and US croplands (Pan, 2009). Increasing SOC is still a challenge for China's agricultural sector, for which more effective practices should be pursued, potentially by the amendment of biochar from crop straw (Pan et al., 2011). Nevertheless, the government of China has been sponsoring incentives for enhancing SOM storage through straw return and combined fertilization using organic and inorganic fertilizers since 2003. Meanwhile, a national project on conservation tillage has been initiated, with the aim of extending conservation tillage to 20% of China's croplands by 2050 (MA-SCDRC, 2009).

Benefits of Soil Organic Matter for Crop Production and Agroecosystem Functioning

As pointed out by Sohi (2012), SOM, particularly stable organic matter, may have a number of benefits for crop production and

Table 27.1. Changes in topsoil (0–20 cm) SOC content of China's croplands over 1985–2006, as revealed by data from different data sets.

Data set	Land use[a]	Initial (g kg^{-1})	Final (g kg^{-1})	Reference
National soil fertility network	Rice paddy (112)	18.43 ± 7.72	19.82 ± 7.47	Cheng et al., 2009
	Dry croplands (187)	10.46 ± 5.89	11.05 ± 6.00	
Soil quality change studies	Rice paddy (404)	15.74 ± 6.06	17.37 ± 6.31	Pan et al., 2009c
	Dry croplands (677)	10.06 ± 6.72	10.83 ± 5.86	
Long-term fertilization experiments	Rice paddy (135)	16.28 ± 6.00	18.83 ± 7.94	Wang et al., 2010
	Dry croplands (346)	8.99 ± 6.02	10.61 ± 6.13	
Long-term tillage experiments	Rice paddy (37)	13.02 ± 5.18	14.15 ± 5.44	Wang et al., 2009
	Dry croplands (51)	9.29 ± 5.53	10.01 ± 5.36	

[a]The number in brackets is the observation number.

plant health, as well as for ecosystem functioning. Several studies conducted on field soils have indicated that SOC sequestration may have a number of co-benefits for crop production and agroecosystem functioning. These benefits may include the following.

Enhanced crop productivity and stability

As mentioned in previously, the role of SOM in sustaining crop production is well recognized. This has been addressed in many long-term experiments. In a study of a long-term fertilization, Pan et al. (2009b) demonstrated an increasing yield with SOM accumulation under combined organic/inorganic fertilization in east China, which was attributed to the increased nitrogen (N) use efficiency with SOM accumulation. This is further supported by observations of yield changes with SOM accumulation in rice paddies under long-term fertilization studies in south China (Yuan et al., 2004; Pan and Zhao, 2005). In one of our long-term experiments with conservation tillage, a significant correlation between relative changes in crop yields and relative changes in SOC were observed (Fig. 27.1). However, some of our long-term field studies have shown further benefits for the enhancement and sustainability of crop productivity. Much lower yield variability over years was observed in plots with high SOC contents under combined organic/inorganic fertilization from long-term fertilization experiments from the Tai Lake region of Jiangsu and from the red soil region of Jiangxi (Pan and Zhao, 2005). Data from the experiments of the long-term monitoring system for soil fertility and fertilizer efficiency, managed by the Chinese Academy of Agricultural Sciences, showed a wider variability of rice and wheat in sites with low background SOM levels than with high levels (Table 27.2). This finding highlights the role SOM enhancement can play in sustaining crop production in the face of environmental disturbance such as climate variability. We argue that higher SOC sequestration with SOM accumulation could have not only a synergic effect on increasing rice yield but also on sustaining productivity against environmental stresses (Lal, 2004a; Pan et al., 2009b). In particular, recent field experiments with biochar soil amendments to enhance the stable organic matter pool support the hypothesis that biochar has a stimulating effect on SOM by encouraging root system development of dry crops, including maize, growing in SOM-depleted soils in arid regions (Zhang et al., 2010, 2012).

Retention of N and P, and reduction of N$_2$O emission in croplands

There have been many studies demonstrating a controlling role of SOM in the retention of N in soils. This leads to a reduction of N release to waters and enhances N use efficiency in agricultural soils. It is already well known that soil N contents are well correlated with SOM levels in natural and agricultural soils. Increase in SOM content generally leads to an increase in the soil's

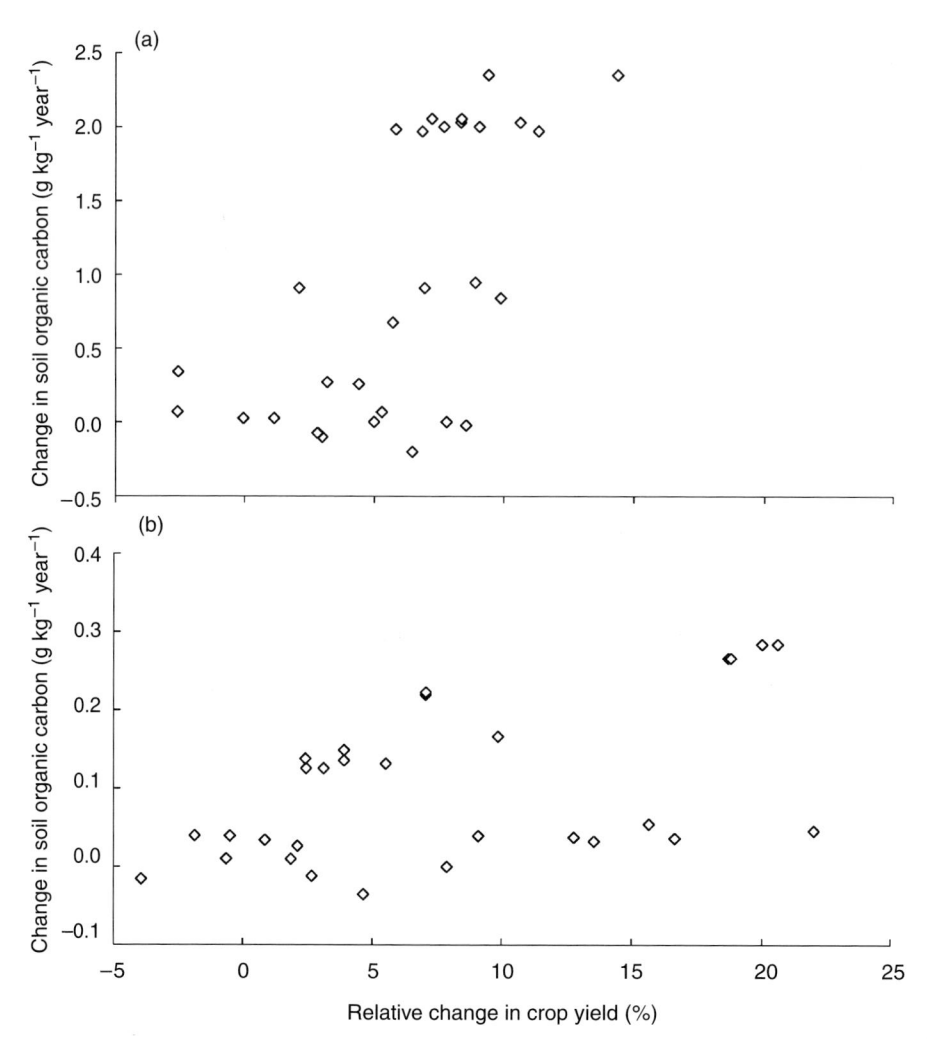

Fig. 27.1. A coupling of soil organic carbon content change with crop yield change under conservation tillage experiments of China (a, rice paddies; b, dry croplands) (Wang et al., 2009).

Table 27.2. Crop yield variability over years, from different experiment sites with various SOM levels. (Courtesy of Dr Xu Mingang for the yield and SOC data from the long-term experiments sites between the late 1980s and early 2000s.)

Agro/ecoregion	Cropping system	Yield variability (%) under different treatments		Topsoil SOM (g C kg^{-1})
		Min	Max	
Tai Lake Plain	Rice	8	30	18.0
Northern Zhejiang Plain	Rice	12	30	16.0
Purple basin of Sichuan	Summer rice	13	20	16.0
Rolling area of red soils, Jiangxi	Double rice	8	17	16.5
Central China Plain	Wheat–maize	30	45	7.5
Rolling area of red soils, Hunan	Winter wheat	30	40	10.0
Purple basin of Sichuan	Winter wheat	15	30	16.0

capacity to protect N. Data collected from ten provincial demonstration farms of high rice productivity in Jiangsu in the early 1990s suggested a high level of total soil N capacity of 1–2 g kg^{-1} under SOM contents over 10 g kg^{-1} (Fig. 27.2). In a study of soil fertility changes with different fertilization from a long-term trial from the Tai Lake region in Jiangsu, topsoil N was found to increase significantly with SOM accumulation. Total N was found to increase by 0.2 g kg^{-1} of topsoil and 0.1 g kg^{-1} of topsoil at a depth of 0–5 cm and 5–15 cm under combined fertilization of organic/inorganic fertilizers compared to under inorganic fertilizer alone, in accordance with an increase in SOC by almost 1.5 and 0.8 g kg^{-1}, respectively (Qiu et al., 2005). The relationship between changes in soil phosphorus (P) retention and SOM accumulation are less well documented. Generally, P becomes more mobilized in OM-rich soils, a phenomenon of organically activated mobilization. However, soil aggregation with SOM accumulation may offer physical protection for P released from dispersed soil particles. One of our earlier studies on P balancing at the plot scale showed less P loss from soils with high SOM contents (accumulated under combined organic/inorganic fertilization with straw return or pig manure) (Jiao et al., 2007). Thus, SOM accumulation may also have benefits for P retention in agricultural soils, as P is generally deficient worldwide.

The question of whether N_2O emission would be increased with increase of soil N is still poorly understood. Although N_2O emissions from croplands are dependent on soil moisture regimes rather than solely on soil N contents, the emission factor of N_2O from N fertilizers is generally smaller in soils high in SOM as compared to those low in SOM. In our database of soil GHG emissions from China's croplands, an increase in N_2O emission from croplands with increasing SOM contents was not visible, nor of CH_4 (Figs 27.3 and 27.4). This indicates that soil C sequestration with SOM accumulation would not necessarily lead to increased GHG emissions, a concern voiced by Schlesinger (2010) and Powlson et al. (2011). This finding also suggests possible changes in soil (micro-)biochemical processes with SOM accumulation in croplands.

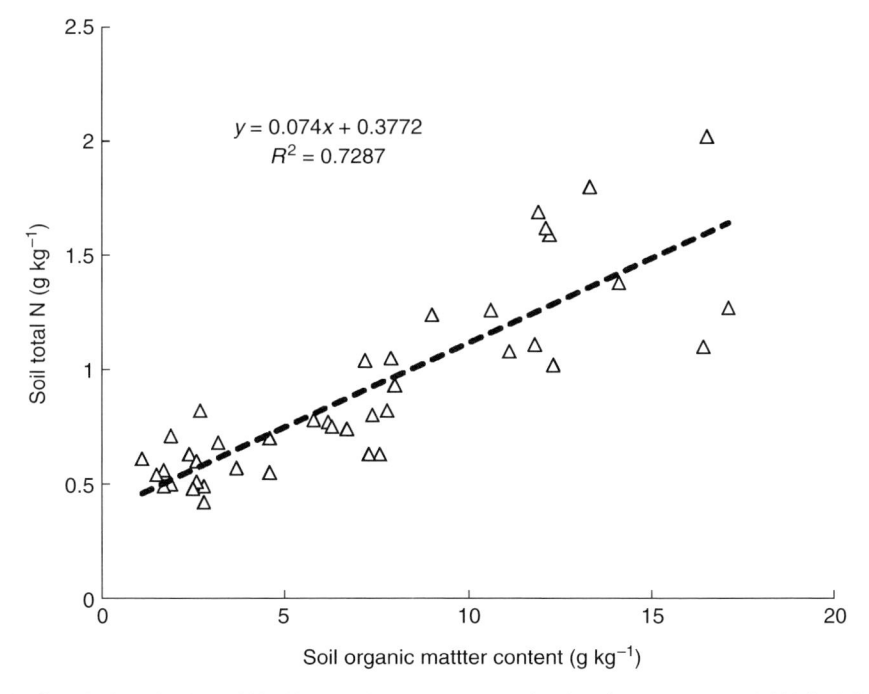

$$y = 0.074x + 0.3772$$
$$R^2 = 0.7287$$

Fig. 27.2. Correlation of soil total N with organic matter content in plots from ten provincial high-yielding demonstration farms in Jiangsu in 1992. (Data collected from farm archives, unpublished.)

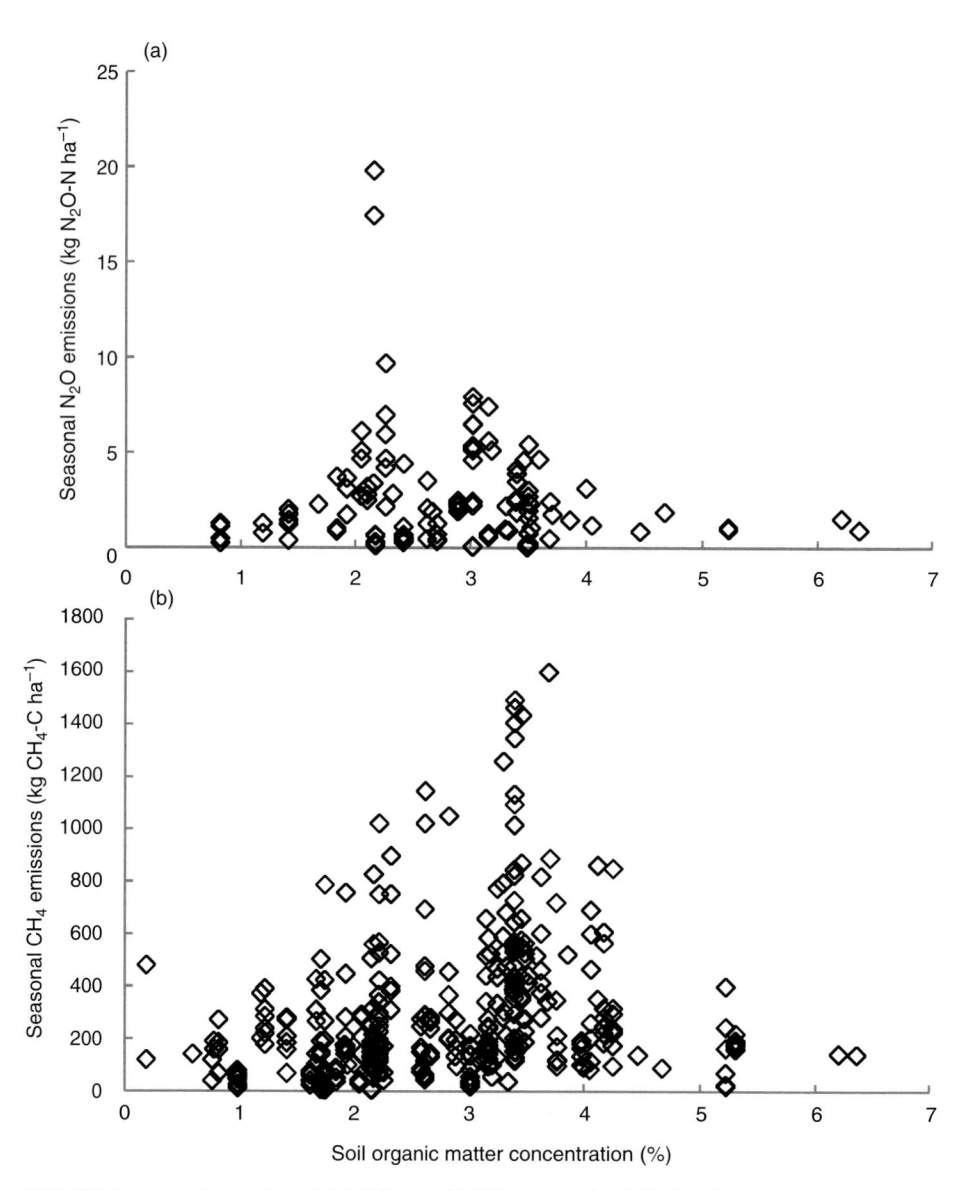

Fig. 27.3. Total seasonal emission of (a) N_2O and (b) CH_4 across the SOM level from rice paddies of China measured in the time span of 1990–2010.

Healthy microbial communities and ecosystem functioning with SOM accumulation

A healthy ecosystem can be characterized by low levels of pollutants and reduced GHG emissions, especially in relation to the C intensity of agricultural production. The control of SOM on metal mobility, particularly of Cd, has been well addressed in experiments and modelling (Sauve, 1999; Jansen et al., 2001). The effect of SOM on organic pollutants has been investigated less frequently in field soils. In the same long-term experiment as reported by Pan (2009), the contents of soil polycyclic aromatic hydrocarbons (PAHs) were significantly lower in plots receiving organic amendments of

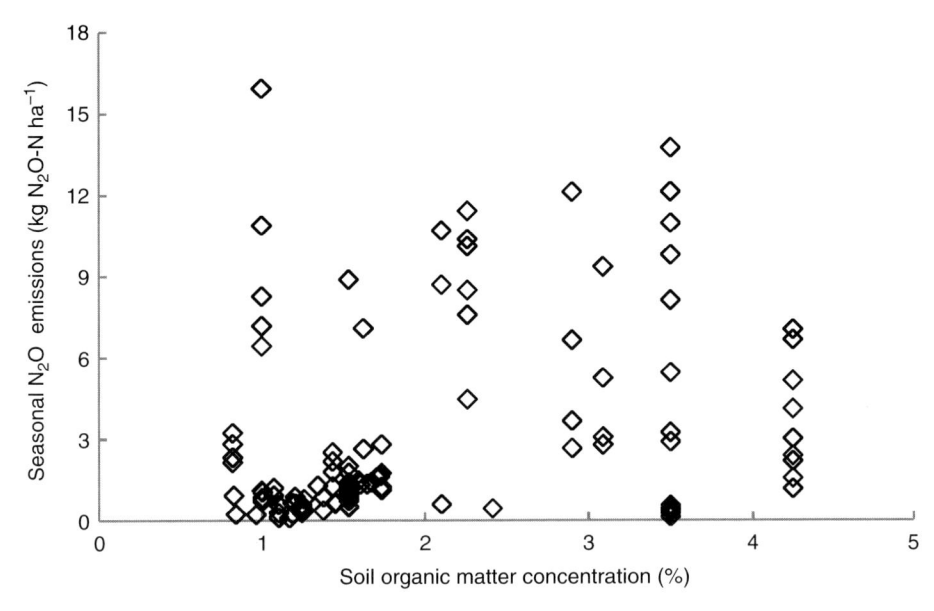

Fig. 27.4. Total seasonal N_2O emission across the SOM level from dry croplands of China.

straw and manure than those receiving chemical fertilizers only (Han et al., 2009). A further study which carried out a laboratory incubation with topsoil samples from different fertilized plots, revealed an enhanced degradation of spiked pyrene from the soil with organic amendments, which showed a significant linear correlation with both SOM content and soil microbial abundance (Han et al., 2009). Another study by Wang et al. (2009) reported reduced extractability of soil PAHS with increasing SOM content for the same experiment. This is the first strong indication that SOM accumulation in soils helps to build up a healthy ecosystem.

Over the past decade, an increasing amount of evidence has been gathered to suggest that SOC sequestration can lead to a healthy soil microbial community. With SOC increase, enhanced microbial abundance and gene diversity were observed under combined organic/inorganic fertilization in a rice paddy (Zhang et al., 2004). One of our recent studies on the changes in microbial activity with rice cultivation in sediment-derived wetland soils from the Yangtze River Valley demonstrated an enhancement of the soil microbial community and diversity, together with an increase in soil enzyme activity with an increase in SOM levels in the rice paddy soils (Huang et al., 2006). It could be the case that SOC accumulation enhances soil productivity through the build-up of an active and healthy soil microbial community in rice paddies.

Changes in soil respiration, and in turn efflux of CO_2 (and CH_4 in rice paddies), with SOM accumulation have been widely argued (Zheng et al., 2007, 2008). A healthy soil may have high microbial abundance but not necessarily high respiration rates, especially the respiration intensity of SOM. In a case study by Zheng et al. (2006), in a situation of SOC accumulation under good fertilization, the metabolic quotient of the soil microbial community and the respiratory quotient of SOC were both reduced, while total soil CO_2 evolution was higher. In many field studies, the correlation coefficient of soil respiration with SOC content in rice soils without N being limiting was generally low. Instead, soil respiratory activity and respired CO_2 flux were often shown to be lower in plots rich in SOC under a well-designed fertilizer scheme when compared with relatively

SOC-poor plots under long-term agroecosystem experiments (Williams et al., 2007). Decreases in soil respiratory activity with increase in SOC have been observed in dry croplands under organic fertilization in comparison with those under non-organic amendments (Meng et al., 2005; Yin and Cai, 2006). An integrated field study using a number of long-term experiments from rice paddies in south China demonstrated an increasing dominance of the fungal over the bacterial community with increasing SOC accumulation. This in turn supports a reduction both in soil respiratory quotient and microbial metabolic quotient under good agro-management (Liu et al., 2011). A laboratory incubation study of methane production from rice paddies differentiating in SOC contents also characterized a reduction in C intensity from methane emission in rice paddy containing high SOC under combined organic/inorganic fertilization. This is further proven by another study with enhanced diversity of methanotrophs, which is responsible for methane exhaustion in rice paddies (Zheng et al., 2008). A soil fauna study by Xiang et al. (2006) documented an enhanced size of the soil fauna community and enhanced diversity, especially those of soil earthworms, thus favouring soil aggregation and nutrient accumulation (Wang et al., 2009) in a long-term experiment site from the Tai Lake region, China.

A new insight into SOC sequestration and GHG emission was that a total global warming potential calculated from all the GHG fluxes in a plot continuously receiving compound fertilizers seemed smaller than from the one receiving the chemical fertilizer only (Li et al., 2009a,b; Liu et al., 2009). Some studies have shown that the net C sink can be 1.5–3 times more under a combined organic/inorganic fertilizer regime than that under chemical fertilization in rice paddies from Jiangxi and Jiangsu, China (Li et al., 2009a,b). A similar study indicated that there was a higher net C sink (by 1.1- to 1.7-fold) under organic amendments compared to under chemical fertilization only (Peng et al., 2009).

In addition, as mentioned above, increase in N efficiency could help to decrease fertilizer-induced C emission per unit of rice production, in turn resulting in a higher net C sink (Li et al., 2009a,b). And this has also been true for the rice paddy in a site of Jiangxi where a higher rate of SOC sequestration has been observed (Li et al., 2010). Thus, SOC sequestration in croplands, particularly in rice paddies, would offer multiple win–win effects for crop productivity and mitigation of GHG emission, as well as ecosystem health in agriculture, so ensuring food security and sustainability of agriculture for China (Pan and Zhao, 2005).

Benefits for soil chemical buffering

In a recent study of soil acidity changes over 1985–2006 in croplands of soil-monitoring sites over China, the extent of soil pH change was shown to be largely dependent on SOM level (Fig. 27.5). Thus, the role of SOM accumulation in buffering chemical processes of croplands is clearly demonstrated. However, this has been poorly assessed in soil or geosciences research.

Conclusion

In conclusion, increase in SOM content would lead to an increase in soil's capacity to act in an integrated and interactive manner for cropland productivity and ecosystem health. In the case of China, accumulation of SOM may have a number of benefits for agricultural production and nutrient use and in sustaining a good quality of agroenvironment. We hypothesized that input of OM to soil might progressively lead to a build-up of soil aggregates with diverse soil microhabitats, a healthy soil microbial community with greater fungal dominance and higher diversity for ecological redundancy. In this way, functional entity is developed with mutual interaction of mineral, chemical, biological and ecological forces. Therein, soil processes and functioning could be magnified to enhance soil capacity for nutrient, moisture and biotic

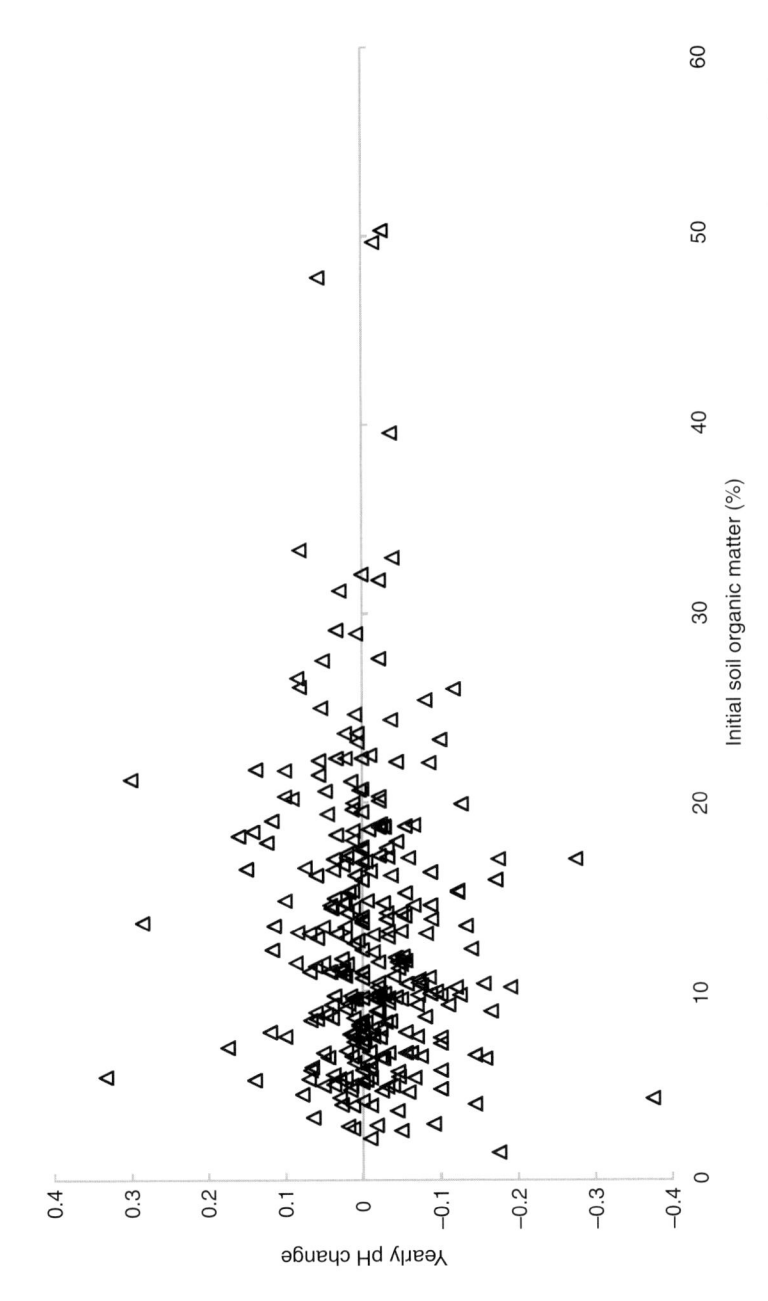

Fig. 27.5. Relative yearly pH change versus initial SOC content of topsoil over 1985–2006 from monitoring sites of dry croplands in China.

conservation. Nevertheless, soil activity with the accumulation of SOM in the topsoil would not be a linear response to the content of SOM but an interaction of SOM with the soil attributes. Finally, soil crop productivity and ecosystem functioning can be harmonized to ensure sustainability.

Acknowledgements

This work was partly funded with a project of the Natural Science Foundation of China under grant 40830528 and a project of the Priority Academic Program Development of Jiangsu Higher Education Institutions.

References

Cheng, K., Pan, G.X., Tian, Y.G. and Li, L.Q. (2009) Changes in topsoil organic carbon of China's crop-land evidenced from the national soil monitoring network. *Journal of Agro-Environment Science* 28, 2476–2481 (in Chinese).

Dawe, D., Dobermann, A., Ladha, J.K., Yadav, R.L., Bao, L., Gupta, R.K., Lal, P., Panaullah, G., Sariam, O., Singh, Y., Swarup, A. and Zhen, Q.-X. (2003) Do organic amendments improve yield trends and profitability in intensive rice systems? *Field Crops Research* 83, 191–213.

EC (European Commission) (2002) Towards a Thematic Strategy for Soil Protection. Communication of 16 April 2002 from the Commission to the Council, the European Parliament, the Economic and Social Committee and the Committee of the Regions. COM (2002), 179, 35 pp. http://www.europarl.europa.eu/sides/getDoc.do?type=REPORT&reference=A5-2003-0354&language=ET(accessed 21 August 2014).

Feller, C., Blanchart, E., Bernoux, M., Lal, R. and Manlay, R. (2012) Soil fertility concepts over the past two centuries: the importance attributed to soil organic matter in developed and developing countries. *Archives of Agronomy and Soil Science* 58, S3–S21.

Han, X.J., Pan, G.X. and Li, L.Q. (2009) Effects of the content of organic matter on the degradation of PAHs: a case of a paddy soil under a long-term fertilization trial from the Tai Lake Region, China. *Journal of Agro-Environment Science* 28, 2533–2539 (in Chinese).

Huang, Y. and Sun, W.J. (2006) Changes in topsoil organic carbon of croplands in mainland China over the last two decades. *Chinese Science Bulletin* 51(15),1785–1803.

Huang, Q.R., Hu, F., Li, H.X., Lai, T. and Yuan, Y.H. (2006) Crop yield response to fertilization and its relations with climate and soil fertility in red paddy soil. *Acta Pedologica Sinica* 43(6), 926–933. [In Chinese with English abstract.]

IPCC (Intergovernmental Panel on Climate Change) (1995) *Impacts, Adaptations and Mitigation of Climate Change: Scientific Technical Analysis*. Working Group II. Cambridge University Press, Cambridge, UK.

Jansen, B., Kotte, M.C., van Wijk, A.J. and Verstraten, J.M. (2001) Comparison of diffusive gradients in thin films and equilibrium dialysis for the determination of Al, Fe(III) and Zn complexed with dissolved organic matter. *Science of the Total Environment* 277, 45–55.

Jiao, S.J., Hu, X.M., Pan, G.X., Zhou, H.J. and Xu, X.D. (2007) Effects of fertilization on nitrogen and phosphorus run-off loss from Qingzi paddy soil in Taihu Lake region during rice growth season. *Chinese Journal of Ecology* 26(4), 495–500 (in Chinese).

Lal, R. (1999) Soil management and restoration for C sequestration to mitigate the accelerated greenhouse effect. *Progress in Environmental Science* 1, 307–326.

Lal, R. (2002) Soil C sequestration in China through agricultural intensification and restoration of degraded and desertified soil. *Land Degradation and Development* 13, 469–478.

Lal, R. (2004a) Soil C sequestration impacts on global climatic change and food security. *Science* 304, 1623–1627.

Lal, R. (2004b) Offsetting China's CO_2 emissions by soil carbon sequestration. *Climatic Change* 63, 263–275.

Li, C.S. (2000) Loss of soil carbon threatens Chinese agriculture: a comparison on agro-ecosystem carbon pool in China and the U.S. *Quaternary Sciences* 20, 345–350 (in Chinese).

Li, J.J., Pan, G.X., Zhang, X.H., Fei, Q.H., Li, Z.P., Zhou, P., Zheng, J.F. and Qiu, D.S. (2009a) An evaluation of net carbon sink effect and cost/benefits of a rice–rape rotation ecosystem under long-term fertilization from Tai Lake region of China. *Chinese Journal of Applied Ecology* 20, 1664–1670 (in Chinese).

Li, J.J., Pan, G.X., Li, L.Q. and Zhang, X.H. (2009b) Estimation of net carbon balance and benefits of rice–rice cropping farm of a red earth paddy under long term fertilization experiment from Jinx, China. *Journal of Agro-Environnent Science* 28(12), 2520–2525 (in Chinese).

Li, L., Xiao, H.A. and Wu, J.S. (2007a) Decomposition and transform of organic substrates in upland and paddy soils red earth region. *Acta Dedolo Sínica* 44, 669–674 (in Chinese).

Li, Z.P., Han, F.X., Su, Y., Zhang, T.L., Sun, B., Monts, D.L. and Plodinec, M.J. (2007b) Assessment of soil organic and carbonate carbon storage in China. *Gendarme* 138, 119–126.

Li, Z.P., Pan, G.X. and Zhang, X.H. (2007c) Top soil organic carbon pool and ^{13}C natural abundance changes from a paddy after 3 years corn cultivation. *Acta Pedologica Sinica* 44, 244–251 (in Chinese).

Li, Z.P., Liu, M., Wu, X.C., Han, F. and Zhang, T. (2010) Effects of long-term chemical fertilization and organic amendments on dynamics of soil organic C and total N in paddy soil derived from barren land in subtropical China. *Soil and Tillage Research* 106(2), 268–274.

Lindert, P.H., Lu, J.A. and Wu, W.L. (1996) Trends in the soil chemistry of South China since the 1930s. *Soil Science* 161, 329–342.

Liu, D.W., Liu, X.Y., Liu, Y.Z., Pan, G., Crowley, D. and Tippkotter, R. (2011) SOC accumulation in paddy soils under long-term agro-ecosystem experiments from South China. VI. Changes in microbial community structure and respiratory activity. *Biogeosciences Discuss* 8, 1–26.

Liu, X.Y., Pan, G.X., Li, L.Q. and Zhang, X.H. (2009) CO_2 emission under long-term different fertilization during rape growth season of a paddy soil from Tailake region, China. *Journal of Agro-Environment Science* 28(12), 2506–2511.

Manlay, R.J., Feller, C. and Swift, M.J. (2007) Historical evolution of soil organic matter concepts and their relationships with the fertility and sustainability of cropping systems. *Agriculture, Ecosystem and Environment* 119, 217–233.

Meng, L., Cai, Z.C. and Ding, W.X. (2005) Carbon contents in soils and crops as affected by long-term fertilization. *Acta Pedologica Sinica* 42, 769–776 (in Chinese).

MA-SCDRC (2009) *A National Planning of Conservation Tillage Project*. Ministry of Agriculture and State Commission of Development and Reform, China. (http://www.moa.gov.cn/zwllm/zcfg/nybgz/200908/t20090828_1340481.htm, accessed 23 October 2014).

Pan, G. (2009) Stock, dynamics of soil organic carbon of China and the role in climate change mitigation. *Advances in Climate Change Research* 5 (Suppl.), 11–18.

Pan, G., Zhou, P., Li, L.Q. and Zhang, X.H. (2007) Core issues and research progresses of soil science of C sequestration. *Acta Pedologica Sinica* 44(2), 327–337.

Pan, G., Smith, P. and Pan, W.W. (2009a) The role of soil organic matter in maintaining the productivity and yield stability of cereals in China. *Agriculture Ecosystems and Environment* 129, 344–348.

Pan, G., Zhou, P., Li, Z.P., Smith, P., Li, L.Q., Qiu, D.S., Zhang, X.H., Xu, X.B., Shen, S.Y. and Chen, X.M. (2009b) Combined inorganic/organic fertilization enhances N efficiency and increases rice productivity through organic carbon accumulation in a rice paddy from the Tai Lake region, China. *Agriculture Ecosystems and Environment* 131, 274–280.

Pan, G.X., Xu, X.W., Smith, P., Pan, W.W. and Lal, R. (2009c) An increase in topsoil SOC stock of China's croplands between 1985 and 2006 revealed by soil monitoring. *Agriculture, Ecosystems and Environment* 136, 133–138.

Pan, G., Lin, Z., Li, L.Q., Zhang, A.F., Zheng, J.W. and Zhang, X.H. (2011) Perspective on biomass carbon industrialization of organic waste from agriculture and rural areas in China. *Journal of Agricultural Science and Technology* 13, 75–82.

Pan, G., Cheng, K., Zheng, J.F., Li, L.Q., Zhang, X.H. and Zheng, J.W. (2013) Organic carbon sequestration potential and the co-benefits in China's cropland. In: Lal, R. and Stewart, A. (eds) *Principles of Sustainable Soil Management in Agroecosystems*. CRC Press, Boca Raton, Florida, pp. 501–520.

Pan, G.X. and Zhao, Q.G. (2005) Study on evolution of organic carbon stock in agricultural soil of China: facing the challenge of global change and food security. *Advanced Earth Sciences* 20(4), 384–392 (in Chinese).

Pan, G.X., Li, L.Q., Wu, L.S. and Zhang, X.H. (2003) Storage and sequestration potential of topsoil organic carbon in China's paddy soils. *Global Change Biology* 10, 79–92.

Peng, H., Ji, X.H., Liu, Z.B., Shi, L.H., Tian, F.X. and Li, H.S. (2009) Evaluation of net carbon sink effect and economic benefit in double rice field ecosystem under long-term fertilization. *Journal of Agro-Environment Science* 28, 2526–2532 (in Chinese).

Powlson, D.S., Whitmore, A.P. and Goulding, W.T. (2011) Soil carbon sequestration to mitigate climate change: a critical re-examination to identify the true and the false. *European Journal of Soil Science* 62(1), 42–55.

Qiu, D.S., Li, L.Q., Jiao, S.J., Pan, G.X. and Zhang, Y. (2005) Change of soil fertility under long-term different fertilization practices in a paddy soil from the Tai Lake region. *Soils and Fertilizers* 4, 28–32 (in Chinese).

Sauve, S. (1999) Speciation and complexation of cadmium in soil solutions. In: Wenzel, W.W., Adirano, D.C., Alloway, B., Doner, H.E., Keller, C., Lepp, N.W., Mench, M., Naidou, R. and Pierzynski, G.M. (eds) *Proceedings of 5th International Conference on the Biogeochemistry of Trace Elements*. ICOBTE, Vienna, pp. 1098–1099.

Schlesinger, W.H. (2010) On fertilizer-induced soil carbon sequestration in China's croplands. *Global Change Biology* 16, 849–850.

Sohi, S. (2012) Carbon storage with benefits. *Science* 338, 1034–1035.

Song, G.H., Li, L.Q., Pan, G. and Zhang, Q. (2005) Topsoil organic carbon storage of China and its loss by cultivation. *Biogeochemistry* 74, 47–62.

Tiessen, H., Cuevas, E. and Chacon, P. (1994) The role of soil organic matter in sustaining soil fertility. *Nature* 371, 783–785.

Wang, C.J., Pan, G.X. and Tian, Y.G. (2009) Characteristics of cropland topsoil organic carbon dynamics under different conservation tillage treatments based on long-term agro-ecosystem experiments across mainland China. *Journal of Agro-Environment Science* 28, 2464–2475 (in Chinese).

Wang, C.J., Pan, G., Tian, Y.G., Li, L.Q., Zhang, X.H. and Han, X.J. (2010) Changes in cropland topsoil organic carbon with different fertilizations under long-term agro-ecosystem experiments across mainland China. *Science in China Series C, Life Sciences* 153(7), 858–867.

Williams, M.A., Myrold, D.D. and Biolltomley, P.J. (2007) Carbon flow from ^{13}C-labelled clover and ryegrass residues into a residue-associated microbial community under field conditions. *Soil Biology and Biochemistry* 39, 819–822.

Wu, H.B., Guo, Z.T. and Peng, C.H. (2003) Land use induced changes of organic carbon storage in soils of China. *Global Change Biology* 9, 305–315.

Xiang, C.G., Zhang, P.J., Pan, G.X., Qiu, D.S. and Chu, Q.H. (2006) Changes in diversity, protein content and amino acid composition of earthworms from a paddy soil under long-term different fertilizations in the Tai Lake Region, China. *Acta Ecologica Sinica* 26, 1667–1674 (in Chinese).

Yin, Y.F. and Cai, Z.C. (2006) Effect of fertilization on equilibrium levels of organic carbon and capacities of soil stabilizing organic carbon for Fluvo-aquic soil. *Soils* 38, 745–749 (in Chinese).

Yuan, Y.H., Li, H.X., Huang, Q.R., Feng, H. and Pan, G.X. (2004) Effects of different fertilization on soil organic carbon distribution and storage in micro-aggregates of red paddy topsoil. *Acta Ecologica Sinica* 24, 2961–2966.

Zhang, A.F., Cui, L.Q., Pan, G.X., Li, L.Q., Hussain, Q., Zhang, X.H., Zheng, J.W. and Crowley, D. (2010) Effect of biochar amendment on yield and methane and nitrous oxide emissions from a rice paddy from Tai Lake plain, China. *Agriculture, Ecosystems and Environment* 139, 469–475.

Zhang, A.F., Liu, Y.M., Pan, G.X., Hussain, Q., Li, L.Q., Zheng, J.W. and Zhang, X.H. (2012) Effect of biochar amendment on maize yield and greenhouse gas emissions from a soil organic carbon poor calcareous loamy soil from Central China Plain. *Plant Soil* 351, 263–275.

Zhang, P.J., Li, L.Q., Pan, G.X. and Zhang, J.W. (2004) Influence of long-term fertilizer management on topsoil microbial biomass and genetic diversity of a paddy soil from the Tai Lake region, China. *Acta Ecologica Sinica* 24, 2819–2824 (in Chinese).

Zheng, J.F., Zhang, X.H., Pan, G.X. and Li, L.Q. (2006) Diurnal variation of soil basal respiration and CO_2 emission from a typical paddy soil after rice harvest under long-term different fertilizations. *Plant Nutrition and Fertilizer Science* 12, 485–494 (in Chinese).

Zheng, J.F., Zhang, X.H., Li, L.Q., Zhang, P.J. and Pan, G.X. (2007) Effect of long-term fertilization on C mineralization and production of CH_4 and CO_2 under anaerobic incubation from bulk samples and particle size fractions of a typical paddy soil. *Agriculture, Ecosystems and Environment* 120, 129–138.

Zheng, J.F., Zhang, P.J., Pan, G.X., Li, L.Q. and Zhang, X.H. (2008) Effect of long-term different fertilization on methane oxidation potential and diversity of methanotrophs of paddy soil. *Acta Ecologica Sinica* 28, 4864–4872 (in Chinese).

28 Assessment of Organic Carbon Status in Indian Soils

Tapas Bhattacharyya*

Abstract

Soil organic carbon (SOC) content in Indian soils has been reported as low, which is in tune with the fact that nearly 60% of the area in India represents the typical tropical climate, which does not permit SOC accumulation. Recent evaluation with the help of more soil and site data, model approaches and long-term fertilizer experiments (LTFEs) show an increasing trend of SOC, as detailed in this chapter through different case studies in two important food growing zones of India, namely the Indo-Gangetic Plains (IGP) and the black soil region (BSR). The study shows the evaluation of Century, RothC and InfoCrop models in LTFEs with contrasting bioclimate in the IGP and the BSR. The Century model experience necessitates the modification of crop information to suit the tropical conditions found in India. The RothC output has been found to be useful to arrive at the threshold limit of the mean annual rainfall as an indicator of organic carbon storage in soil. The InfoCrop cotton model in the BSR indicated that the interaction of increased temperature and CO_2 concentration had a compensatory effect on crop yield. A methane emission study on Indian agricultural soils has been computed as 2.54 Tg C-CH_4 year^{-1} and constitutes 0.23% of global warming. However, although an increase of SOC has been found in IGP and BSR soils, even in arid and semi-arid environments, the status of soil inorganic carbon will require attention by planners and resource managers in view of soil degradation.

Introduction

The restoration of soil health through soil organic carbon (SOC) management is a major concern for tropical soils. Barring its importance for sustainable crop production, the accelerated decomposition of SOC due to agriculture, resulting in loss of carbon to the atmosphere and its contribution to the greenhouse effect, is a serious global problem. The contribution of SOC for improving soil physical, chemical and biological properties, and thus maintaining its health in order to sustain crop productivity, has been well known since the dawn of human civilization. Important factors controlling SOC status include climate (especially rainfall and temperature), hydrology, parent material, soil fertility, biological activity, vegetation and land use. SOC is sensitive to human activities such as deforestation, biomass burning, land-use changes and environmental pollution. It is estimated that land-use change results in the transfer of 1–2 Pg C year^{-1} from

*E-mail: tapas11156@yahoo.com

the terrestrial ecosystem to the atmosphere, of which 15–17% C is contributed by the decomposition of SOC (Houghton and Hackler, 1994). It is important to note that organic matter (OM) preferentially accumulates in submerged rice systems, which store relatively more SOC as compared to their upland cropping system (Sahrawat et al., 2005).

To sustain the quality and productivity of soils, information on SOC in terms of its amount and quality is essential. In recent years, global warming has created awareness on the role of C cycling in agroecosystems in storing atmospheric C in Indian soils. The first comprehensive study on SOC status in Indian soils was conducted by Jenny and Raychaudhuri (1960), followed by an estimation of soil organic C stock of 24.3 Pg by Gupta and Rao (1994), which was later revised by Bhattacharyya et al. (2008) (Table 28.1). The total C stock in Indian soils is low, as compared to tropical regions and the rest of the world (Table 28.2). Comprehensive research efforts on SOC in Indian soils have provided interesting information. For brevity, a few case studies from the two important food-growing zones of the country, namely the Indo-Gangetic Plains (IGP) and the black soil region (BSR), are discussed here.

Dynamics of SOM in India's Croplands over the Past Two Decades

Reduced productivity of the rice–wheat (R/W) system in the IGP has been linked with declining SOM (Bhandari et al., 2002). The earlier report on declining productivity in Haryana and Punjab by Sinha et al. (1998) hinted at a decrease in SOC from 0.5% in the 1960s to 0.2% in 1998 in major R/W regions of the Indian IGP. A key question being raised often is whether the R/W system of the IGP is sustainable in terms of soil health, with SOC as the key soil quality parameter. As compared to the IGP, the BSR has been experiencing less intensive agriculture, as is evidenced by relatively less potassium (K) stock in the IGP due to excessive mining (Bhattacharyya et al., 2007a). Besides, the high annual temperature in the central, western and the peninsular region of the country, representing the BSR, does not permit SOC build-up either (Bhattacharyya et al., 2000).

Depletion of SOM in croplands in history of agricultural development of India

The IGP, with about 13% geographical coverage in India, produces nearly 50% of the food grains for 40% of the total population of India. However, recent reports of the land use and soils of the IGP indicate a general decline in soil fertility (Bhandari et al., 2002). Soils that earlier rarely showed any nutrient deficiency symptoms are now deficient in many nutritional elements. Long-term soil fertility studies have shown a reduction in soil organic matter content, as well as in the other essential nutrients that had higher levels of nutritional elements in the earlier years (Abrol and Gupta, 1998; Bhandari et al., 2002). The biological activity of soils has declined gradually, resulting in the reduced efficiency of the inputs applied (Abrol and Gupta, 1998). As a consequence, parts of the IGP have an aridic environment at present (Eswaran and van den Berg, 1992). The sustainability ratings of some soils of the IGP for the R/W system indicate many soil constraints, including low SOC (Bhattacharyya et al., 2004). It is in this context that the soils of the IGP of the Indian subcontinent require focused attention.

The BSR, constituting nearly 74.6 million hectares (Mha) in area, is known for low average SOC content (Bhattacharyya et al., 2008); however, the occurrence of high-quality brown forest soils (mollisols), with nearly 3–5% SOC, is not rare (Bhattacharyya et al., 2006a). The current arid and semi-arid environment prevailing in central and southern peninsular India, representing the major BSR, has been ascribed to the global warming phenomenon (Eswaran and van den Berg, 1992). This has been indicated as the causative factor for the low SOC level in the shrink–swell soils of India (Bhattacharyya et al., 2000).

Table 28.1. Soil carbon stocks in different bioclimatic systems in India. (From Bhattacharyya et al., 2008.)

Bioclimatic systems[a]	Area		SOC			SIC			TC		Pg Mha^{-1}	
	Coverage (Mha)	TGA (%)	Stock Pg	Per cent of total SOC		Stock Pg	Per cent of total SIC		Stock Pg	Per cent of total TC	SOC	SIC
Arid cold	15.2	4.6	0.6	6		0.7	17		2.7	20	0.0192	0.0327
Arid hot	36.8	11.2	0.4	4		1.0	25					
Semi-arid	116.4	35.4	2.9	30		1.9	47		4.8	35	0.0249	0.0163
Subhumid	105	31.9	2.5	26		0.3	8		2.8	20	0.0238	0.0029
Humid to per humid	34.9	10.6	2.1	21		0.04	1		2.14	15	0.0602	0.0011
Coastal	20.4	6.2	1.3	13		0.07	2		1.37	10	0.0637	0.0034

[a]Ranges in rainfall: arid = <550 mm; semi-arid = 550–1000 mm; subhumid = 1000–1500 mm; humid to per humid = 1200–3200 mm; coastal = 900–3000 mm. SOC, soil organic carbon; SIC, soil inorganic carbon; TC, total carbon; TGA, total geographical area.

Table 28.2. Carbon stocks in the Indo-Gangetic Plains, India, and other parts of the world (values in Pg).

Region	Soil depth (m)	
	0–0.3	0–1.5
IGP, India[a]	0.63	2.00
India[b]	9.77	29.97
Tropical Regions[c]	201–213	616–640
World[c]	684–724	2376–2456[d]

[a]Bhattacharyya et al., 2004; [b]Bhattacharyya et al., 2008; [c]Batjes, 1996; [d]for 0–2.0 m soil depth.

The increasing trend of soil organic matter over the past two decades in India

Recent studies on SOC over a long period indicate an increasing trend in both the IGP (Benbi and Brar, 2009) and the BSR (Bhattacharyya et al., 2007b). The evaluation of large soil test data for 25 years (1981/82 to 2005/06) has shown improvement in the SOC status under intensive agriculture in Punjab. On average, the SOC increased by 38% for the whole Punjab, from 2.9 g kg^{-1} in 1981/82 to 4 g kg^{-1} in 2005/06. The increase in SOC content was significantly related to the increase in R/W performance. This increase was linear, from 5.9 t ha^{-1} in 1981/82 to 8.1 t ha^{-1} in 1999/2000. The increased productivity of R/W has resulted in an increased sequestration of carbon in the plough layer by 0.8 t C ha^{-1} of increased grain production (Benbi and Brar, 2009). The All India Coordinated Research Project (AICRP) on long-term fertilizer experiments (LTFEs) showed that SOC contents were improved over years under the recommended level of fertilizer application (Biswas and Benbi, 1997; Manna et al., 2006; Singh and Wanjari, 2009). The increased sequestration of C in the R/W system under intensive agriculture is due to improved crop productivity, greater turnover of biomass (both above ground and below ground), greater transport of carbon into roots (rhizosphere) and reduced organic matter decomposition under anaerobic wetland rice.

An increase in carbon stocks in soils under continuous wetland rice has also been reported (Bronson et al., 1998; Sahrawat et al., 2005; Kukal and Rehana-Rasool Benbi, 2008). A Global Environment Facility (GEF) financed project on the assessment of SOC stocks and the change at the national scale in four countries including India (Milne et al., 2006) indicated an increase in the SOC stock in the IGP by about 4.5% from 1990 to 2000, as per the Century model, due mainly to the increase in productivity in rice-based systems (Bhattacharyya et al., 2007c). SOC stocks in 2030 are predicted to decline to the level of 1990 in the wake of presumed land management changes in near elimination of the fallow-rice system and doubling of the area under triple-cropped rotations. An Intergovernmental Panel on Climate Change (IPCC) model showed essentially no change in SOC stocks from 1990 to 2030. Studies carried out under the Global Environment Facility Soil Organic Carbon (GEFSOC) project and the National Agricultural Technology Project (NATP) on SOC for the IGP and BSR regions have further strengthened the viewpoint that there is an improvement in the SOC of soils under the existing agricultural management practice (Tables 28.3 and 28.4). Continuation of the same cropping pattern, therefore, rings no alarm bells on the drop of SOC levels in the IGP and the BSR in the near future, although the increase in $CaCO_3$ in soils will remain a concern (Bhattacharyya et al., 2007b).

Benefits of SOC for Crop Production and Agroecosystem Functioning

SOC and its influence has been studied by different models, namely Century, RothC and InfoCrop, with the help of different LTFEs. Many long-term fertilizer experimental data sets have been documented for the IGP and the BSR (Bhattacharyya et al., 2007c, 2011, 2012). Besides, soil survey data sets have also been used to find out the benefits of SOC in different land-use systems.

Table 28.3. Changes in carbon stock over years in selected benchmark spots in the Indo-Gangetic Plains, India (0–150 cm). (From Bhattacharyya et al., 2007b.)

Bioclimatic systems	Soil series	SOC stock (Tg 100,000 ha^{-1})		SOC change over 1980 stock (%)	SIC stock (Tg 100,000 ha^{-1})[a]		SIC change over 1980 stock (%)
		1980	2005		1980	2005	
Semi-arid	Phaguwala	3.36	5.48	63	13.10	26.14	99
	Ghabdan	2.63	7.04	167	18.95	7.71	−59
	Zarifa Viran	4.13	5.38	30	22.36	16.98	−24
	Fatehpur	1.11	5.50	395	0	58.13	100
	Sakit	4.05	8.55	111	51.03	5.37	−89
	Dhadde	4.47	5.84	31	0	10.15	100
Subhumid	Bhanra	1.81	5.34	197	0	0.58	100
	Jagjitpur	2.52	8.76	248	2.52	8.86	251
	Haldi	8.55	6.28	−26	0	2.84	100
Humid	Hanrgram	6.93	11.02	59	0	3.68	100
	Madhpur	3.99	4.97	25	4.03	15.98	296
	Sasanga	5.25	8.42	61	0.88	4.45	405

[a]SIC indicates contribution of CaCO$_3$ in soils.

Table 28.4. Changes in carbon stock over years in selected benchmark spots of the black soil region, India (0–150 cm). (From Bhattacharyya et al., 2007b.)

Bioclimatic systems	Soil series	SOC stock (Tg 100,000 ha^{-1})		SOC change over 1980 (%)	SIC stock (Tg 100,000 ha^{-1})[a]		SIC change over 1980 (%)
		1980	2005		1980	2005	
Arid	Sokhda	11.19	9.20	−18	23.63	60.92	158
Semi-arid	Asra	6.29	13.59	116	2.00	2.00	0
	Teligi	7.41	15.20	105	21.01	29.60	41
	Semla	15.78	13.28	−16	73.82	46.11	−37
	Vijaypura	7.70	7.70	0	0	0	0
	Kaukuntla	4.71	10.25	118	0	12.52	100
	Patancheru	8.39	16.72	101	0	11.78	100
Subhumid	Kheri	5.62	10.51	87	8.32	9.71	17
	Linga	9.66	12.92	34	15.41	21.66	40

[a]SIC indicates contribution of CaCO$_3$ in soils.

Identifying systems for organic C sequestration

SOC, SIC (soil inorganic carbon) and TC (total carbon) stocks were estimated while identifying systems for carbon sequestration in the semi-arid tropics (SAT) in India. The database generated was used to identify systems keeping in view the fact that effective carbon management can help not only in building the SOC stock to a level of 10.5 Pg, from their existing level of 2.5 Pg, but also in reducing the SIC stock to the tune of 1.9 Pg, much to the benefit of growing plants in terms of the better physical and chemical environment of soils (Bhattacharyya et al., 2006b). In SAT, benchmark spots of agriculture and horticulture were identified as suitable management under two distinctly different conditions (Fig. 28.1). The minimum threshold limit of SOC is 0.63% (0–30 cm soil depth), which is associated with a maximum threshold value of bulk density (BD) (1.6 Mg m^{-3}). These threshold limits were arrived at since they (of SOC and BD) correspond to approximately 10 Pg SOC stock in India (0–30 cm soil depth) (Bhattacharyya et al., 2000). These minimum SOC and maximum BD values correspond to an average value of 1.19% SIC. The minimum threshold value of SOC may, therefore, increase SIC and BD effecting poor soil drainage. The maximum threshold limit of SOC is 2.42%, which corresponds to a minimum threshold limit of BD (1.22 Mg m^{-3}). This is the maximum SOC obtained in the forest ecosystem in SAT, India.

Century model experience

LTFE data sets were used to evaluate the performance of the Century ecosystem model in contrasting ecoregions of India, viz. Mohanpur (humid) and Akola (semi-arid), with a mean annual rainfall of 1619 mm and 793 mm, respectively. At the humid site, the modelled data simulated the measured data reasonably well for all the treatments, with control and treatments with fertilizer alone and in combination with organic inputs. Century modelled changes in SOC more successfully at Mohanpur than at Akola. The SOC levels were low to start with at Mohanpur and gradually showed an increasing trend when external inputs (fertilizers and manures) were added. Century closely predicted the SOC levels measured at Mohanpur for all the treatments. In Akola, measured SOC exhibited a sharp increase when both inorganics and organics were added in combination or organics applied separately. At the semi-arid site, Century performed well for the early years and less well during the end

```
                    Conditions
                   /          \
                  /            \
    Minimum threshold of SOC (0.63%)   Maximum threshold of SOC (2.42%)
    Maximum threshold of BD (1.6 Mg m⁻³)   Minimum threshold of BD (1.22 Mg m⁻³)

             Corresponds to ~10 Pg SOC stock in 0–30 cm soil
                            depth in India
```

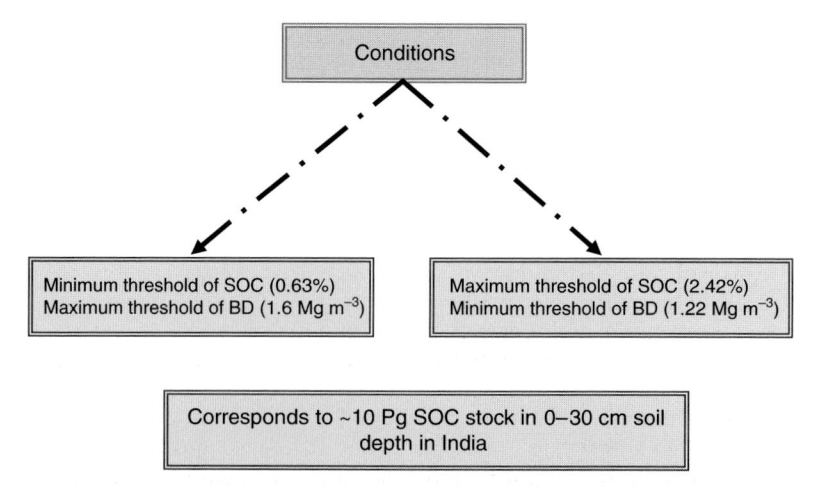

Fig. 28.1. Conditions of identifying systems for organic carbon sequestration in soils of SAT, India. (From Bhattacharyya et al., 2006b.)

of the experiment. The comparison between the reported and the modelled yield for the three crops (rice, wheat and sorghum) showed a reasonably good correlation ($r = 0.8$) (Bhattacharyya et al., 2010). Earlier experience of evaluation of the Century model in the IGP showed overestimation by Century at the humid site (Bhattacharyya et al., 2007c). More recently, Milne et al. (2008) suggested that Century might be less suited to estimate the carbon dynamics of soils under rice flooded every year. In contrast, in the present study at Mohanpur, flooded rice (flooded for 6 months over a period of 19 years of experimentation) yielded a good Century output from the humid bioclimatic system. It appears that further development of the Century model is needed to improve its performance when modelling carbon return to the soil under extremely clayey soils (>50% clay), as in the BSR (Akola). The crop files obtained along with the Century model were modified to suit the tropical conditions (Table 28.5). In general, Century appeared to simulate crop yield reasonably well for all the crops (Fig. 28.2). With the proper relationships, Century simulation of measured yields is a gauge of the model's performance in replicating the actual crop residue returns of carbon and nitrogen to the soil and in replicating crop–water relations.

RothC model experience

RothC could simulate changes in total organic C (TOC) in two contrasting ecosites. Observed trends in TOC consist of an increase in the subhumid sites, while manures alone or in combination increase TOC appreciably in arid and semi-arid sites. TOC remained, however, almost similar over years for the control (no fertilizer or manure) and nitrogen, phosphorus and potassium (NPK) treatments in all the four sites. In the subhumid site, TOC increased marginally when a less than recommended dose of fertilizer was added. The addition of a higher dose of fertilizer and manure brought a rapid increase in TOC. The addition of fertilizer only did not increase TOC but increased the crop yield, while the application of fertilizers in combination with organic materials brought an increase in TOC as well as in crop yield. Usually, the harvesting of annual crops is done from almost ground level. Grain and straw are both used, although for different purposes. Stubble is removed before sowing the next crop. This gives very little scope of returning the biomass back to the soil. This happens in cases when higher fertilizer is added: yield increased but SOC did not change much. This is in contrast to the treatments where organic amendments are applied along with inorganics. The manures, since they are mixed with the soil, help to increase the SOC content. This model has been found useful as a tool to arrive at different threshold values of rainfall, which influence the decomposition rate modifier and also the rate of organic carbon sequestration in various bioclimatic systems (Fig. 28.3). The rate modifier for the moisture sum gradually decreases from a subhumid moist to an arid bioclimatic system, with a clear distinction at Sarol, Akola and Zarifa Viran. Various other details of these sites vis-à-vis the rate modifiers indicate that rainfall of nearly 850, 550 and 500 mm during the wet months appears to be the three threshold limits in these five bioclimatic systems in deciding the rate modifier for the moisture sum. This means that below these wet months' rainfall, the organic carbon turnover rate will be reduced, causing less organic carbon storage in soils. While explaining inorganic carbon sequestration and its consequences on soil sodicity, a threshold limit of 850 mm mean annual rainfall (MAR) was reported (Bhattacharyya et al., 2000, 2004), below which the soils became more calcareous, alkaline and sodic. RothC model output can thus help to find the threshold limits of the climatic parameters that influence SOC decomposition and its content, which can serve as a good indicator for soil quality and health. The model showed that the addition of fertilizers only should maintain the TOC level in soils. To increase the level of organic carbon sequestration for posterity, the addition of organics along with fertilizers may be recommended. Recent findings on the effects of organics on shrink–swell soils

Table 28.5. Crop files modified in the Century model for the Indo-Gangetic Plains (IGP) and the black soil region (BSR), India. (From Bhattacharyya et al., 2010.)

Rice				
Abbreviation	IGPRM[a]	IGPRM[b]	RICL[c]	
Description	Rice monsoon	Rice monsoon	Lowland rice	
Sowing time	June–July	June–July	–	
Duration (months)	5	5	5	
Irrigation	Rain fed	Rain fed	Rain fed	
PPDF (2)[d]	45	42	45	

Wheat				
Abbreviation	IGPWE[a]	IGPWEM[e]	WW3S[c]	BSRW[f]
Description	Irrigated wheat eastern IGP	Irrigated wheat eastern IGP (Mohanpur)	Soft winter wheat high harvest index	Irrigated wheat black soil region
Sowing time	3rd week December	November		November
Duration (months)	4–5	6	6	5
Irrigation	4–5	6	–	5
PRDX (1)[g]	300	325	450	475
PPDF (1)[h]	30	18	18	18
PPDF (2)[d]	35	35	33	40

Sorghum		
Abbreviation	BSRS[f]	SORG[c]
Description	Sorghum black soil region	Grain sorghum
Sowing time	June	June–July
Duration (months)	4	
Irrigation	–	
PRDX (1)[g]	375	680
PPDF (1)[h]	30	30
PPDF (2)[d]	45	45

[a]Please see Table 3 in Bhattacharyya et al., 2007c; [b]modified for rice in this study; [c]original crop file Century; [d]PPDF (2): maximum temperature for production for parameterization of Poisson density function curve to simulate temperature effect on growth; [e]modified for this study for IGP (Mohanpur); although wheat is grown during late December in eastern part of the IGP, the present experiment (our study area of Mohanpur) reports wheat sowing during November; [f]modified for the study of the BSR (Akola); [g]PRDX (1): potential aboveground monthly production for crops (g C m^{-2}); [h]PPDF (1): optimum temperature for production for parameterization of a Poisson density function curve to simulate temperature effect on growth.

(vertisols) indicate an increase in the active pool of organic carbon (Chivhane and Bhattacharyya, 2010). Although such pools may not influence crop yield immediately, in the course of time crop yield and soil health will be bettered by such practice. Among organic manures, farmyard manure and wheat straw was, by far, the best combination for the maximum influence on organic carbon sequestration. The other options are cow dung slurry and urban compost and farmyard manure. Earlier, Guo et al. (2006) reported RothC to simulate changes in the SOC accurately across a wide area of northern China. They indicated that manures applied at an appropriate rate were more effective in increasing and maintaining SOC than fertilizers, which, in turn, were more effective in increasing crop yield.

InfoCropmodel studies

The validated InfoCrop cotton model was tested for its sensitivity to atmospheric CO_2

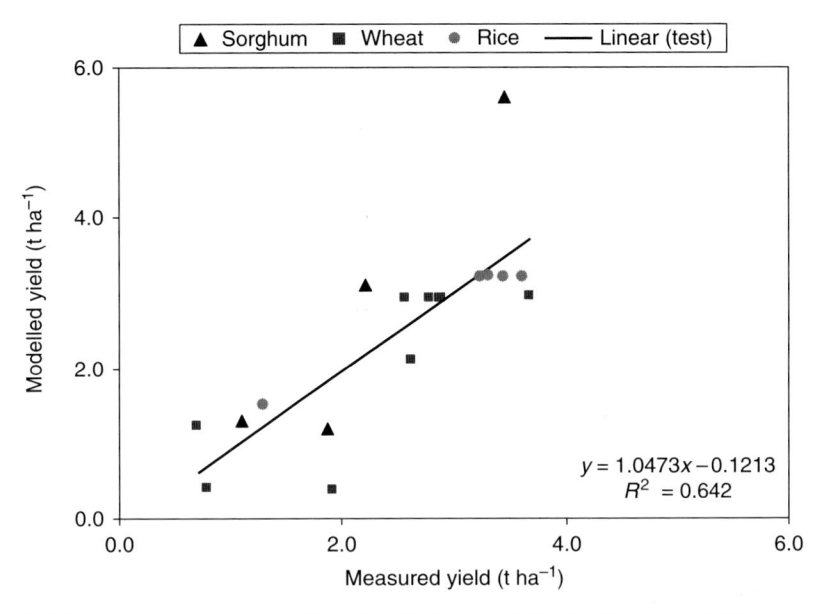

Fig. 28.2. Modelled versus measured yield from the Century model for two long-term fertilizer experiments in India. (From Bhattacharyya et al., 2010.)

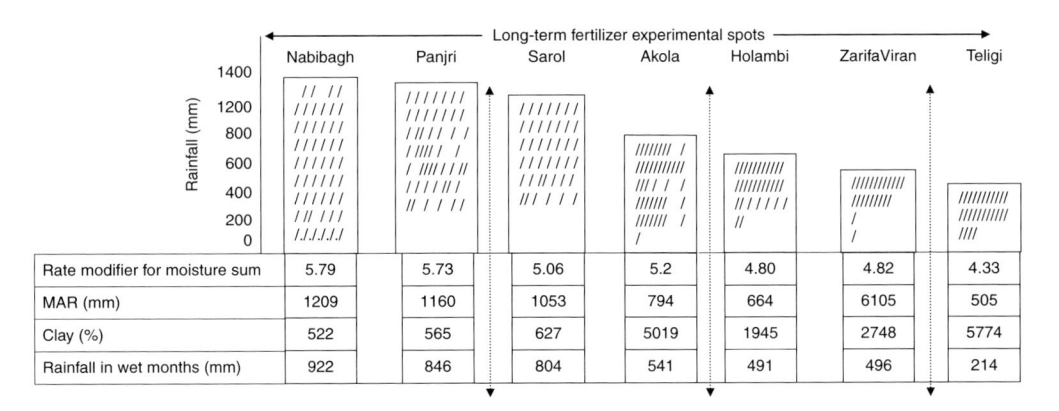

	Nabibagh	Panjri	Sarol	Akola	Holambi	ZarifaViran	Teligi
Rate modifier for moisture sum	5.79	5.73	5.06	5.2	4.80	4.82	4.33
MAR (mm)	1209	1160	1053	794	664	6105	505
Clay (%)	522	565	627	5019	1945	2748	5774
Rainfall in wet months (mm)	922	846	804	541	491	496	214

Fig. 28.3. Schematic diagram showing the decomposition rate modifiers in different long-term fertilizer experiments. The LTFE spots of Akola, Holambi and Zarifa Viran are shown for comparison to find the threshold limits for rainfall, but are not discussed in this paper (Bhattacharyya et al., 2011). MAR, mean annual rainfall.

and temperature in four different scenarios of climate change. The effect of change in temperature and CO_2 was studied in rainfed cotton for a gradual increase in CO_2 (369, 543 and 789 ppm) and temperature (0, 1, 2, 3, 4 and 5°C) during the entire crop growth period (Bhattacharyya et al., 2012). Weather and agronomic management practices, which influence crop growth and yield, vary considerably in different parts of India. The simulation results indicate that increased temperature has a pronounced negative effect on cotton crop yields. Across all four selected places, the simulated potential yield of cotton decreased with an increase in temperature. The decrease in potential yield varied from 13 to 32% against a temperature increase of 3°C, and varied from

21 to 40% against a temperature increase of 5°C. In the case of simulated water-limited yields, the decrease varied from 10 to 36% against a temperature increase of 3°C and from 16 to 43% against a temperature increase of 5°C. In all four places, Nagpur in central India experienced a higher decrease in potential yields of 32 and 40%, as well as in water-limited yields of 36 and 43% against a temperature increase of 3°C and 5°C, respectively. Increased CO_2 concentration has a positive effect on cotton crop yields (Figs 28.4 and 28.5). When the CO_2 concentration increased from 369 ppm to 543 ppm, the increase in potential and water-limited yields was in the range of 4–8% and 9–16%, respectively. If the CO_2 concentration increased from 369 ppm to 789 ppm, the increase in potential yields varied from 7 to 11% and from 12 to 24% in water-limited yields. The utilization of increased CO_2 concentration was relatively better in the case of water-limited yields, due to soil moisture consideration. In the simulated potential yields, soil moisture is not considered. It was observed that the interaction of increased temperature and increased CO_2 concentration had a compensatory effect on cotton yields, resulting in a smaller reduction in potential yield and a non-significant reduction in water-limited yield.

Methane Production in Submerged and Paddy Soils

Rice crop area and livestock population are the two major sources of CH_4 emissions from the agriculture sector. India is a major rice growing country, with a very diverse growing environment. There are several studies on the estimation of methane emission from rice fields using various methods (Table 28.6). Methane Campaign 1991 (MC-1991) reported an annual methane emission of 4.0 ± 2.0 Tg (1 Tg = 1012 g) for Indian paddy soils. A state-wise study, conducted in 1994, indicated a methane emission of

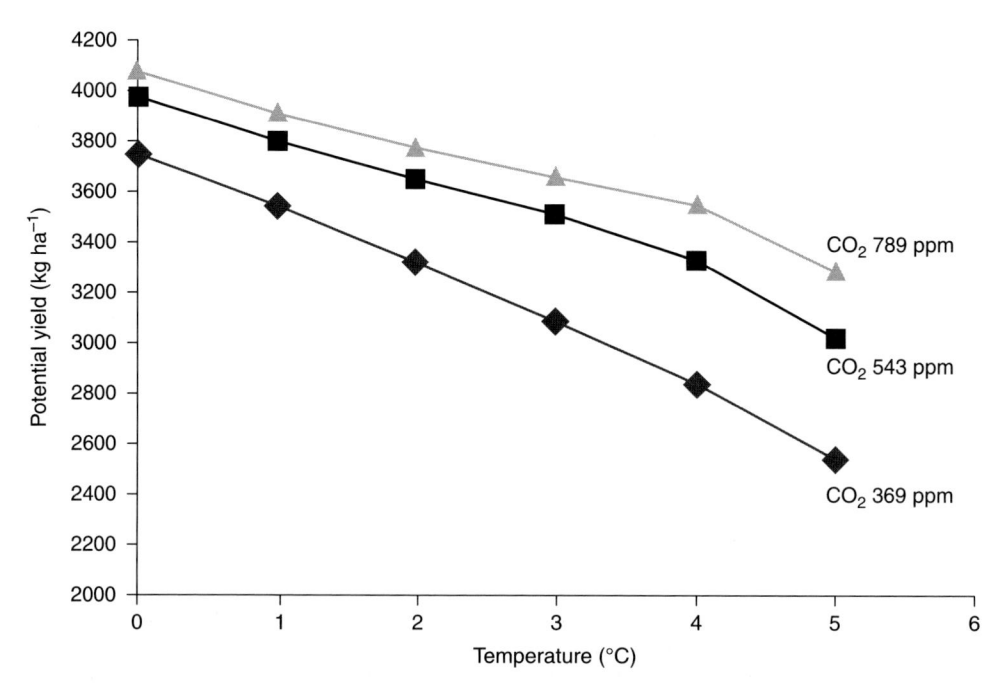

Fig. 28.4. InfoCrop model output showing simulated potential yield of cotton crop in Akola, India. (From Bhatttacharyya et al., 2012.)

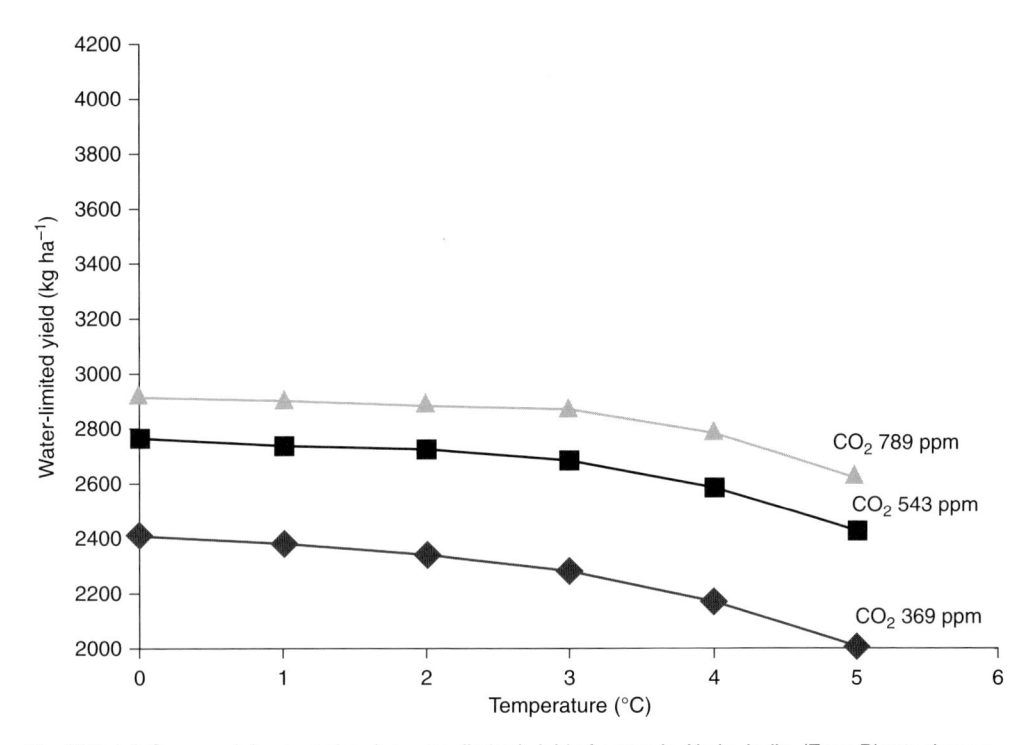

Fig. 28.5 InfoCrop model output showing water-limited yield of cotton in Akola, India. (From Bhatttacharyya et al., 2012.)

Table 28.6. Annual CH_4 emission estimates from rice fields in India. (From Bhatia et al., 2012, with permission from Wiley.)

Reference	Estimate (Tg CH_4 year^{-1})	Methodology used
Ahuja (1991)	37.5	Extrapolated from studies in the USA and Europe to the rice growing regions in India
Cao et al. (1996)	14.4	MEM
Matthews et al. (2000)	2.1	MERES simulation model
Yan et al. (2003)	5.9	Region-specific emission factors
Bhatia et al. (2004)	2.9	IPCC methodology and measured CH_4 emission coefficients
Pathak et al. (2005)	1.5	DNDC model
Gupta et al. (2009)	4.1	State specific CH_4 emission coefficients and IPCC approach
Bhatia et al. (2012)	2.1	Validated InfoCrop model

MEM, methane emission model; MERES, methane emission from rice ecosystem; IPCC, Intergovernmental Panel on Climate Change; DNDC, denitrification decomposition.

4.09 ± 1.19 Tg year^{-1} at the national level, and the trend from 1979 to 2006 was in the range of 3.62 ± 1.0 to 4.09 ± 1.19 Tg year^{-1} (Gupta et al., 2009). Bhatia et al. (2012) reported the simulated methane emission (using the InfoCrop model) from 42.21 Mha of rice cultivation as 2.07 Tg year^{-1}, which was very much similar to the estimate obtained by Matthews et al. (2000) using Methane Emission from Rice Ecosystem (MERES).

The reported emissions due to enteric fermentation and rice cultivation were 6.59 Tg C-CH_4 year^{-1} and 3.07 Tg C-CH_4 year^{-1}, respectively (NATCOM, 2004). The average emission coefficient derived from all categories weighted for the Indian rice crop was 74.05 ± 43.28 kg ha^{-1} (Manjunath et al., 2011). The total mean emission from the rice lands of India was computed as 2.54 Tg C-CH_4 year^{-1}. Methane from Indian agricultural soils is only 0.23% of the global warming caused by the world's CO_2 emission (Bhatia et al., 2004).

General Observations

Out of the two important food-growing regions, the IGP has contributed largely to high levels of crop production compared to the BSR. It was observed that during the post-Green Revolution era, cropping intensity in the dominant states of the IGP increased from 137% (1976/77) to 158% (1999/2000). During the same period, the BSR remained less intensively cultivated, with an increase in cropping intensity from 111% to 123%. Despite this difference, SOC stock of both the soils has increased from 1980 to 2005. However, the increase was more in the IGP than in the BSR. This was due to more biomass turnover to the soils, as evidenced from the increased SOC in the fertilized areas of a long-term experiment (30 years) in the IGP. In addition, an exercise through the GEFSOC modelling system also projected an increase in SOC stock using the LTFE data sets of the IGP. SOC stocks in the BSR indicated an increase, albeit more in the double-cropped areas and also in areas where green manuring was practised, indicating that the prevailing agricultural land uses helped in sequestering more organic carbon in soils of both these regions. The mechanisms involved in preferential accumulation of organic matter in wetland soils under paddy may be ascribed mainly to anaerobiosis and the associated chemical and biochemical changes that take place in submerged soils. It has recently been reported that the SIC:nitrogen ratio is relatively narrow in lowland rice–rice systems, which indicates that the pedo environment in rice soils keeps the deteriorating effect of $CaCO_3$ formation and the concomitant sodicity at bay. Pedogenic $CaCO_3$ formation has been linked with the development of soil sodicity. This sodicity causes chemical soil degradation, indicating poor content of SOC. Pan et al. (1997) indicated that both SOC and SIC were equally important for C transfer and potential CO_2 sequestration. They found a negative correlation between SOC and SIC in arid ecoregions of China. It is now known that, with the adversity of the climate, there will be depletion of organic carbon and C will be sequestered as $CaCO_3$ in the soils (Pan and Guo, 2000; Bhattacharyya et al., 2004). Despite this, the system of intensive agriculture in the IGP and the BSR has increased the SOC stock. In spite of the formation of $CaCO_3$ in the soils, the SOC increase suggests that the prevailing agricultural land uses have been able to enhance or maintain the level of organic carbon in the soils of these two food production zones of the country. Despite the fact that the increase in $CaCO_3$ is a bane for farmers, the increase in SOC has always been possible due to the adoption of suggested management interventions, even in arid and semi-arid environments. However, the rise in $CaCO_3$ (Tables 28.3 and 28.4) warrants a fine-tuning of the existing management interventions. Until then, the status of inorganic C in soils will remain a warning signal for potential soil degradation (Bhattacharyya et al., 2007b).

Acknowledgements

The present work is a compilation of findings of different projects funded by the Global Environment Facility (GEF), the Department of Science and Technology (DST), India, the National Agricultural Technology Project (NATP) and the Indian Council of Agricultural Research (ICAR), New Delhi, India.

References

Abrol, I.P. and Gupta, R.K. (1998) Indo-Gangetic Plains – Issues of Changing Land Use. *LUCC Newsletter*, March 1998, No 3.

Ahuja, D.R. (1991) Estimating regional anthropogenic emissions of greenhouse gases. In: Khoshoo, T.N. and Sharma, M. (eds) *The Indian Geosphere-Biosphere Programme*. Vikas, New Delhi.

Batjes, N.H. (1996) Total carbon and nitrogen in the soils of the world. *European Journal of Soil Science* 47, 151–163.

Benbi, D.K. and Brar, J.S. (2009) A 25 year record of carbon sequestration and soil properties in intensive agriculture. *Agronomy for Sustainable Development* 29, 257–265.

Bhandari, A.L., Ladha J.K., Pathak, H., Padre, A.T., Dawe, D. and Gupta, R.K. (2002) Yield and soil nutrient changes in a long-term rice–wheat rotation in India. *Soil Science Society of America Journal* 66, 162–170.

Bhatia, A., Pathak, H. and Aggarwal, P.K. (2004) Inventory of methane and nitrous oxide emissions from agricultural soils of India and their global warming potential. *Current Science* 87, 317–324.

Bhatia, A., Aggarwal, P.K., Jain, N. and Pathak, H. (2012) Greenhouse gas emission from rice- and wheat-growing areas in India: spatial analysis and upscaling. *Greenhouse Gas Science and Technology* 2, 115–125.

Bhattacharyya, T., Pal, D.K., Velayutham, M., Chandran, P. and Mandal, C. (2000) Total carbon stock in Indian soils: issues, priorities and management. In: *Special Publication of the International Seminar on Land Resource Management for Food, Employment and Environmental Security (ICLRM)*, New Delhi, 8–13 November 2000. ICLRM, All India Soil & Land Use Survey, New Delhi, pp. 1–46.

Bhattacharyya, T., Pal, D.K., Chandran, P., Mandal, C., Ray, S.K., Gupta, R.K. and Gajbhiye, K.S. (2004) *Managing Soil Carbon Stocks in the Indo-Gangetic Plains*, India. Rice–Wheat Consortium for the Indo-Gangetic Plains, New Delhi.

Bhattacharyya, T., Pal, D.K., Lal, S., Chandran, P. and Ray, S.K. (2006a) Formation and persistence of Mollisols on zeolitic Deccan basalt of humid tropical India. *Geoderma* 136, 609–620.

Bhattacharyya, T., Chandran, P., Ray, S.K., Mandal, C., Pal, D.K., Venugopalan, M.V., Durge, S.L., Srivastava, P., Dubey, P.N., Kamble, G.K. et al. (2006b) *Estimation of carbon stocks in red and black soils of selected benchmark spots in semi-arid tropics of India*. Global Theme on Agroecosystems Report No 28. International Crops Research Institute for the Semi-Arid Tropics (ICRISAT), Patancheru, Andhra Pradesh, India.

Bhattacharyya, T., Pal, D.K., Chandran, P., Ray, S.K., Durge, S.L., Mandal, C. and Telpande, B. (2007a) Available K reserve of two major crop growing regions (Alluvial and shrink-swell soils) in India. *Indian Journal of Fertilizers* 3, 41–52.

Bhattacharyya, T., Chandran, P., Ray, S.K., Pal, D.K., Venugopalan, M.V., Mandal, C. and Wani, S.P. (2007b) Changes in levels of carbon in soils over years of two important food production zones of India. *Current Science* 93, 1854–1863.

Bhattacharyya, T., Pal, D.K., Easter, M., Williams, S., Paustian, K., Milne, E., Chandran, P., Ray, S.K., Mandal, C., Coleman, K. et al. (2007c) Evaluating the Century C model using long-term fertilizer trials in the Indo-Gangetic Plains, India. *Agriculture, Ecosystems and Environment* 122, 73–83.

Bhattacharyya, T., Pal, D.K., Chandran, P., Ray, S.K., Mandal, C. and Telpande, B. (2008) Soil carbon storage capacity as a tool to prioritise areas for carbon sequestration. *Current Science* 95, 482–494.

Bhattacharyya, T., Pal, D.K., Williams, S., Telpande, B., Deshmukh, A.S., Chandran, P., Ray, S.K., Mandal, C., Easter, M. and Paustian, K. (2010) Evaluating the Century C model using two long-term fertilizer trials representing humid and semi-arid sites from India. *Agriculture, Ecosystems and Environment* 139, 264–272.

Bhattacharyya, T., Pal, D.K., Deshmukh, A.S., Deshmukh, R.R., Ray, S.K., Chandran, P., Mandal, C., Telpande, B., Nimje, A.M. and Tiwary, P. (2011) Evaluation of RothC model using four long term fertilizer experiments in black soils, India. *Agriculture, Ecosystems and Environment* 144, 222–234.

Bhattacharyya, T., Chandran, P., Venugopalan, M.V., Pal, D.K., Ray, S.K., Mandal, C., Sarkar, D., Tiwary, P., Nimje, A., Dasgupata, D., Balbuddhe, D. and Shaikh, S. (2012) Changes in soil carbon reserves as influenced by different ecosystems and land use in India. Project Completion Report – Network Project on Climate Change (NPCC) ICAR on 'Changes in soil carbon reserves as influenced by different ecosystems and landuse in India'. NBSS&LUP, Nagpur, India.

Biswas, C.R. and Benbi, D.K. (1997) Sustainable yield trends of irrigated maize and wheat in a long-term experiment on loamy sand in semi-arid India. *Nutrient Cycling in Agroecosystems* 46, 225–234.

Bronson, K.F., Cassman, K.G., Wassman, R., Olk, D.C., Noordwijk, M. and van Garrity, D.P. (1998) Soil carbon dynamics in different cropping systems in principal eco-regions of Asia. In: Lal, R.,

Kimble, J.M., Follett, R.F. and Stewart, B.A. (eds) *Management of Carbon Sequestartion in Soil*. CRC, Boca Raton, Florida, pp. 35–57.

Cao, M., Gregson, K., Marshall, S., Dent, J.B. and Heal, O. (1996) Global methane emissions from rice paddies. *Chemosphere* 55, 879–897.

Chivhane, S.P. and Bhattacharyya, T. (2010) Effect of land use and bio-climatic system in organic carbon pool of shrink-swell soils in India. *Agropedology* 20, 145–156.

Eswaran, H. and van den Berg, E. (1992) Impact of building of atmospheric CO_2 on length of growing season in the Indian sub-continent. *Pedologie* 42, 289–296.

Guo, L., Falloon, P., Coleman, K., Zhou, B., Li, Y., Lin, E. and Zhang, F. (2006) Application of the RothC model to the results of long term experiments on typical upland soils in Northern China. *Soil Use Management* 23, 63–70.

Gupta, P.K., Gupta, V., Sharma, C., Das, S.N., Purkait, N., Adhya, T.K., Pathak, H., Ramesh, R., Baruah, K.K., Venkatratnam, L., Singh, G. and Iyer, C.S.P. (2009) Development of methane emission factors for Indian paddy fields and estimation of national methane budget. *Chemosphere* 74, 590–598.

Gupta, R.K. and Rao, D.L.N. (1994) Potential of wastelands for sequestering carbon by reforestation. *Current Science* 66, 378–380.

Houghton, R.A. and Hackler, J.L. (1994) The net flux of carbon from deforestation and degradation in South and Southeast Asia. In: Dale, V. (ed.) *Effects of Land-Use Change on Atmospheric CO_2 Concentrations: South and Southeast Asia as a Case Study*. Springer, New York, pp. 301–327.

Jenny, H. and Raychaudhuri, S.P. (1960) *Effect of Climate and Cultivation on Nitrogen and Organic Matter Reserves in Indian Soils*. ICAR, New Delhi.

Kukal, S.S. and Rehana-Rasool Benbi, D.K. (2008) Soil organic carbon sequestration in relation to organic and inorganic fertilization in rice–wheat and maize–wheat systems. *Soil and Tillage Research* 102, 87–92.

Manjunath, K.R., Panigrahy, S., Adhya, T.K., Beri, V., Rao, K.V. and Parihar, J.S. (2011) Methane emission pattern of Indian rice-ecosystems. *Journal of Indian Society of Remote Sensing* 39, 307–313.

Manna, M.C., Swarup, A., Wanjari, R.H., Singh, Y.V., Ghosh, P.K., Singh, K.N., Tripathi, A.K. and Saha, M.N. (2006) Soil organic matter in a West Bengal Inceptisol after 30 years of multiple cropping and fertilizers. *Soil Science Society of America Journal* 70, 121–129.

Matthews, R.B., Wassmann, R., Knox, J. and Buendia, L.V. (2000) Using a crop/soil simulation model and GIS techniques to assess methane emissions from rice fields in Asia. IV. Up scaling to national levels. *Nutrient Cycling in Agroecosystems* 58, 201–217.

Milne, E., Bhattacharyya, T., Pal, D.K., Paustian, K., Easter, M. and Williams, S. (2006) A system for estimating soil organic carbon stocks and changes at the regional scale – a case study from the Indo-Gangetic Plains, India. In: *Proceedings of the International Conference on Soil, Water and Environmental Quality*. Indian Society of Soil Science, New Delhi, pp. 303–313.

Milne, E., Williams, S., Brye, K.R., Eastern, M., Killian, K. and Paustian, K. (2008) Simulation soil organic carbon in a rice–soybean–wheat–soybean chronosequence in Prairie Country, Arkansas using the Century model. *Journal of Integrated Bioscience* 6, 41–52.

NATCOM (2004) *India's Initial National Communication to the United Nations Framework Convention on Climate Change, Ministry of Environment and Forest (MoEF)*. Government of India, New Delhi.

Pan, G. and Guo, T. (2000) Pedogenic carbonate of acidic soils in China and its significance in carbon sequestration in terrestrial systems. In: Lal, R., Kimble, J.M., Eswaran, H. and Stewart, B.A. (eds) *Global Climate Change and Pedogenic Carbonates*. CRC Press, Boca Raton, Florida, pp.135–148.

Pan, G., Tao, Y.X., Shun, Y.H., Xn, S.Y., Teng, Y.Z. and Han, F.S. (1997) Some features of carbon cycling in humid subtropical Karst region: an example of Guilin Yaji Karst experiment site. *Journal of Chinese Geography* 7, 48–57.

Pathak, H., Li, C. and Wassmann, R. (2005) Greenhouse gas emissions from Indian rice fields: calibration and upscaling using the DNDC model. *Biogeosciences* 2, 113–123.

Sahrawat, K.L., Bhattacharyya, T., Wani, S.P., Chandran, P., Ray, S.K., Pal, D.K. and Padmaja, K.V. (2005) Long-term lowland rice and arable cropping effects on carbon and nitrogen status of some semi-arid tropical soils. *Current Science* 89, 2159–2163.

Singh, M. and Wanjari, R.H. (2009) Annual Report 2008–2009. *All India Coordinated Research Project on Long Term Fertilizer Experiments to Study Changes in Soil Quality, Crop Productivity and Sustainability*. IISS, Bhopal, India, pp. 1–95.

Sinha, S.K., Singh, G.B. and Rai, M. (1998) *Decline in Crop Productivity in Haryana and Punjab: Myth or Reality*. Indian Council of Agricultural Research, New Delhi.

Yan, X., Ohara, T. and Akimoto, H. (2003) Development of region specific emission factors and estimation of methane emission from rice fields in the East, Southeast and South Asian countries. *Global Change Biology* 9, 1–18.

29 Policy Frameworks

Luca Montanarella*, Francesca Bampa and Delphine de Brogniez

Abstract

Policy frameworks concerning soil carbon are rapidly evolving, both in Europe and at the global level. Within Europe, the Roadmap to Resource-Efficient Europe (RRE) (COM (2011) 571 final; EC, 2011a), as well as the implementation of the Soil Thematic Strategy (COM (2012) 46 final; EC, 2012c), highlight the relevance of soil organic carbon (SOC) and the need to reverse its decline in many parts of the European Union (EU). Integration of this concern into several related policies, such as the Common Agricultural Policy (CAP) or the Climate Change Policy in relation to the LULUCF (land use, land-use change and forestry) negotiation process, shows a potential for reverting the current negative trends. The recognition that SOC played a crucial role in the current Multilateral Environmental Agreements negotiated in Rio de Janeiro (United Nations Framework Convention on Climate Change – UNFCCC; United Nations Convention on Biological Diversity – UNCBD; and United Nations Convention to Combat Desertification – UNCCD) was clearly identified in the recent Rio+20 Sustainable Development Conference. Soil, as an important global terrestrial C pool, as well as a large biodiversity reservoir, is gaining attention within the UNFCCC and the UNCBD, while remaining a focus for the UNCCD, traditionally the global convention dealing with soil-related issues. The proposed Sustainable Development Goal of Zero Net Land and Soil Degradation paves the way towards a renewed global effort of soil protection and restoration activities. The framework of the new Global Soil Partnership (GSP) of the Food and Agriculture Organization of the United Nations (FAO) will certainly contribute towards facilitating these recent developments.

Introduction

Why soil organic carbon needs a policy framework

Soils are recognized after oceans as the second largest carbon (C) pool at the global scale. Globally, they contain $c.$2300 Gt of organic carbon (OC) in the top 3 m (Batjes, 1996; Jobbágy and Jackson, 2000; Stockmann et al., 2013). Estimates vary for different soil depths: 1500 Gt of OC are estimates for the first metre and about 615 Gt of OC for the top 20 cm. The distribution of OC along the soil profile depends on plant production, depth, microbial activity and climatic conditions

*E-mail: luca.montanarella@jrc.ec.europa.eu

(Jobbágy and Jackson, 2000; Guo and Gifford, 2002). In turn, soils depend largely on the levels of SOC for delivering major ecosystem services and functions (Schmidt et al., 2011). A soil rich in organic matter is a habitat for a rich biodiversity pool, a buffer for contaminants, a store of water and major nutrient elements for food and biomass production and a regulator layer for gas exchanges with the atmosphere influencing the global greenhouse gases (GHGs) balance (Stockmann et al., 2013). Soils are a source of GHGs (CO_2, CH_4 and N_2O), especially since the shift of attention from drylands to a more global approach to land degradation, usually interlinked with SOC depletion (FAO, 2004).

In June 2012, world leaders gathered at the United Nations Conference on Sustainable Development (Rio+20) to agree on a sustainable goal on land. The proposed Sustainable Development Goal of a 'Zero Net Land and Soil Degradation World' paves the way towards a renewed global effort on land, soil protection and restoration activities, including food security and poverty eradication. The goal needs to be achieved by 2030 and will require the commitment of both public and private sectors (Ashton, 2012).

UNFCCC (LULUCF sector)

Climate change negotiations have been focusing mainly on the reduction of GHG emissions, especially from industrial activities, but also from deforestation and forest degradation (IPCC, 2003). The recent shift to a more holistic approach towards accounting for LULUCF has brought the issue of SOC and its possible role in the negotiation process to the attention of policy makers. In 2011, at COP 17 of the UNFCCC in Durban, South Africa, two major decisions related to SOC were approved. First, Decision 2/CMP.7 has included the new activity, 'Wetland drainage and rewetting' under Article 3.4. of the Kyoto Protocol, which enables the rewetting of organic soils under all land-use categories (independent of whether associated activities – such as cropland management and grazing land management – have been chosen), as an activity accountable for reaching committed emission reductions. Second, Decision 15/CMP.7 explicitly calls on parties to support developing countries in capacity-building activities for establishing the necessary expertise to estimate changes in carbon stock in soils. Indeed, one of the major concerns in including SOC in any global C accounting system has been the difficulty in the monitoring, reporting and verification (MRV) of SOC levels over relatively short periods, especially in countries lacking the necessary technology and knowledge.

The dynamics of OC in soils are complex and slow. Changes in concentration and stocks are therefore difficult to detect and understand. The spatial variability of soil properties, the effect of soil and vegetation management practices and general environmental conditions are important factors to consider when designing a monitoring network (Post et al., 2001; Goidts et al., 2009). In a simulation study, Saby et al. (2008) showed that a time interval of about 10 years would enable the detection of large changes in SOC content in most European countries. The complex feasibility to assess and verify the effect of changes in OC has made the application of stringent verification procedures difficult (Smith, 2004). Effective SOC monitoring is indeed a very costly and time-consuming activity. New emerging technologies, like spectral reflectance methods and remote sensing, are still in the development phase and not yet fully operational (Brown et al., 2006; Stockmann et al., 2013). A more realistic approach is to base verification and accounting on the indirect assessment of SOC by monitoring land-use changes and applying robust SOC modelling techniques (Smith et al., 2005; Álvaro-Fuentes et al., 2009; van Wesemael et al., 2010). First results are promising, and may allow for consistent reporting of SOC at national and global scales. A supplement to the 2006 IPCC guidelines, drafted in 2013, is expected to provide major guidance in addressing specifically organic soils (drained, rewetted and wet, be it under forest, cropland, grassland, wetlands or whatever land-use category), and may be adopted at UNFCCC

COP 19. Together with the supplement to the 2003 IPCC Good Practice Guidance, translating the decisions made by the UNFCCC, it will provide a new formal reference for future SOC accounting, if adopted at UNFCCC COP 19.

CBD

Many areas of the world host a richer biodiversity below ground than above ground in terms of abundance, numbers of species and functions of organisms. The dynamic equilibrium of SOC under unchanged land use is driven by the active soil biodiversity pool. For this reason, any disturbance of SOC levels immediately affects soil biota. In tropical latitudes, Midgley et al. (2010) demonstrated that high C levels corresponded to a high biodiversity system. However, no clear global relationship between soil biodiversity and C sequestration has yet been established. Given that many soil organisms remain unknown or unclassified, there is a possibility that a large unknown part of the soil biodiversity has already been irreversibly lost worldwide due to rapid SOC depletion. Soil biota perform a wide variety of processes and functions, including organic matter decomposition, nutrient cycling, soil structure formation, pest regulation and bioremediation of contaminants. In turn, various ecosystem services such as food production, climate regulation and water provision benefit from a rich soil biodiversity pool (Dominati et al., 2010; Pulleman et al., 2012). It is therefore fundamental to complete the full assessment of global soil biodiversity resources as well as to implement an effective strategy for protecting endangered soil species. The CBD, as a contribution to the achievement of the Millennium Development goals, has already recognized the importance of an *Agricultural biological diversity*[1] in the Fifth Meeting of the Conference of the Parties held in Kenya, Nairobi (COP5 Decision V/5) and has established in the Eighth Meeting held in Brazil, Curitiba the *International Initiative for the Conservation and Sustainable Use of Soil Biodiversity*[2] (COP8 Decision VIII/23), coordinated by the FAO. The Global Soil Biodiversity Initiative,[3] launched in September 2011, should further support the work of the FAO in this sense in order to develop a platform for promoting soil biodiversity into environmental policy and sustainable land management for the protection and enhancement of ecosystem services.

UNCCD

The UNCCD was originally negotiated, in 1994, to address land degradation in drylands (arid, semi-arid and dry subhumid areas), especially in sub-Saharan Africa, one of the most vulnerable ecosystems (FAO, 2004). Only in recent years has the Convention been realigning its strategy towards addressing land degradation in other parts of the world too (*10-Year Strategy of the UNCCD, 2008–2018*). The transition from a regional towards a global focus was well reflected by the ratification of nearly all countries in the world, all aiming to reverse global land degradation trends. The SOC pool tends to decrease exponentially with temperature. Consequently, SOC content in drylands (usually smaller than 1%) is recognized as a parameter that reflects degradation and desertification trends (Lal, 2000, 2002). However, despite the relevance of SOC in monitoring and assessing desertification, only little attention has been given to this parameter in recent policy assessments. Including SOC as a relevant indicator within the regular reporting system of the UNCCD would certainly raise the attention to soils in the Convention negotiation process, and also allow for developing synergies with the CBD and UNFCCC. The potential for such synergies has already been well recognized within the Millennium Ecosystem Assessment (Texier, 2005).

Global Soil Partnership of the FAO

With their lengthy negotiation processes, binding multilateral environmental agreements have shown their limits during the past 20 years. In the *Rio+20 Sustainable Development Conference*,[4] a new approach to sustainability and environmental issues has been put forward, based on partnerships and 'coalitions of the willing'. A large community of stakeholders and decision makers is in favour of more effective measures to protect natural resources for future generations, and is willing to form voluntary partnerships to move forward with more ambitious agendas.

A new initiative, a Global Soil Partnership[5] (GSP), has been put forward by the FAO, with the strong support of the European Commission, to address the sustainable management of global soil resources and federate all stakeholders and parties that are willing to move on with effective soil protection measures. The partnership should establish a more effective science–policy interface, addressing policy-relevant scientific and technical issues related to soils (Montanarella and Vargas, 2012).

SOC is a key element of the GSP. Sustainable soil management practices promoted by the GSP will protect current SOC resources from further depletion and will allow for increased SOC levels in the long term. The GSP is in the beginning of its implementation, following its official approval at the end of 2012 by the FAO Council. The planned Intergovernmental Panel on Land and Soil (ITPS) will become the main scientific reference for future policy making in relation to soils and SOC management. Full implementation of the GSP, started in 2013, will hopefully lead to a new approach to soil protection at global and regional scales.

SOC in European Legislative Frameworks

SOC is a crosscutting issue entering many different EU policy frameworks. The EU counts for more than 70 billion t of OC in soils. For this reason, it is necessary to promote practices that favour maintaining or even increasing SOM levels (SoCo, 2009). In Europe, we have observed a decline of organic matter levels and the need to reverse this negative trend is recognized (Rusco et al., 2001; EC, 2006b).

The Soil Thematic Strategy (COM (2006) 231 final, COM (2006) 232 and COM (2012) 46)

The Thematic Strategy for Soil Protection (COM (2006) 231 final; EC, 2006b) outlines the overall strategy concerning soils in the EU. The Thematic Strategy and its related proposal for an EU Soil Framework Directive (COM (2006) 232; EC, 2006a) have been reviewed recently by the European Commission (COM (2012) 46; EC, 2012c). The Strategy states that soils are a non-renewable resource subject to a series of degradation processes or threats: erosion, decline in OM, local and diffuse contamination, sealing, compaction, decline in biodiversity, salinization, landslides (EC, 2006a,b, 2012).

SOC management is at the core of the strategy to protect SOC as one of the main soil properties. The proposed Soil Framework Directive calls for the delineation of areas in Europe threatened by SOC decline and for the establishment of appropriate measures to reverse the negative trend. Around 45% of the soils in Europe have low or very low OC content (from 0 to 2%) and 45% have a medium content (from 2 to 6%). Several factors are responsible for the decline in SOM, and many of them relate to human activity (Van-Camp et al., 2004; Zdruli et al., 2004). According to the report from the EEA (2010), these factors include conversion of grassland, forests and natural vegetation to arable land; deep ploughing of arable soils; drainage, liming, nitrogen (N) fertilizer use; and tillage of peat soils and unsustainable crop rotations without temporary grasslands. The directive is still in its discussion phase within the Council, with some EU Member States still voicing strong opposition to the proposed legislative framework. Nevertheless, the Soil Framework Directive is one of the elements of the EU Thematic Strategy that also includes the systematic integration of soil protection elements in other related EU legislative instruments, such as the Common

Agricultural Policy (CAP), the climate change policy (Adaptation Strategy, LULUCF Accounting), natural resource management (Roadmap for a Resource Efficient Europe (EC, 2011a)) and waste policy (Biowaste).

Common Agricultural Policy (CAP)

A substantial proportion of land in Europe is occupied by agriculture, and consequently this sector plays a crucial role in natural resources protection. Adopting inefficient and non-sustainable management practices, including land abandonment, has an adverse impact on natural resources. Efficient agricultural management can maintain the same rate in food production while having a positive effect on the state of soil. By adopting C sequestration practices, SOC concentration can be maintained and sometimes enhanced (Lal, 2006, 2011). Agricultural production is very sensitive to climatic variations and its management needs to adapt accordingly in order to ensure sufficient food production.

The Common Agricultural Policy (CAP) is the agricultural and rural development policy of the EU concerned with ensuring sufficient food at reasonable and stable prices. Its main goal is to ensure food security, but nowadays the CAP is designed to meet a wide range of needs, including the maintenance of farm incomes and good farming practices, enhancing food quality and promoting animal welfare. In particular, the interaction between agriculture and the environment is integrated by the concept of sustainable agriculture. In this way, the management of agroecosystems ensures benefits also for the future. With respect to soil protection, the CAP contributes to preventing and mitigating soil degradation in order to build up SOM, enhance soil biodiversity and reduce soil erosion, contamination and compaction. Due attention has been given to SOC in the definition of Good Agricultural and Environmental Conditions (GAEC), as referred to in Article 6 of Regulation No 73/2009 establishing the rules for direct support scheme payments for farmers (EC, 2009). These GAEC criteria explicitly include a reference to the need to maintain OM levels through appropriate good agricultural and environmental practices. Farmers receiving direct support through the CAP need to comply with these GAECs (commonly referred to as Cross-compliance Scheme). Monitoring and verification of the impact of this scheme on European agricultural soils has proven difficult, certainly with respect to organic soils, for which the proposed monitoring procedures are unsuitable, as they focus on the topsoil layer only. New tools are needed for cost-effective SOC monitoring from the European to farm level (EC, 2011b, 2012a). The EC is actually preparing the ground for the new CAP towards 2020. The new policy for the period 2014–2020 wants to assure: viable food production, sustainable management of natural resources and climate action, and a balanced territorial development. The last Communication on the CAP towards 2020 remarks how farming practices could limit soil depletion, water shortages, pollution, C losses and loss of biodiversity.

Roadmap to resource-efficient Europe (COM (2011) 571 final)

The Resource Efficiency Roadmap is part of the Resource Efficiency Flagship of the *Europe 2020 Strategy* (EC, 2011a). The Roadmap aims to set out a framework for the design and implementation of future actions in resource efficiency. It also outlines the structural and technological changes in the EU's economy needed by 2050, including sustainability milestones to be reached by 2020. It proposes ways to increase resource productivity, to decouple economic growth from resource use and its environmental impact, and it investigates how policies interrelate and build on each other. Areas where policy action can make a real difference are of particular focus, and specific bottlenecks, like inconsistencies in policy and market failures, are tackled to ensure that policies are all going in the same direction. The framework for actions comprehends many policy areas, such as climate change, energy, transport, industry, raw materials, agriculture, fisheries, biodiversity and regional development. The soil and land related milestone indicates that by 2020 EU policies must take into account their direct and indirect impact on land use in the EU and globally and that the rate of land take is on track with the aim to achieve zero net land take by 2050, soil erosion is reduced and soil organic matter increased, with remedial work on contaminated sites well under way (EC, 2011a).

The European environment – state and outlook 2010 (EEA, 2010)

In 2010, the European Environment Agency (EEA) published a report on the state of the European Environment, outlining the influence of climate on European land. As regards the EU-27 and other industrialized countries, the EEA declared that GHG emission cuts of 25–40% by 2020 and 80–95% by 2050 were needed. The main sources of GHG emissions globally are the burning of fossil fuels for electricity generation, transport, industry and households – which together account for about two-thirds of total global emissions. Other sources include deforestation – which contributes about one-fifth – agriculture, land-filling of waste and the use of industrial fluorinated gases. The projected impacts of climate change are expected to vary considerably across Europe, with pronounced events expected in the Mediterranean basin, northwestern Europe, the Arctic and mountainous regions. Regarding soil ecosystems, information on the impacts of climate change is very limited, but changes will likely be due to projected rising temperatures, changing precipitation intensity and frequency and more severe droughts. Such changes can lead to a decline in SOM content and an increase in CO_2 emissions (EEA, 2010).

The main messages from the EEA 2010 State of Environment Report concerning SOC were:

- State of soil organic carbon levels: around 45% of the mineral soils in Europe have low or very low OC content (0–2%), and 45% have a medium content (2–6%) (Rusco et al., 2001). The problem exists in particular in the southern countries of Europe, where 74% of the land is covered by soils that have less than 2% of OC in the topsoil (0–30 cm) (Zdruli et al., 2004). However, low levels of OM are not restricted to southern Europe, as they are also observed in France, the UK, Germany, Norway and Belgium.
- There is growing realization of the role of soil, in particular peat, as a store of C and its role in managing terrestrial fluxes of atmospheric CO_2.
- Other than in tropical ecosystems, soil contains about twice as much OC as aboveground vegetation. SOC stocks in the EU-27 are estimated to be around 75 billion t, of which about 50% is in Sweden, Finland and the UK because of their large areas of peatlands and forest soils (EC, 2008b).
- Peat soils represent the highest concentration of organic matter in all soils. Peatlands are currently under threat from unsustainable practices such as drainage, clearance for agriculture, fires and extraction.

The European peatlands are estimated to cover about 52 Mha (Joosten and Clarke, 2002; Joosten, 2009), of which about 31 Mha occur in the northern latitudes (EC, 2008b). The total C storage of European peatlands is estimated at 42 billion t, accounting for 10–15% of the C stock in northern peatlands and about 20% of the European SOC stock (Aertsens et al., 2013). Almost one-third of European peatlands are in Finland, and more than one-quarter are in Sweden. The remainder are in Poland, the UK, Norway, Germany, Ireland, Estonia, Latvia, the Netherlands and France. Small areas of peat and peat-topped soils also occur in Lithuania, Hungary, Denmark and the Czech Republic (Montanarella et al., 2006). There has been an estimated loss of 50% of wetlands globally since 1900, due mainly to intensive management (EEA, 2010). There is no harmonized exhaustive inventory of peat stocks in Europe; for example, in the

CLIMSOIL report it is estimated that more than 65,000 km² or 20% of all peatlands have been drained for agriculture, almost 90,000 km² or 28% for forestry and 2273 km² or 0.7% for peat extraction.

The lack of availability of reliable data on the area of peat soils in agricultural use and whether it is grassland or arable land is caused primarily by land-use changes and degradation of peat soils that have turned into mineral soils following oxidation of the peat (EC, 2008b). The large amount of C stored in peatland justifies a prominent place of this specific landform in the European climate change debate. The result is that the most effective option to manage soil C is to preserve the existing rich stocks of OM in peat soils. Peatlands degradation and drainage, due to climate change and land management, has become a relevant concern because of the deleterious effects on this specific ecosystem and the associated GHG release (CO_2, N_2O and CH_4).

Virgin peatlands, called *mires*, accumulate atmospheric C and N_2O, but emit CH_4 (Jandl et al., 2011). The extensive intact European mires are still functioning as sinks for C. For this reason, a main goal for Europe is to monitor these organic soils and plan their effective protection. The difficulties are mainly in collecting data on land use in peatlands, and particularly on their drainage status for forestry and agriculture. The debate about the temperature sensitivity of SOC decomposition should be broadened specifically to include peatlands. Because of their high sensitivity to temperature changes, during the next few decades these lands may mobilize large stocks of C. For this reason, high research priority should be given to decomposition constraints in these environments and their feedbacks with climate (Davidson and Janssens, 2006).

Conclusions

Policymakers recognize SOC as relevant for several policy areas: climate change, agriculture, environmental protection, waste management, biodiversity, energy, etc. There is no single framework addressing SOC at the global or regional scale. Existing multilateral environmental agreements (MEAs) address SOC from different perspectives. There is clear need for a more coordinated and coherent approach to SOC management and related policies. The recently proposed GSP by the FAO has the potential to allow for voluntary coordination on SOC management at the global scale. Nevertheless, without proper financial incentives, it will be difficult to implement an effective SOC management policy at the local scale. Within the EU, there could be scope for the introduction of an effective incentive system, but it will require a reliable monitoring, verification and reporting system at the farm level. Given the current cost associated with effective SOC measurements *in situ*, it may be difficult to implement such a system at the continental scale. A more realistic approach could be to consider land-use changes and the related effects on SOC levels through a series of standardized conversion factors, potentially derived locally from a network of benchmark sites. Some ongoing projects could lead in the long term to the establishment of such a network of reference sites for the European Union.

Notes

[1] http://unfccc.int/
[2] http://www.cbd.int
[3] http://www.unccd.int/
[4] http://www.cbd.int/decision/cop/?id = 7147
[5] http://www.cbd.int/decisions/cop/?m = cop-08
[6] http://www.globalsoilbiodiversity.org/
[7] http://www.uncsd2012.org/
[8] http://www.fao.org/globalsoilpartnership/home/en/

References

Aertsens, J., De Nocker, L. and Gobin, A. (2013) Valuing the carbon sequestration potential for European agriculture. *Land Use Policy* 31, 584–594.

Alvaro-Fuentes, J., López, M.V., Arrúe, J.L., Moret, D. and Paustian, K. (2009) Tillage and cropping effects on soil organic carbon in Mediterranean semiarid agroecosystems: testing the Century model. *Agriculture, Ecosystems and Environment* 134, 211–217.

Ashton, R. (2012) *Zero Net Land Degradation. A Sustainable Development Goal for Rio+ 20.* UNCCD, Bonn, Germany.

Batjes, N.H. (1996) Total carbon and nitrogen in the soils of the world. *European Journal of Soil Science* 47, 151–163.

Brown, D.J., Shepherd, K.D., Walsh, M.G., Dewayne Mays, M. and Reinsch, T.G. (2006) Global soil characterization with VNIR diffuse reflectance spectroscopy. *Geoderma* 132, 273–290.

Davidson, E.A. and Janssens, I.A. (2006) Temperature sensitivity of soil carbon decomposition and feedbacks to climate change. *Nature* 440, 165–173.

Dominati, E., Patterson, M. and Mackay, A. (2010) A framework for classifying and quantifying the natural capital and ecosystem services of soils. *Ecological Economics* 69, 1858–1868.

EC (European Commission) (2006a) (COM (2006) 232) Proposal for a Directive of the European Parliament and of the Council Establishing a Framework for the Protection of Soil and Amending Directive 2004/35/EC. EC, Brussels.

EC (2006b) (COM (2006) 231 final) Communication from the Commission to the Council, the European Parliament, the European Economic and Social Committee of the Regions, Thematic Strategy for Soil Protection. EC, Brussels.

EC (2008a) (COM (2008a) 19 final) Proposal for a Directive of the European Parliament and of the Council on the Promotion of the Use of Energy from Renewable Sources. EC, Brussels.

EC (2008b) Review of Existing Information on the Interrelations between Soil and Climate Change (CLIMSOIL) – Final Report. Contract number 70307/2007/486157/SER/B1:208. R. Schils. EC, Brussels.

EC (2009) (EC No 73\2009) Council Regulation establishing common rules for direct support schemes for farmers under the common agricultural policy and establishing certain support schemes for farmers, amending Regulations (EC) No 1290/2005 (EC) No 247/2006 (EC) No 378/2007 and repealing Regulation (EC) No 1782/2003. Official Journal of the European Union, Brussels.

EC (2011a) (COM (2011a) 571 final) Communication from the Commission to the European Parliament, the Council, the European Economic and Social Committee and the Committee of the Regions, Roadmap to a Resource Efficient Europe. EC, Brussels.

EC (2011b) (SEC (2011) 1153 final/2) Commission staff working paper. Impact assessment Common Agricultural Policy towards 2020 accompanying the document proposals for a Regulation of the European Parliament and of the Council. EC, Brussels.

EC (2012a) (COM (2010) 672 final) Communication from the Commission to the European Parliament, the Council, the European Economic and Social Committee and the Committee of the Regions, The CAP towards 2020: meeting the food, natural resources and territorial challenges of the future. EC, Brussels.

EC (2012b) (COM (2012a) 93 final) Proposal for a Decision of the European Parliament and of the Council on accounting rules and action plans on greenhouse gas emissions and removals resulting from activities related to land use, land use change and forestry. EC, Brussels.

EC (2012c) (COM (2012b) 46 final) Report from the Commission to the European Parliament, the Council, the European Economic and Social Committee and the Committee of the Regions, The implementation of the Soil Thematic Strategy and on going activities. EC, Brussels.

EC (2012d) (COM (2012c) 595 final) Proposal for a directive of the European Parliament and of the Council amending Directive 98/70/EC relating to the quality of petrol and diesel fuels and amending Directive 2009/28/EC on the promotion of the use of energy from renewable sources. EC, Brussels.

EC (2012e) (COM (2012d) 94 final) Communication from the Commission to the European Parliament, the Council, the European Economic and Social Committee and the Committee of the Regions, Accounting for land use, land use change and forestry (LULUCF) in the Union's climate change commitments. EC, Brussels.

EC (2012f) (SWD (2012e) 41 final) Commission Staff Working Document, Impact Assessment on the role of land use, land use change and forestry (LULUCF) in the EU's climate change commitments, Accompanying the document Proposal for a Decision of the European Parliament and of the Council on

Accounting Rules And Action Plans on Greenhouse Gas Emissions and Removals Resulting from Activities Related to Land Use, Land Use Change and Forestry. EC, Brussels.

EEA (European Environment Agency) (2010) *The European Environment — State and Outlook 2010: Synthesis*. European Environment Agency, Copenhagen.

FAO (Food and Agriculture Organization) (2004) *Carbon Sequestration in Dryland Soils*. World Soils Resources Reports, 102. Food and Agriculture Organization of The United Nations, Rome.

Goidts, E., van Wesemael, B. and Crucifix, M. (2009) Magnitude and sources of uncertainties in soil organic carbon (SOC) stock assessments at various scales. *European Journal of Soil Science* 60, 723–739.

Guo, L.B. and Gifford, R.M. (2002) Soil carbon stocks and land use change: a meta analysis. *Global Change Biology* 8, 345–360.

IPCC (Intergovernmental Panel on Climate Change) (2003) In: Penman, J., Gytarsky, M., Krug, T., Kruger, D., Pipatti, R., Buendia, L., Miwa, K., Ngara, T., Tanabe, K. and Wagner F. (eds) *Good Practice Guidance for Land Use, Landuse Change and Forestry (GPG-LULUCF)*. IPCC-IGES, Kanagawa, Japan.

Jandl, R., Rodeghiero, M. and Olsson, M. (2011) *Introduction to Carbon in Sensitive European Ecosystems: From Science to Land Management*, Wiley, Chichester, UK.

Jobbágy, E. and Jackson, R.B. (2000) The vertical distribution of soil organic carbon and its relation to climate and vegetation. *Ecological Applications* 10, 423–436.

Joosten, H. (2009) *The Global Peatland CO_2 Picture, Peatland Status and Drainage Related Emissions in All Countries of the World*. Wetlands International, Ede, the Netherlands.

Joosten, H. and Clarke, D. (2002) *Wise Use of Mires and Peatlands*. International Mires Conservation Group and the International Peat Society, Finland.

Lal, R. (2000) Carbon sequestration in drylands. *Annals of Arid Zone* 39, 1–10.

Lal, R. (2002) Carbon sequestration in dryland ecosystems of West Asia and North Africa. *Land Degradation and Development* 13, 45–59.

Lal, R. (2006) Enhancing crop yields in the developing countries through restoration of the soil organic carbon pool in agricultural lands. *Land Degradation and Development* 17, 197–209.

Lal, R. (2011) Sequestering carbon in soils of agro-ecosystems. *Food Policy* 36(Suppl. 1), S33–S39.

Midgley, G.F., Bond, W.J., Kapos, V., Ravilious, C., Scharlemann, J.P.W. and Woodward, F.I. (2010) Terrestrial carbon stocks and biodiversity: key knowledge gaps and some policy implications. *Current Opinion in Environmental Sustainability* 2, 264–270.

Montanarella, L. and Vargas, R. (2012) Global governance of soil resources as a necessary condition for sustainable development. *Current Opinion in Environmental Sustainability* 4, 559–564.

Montanarella, L., Jones, R.J.A. and Hiederer, R. (2006) Distribution of peatland in Europe. *Mires and Peat* 1, 1–10.

Post, W.M., Izaurralde, R.C., Mann, L.K. and Bliss, N. (2001) Monitoring and verifying changes of organic carbon in soil. *Climatic Change* 51, 73–99.

Pulleman, M., Creamer, R., Hamer, U., Helder, J., Pelosi, C., Pérès, G. and Rutgers, M. (2012) Soil biodiversity, biological indicators and soil ecosystem services – an overview of European approaches. *Current Opinion in Environmental Sustainability* 4, 529–538.

Rusco, E., Jones, R.J. and Bidoglio, G. (2001) *Organic Matter in the Soils of Europe: Present Status and Future Trends*. EUR 20556 EN. Publications Office of the European Union, European Commission Joint Research Centre, Luxembourg.

Saby, N.P.A., Bellamy, P.H., Morvan, X., Arrouays, D., Jones, R.J.A., Verheijen, F.G.A., Kibblewhite, M.G., Verdoot, A., Uveges, J.B., Freudenschuss, A. and Simota, C. (2008) Will European soil-monitoring networks be able to detect changes in topsoil organic carbon content? *Global Change Biology* 14, 2432–2442.

Schmidt, M.W., Torn, M.S., Abiven, S., Dittmar, T., Guggenberger, G., Janssens, I.A., Kleber, M., Kögel-Knabner, I., Lehmann, J., Manning, D.A. *et al*. (2011) Persistence of soil organic matter as an ecosystem property. *Nature* 478, 49–56.

Smith, P. (2004) Monitoring and verification of soil carbon changes under Article 3.4 of the Kyoto Protocol. *Soil Use and Management* 20(Suppl.), 264–270.

Smith, P., Martino, D., Cai, Z., Gwary, D., Janzen, H., Kumar, P., McCarl, B., Ogle, S., O'Mara, F., Rice, C. *et al*. (2008) Greenhouse gas mitigation in agriculture. *Philosophical Transactions of the Royal Society B: Biological Sciences* 363, 789–813.

SoCo (2009) *Addressing Soil Degradation in EU Agriculture: Relevant Processes, Practices and Policies*. Report on the project Sustainable Agriculture and Soil Conservation (SoCo). European Communities, Luxembourg.

Stockmann, U., Adams, M.A., Crawford, J.W., Field, D.J., Henakaarchchi, N., Jenkins, M., Minasny, B., McBratney, A.B., de Remy de Courcelles, V., Singh, K. *et al*. (2013) The knowns, known unknowns and unknowns of sequestration of soil organic carbon. *Agriculture, Ecosystems and Environment* 164, 80–99.

Texier, C. (2005) *Millennium Ecosystem Assessment. Ecosystems and Human Well-being: Synthesis*. Island Press, World Resources Institute, Washington, DC.

Van-Camp, L., Bujarrabal, B., Gentile, A.R., Jones, R.J.A., Montanarella, L., Olazabal, C. and Selvaradjou, S.-K. (2004) *Reports of the Technical Working Groups Established under the Thematic Strategy for Soil Protection*. EC, 872. Office for Official Publications of the European Communities, Luxembourg.

van Wesemael, B., Paustian, K., Meersmans, J., Goidts, E., Barancikova, G. and Easter, M. (2010) Agricultural management explains historic changes in regional soil carbon stocks. *Proceedings of the National Academy of Sciences of the United States of America* 107, 14926–14930.

Zdruli, P., Jones, R.J.A. and Montanarella, L. (2004) *Organic Matter in the Soils of Southern Europe*. European Soil Bureau Technical Report, EUR 21083 EN. Office for Official Publications of the European Communities, European Commission, Luxembourg.

30 National Implementation Case Study: China

Yongcun Zhao*

Abstract

As a developing country with limited arable land and a large population, governance of soil carbon in China has to face a dual challenge, where both maintaining a steady increase in crop production for ensuring adequate food supplies and addressing environmental problems raised by rapid industrialization and agronomic development must be satisfied simultaneously. In this chapter, the possible approaches for soil carbon governance in China such as land management, agricultural management practice, forestry activity and pasture management and recovery of degraded land are reviewed, and the implementation of a soil testing and fertilizer recommendation project, a fertile soil project, conservation tillage and crop residue returning, as well as an ecological construction project for sequestrating carbon in the soils of China is explored. Moreover, funding and technology limitation, notion and knowledge gap and policy challenge are also discussed in the chapter.

Introduction

The terrene environment is the most important living space for humans, and soils are at the core of terrestrial ecosystems. In terrestrial ecosystems, the soil carbon reservoir is nearly three times as large as carbon storage for vegetation and twice as large as atmospheric carbon storage. In addition, a slight change in soil carbon may affect the concentration of greenhouse gases in the atmosphere, amplifying global change. Therefore, governance of carbon in soils is extremely important for mitigating global climate change.

So far, the most accurate estimate of soil organic carbon (SOC) storage for China is 89.14 Pg C (upper 1 m; Yu *et al.*, 2007), accounting for approximately 6% of global SOC storage (1500–1550 Pg C; Jobbagy and Jackson, 2000; Lal, 2004). As a developing country with limited arable land and a large population, however, rational governance of soil carbon for China is extremely important, because the dual challenge of maintaining a steady increase in crop production for ensuring adequate food supplies and addressing environmental problems raised by rapid industrialization and agronomic development must be satisfied simultaneously. This chapter reviews the possible approaches for soil carbon governance in China and the current national action and policy for soil carbon governance, and discusses the difficulties and challenges for soil carbon sequestration in China.

*E-mail: yczhao@issas.ac.cn

Approaches for Soil Carbon Governance in China

Land use and land cover

Land-use/land-cover change is a critical factor affecting the storage of carbon in terrestrial ecosystems. Transformation from natural woodland, meadow and wetland to cropland may cause soil carbon losses, while conversion of cropland to vegetable fields in China sequestrates SOC due to the high inputs of fertilizer and manure in vegetable fields (Zhang et al., 2006, 2007; Kong et al., 2006). In China, vegetable fields covered only 2.8% of the arable land in 1978, whereas this figure increased to 16.4% in 2007, according to the *China Statistical Yearbook* (2008) (National Bureau of Statistics of China, 2008). This increase has turned into an important driving factor affecting the sequestration of SOC (Liu et al., 2012). Conversion of uplands to paddy fields can also increase soil carbon content. For example, the acreage of paddy fields in the Jiangsu Province of China increased by 1.73 million ha (Mha) from 1949 to 1998, and 17 Tg C was sequestrated in soils (Pan et al., 2003).

Agricultural management practice

Conservation tillage, fertilization and water management and the optimization of cropping systems are major approaches for sequestrating carbon in cropland soils of China. Conservation tillage is a tillage system that reduces loss of soil or water relative to conventional tillage (Mannering and Fenster, 1983), and it is usually a combination of reduced tillage, no tillage and straw mulching. Utilizing the DNDC model (denitrification–decomposition model; for additional information on the DNDC model, please refer to Li et al., 1994), the carbon sequestration potential (years 2009–2050, with 2008 as the baseline) for China's paddy soils is estimated at 239 Tg C under the condition of reduced tillage and 415 Tg C with no tillage, 437 Tg C with a crop residue return rate of 50%, 91 Tg C with a farmyard manure incorporation rate of 200% and 465 Tg C with a crop residue return rate of 30% combined with reduced tillage (Xu et al., 2011). Moreover, fertilization and water management and the optimization of rotation and cropping systems have also often been used jointly with the conservation tillage technique, which may amend the soil's physical properties, reduce possible soil erosion and thereby improve soil fertility and SOC content.

Forestry activity and pasture management

Forestry activities such as forestation, reforestation, restoration of degraded ecosystems and establishment of agroforestry ecosystems are the most effective approaches to increase carbon sequestration in the vegetation and soils in China (Zhang et al., 2005). The percentages of forest cover in China increased from 16.55% in 1998 to 18.21% in 2003 (State Forestry Administration of China, 2005), which largely benefited the accumulation of carbon in forestry soils. Moreover, the rapid restoration of the shrub-covered area also made important contributions to soil carbon sequestration (Huang et al., 2010). Fenced-in grazing (the pasture is fenced into many subpastures during grazing, each grazed for a short period and then given adequate rest periods for regrowth), artificial planting of grass, returning farmland to grassland, fertilization, irrigation and grazing management are the main methods for carbon sequestration in pasture soils. In Inner Mongolia, Tibet and Xinjiang (China), SOC storage can be increased by 4561 Tg C through reducing grazing pressure, on the assumption that all degraded grassland (55% grassland in the area) can be fully recovered, and, utilizing similar methods, the annual sequestration rate of SOC under the condition of artificial planting of grass, returning farmland to grassland and fenced-in grazing can be estimated at 25.6, 1.5 and 12.0 Tg C year^{-1}, respectively (Guo et al., 2008).

Recovery of degraded land

Vegetation recovery is still the most effective and feasible method for enhancing the carbon sequestration potential of degraded land in China (Peng et al., 2005). Preliminary estimates by Lin et al. (2005) showed that the annual sequestration rate of soil carbon through erosion control could achieve 313 Tg C year^{-1} in China, and the restoration of degraded land and the control of desertification could reach 1.8 Tg C year^{-1}. Moreover, with the expectation of agricultural carbon trading and compensation mechanisms, the application of carbon capture and biochar techniques in agriculture also seems economical and feasible.

Current National Action and Policy for Soil Carbon Governance

Soil testing and fertilizer recommendation (STFR) project

The STFR project initiated by the Ministry of Agriculture of China (MOA) aims at improving the soil organic matter (SOM) content in China's croplands through a rational formula of fertilizer, deep fertilization instead of surface fertilization and the joint use of organic fertilizer and chemical fertilizer. The STFR project was started in 2005, and the 'Technical specification of balanced fertilization by soil testing' (NY/T 1118-2006) was distributed in 2006 (MOA, 2006) for guiding and normalizing the STFR project that was implemented in each county. A total of 2498 counties (all agricultural counties in China) were covered by the STFR project until the end of 2009, and over 90% of grain crops in the counties were grown using the STFR technology. Unsuitable levels of fertilization have been decreased by 3,000,000 t and the efficiency of fertilizer use has increased by 3% due to the implementation of the STFR project over the past 5 years (MOA, 2009). The STFR project also largely benefited the sequestration of carbon in agricultural soils through the improvement of SOM content and the reduction of N_2O emissions caused by excessive fertilization (Huang, 2006).

The fertile soil (FS) project

Until the 1950s, agriculture in China depended mainly on organic fertilizer to provide nutrients. The FS project (2003–2007) proposed by the MOA is a soil fertility project that aims at ensuring the security of national food supply through the increase of organic fertilizer application, improvement of fertilizer use efficiency, soil amelioration and the prevention of soil degradation in basic farmland protection and demonstration areas of China. Balanced fertilization, comprehensive utilization of organic fertilizer and soil fertility improvement are the main agricultural extension technologies that are used in the project (Lin et al., 2005). The SOM content was increased largely through the implementation of the FS project, because of the increase in organic fertilizer input (annual increasing rate for organic fertilizer input is 5% and for areas with crop residue returning it is 10%), the improvement of organic fertilizer quality and the extension of the above-mentioned soil fertility improvement technology (Lin et al., 2005).

Conservation tillage (ConsT) and crop residue returning (CR)

The long-term use of ConsT practices would be beneficial to improve SOC content and soil fertility. Experimental research and extension of ConsT in China was started in 1960s (Wang et al., 2006). The promotion of ConsT in dry farmland demonstration areas has been extended by the MOA since 2002. A total of 167 ConsT demonstration counties in 15 provinces of northern China had been set up by the central government and 262 demonstration counties had been set up by the provincial government by the end of 2006 (MOA, 2007). The total area of ConsT demonstration and extension is as large as

1,360,000 ha. There are around 67 Mha of dry farmland in China suitable for using ConsT technology. The ConsT-covered area, however, is only about 1.36 Mha, accounting for only 2% of ConsT-suitable area. In April 2007, the MOA delivered an official report to promote the extension of ConsT technology. This plan will cover 4 Mha of farmland for ConsT by the end of the Eleventh Five-Year Plan (MOA, 2007). Up to the end of 2008, China had set up 226 national demonstration counties and 365 provincial demonstration counties for ConsT. The implemented area of ConsT is 3.33 Mha. ConsT technology was pushed further forward in 2012, when the central government spent 30 million Yuan to support the extension of ConsT in 204 counties of China and 300 million Yuan to support the establishment of the ConsT demonstration base in 80 additional counties (Department of Agricultural Mechanization Management, MOA, 2012).

The implementation of CR in China was mainly started in the 1980s, and the CR-covered area was about 3.5 Mha by the end of 1999, with levels of CR in the field at around 3.5 t ha^{-1} (Huang *et al.*, 2010). The implementation of CR through agricultural machinery for primary crops such as wheat, maize and paddy rice is currently about 23.8 Mha, and this accounts for about 15% of the crop-sown area in China (Department of Science and Education, MOA, 2010).

Ecological construction (EC) project

The EC project of China officially originated from the 'National Ecological Environment Construction Plan' issued by the former State Planning Committee of China at the end of 1998 (Shen, 2007). A total of six subplans were included in the plan, which were nature conservation (ecosystem, wildlife and natural landscape), afforestation, soil and water conservation, desertification control, grassland protection and ecological agriculture development. Subsequently, several EC projects, such as the Natural Forest Protection Programme (NFP), the 'Grain for Green' Project (GGP, e.g. returning farmland to forest/pasture) and fenced-in grazing were also proposed and implemented in China. After the 6-year effort from 1999 to 2004, a total of 16.71 Mha of grain-planted area was returned successfully to forest/grassland, including 7.00 Mha of forest returned from cropland and 9.71 Mha of barren mountains used for grass growing (Wang and Chen, 2006). The desertification area decreased by 6416 km^2, with an average decrease rate of 1283 km^2 $year^{-1}$ (Shen, 2007), and approximately 18% of the soil erosion area was also controlled (Lin *et al.*, 2005). Moreover, the policy to restrict and ban biomass cutting, and the six forestation projects (State Forestry Administration of China, 2005), also benefited the accumulation of carbon in the vegetation and soils of China. The six forestation projects are: the NFP; the GGP; the Shelterbelts Project in northern, north-eastern and north-western China and in the Yangtze River basin; the desertification prevention and control around Beijing project; the national wildlife conservation and nature reserve construction project; and the fast-growing and high-yielding timber base construction programme (State Forestry Administration of China, 2005).

The Difficulty and Challenge of Soil Carbon Sequestration in China

The carbon sequestration potential in the soils of China is large and promising, which may offset the emission of greenhouse gases to a certain extent, but China still has to face many difficulties and challenges in the governance of soil carbon.

Funding and technology limitation

China, as a developing country, is still under conditions of traditional agricultural production. Although the extension and demonstration of the STFR, FS, ConsT, CR and EC projects have been implemented in the country, funding and technology are still the predominant limiting factors. Moreover, the establishment of compensative mechanisms is also necessary for sequestrating carbon in soils, which may need a rapid and feasible assessment system so that a compensative strategy can be implemented and evaluated dynamically. To overcome these challenges, the Chinese government has taken a series of actions to respond to these issues. Biological carbon sequestration technologies and carbon sequestration projects have been incorporated into the *Guidelines on National Medium- and Long-term Program for Science and Technology Development (2006–2020)* (State Council of PR China, 2006), and several research projects at national level have also been started to solve carbon-related issues. These projects include, for example, the 'Global climate change and ecological issue' projects supported by the Chinese Academy of Science (CAS) in 2009 (Ding *et al.*, 2009), 'Carbon sequestration function and potential in natural forest and grassland of China' project supported by the Ministry of Science and Technology of China in 2010 and the strategic priority research programme – 'Climate change: carbon budget and related issues' – hosted by CAS in 2011 (Wang *et al.*, 2012). These research projects concentrate mainly on the evaluation of carbon sequestration potential originated from forest, pasture, shrub, cropland management, the EC project and the experiment on optimal configuration, integration and demonstration of current carbon sequestration technology. However, carbon sequestration-related new technology, such as CO_2 capture and storage, CR/ConsT machinery, application of biochar and biological inhibitors, crop breeding and selection and the development of controlled-release (CRF) and slow-release (SRF) fertilizers are still not perfect and need to be improved. This improvement may demand huge financial support or technical support imported from developed countries. For example, the agricultural machinery used in the ConsT/CR is still the primary limiting factor for the adoption of these techniques in China.

Notion and knowledge gap

Chinese farmers usually prefer to pursue their traditional tillage methods as they think these are the most effective for high productivity. Traditional tillage, however, may deteriorate the physical properties and organic matter of agricultural soils, and make the soils subject to erosion. Consequently, soil fertility is decreased dramatically. Traditional tillage in China is essentially not suitable for establishing a sustainable development model of agriculture. Therefore, the efforts for bridging the notion gaps between scientific research and farmers, administrators and policy makers are indispensable. For example, most Chinese farmers believe that the more the fertilizer input, the higher will be the crop yield. Consequently, excessive chemical fertilizers and pesticides have often been applied, and the optimal balance between cost and benefit missed. Moreover, Chinese farmers rarely accept the recommendation of balanced fertilizations, because the application of organic fertilizer such as manure is now considered to be hard and time-consuming work. Therefore, the change of current fertilization notions is still difficult, as long as there are no incentivizing policies. Moreover, during the extension of ConsT technology, extension staff should introduce effective tillage methods such as ridge-till, strip-till and mulch-till to farmers, and also the development of ConsT technology that is suitable for specific ecological environment conditions. Acceptance by Chinese farmers is also necessary because this could bridge the knowledge gaps between them and new technologies and practices.

Policy challenge

During the transition period from traditional agriculture to low-carbon as well as cost- and resource-effective agriculture in China, the promotion and extension of carbon-sequestering agriculture are also constrained by the lack of sufficient policy support and encouragement. The uncertainties in farmers' income in the extension of low-carbon agriculture are still the main obstacle. Therefore, encouraging, compensative and income-guaranteed policies, such as free technological service, training and financial support, and a free/low-interest agriculture loan policy must be made the highest priority, so that the governance of soil carbon in China can be pushed forward persistently and effectively.

Acknowledgements

This chapter was partly supported by the Strategic Priority Research Program – Climate Change: Carbon Budget and Related Issues of the Chinese Academy of Sciences (XDA05050509) – and the National Basic Research Program of China (973 Program) (2010CB950702).

References

Department of Agricultural Mechanization Management, MOA (2012) Website of China Agricultural Mechanization Information Network, December, 2012. http://www.amic.agri.gov.cn/nxtwebfreamwork/ztzl/js_bhxgz/detail.jsp?articleId = ff8080813bcb2b00013bd12657e905a (accessed 7 June 2013).

Department of Science and Education, MOA (2010) National Crop Straw Resource Investigation and Evaluation Report (in Chinese). Official investigation report released by Department of Science and Education, MOA, in December 2010. http://www.fjagri.gov.cn/upload/File/20110217103019.pdf (accessed 10 August 2014).

Ding, Z.L., Fu, B.J., Han, X.G. and Ge, Q.S. (2009) Brief introduction to a cluster of projects of 'Research in Key Issues of International Negotiation with Regard to Coping with Climate Change' by CAS. *Bulletin of Chinese Academy of Sciences* 1, 8–17 (in Chinese).

Guo, R., Wang, X.K., Lu, F., Duan, X.N. and Ouyang, Z.Y. (2008) Soil carbon sequestration and its potential by grassland ecosystems in China. *Acta Ecologica Sinica* 28(2), 862–867 (in Chinese).

Huang, Y. (2006) Emissions of greenhouse gases in China and its reduction strategy. *Quaternary Sciences* 26(5), 722–732 (in Chinese).

Huang, Y., Sun, W.J., Zhang, W. and Yu, Y.Q. (2010) Changes in soil organic carbon of terrestrial ecosystems in China: a mini-review. *Science China Life Sciences* 53, 766–775.

Jobbagy, E.G. and Jackson, R.B. (2000) The vertical distribution of soil organic carbon and its relation to climate and vegetation. *Ecological Applications* 10, 423–436.

Kong, X.B., Zhang, F.R., Wei, Q., Xu, Y. and Hui, J.G. (2006) Influence of land use change on soil nutrients in an intensive agricultural region of North China. *Soil and Tillage Research* 88, 85–94.

Lal, R. (2004) Soil carbon sequestration impacts on global climate change and food security. *Science* 304, 1623–1627.

Li, C.S., Frolking, S. and Harriss, R.C. (1994) Modeling carbon biogeochemistry in agricultural soils. *Global Biogeochemical Cycles* 8, 237–254.

Lin, E.D., Li, Y.E. and Guo, L.P. (2005) *Carbon Sequestration Potentials in Argicultural Soils of China and Climate Change* (in Chinese). Science Press, Beijing.

Liu, Y., Yu, D.S., Shi, X.Z., Zhang, G.X. and Qin, F.L. (2012) Influence of vegetable cultivation methods on soil organic carbon sequestration rate. *Acta Ecologica Sinica* 32(9), 2953–2959 (in Chinese).

Mannering, J.V. and Fenster, C.R. (1983) What is conservation tillage. *Journal of Soil and Water Conservation* 38(3), 140–143.

MOA (2006) Technical specification of balanced fertilization by soil testing. Agriculture standard of China, NY/T 1118-2006. Chinese Ministry of Agriculture, Beijing.

MOA (2007) Central government of China web portal, website of Ministry of Agriculture, April 2007. http://www.gov.cn/gzdt/2007-04/25/content_596175.htm (accessed 7 June 2013).

MOA (2009) MOA press conference held 21 December 2009. Chinese Ministry of Agriculture, Beijing.

National Bureau of Statistics of China (2008) *China Statistical Yearbook (2008)*. China Statistics Press, Beijing.

Pan, G.X., Li, L.Q., Zhang, X.H., Dai, Y.J., Zhou, Y.C. and Zhang, P.J. (2003) Soil organic carbon storage of China and the sequestration dynamics in agricultural lands. *Advance in Earth Science* 18(4), 609–618 (in Chinese).

Peng, W.Y., Zhang, K.L., Chen, Y. and Yang, Q.K. (2005) Research on soil quality change after returing farmland to forest on the Loess sloping croplands. *Journal of Natural Resources* 20(2), 272–278 (in Chinese).

Shen, G.F. (2007) China's ecological construction projects: concept, scope and achievements. *Forestry Economics* 11, 3–5 (in Chinese).
State Council of PR China (2006) *Guidelines on National Medium- and Long-term Program for Science and Technology Development (2006–2020)*. Issued by the State Council of PR China on 9 February 2006. http://www.most.gov.cn/mostinfo/xinxifenlei/gjkjgh/200811/t20081129_65774.htm (accessed 10 August 2014).
State Forestry Administration of China (2005) *China Forestry Development Report* (in Chinese). Beijing.
Wang, Q.F., Liu, Y.H., He, N.P., Fang, H.J., Fu, Y.L. and Yu, G.R. (2012) Demands and key scientific issues in the synthesis research on regional terrestrial ecosystem carbon budget in China. *Progess in Geography* 31(1), 78–87 (in Chinese).
Wang, R.P. and Chen, K. (2006) Analysis of the situation and problems in reverting farmland to forests and grassland in China. *Chinese Agricultural Science Bulletin* 22(2), 404–409 (in Chinese).
Wang, X.B., Cai, D.X., Hua, L., Hoogmoed, W.B., Oenema, O. and Perdok, U.D. (2006) Soil conservation tillage – the highest priority for global sustainable agriculture. *Scientia Agricultura Sinica* 39(4), 741–749 (in Chinese).
Xu, S.X., Shi, X.Z., Zhao, Y.C., Yu, D.S., Li, C.S. and Wang, S.H. (2011) Carbon sequestration potential of recommended management practices for paddy soils of China, 1980–2050. *Geoderma* 166, 206–213.
Yu, D.S., Shi, X.Z., Wang, H.J., Sun, W.X., Chen, J.M., Liu, Q.H. and Zhao, Y.C. (2007) Regional patterns of soil organic carbon stocks in China. *Journal of Environmental Management* 85, 680–689.
Zhang, H.B., Luo, Y.M., Wong, M.H., Zhao, Q.G. and Zhang, G.L. (2007) Soil organic carbon storage and changes with reduction in agricultural activities in Hong Kong. *Geoderma* 139(3/4), 412–419.
Zhang, X.Q., Wu, S.H., He, Y. and Hou, Z.H. (2005) Forests and forestry activities in relations to emission mitigation and sink enhancement. *Scientia Silvae Sinicae* 41(6), 150–156 (in Chinese).
Zhang, X.Y., Chen, L.D., Fu, B.J., Li, Q., Qi, X. and Ma, Y. (2006) Soil organic carbon changes as influenced by agricultural land use and management: a case study in Yanhuai Basin, Beijing, China. *Acta Ecologica Sinica* 26(10), 3198–3204 (in Chinese).

31 Avoided Land Degradation and Enhanced Soil Carbon Storage: Is There a Role for Carbon Markets?

Meine van Noordwijk*

Abstract

Avoidance of depletion of soil organic matter as part of land degradation and enhanced restoration in depleted soils are of direct importance to agriculture, ranching, forestry and other land uses, but are also a relevant part of national C accounting and the global C cycle. 'Carbon markets' imply economic, performance-based incentives that relate global climate and greenhouse gas concerns, via national commitments to reduce overall CO_2 emissions, to incentives at the level of land users to increase net C storage. We focus on three groups of questions:

1. What is the value chain involved? Can soil C be separated from aboveground land-use effects?
2. Are market-based solutions feasible? What can we learn from the pilots?
3. Will the prices be worth it for land managers once transaction costs are accounted for? Are there better ways to provide effective performance-based incentives from the public perspective?

We conclude that a combination of the commodification, compensation and co-investment versions of the broader payments or rewards for environmental services (ES) debate is probably needed to achieve the effects desired at both ends of the carbon value chain. Commodification, the purest market-based paradigm, by itself remains controversial and can be counterproductive. Synergy between private and public sectors is needed to make progress, while the primary attention will have to remain with the primary sources of anthropogenic carbon emissions.

Introduction

The preceding chapters have established the many functions that soil organic matter (SOM) plays in soil health (Chapter 14, this volume), buffering of water and nutrient supply to plants (Chapters 7 and 12, this volume), soil aggregation and infiltration (Chapters 8 and 22, this volume), filtering of contaminants (Chapter 8, this volume) and in support of agricultural production systems that minimize environmental impacts per unit harvested product. Yet, in the short-term, agricultural practices that deplete SOM, for example through intensive tillage, utilization of all crop residues, drainage of wetland and peat soils, may appear to be profitable with the current prices of agricultural inputs and outputs and in the absence of a direct market valuation of soil carbon

*E-mail: m.vannoordwijk@cgiar.org

(Chapters 15 and 18, this volume). There is thus a challenge of balancing the long-term on-farm benefits of enhancing SOM with short-term opportunity costs of foregoing more profitable, SOM-depleting practices. In economic analysis, this temporal trade-off depends on discount rates of future costs and benefits, and thus on security of tenure, risk preferences and farmers' future outlook. Discussions on this trade-off in terms of soil fertility go back at least 2000 years to the Roman books on agriculture by Marcus Porcius Cato and Marcus Terentius Varro (Winiwarter, 2000; even though soil organic matter and carbon were not known in current terms, the terms used, calidus (hot) or frigidus (cold), as descriptors of fertile and infertile soil were the same as described from local ecological knowledge in South-east Asia, but with opposite meanings). The recent concern on increases in the greenhouse gas effect may, however have opened new ways to resolve this age-old issue by providing a basis for economic incentives to let the long-term farmer benefits of enhanced soil C storage coincide with the public interest in mitigating climate change.

The carbon content of SOM is relevant in the global C balance and in efforts to contain the rapid increase of atmospheric CO_2 concentrations that contributes to global climate change (Chapters 9 and 20, this volume). The economic value of maintaining and enhancing soil carbon derives from a comparison with other activities that increase or decrease atmospheric CO_2 concentrations but are constrained by the institutional framework of rights and obligations that is formed to implement the ultimate objective of the UN Framework Convention on Climate Change (UNFCC) (UN, 1992). This objective is to:

> stabilize greenhouse gas concentrations in the atmosphere at a level that would prevent dangerous anthropogenic interference with the climate system. Such a level should be achieved within a time-frame sufficient to allow ecosystems to adapt naturally to climate change, to ensure that food production is not threatened and to enable economic development to proceed in a sustainable manner.
>
> (UN, 1992)

In this chapter, we review to what extent 'markets' can and already are linking the global significance of soil C storage to the land-use decisions that otherwise might opt for degradation of soil carbon. Soil C loss is related to between-categories changes in land cover (e.g. forest-to-agriculture conversion), and to details of land management within a category. Any economic incentives for soil carbon management interacts with rights-based approaches to nudge the decisions of land users in a direction desirable by external stakeholders, as well as with the direct economic benefits derived from land use. Incentives can target the primary actors ('land users', 'farmers'), as well as the underlying drivers and conditioning factors (Fig. 31.1).

We will discuss the concept of markets in the broader context of economic incentives that can apply at multiple scales, linking plot-level land-use decisions by farmers and other land managers to global stakeholders. Markets essentially are feedback mechanisms that relate current and expected future supply and demand, and the elasticity of production and demand, to prices, perceived risk and transaction costs. Markets require buyers and sellers, and often involve brokers, as basic levels of trust are needed to link 'pay-for-what-you-get'

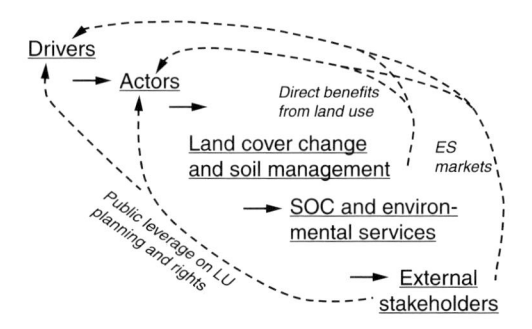

Fig. 31.1. External stakeholders of soil carbon, beyond the on-farm effects of soil organic matter, can use rights-based approaches (land-use planning, access rights) and economic instruments ('environmental service markets') to influence decision making by the primary actors and/or the underlying drivers. (From van Noordwijk et al., 2011.) LU, land use; ES, ecosystem services; SOC, soil organic carbon.

and 'get-what-you-pay-for'. The less directly visible the product quality is, the more trust and brokers play a role. Soil carbon is not directly visible, so we can expect trust and brokers to play a large role. We will structure this chapter around the following three groups of questions.

1. What is the value chain involved? Can soil C be separated from aboveground land-use (LU) effects?
2. Are market-based solutions feasible? What can we learn from the pilots?
3. Will the prices be worth it for land managers once transaction costs are accounted for? Are there better ways to provide effective performance-based incentives from the public perspective?

We will interpret 'markets' as referring to two out of four paradigms that are recognized in the broader payments or rewards for the environmental services (ES) debate (van Noordwijk et al., 2012a): (ia) commodification of ecosystem services (ES) as such; (ib) coupling ES to marketed commodities; (ii) compensation for opportunity costs of voluntary or mandatory LU restrictions that enhance ES but reduce profitability; and (iii) co-investment in environmental stewardship. In earlier discussions, (ia) and (ib) were jointly implied under commodification (van Noordwijk and Leimona, 2010). Only the first paradigm is fully aligned with the payments for environmental services (PES) definition of Wunder (2005).

Q1: What Value Chain is Involved? Basics of Soil C Accounting

Before answering the market-related questions 2 and 3, a basic understanding is needed of the actors and multiple scales, and the many cross-scale interactions, involved in the value chain:

A. Global citizens, experiencing climate change, consuming goods derived from international trade and, through elections, determining the immediate political relevance of environmental issues and willingness to commit to serious overall emission reduction targets of the level needed to contain global climate change.
B. Sovereign states as negotiation and implementation parties in the UNFCCC, balancing economic growth and demographic and economic transitions with environmental concerns and voter satisfaction.
C. Subnational governments (provinces, districts), implementing national policies while meeting local ambitions for economic growth, social movements and environmental integrity.
D. Private sector meeting obligations for emission reduction in industrialized countries only, but increasingly held responsible for the global footprints of their value chains.
E. Concession holders (forestry, plantation agriculture, mining), managing large tracts of land for economic benefit within rules set by the government.
F. Farmers and other land users, deciding on land cover conversion and land management affecting soil C storage.
G. Plants responding to land management by modified allocations to root and associated symbiont turnover and litterfall, and soil biota responding to management by modifying rates of decomposition and CO_2 release.

Combinations of the four ES reward paradigms (ia, ib, ii and iii) across scales (between E/F and A) are possible and form an important part of the discourse, as has emerged at the interface of Integrated Conservation Development Projects (ICDPs) and efforts to Reduce Emissions from Deforestation and Degradation (REDD) (Minang and van Noordwijk, 2013).

Within the UNFCCC, however, national borders have special significance for accounting rules of both area-based change in C stocks and trade-based use of fossil energy, fertilizers, cement and other elements of the national greenhouse gas (GHG) inventories (IPCC, 2006). Such reporting, on a 5-year cycle for developing countries, requires different types of data to be combined (Fig. 31.2) and has a major challenge in harmonizing operational definitions across

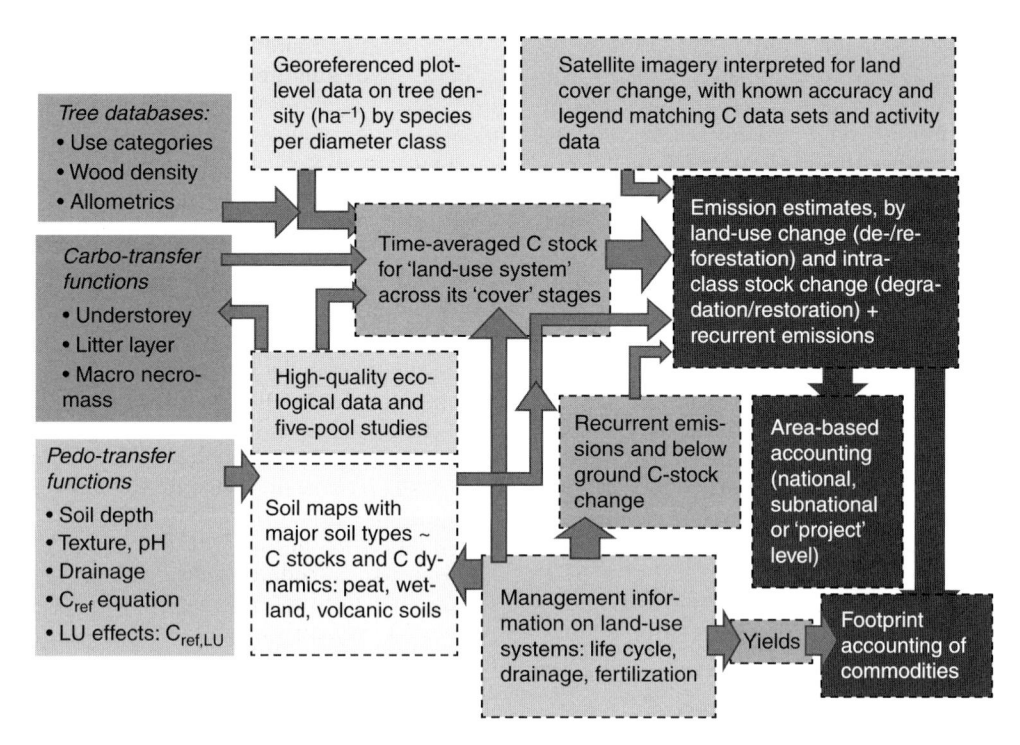

Fig. 31.2. Data requirements for a national C stock accounting system compliant with IPCC (2006) standards and its derivatives for commodity footprints. (From van Noordwijk et al., 2012b.)

sectors. While there has been a tendency for 'voluntary market' approaches to set up their own accounting rules that do not relate directly to national GHG inventories and reporting, harmonization will be essential for achieving the impacts at the scale required for the current actions to make sense (van Noordwijk et al., 2012b).

The same data sources plus yield data are needed for the footprint accounting of commodities in international trade, which are emerging as options to break the gridlock of country-based negotiations (Minang et al., 2010; Khasanah et al., 2011). While footprints have been quantified for average production conditions, recent interest in the swing potential or the variation in footprint within a realistic and plausible range of management practices (Davis et al., 2013) provides new perspectives. In the current global economy, consumers influence land-use decisions through selective use of cross-border trade, through regulations and through 'green' investment (Plate 15).

Can land use C storage effects be compared to fossil energy C emissions?

A long-standing debate relates to the issue of the timescale at which changes in terrestrial carbon stock can be compared to the return to the atmosphere of carbon captured in geological history of the planet and currently stored in fossil fuel (coal, oil, gas, peat). While return of such fossil carbon to the atmosphere is 'permanent', apart from its temporary absorption by oceans and terrestrial systems, the reductions in current use leave economically attractive options in the hands of market players. This is similar to what happens when forests as terrestrial C stocks are protected from immediate pressures. Somehow, however, the 'permanence' discussion has applied higher standards to avoided emissions of terrestrial C stocks than to avoided emissions of fossil fuel carbon (Swallow et al., 2007). In fact, neither of them is permanent at the timescale relevant to the climate change debate. The only option to achieve permanence is in the continuity and integrity of national accounts and international commitments and accountability – the track record of the latter is not very good, with the failure of the UNFCCC parties to agree on full implementation and a seamless sequel to the Kyoto Agreement (2008–2012).

Land-use change as part of total anthropogenic emissions

In terms of quantity, the current changes in terrestrial C stocks are dwarfed by fossil fuel emissions of CO_2. Historically, anthropogenic land cover change is estimated to have released 156 petagrams (Pg) of C to the atmosphere in the period 1850–2000, equivalent to 57% of fossil fuel emissions over the period (DeFries et al., 1999; Houghton, 2003). Despite shifts in geographic focus ('hotspots'), the net terrestrial C emissions have not changed much between the 1960s and now, at about 1.1 Pg C year^{-1}; in the 1960s, this represented about 30% of total anthropogenic emissions, in the 1990s, 18% and in 2010, 9% of the total, due to the large increase of fossil fuel emission (Canadell et al., 2007; Le Quéré et al., 2009; Peters et al., 2012).

Changing above- plus belowground terrestrial C stocks from a net source to become a net sink of atmospheric CO_2 is an essential element of strategies to keep global temperature rise below the +2°C threshold, which is considered a boundary of the safe planetary operating space (Rockström et al., 2009), alongside shifts in energy use. Estimates of emissions from soil due to land-use change are, however, lacking or highly uncertain, while subnational and project-based emission reduction efforts face substantive measurement costs if they want to include soil C in the considerations at required precision levels; costs of measurement that are generally not justified by the economic compensation for enhanced C storage (Cacho et al., 2008), except for the special cases where high emission rates can be avoided.

Peatland soils as hotspots of soil C emissions

Peat soils store more than half of the world's soil carbon on less than 10% of the area, and land conversion on peat contributes a disproportionately large fraction of total soil-based emissions (van Noordwijk et al., 2014). Peat soils, however, are more frequent in cool and subarctic climates, and in the tropics the equivalence of peat and mineral soils in terms of carbon stock is only true in the South-east Asia region, with Indonesia and Malaysia holding about 80% of global tropical peatland carbon. The high fluxes due to land use and land-use change on peat are associated with high uncertainty; a recent effort to establish emission factors for tropical peatlands in the IPCC has failed to reach consensus. Peatlands are hotspots of conflict over access rights as well (Galudra et al., 2011).

Erosion and sedimentation as fractal dimension

The net contribution of belowground carbon losses to the total of terrestrial C losses is uncertain, as part of the on-site loss measured in agricultural lands has been due to erosion and lateral transport followed by deposition elsewhere in the landscape, rather than direct release to the atmosphere (Paustian et al., 1997). Due to the lateral flow interactions, the net loss of soil + organic matter from any area has a fractal dimension, with deposition taking a larger and larger share of the plot-level erosion values, the larger the area considered (van Noordwijk et al., 1998a; Verbist et al., 2010). Incorporating the fate of deposited materials, van Oost et al. (2012) concluded that historical erosion in the landscape they studied had been at least neutral in terms of atmospheric CO_2 emissions. These insights are yet to become mainstream in accounting, as critical data for deposition sites are not normally collected.

Topsoil only or land-use effects on deeper soil layers

Most of the data refer to changes in the topsoil, which generally are more readily observable and where changes may occur more rapidly after land cover and land-use change. For the total soil–atmosphere flux, deeper soil layers may be at least as important (Nepstad et al., 2002), depending on the presence of deep-rooted trees that stay green in dry periods.

In a recent study in Sumatra, an area where deforestation has almost run its course over the past three decades, statistically significant differences in soil C stock with depth could only be ascertained for the top 10 cm, once the depth data were corrected for the bulk density under the influence of land use (Fig. 31.3). The bulk density correction removed about half of the apparent difference in soil C stocks if only C_{org} data were available. Correction for differences in soil texture using the Cref equation of van Noordwijk et al. (1997) and Hairiah et al. (2011) helps in separating true land-use effects from the covariance of land use, positions in the landscape and the soil texture typical for these positions. An example is the rubber agroforest that tends to occur close to rivers in soils with higher clay + silt content than the locations where natural forest is left in the landscape; higher C_{org} data for rubber agroforest relative to natural forest turn into lower C_{org}/C_{ref} ratios once corrected for texture.

A change in the intensity and depth of soil tillage affects C_{org} distribution in the soil, rather than the total C stock, and initial claims that reduced tillage is a C storage option had to be reconsidered when data for the whole soil profile were included in the analysis. When assessed across nitrous oxide, CO_2 and methane, there may still be positive effects (Ruan and Robertson, 2013).

Default estimates for land-use effects on soil C stocks

As part of the second IPCC review, Paustian et al. (1997) summarized known effects of land-use change on soil carbon across climatic zones and soil types. Subsequent literature has lead to some refinement. Don et al. (2011) in a global meta-analysis of 385

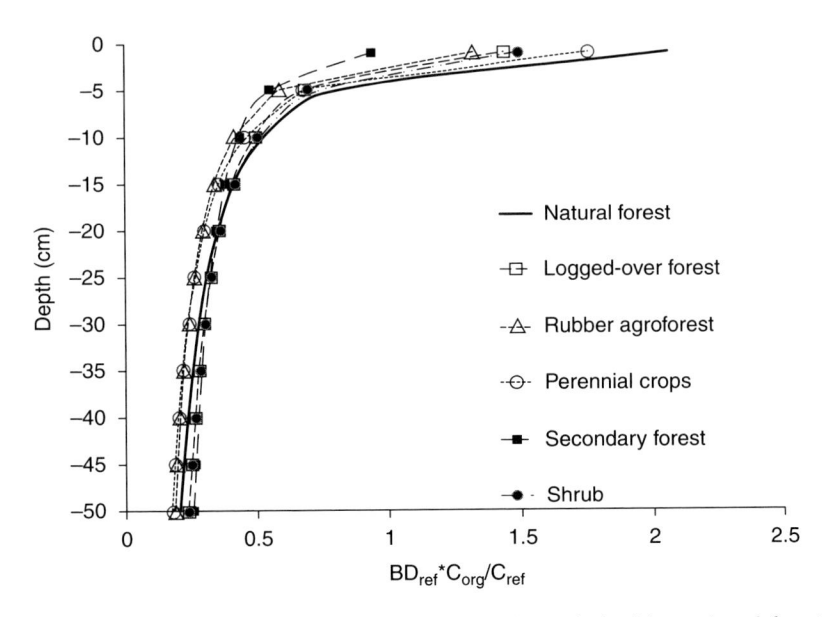

Fig. 31.3. Relationship between C_{org}/C_{ref} and the mean depth of sampled soil layers in a deforestation landscape in Sumatra (Indonesia); the regression lines shown represent 54–85% of variation, except the secondary forest line, which represents only 31% of variation at sample level.

studies on land-use change in the tropics, expanding substantively from Guo and Gifford (2002), found that the highest C_{org} losses were caused by conversion of primary forest into cropland (−25%) and perennial crops (−30%), but forest conversion into grassland also reduced C_{org} stocks by 12%. If it were a simple additive system, one might thus expect conversion of grasslands to perennial crops to lead to a decrease of C_{org} by about 18%. The discrepancy may be due to the non-average starting points for the land-use transitions measured. The time frame over which land-use effects are measured is unclear, and can be responsible for at least part of the discrepancies. Another recent meta-analysis (Powers et al., 2011) focused on 'paired plot' literature and found little consistency in C_{org} change, with both 'forest to grassland' and 'grassland to forest' conversions leading to statistically significant C_{org} gain; this may raise doubts on the selection bias in the results that get published. Both reviews confirm that complete data sets that combine soil bulk density and soil organic carbon are scarce, and that spatial extrapolation is affected by unbalanced representation of tropical soils. Bruun et al. (2009) exposed common misunderstandings in the role of swiddening in this respect.

These numbers provide a perspective on the possibility of soil C storage beyond peatlands to be a significant component of plans to reduce net anthropogenic C emissions. Net effects can be a modest, but significant, contribution to agreed emission reduction (which was a modest 5% for the first Kyoto commitment period), but only if the efforts and associated finance used to invest in terrestrial C is 'additional'.

Can soil C be separated from aboveground land-use effects?

The development of economic incentives for land use-related C stock change has so far focused on changes in tree cover, with rules focused on tree planting, deforestation and forest degradation, with some attention to peatlands as hotspots of belowground C stock change. The changes in aboveground C stock involved in deforestation can be of the order of 200 Mg C ha^{-1}, while the management swing potential of soil C stocks in agriculturally managed soils is an order of magnitude smaller, around 20 Mg C ha^{-1} (Buysse et al., 2013), unless peat soils are involved. Given these orders of magnitude, it is unlikely that soil C attracts the attention of market-based mechanisms separate from changes in aboveground C stocks. If soil C protecting practices would reduce yields, the indirect negative effects of such land management on anthropogenic atmospheric CO_2 concentrations by increasing rates of forest-to-agriculture conversion can easily surpass the positive effects on local C storage. If the practices increase yields, it probably does not require support to be readily adopted.

A further distinction is usually made between 'avoided degradation' and 'assisted restoration', recognizing that changes in above- and belowground C storage tend to be in the same direction. Comparing aboveground C storage transition curves (linked to forest and tree cover transitions) to the dynamics of the belowground C storage in (agro-)ecosystems, we expect a reduced 'management swing potential', a more temporally buffered dynamics and a time lag for the effects of tree cover change to impact on soil organic matter via root and associated symbiont turnover (van Noordwijk et al., 1998b) (Fig. 31.4). The concept of management swing potential was recently introduced as the difference in footprint of the best and worst modes of production (Davis et al., 2013). There is empirical evidence of a recovery of soil carbon in intensive rice-based cropping systems in East and Southeast Asia that is not related to the use of trees but to an increase in the number of crops per year and associated increase in root biomass inputs to the soil (Minasny et al., 2011).

Offsets and additionality

Additionality is usually discussed on the supply side, with the burden of evidence on

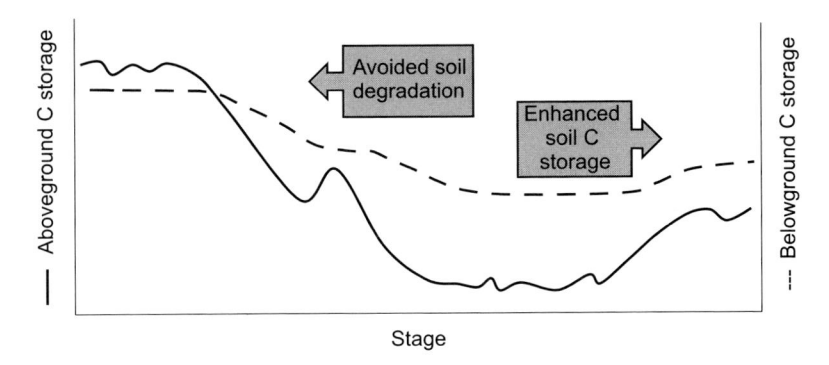

Fig. 31.4. Schematic dynamics of above- and belowground C storage in relation to a forest or tree cover transition curve; the translation of 'stage' to 'time' can vary with the systems studied.

C emission reduction proponents that the emission reduction claimed relative to a baseline or counterfactual is truly additional and would not have happened without their intervention. If the proposed activities are economically feasible, additionality arguments have to involve investment or regulatory bottlenecks that their project will overcome but that otherwise prevent action. Proof of additionality is simpler for activities that are not economically feasible without intervention – but that raises questions on attractiveness and permanence. The observation of 'spontaneous' soil C recovery discussed above is both good and bad news from a C market perspective: it suggests that restoration is possible, but also that it may happen without specific targeting, as part of ecological intensification of land-use practices.

However, additionality in terms of total anthropogenic GHG emissions needs to involve the demand side as well. Offset markets that exchange 'emission rights' in the form of 'carbon credits' do not lead to net global emission reduction, unless they are directly associated with commitments to deeper overall cuts compared to business-as-usual practice. Evidence for this type of additionality in global climate negotiations is weak. Although there have been, and still are, experiments in 'terrestrial carbon projects' that target offset finance, it is unlikely, at the current state of negotiations, to contribute substantively to global climate change mitigation. Where 'new finance' is used, however, net positive effects are feasible.

Q2: Are Market-based Solutions Feasible? What Can We Learn From The Pilots?

Certified emission reduction units as marketable commodities

The term 'carbon market' as commonly used is, in fact, a market for Certified Emission Reduction Units, where each word has a specific meaning; **certified** (by a legitimate, transparent and credible institution that can be trusted in its quality control); **emission** (of all major GHGs, specifically acknowledging the different dynamics of CO_2, CH_4 and N_2O in their interactions with soil fertility and soil management, and the different accounting methods that apply to the different gases); **reduction** (relative to a prior agreed baseline of 'emission rights'); **unit** (the different gases can be combined on a CO_2 equivalent basis, but can refer to a unit of space and time and/or a unit of product that enters the value chains for food, fibre or bioenergy). It should be noted that the abbreviation CER is used in the context of the Clean Development Mechanism of the Kyoto protocol and synonyms exist for other formal and voluntary frameworks.

The various market-based mechanisms reviewed by Jindal and Namirembe (2012) differ in the authority that provides certification, the fungibility (or degree to which land use-related emission reduction can offset ongoing fossil energy emissions) and rules for setting baselines. We will discuss three groups of market rules.

Instruments under the Kyoto protocol and its post-2012 extension

Propositions to include all land-based emissions under the framework of the Clean Development Mechanism of the Kyoto Protocol were stranded at the negotiation table as fears that flexible mechanisms would distract from the need for reducing fossil energy-based emissions.

Limited opportunities have been created for 'assisted restoration' of terrestrial C stocks based on offset finance (generating 'carbon credits') under the heading of Afforestation/Reforestation Clean Development Mechanisms (A/R-CDM) (Jindal and Namirembe, 2012). Stringent safeguards and rules regulate the application of these mechanisms, and so far their implementation has been far below the allowable part of global mitigation efforts. The discussion and emergence of operational rules has provided significant learning opportunities for all involved. However, in the context of continuously increasing net anthropogenic emissions, the time and mental energy spent might have had better uses (Murdiyarso et al., 2008; van Noordwijk et al., 2008). Jindal et al. (2008) described 23 projects in 2008, before the REDD experimentation phase started.

Meanwhile, the agreed international accounting procedures for Land Use, Land Use Change and Forestry (LULUCF) or the updated but yet-to-be adopted protocols on Agriculture, Forestry and Other Land Uses (AFOLU) in industrialized countries continue to be debated, which so far has prevented their inclusion as fungible parts of the commitment to net emission reduction. The European carbon market does not yet include forest, tree or soil carbon-related emission reduction to be traded in equivalence with fossil fuel emissions reduction rights.

REDD+, NAMA not yet covered in binding agreements

The conference of parties of the UNFCCC in Bali in 2007 opened the door for experiments both with efforts to reduce emissions from deforestation and forest degradation (REDD+) and nationally appropriate mitigation actions (NAMA) in developing countries. The NAMA provides for comprehensive emissions reduction across all sectors.

The specific restriction of REDD+ to 'forest' has caused significant complications in the absence of a globally accepted and agreed forest definition – with the existing definition referring to 'forest institutional frameworks' (including 'temporarily unstocked forest'), rather than actual tree cover (van Noordwijk et al., 2008). Over the past 5 years, expectations of 'avoided emissions' in the forest sector have increased, peaked and crashed with reference to REDD+ (reducing emissions from forest degradation and deforestation and increased storage by forest restoration). Although there is still some hope that workable rules can be agreed on, the combination of safeguards that are aimed at avoiding negative externalities on local livelihoods, rights and biodiversity has made 'transaction costs' an important part of the total cost. The initial observation of low opportunity costs of a large share of historical emissions (i.e. much of the emissions has not lead to economically profitable land use) (Stern, 2006) has not been turned into implementable abatement programmes (as abatement costs include opportunity, transaction plus implementation costs). One of the challenges to REDD has been that emissions from peatland, while economically avoidable, transcend the 'forest definition' restriction of REDD+, as does the debate on swiddens (Ziegler et al., 2012; van Noordwijk et al., 2015).

More holistic landscape approaches are needed and have been tried – not limited yet by narrow definitional issues of agreed scope. More generally, landscape approaches that target emission reduction from all land uses have found increasing support – even though the international negotiations have not taken this route as yet. The slow emergence of an 'agricultural mitigation' interest, however, is to be mentioned. Currently, it appears that approaches that explicitly reduce human vulnerability through a combination of mitigation and adaptation at the landscape scale ('climate smart agriculture') are in the focus of attention. Avoided loss of soil C (associated with land degradation) and support for increased soil C storage is a key component. As to the finance of 'climate smart agriculture', however, the options for offset finance (tradeable carbon credits) and/or new investment are barely discussed.

Voluntary markets

Jindal and Namirembe (2012) discuss the Verified Carbon Standard (VCS, formerly Voluntary Carbon Standard), the Gold Standard, the Plan Vivo Standard and others that have emerged to help the voluntary C market obtain the level of trust needed to function. As voluntary C credits are not fungible under formal C emission commitments, they do pass the demand-side additionality test, and can focus on supply-side additionality and the possibility of leakage (negative effects outside the project area that are attributable to project interventions). Planting trees remains an emotionally satisfying activity that is popular for voluntary compensation of fossil energy-based emissions – even though the temporary crediting rules of A/R-CDM have made this a rather unattractive option in the compliance market.

Compliance markets may operate under 'offset' rules, which imply that emission reduction in a certified project can be used to meet formal obligations to reduce emissions elsewhere. The credited emission reduction can be sold as a right to emit – which means it may have zero effect on global emissions. The separation in voluntary and compliance markets is fluid in time, with efforts to regularize and gain recognition for what started as voluntary actions, additional to government commitments.

Price volatility as a primary challenge to market-based mechanisms

By the end of 2012, both the EU trading scheme and CER prices crashed, in the absence of a clear sequel to the Kyoto protocol. The CER prices came down to US$0.5 t^{-1} of CO_2 (Ecosystem Market Place, 2013). Fluctuations on the demand side for carbon credits has a major effect on the volatility of prices under a market mechanism, as existing allowances of carbon rights are independent of ups and downs in the global economy, and the interest in buying credits relates to the margin between actual emissions and allowances. Carbon market prices are far more dependent on global economic moods than commodity prices that determine the opportunity costs of emission-reduction actions.

Product-based emission accounting in international markets

The enthusiasm for biofuels that emerged a decade ago could probably be understood from the attractiveness of an option that required minimum adjustments in the means of transport and associated lifestyles, while agriculture was challenged by surpluses in grain and other commodities that exceeded financially viable demand (even though hunger persisted for those without sufficient income). Importing feedstocks of biofuel, with sugarcane and palm oil the leaders for bioethanol and biodiesel, respectively, appeared to be attractive – until public opinion was made aware of the question of the emissions and loss of biodiversity associated with the rapid expansion of export-oriented crops in tropical rainforests. Subsequent regulations of biofuels in the EU and USA have tried to guarantee that at least some (20–40%) net global emission reduction is achieved – even though the fossil fuel substitution is fully credited by the industrialized countries and the emissions caused by the production of biofuels (up to 20–40% of the substitution) that occur in developing countries remain outside of the purview of the C accountants. This discussion on biofuels has spilled over to palm oil grown for other uses, and a voluntary standard (Roundtable for Sustainable Palm Oil, RSPO) aims to mainstream best agroecological practice, associated with low emissions, by the avoidance of peatland and forest clearing, as well as fine-tuned fertilizer applications. Technically, a carbon-neutral conversion to palm oil production is feasible and exists in some 10–20% of current production (Khasanah et al., 2011).

Exploiting the loopholes in international carbon accounting, the ideas of biological atmospheric carbon dioxide removal through bioenergy with industrial carbon capture and storage in importing countries might allow industrialized countries not only to shift from GHG-emitting fossil energy sources to carbon neutral ones but also even to claim a net atmospheric C capture, regardless of the ecological and social consequences in the areas from which bioenergy would be sourced (Smith and Torn, 2013). Such approaches have become the target for the strategies of industrialized countries with strong emission-reduction commitments, but there are many unresolved issues surrounding this, some technical, others institutional on the integrity of international accounting systems (Table 31.1).

Overview of options

In the interaction between the above options and national GHG accounting (Fig. 31.5), we see that national-scale emissions interact with area-based efforts of limited spatial extent ('projects'), as well as with efforts to modify the footprint of commodities in international trade. The interaction between these two and comprehensive NAMA remains largely unresolved.

Q3: Will The Prices Be Worth It Once Transaction Costs Are Accounted For?

With agriculture finally finding its place on the global climate change agenda at the Durban UNFCCC Conference of Parties in 2011, the level of uncertainty in changes in soil carbon stocks linked to land-use change is likely to get renewed attention. Lipper et al. (2011) suggested that soil carbon sequestration benefits could be 'harvested'. The question remains at what scale this can happen.

The analysis of Cacho et al. (2008) still stands, that measurement costs at the level of precision required by current C market standards are a major obstacle for including soil carbon in project designs for A/R-CDM and its voluntary market counterparts. The standards require evidence that effects on pools that are not measured will not be negative. Rather than claiming positive effects, it may be more economical to collect just enough data to justify a 'no harm' argument, treating any additional soil C storage as a co-benefit. Positive effects of soil organic matter on reducing vulnerability to climate variability may be a further co-benefit.

In the special case of peatlands, the emissions linked to land-use change are large (tens of tonnes of CO_2 eq ha^{-1} $year^{-1}$ over decades) and attention is warranted and forthcoming, as these still are large fluxes with large uncertainties and controversies over prospects of restoration activities. There are some rough edges to the peat issues in terms of definitions: soils with less than 50 cm of peat do not classify as peatlands, yet can cause large emissions. Peatlands gradually merge into other wetland issues of GHG fluxes, which merge into temporarily flooded riparian zones. Mangrove soils with C_{org} levels of around 10% down to several metres in depth have recently gained attention, as little is known about the C dynamics in these soils or of the fate of C-rich sediments under coastal abrasion. Lack of clear definition of such 'special cases' may be an

Table 31.1. Accounting rules and accountability for soil carbon to be included in market-based mechanisms in different countries.

Entity	Accounting	Accountability	Soil C fungibility with other emission types	Voluntary investment
Early-industrialized countries (Annex-I), following Kyoto protocol	National reporting using IPCC standards	Legal + moral	Not within their national boundaries; some in A/R-CDM context	Consumer pressure shifts private sector
Early-industrialized countries (Annex-I), *not* following Kyoto protocol	National reporting using IPCC standards	Moral	Idem	Higher voluntary commitment?
Latecomers in economic development (non-Annex-I), with articulated NAMA	National reporting using IPCC standards	Moral	Land-based emissions tend to be a large share of total	Taxing investment not aligned with national priorities
Idem, without NAMA	National reporting using IPCC standards	Ignored	Idem	Various attempts to tax, control and influence
International trade in bioenergy	Ignored beyond Annex-I involvement	Moral, if national goals are achieved on back of global emission increase	Peatland recognized as emission hotspots	Trade self-regulation, e.g. RSPO
International trade in other products	Idem	None	Idem	Idem
Voluntary carbon transactions	No specific relation	None	Flexible	Multiple motivations

A/R-CDM, afforestation/reforestation clean development mechanisms; IPCC, Intergovernmental Panel on Climate Change; NAMA, nationally appropriate mitigation actions; RSPO, Roundtable on Sustainable Palm Oil.

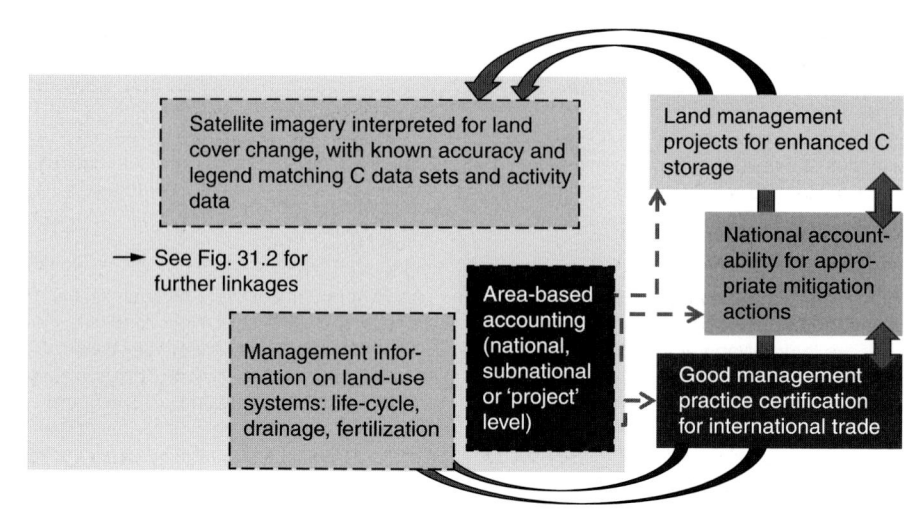

Fig. 31.5. Area-based and trade-based interventions to reduce net emissions from land use (change) interact with the national accounting frame (compare Fig. 31.2) and accountability vis-à-vis nationally appropriate mitigation actions.

argument to include all soils, to avoid the type of definitional confusion that has considerably slowed down REDD efforts.

The IPCC AFOLU accounting frameworks require that changes in soil carbon stocks across all land uses are part of a 5-yearly (non-Annex-I) or annual (Annex-I) reporting cycle at the national scale. Is there scope for monitoring at subnational or 'project' scales as well? Table 31.2 summarizes the pros and cons of the argument.

Integrity of accounting rules

Spatial aggregation shifts determinants of uncertainty in soil C stocks in ways that are poorly recognized as yet (Lusiana et al., 2013). The confidence intervals around national estimates of net soil C change are much narrower than those at project level, due to the number of the at least partially independent replicates involved. The aggregate numbers mostly suffer from possible bias, rather than random error. Biases inherent in the availability of data that are not derived from stratified random designs, but depend on whatever has been collected for other reasons, are a major concern that cannot be addressed easily without new research.

The integrity of the accounting system is challenged by cross-border trade, especially where this involves Annex-I and non-Annex-I countries. Long-term stock change is the primary accounting base for terrestrial carbon and has a diminishing cumulative error, as overestimates of change for a single period tend to be compensated by underestimates for a subsequent period. Flow accounting errors do not diminish by accumulation, and cumulative trade estimates need to be reconciled with the stock changes they lead to. There is, however, little reason to treat carbon in internationally traded wood differently from carbon in other organic produce, be it used as animal feed, human food and fibre, or as a source of bioenergy (van Noordwijk et al., 1997). The inclusion of international trade in wood and wood products in carbon accounting has long been debated (Winjum et al., 1998). If implemented, it allows wood-exporting countries to claim carbon sequestration – but the consequent emissions in importing countries are preferably ignored. Lauk et al. (2012) concluded that increments from 1900 to 2008 in the carbon stocks held in wood and

Table 31.2. The pros and cons of the arguments for inclusion of soil carbon stocks in subnational and project-scale C accounting efforts.

Arguments in favour	Arguments against
Soil carbon stocks have longer residence times, and cumulative changes with time are less vulnerable to change than aboveground C stocks	Small annual changes in a pool that has high spatial variability which is only partially attributable to easily measurable covariates (Don et al., 2011; World Bank, 2012)
Beyond peat and wetlands, the case for recovery of C_{org} in overgrazed and degraded drylands is sufficiently strong to warrant action, as the areas involved are large (Wang et al., 2011)	Evidence of changes in relative C_{org} distribution with depth that are uncorrelated with the more readily observable change in topsoil C_{org}; this relates to changes in soil tillage (VandenBygaart and Angers 2006) and shifts between grasslands and tree-based vegetation (Jobbágy and Jackson, 2000)
Dynamic process-based models of C_{org} continue to improve and can be used for refined and downscales of national estimates (van Wesemael et al., 2011; Smith et al., 2012)	Sample bias in current published data of paired-plot comparisons as explored by Powers et al. (2011) and internal contradictions (both forest => grassland and grassland => forest changes are reported to increase C_{org})
New methods based on spectral analysis reduce the costs of analysis, and correlations with standard (wet chemistry or dry combustion) analysis are fairly good (*no peer-reviewed use for C_{org} temporal monitoring, though*)	Costs of sample analysis with the required levels of replication that can overcome spatial variability can take up >100% of the economic value of increased certainty about C stock changes over short time intervals
Enhancing C_{org} has co-benefits for agricultural productivity and climate change adaptation, so there are win–win opportunities	Soil C stocks have been shown to recover spontaneously with agricultural intensification (Minasny et al., 2011), undermining 'additionality'

other organic products in society corresponded to 2.2–3.4% of global fossil fuel-related carbon emissions for the period; although this is not negligible, the growth over time is not a major climate change mitigation option and there is only modest potential to mitigate climate change by the increase of carbon stocks in society held in wood and other organic products. As quantified by Larson et al. (2012), the significant carbon stocks in harvested wood products in society are associated with substantial fossil fuel-based emissions as well.

Where the importing countries use the imported organic substrate to reduce their fossil fuel use and associated emissions, the integrity of the global C accounting system is at stake if stock change in the exporting country is not properly accounted for. The 'emission transfer' that can cross the boundaries of the part of the world with accountability for net emissions is the basis for seeking 'footprint' numbers that can be associated with international trade.

Are there better ways to provide effective performance-based incentives?

The basic market proposition is that self-interest can generate public goods through Smith's invisible hands (Rothschild, 1994). Current debate on the application of market concepts in REDD+ shows a range of opinions.

On one hand, Venter and Koh (2012) conclude that REDD+ currently is the most promising mechanism driving the conservation of tropical forests, but that if it is to emerge as a true game changer, REDD+ must still demonstrate that it can access low transaction cost and high-volume carbon markets or funds. It will have to do this while also providing or complementing a suite of non-monetary incentives to encourage a developing nation's transition from forest losing to forest gaining, and align with, not undermine, a globally cohesive attempt to mitigate anthropogenic climate change.

On the other hand, Corbera (2012) critiqued the current REDD+ policy framework and its commodification of ecosystems' carbon storage and sequestration functions on a global scale, as part of a 'neoliberalization of nature'. It eases a transition from an ethically informed conservation ethos to a utilitarian one that simplifies nature and undermines socioecological resilience. This approach relies on a single valuation language that may crowd out conservation motivations in the short and long term; while it is sustained on a 'multiple-win' discourse that in practice lacks procedural legitimacy in many developing countries and reproduces existing inequities and forms of social exclusion.

In the contrast between these two perspectives on market-based mechanisms to reduce emissions from forest conversion to other use, generally including changes in soil C stocks, we can see recognition that there are multiple levels of motivation involved. This occurs on both ends of the value chain, and the interaction between them can be negative as well as positive.

Elsewhere (van Noordwijk et al., 2013), we proposed a motivational pyramid for governments' concerns in developing and developed countries (Fig. 31.6) that leads to three entry points in the current debate: (i) REDD+ and similar market instruments offering rents that exceed those for a business as the usual scenario; (ii) concerns about branding and maintenance of market share in international markets; and (iii) genuine interest in global emission reduction.

van Noordwijk et al. (2012a) and Minang and van Noordwijk (2013) argue that along a carbon value chain a combination of PES paradigms (discussed above) can be used: (ia) commodification of verifiable carbon emission credits at national scale; (ii) compensation in the interaction between national governments and sectors or subnational entities; and (iii) co-investment in the interaction with land users. This perspective may address the concerns expressed by

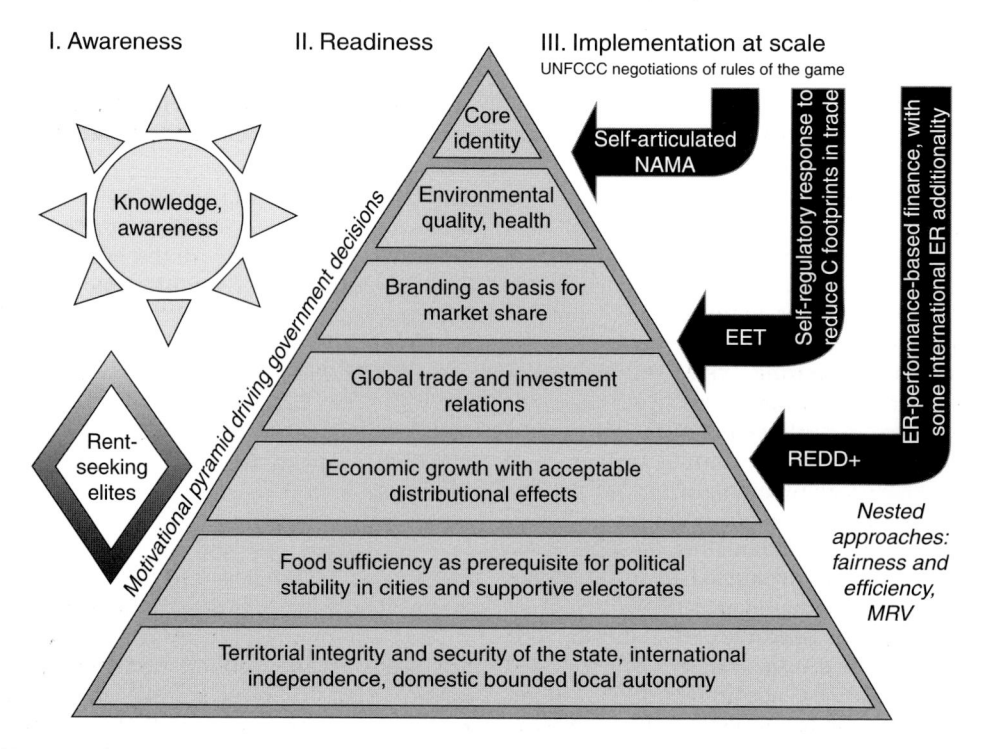

Fig. 31.6. Motivational pyramid of governance decisions and the multiple entry points of market-based interests in reducing C emissions. (From van Noordwijk et al., 2013.)

both Venter and Koh (2012) and Corbera (2012), as cited. Co-investment may well have to start with the clarification of use rights and negotiation of agreements for land use in high C stock parts of the landscape (Akiefnawati et al., 2010). Co-investment will also have to rely on clear local roles in monitoring, rather than full reliance on technical remote sensing approaches (Danielsen et al., 2013).

Meanwhile, the response of private sector entities to critique and footprint concerns of their end-consumers has led to the emergence of paradigm (ib), where ES issues are linked to existing commodity trade, rather than being commoditized separately. So far, the voluntary restriction of further peatland conversion, in the face of public concerns, probably has been the major 'market-based' reduction of net soil carbon emissions. It follows a pathway different from what is usually considered, and has a strong focus on recognized hotspots rather than supporting the large number of smallholders who might make small per hectare gains in carbon stocks on small landholdings, subject to expensive monitoring.

Conclusions

In conclusion, the prospects for carbon markets to support plot- or farm-level performance-based efforts effectively to increase soil C storage are dim. However, there is space for national aggregation to commercially viable scales, by holistic land-use accounting systems that include soil C, with domestic incentive mechanisms that support land-use practices from an approved list, and/or become part of integrated land-use planning for low C emission economies (Dewi et al., 2011).

Avoidance of depletion of soil organic matter as part of land degradation and enhanced restoration in depleted soils are of direct importance to agriculture, ranching, forestry and other land uses, but are also a relevant part of national C accounting and the global C cycle. 'Carbon markets' imply economic performance-based incentives that relate global climate and GHG concerns via national commitments to reduce overall CO_2 emissions to incentives at the level of land users to increase net C storage. If these incentives are based on offset principles, however, they only reduce net anthropogenic emissions if they lead to stronger emission reduction targets at the negotiation tables for an agreed cap overall.

Uncertainty in performance measures at the national scale differs essentially from that at the plot, farm or landscape level. Clear effects in national accounts derive from the aggregation of data with substantial spatial variation in which inherent soil properties and management-dependent soil carbon are not separated as easily as directly visible aboveground C stocks in vegetation. The chances for effective market-based mechanisms to provide farm-level incentives for increasing soil C stocks are limited by: (i) unclear additionality at both the supply and demand level; (ii) lack of adequate international commitments to reduce C emissions to the atmosphere; (iii) technical issues of monitoring at high spatial resolution; and (iv) lack of agreement on modalities for direct incentives and timescales for assessing performance. Currently, and in the foreseeable future, the main market-based mechanisms for avoiding land degradation and enhancing soil C storage in low C emission economies are the use of minimum standards for the sustainability of production systems. This approach would be for commodities acceptable in international trade and opportunities to get market recognition for better-than-average practices. Soil carbon will likely not be targeted separately from aboveground C stocks, and the major changes involved in the loss of tree biomass and the use of peatlands will remain the primary attractor of attention.

Acknowledgements

Comments on an earlier version are acknowledged with thanks from Unai Pascual and other SCOPE workshop participants, as well as from Keith Paustian, Sara Namirembe and Betha Lusiana.

References

Akiefnawati, R., Villamor, G.B., Zulfikar, F., Budisetiawan, I., Mulyoutami, E., Ayat, A. and van Noordwijk, M. (2010) Stewardship agreement to reduce emissions from deforestation and degradation (REDD): case study from Lubuk Beringin's Hutan Desa, Jambi Province, Sumatra, Indonesia. *International Forestry Review* 12, 349–360.

Bruun, T.B., De Neergaard, A., Lawrence, D. and Ziegler, A.D. (2009) Environmental consequences of the demise in Swidden cultivation in Southeast Asia: carbon storage and soil quality. *Human Ecology* 37(3), 375–388.

Buysse, P., Roisin, C. and Aubinet, M. (2013) Fifty years of contrasted residue management of an agricultural crop: impacts on the soil carbon budget and on soil heterotrophic respiration. *Agriculture, Ecosystems and Environment* 167, 52–59.

Cacho, O., Hean, R., Ginoga, K., Wise, R., Djaenudin, D., Lugina, M., Wulan, Y., Subarudi Lusiana, B., van Noordwijk, M. and Khasanah, N. (2008) *Economic potential of land-use change and forestry for carbon sequestration and poverty reduction*. Technical Reports 68. Australian Centre for International Agricultural Research (ACIAR), Canberra. (http://www.aciar.gov.au/publication/TR68, accessed 12 December 2013).

Canadell, J.G., Le Quéré, C., Raupach, M.R., Field, C.B., Buitenhuis, E.T., Ciais, P., Conway, T.J., Gillett, N.P., Houghton, R.A. and Marland, G. (2007) Contributions to accelerating atmospheric CO_2 growth from economic activity, carbon intensity, and efficiency of natural sinks. *Proceedings of the National Academy of Sciences* 104, 18866–18870.

Corbera, E. (2012) Problematizing REDD+ as an experiment in payments for ecosystem services. *Current Opinion in Environmental Sustainability* 4, 612–619.

Danielsen, F., Adrian, T.P., Brofeldt, S., van Noordwijk, M., Poulsen, M., Ruhayu, S., Rutishauser, E., Theilade, I., An, N.T., Budiman, A. *et al.* (2013) Community monitoring for REDD+: international promises and field realities. *Ecology and Society* 18(3), article 41.

Davis, S.C., Boddey, R.M., Alves, B.J.R., Cowie, A., Davies, C., George, B., Ogle, S.M., Smith, P., van Noordwijk, M. and van Wijk, M. (2013) Management swing potential for bioenergy crops. *Global Change Biology Bioenergy* 5, 623–638.

DeFries, R.S., Field, C.B., Fung, I., Collatz, G.J. and Bounoua, L. (1999) Combining satellite data and biogeochemical models to estimate global effects of human-induced land cover change on carbon emissions and primary productivity. *Global Biogeochemistry Cycles* 13, 803–815.

Dewi, S., Ekadinata, A., Galudra, G., Agung, P. and Johana, F. (2011) *LUWES: Land Use Planning for Low Emission Development Strategy*. World Agroforestry Centre (ICRAF), Bogor, Indonesia.

Don, A., Schumacher, J. and Freibauer, A. (2011) Impact of tropical land-use change on soil organic carbon stocks – a meta-analysis. *Global Change Biology* 17, 1658–1670.

Ecosystem Market Place (2013) http://www.ecosystemmarketplace.com/pages/dynamic/article.page.php?page_id=9569&section=news_articles&eod=1 (accessed 9 March 2013).

Galudra, G., van Noordwijk, M., Suyanto, Sardi, I., Pradhan, U. and Catacutan, D. (2011) Hot spots of confusion: contested policies and competing carbon claims in the peatlands of Central Kalimantan (Indonesia). *International Forestry Review* 13, 431–441.

Guo, L.B. and Gifford, R.M. (2002) Soil carbon stocks and land use change: a meta analysis. *Global Change Biology* 8, 345–360.

Hairiah, K., Dewi, S., Agus, F., Velarde, S.J., Ekadinata, A., Rahayu, S. and van Noordwijk, M. (2011) *Measuring Carbon Stocks Across Land Use Systems: A Manual*. World Agroforestry Centre (ICRAF), Bogor, Indonesia.

Houghton, R.A. (2003) Revised estimates of the annual net flux of carbon to the atmosphere from changes in land use and land management 1850–2000. *Tellus B Chemistry, Physics, Meteorology* 55(2), 378–390.

IPCC (Intergovernmental Panel on Climate Change) (2006) *IPCC Guidelines for National Greenhouse Gas Inventories*. The National Greenhouse Gas Inventories Programme, IGES, Japan.

Jindal, R. and Namirembe, S. (2012) *International market for forest carbon offsets: how these offsets are created and traded*. ASB Partnership for the Tropical Forest Margins Lecture Note 14. World Agroforestry Centre (ICRAF) Nairobi.

Jindal, R., Swallow, B. and Kerr, J. (2008) Forestry- based carbon sequestration projects in Africa: potential benefits and challenges. *Natural Resources Forum* 32, 116–130.

Jobbágy, E.G. and Jackson, R.B. (2000) The vertical distribution of soil organic carbon and its relation to climate and vegetation. *Ecological Applications* 10, 423–436.

Khasanah, N., Ekadinata, A., Rahayu, S., van Noordwijk, M., Ningsih, N., Setiawan, A., Dwiyanti, E., Dewi, S. and Octaviani, R. (2011) *Carbon Footprint of Indonesian Palm Oil Production*. Oil Palm Flyer No 1. World Agroforestry Centre (ICRAF) Southeast Asia Program, Bogor, Indonesia.

Larson, C., Chatellier, J., Lifset, R. and Graedel, T. (2012) Role of forest products in the global carbon cycle: from the forest to final disposal. In: Ashton, M.S., Tyrrell, M.L., Spalding, D. and Gentry, B. (eds) *Managing Forest Carbon in a Changing Climate*. Springer, Berlin, pp. 257–282.

Lauk, C., Haberl, H., Erb, K.H., Gingrich, S. and Krausmann, F. (2012) Global socioeconomic carbon stocks in long-lived products 1900–2008. *Environmental Research Letters* 7(3), 034023. doi:10.1088/1748-9326/7/3/034023.

Le Quéré, C., Raupach, M.R., Canadell, J.G. and Marland, G. (2009) Trends in the sources and sinks of carbon dioxide. *Nature Geoscience* 2(12), 831–836.

Lipper, L., Neves, B., Wilkes, A., Tennigkeit, T., Gerber, P., Henderson, B., Branca, G. and Man, W. (2011) *A Guide Book to Harvesting Soil Carbon Sequestration Benefits*. FAO, Rome.

Lusiana, B., van Noordwijk, M., Johana, F., Galudra, G., Suyanto, S. and Cadisch, G. (2013) Implication of uncertainty and scale in carbon emission estimates on locally appropriate designs to reduce emissions from deforestation and degradation (REDD+). *Mitigation and Adaptation Strategies for Global Change* 19, 757–772.

Minang, P.A. and van Noordwijk, M. (2013) Design challenges for achieving reduced emissions from deforestation and forest degradation through conservation: leveraging multiple paradigms at the tropical forest margins. *Land Use Policy* 31, 61–70.

Minang, P.A., van Noordwijk, M., Meyfroidt, P., Agus, F. and Dewi, S. (2010) Emissions embodied in trade (EET) and land use in tropical forest margins. *ASB Policy Brief* 17. ASB Partnership for the Tropical Forest Margins, Nairobi.

Minasny, B., Sulaeman, Y. and McBratney, A.B. (2011) Is soil carbon disappearing? The dynamics of soil organic carbon in Java. *Global Change Biology* 17, 1917–1924.

Murdiyarso, D., van Noordwijk, M., Puntodewo, A., Widayati, A. and Lusiana, B. (2008) District scale prioritization for A/R CDM project activities in Indonesia in line with sustainable development objectives. *Agriculture Ecosystems and Environment* 126, 59–66, doi:10.1016/j.agee.2008.01.008.

Nepstad, D.C., De Carvalho, C.R., Davidson, E.A., Jipp, P.H., Lefebvre, P.A., Negreiros, G.H., Da Silva, E.A., Stone, T.A., Trumbore, S.E. and Vieira, S. (2002) The role of deep roots in the hydrological and carbon cycles of Amazonian forests and pastures. *Nature* 372, 666–669.

Paustian, K., Andrén, O., Janzen, H.H., Lal, R., Smith, P., Tian, G., Tiessen, H., van Noordwijk, M. and Woomer, P.L. (1997) Agricultural soils as a sink to mitigate CO_2 emissions. *Soil Use and Management* 13, 230–244.

Peters, G.P., Marland, G., Le Quéré, C., Boden, T., Canadell, J.G. and Raupach, M.R. (2012) Rapid growth in CO_2 emissions after the 2008–2009 global financial crisis. *Nature Climate Change* 2, 2–4.

Powers, J.S., Corre, M.D., Twine, T.E. and Veldkamp, E. (2011) Geographic bias of field observations of soil carbon stocks with tropical land-use changes precludes spatial extrapolation. *Proceedings of the National Academy of Sciences* 108, 6318–6322.

Rockström, J., Steffen, W., Noone, K., Persson, Å., Chapin, F.S. III, Lambin, E., Lenton, T.M., Scheffer, M., Folke, C., Schellnhuber, H. et al. (2009) Planetary boundaries: exploring the safe operating space for humanity. *Ecology and Society* 14, article 32.

Rothschild, E. (1994) Adam Smith and the invisible hand. *The American Economic Review* 84, 319–322.

Ruan, L. and Robertson, G.P. (2013) Initial nitrous oxide, carbon dioxide, and methane costs of converting conservation reserve program grassland to row crops under no-till vs. conventional tillage. *Global Change Biology* 19, 2478–2489.

Smith, L.J. and Torn, M.S. (2013) Ecological limits to terrestrial biological carbon dioxide removal. *Climatic Change* 118, 89–103, doi:10.1007/s10584-012-0682-3#.

Smith, P., Davies, C.A., Ogle, S., Zanchi, G., Bellarby, J., Bird, N., Boddey, R.M., McNamara, N.P., Powlson, D., Cowie, A. et al. (2012) Towards an integrated global framework to assess the impacts of land use and management change on soil carbon: current capability and future vision. *Global Change Biology* 18, 2089–2101.

Stern, N. (2006) *The Stern Review: The Economics of Climate Change*. Cambridge University Press, Cambridge, UK.

Swallow, B., van Noordwijk, M., Dewi, S., Murdiyarso, D., White, D., Gockowski, J., Hyman, G., Budidarsono, S., Robiglio, V., Meadu, V. et al. (2007) Opportunities for avoided deforestation with sustainable benefits: an interim report by the ASB Partnership for the Tropical Forest Margins. Nairobi, Kenya. (http://www.asb.cgiar.org, accessed 19 August 2014.)

UN (1992) United Nations Framework Convention on Climate Change. (http://unfccc.int/resource/docs/convkp/conveng.pdf, accessed 12 December 2013).

VandenBygaart, A.J. and Angers, D.A. (2006) Towards accurate measurements of soil organic carbon stock change in agroecosystems. *Canadian Journal of Soil Science* 86, 465–471.

van Noordwijk, M. and Leimona, B. (2010) Principles for fairness and efficiency in enhancing environmental services in Asia: payments, compensation, or co-investment? *Ecology and Society* 15, article 17. http://www.ecologyandsociety.org/vol15/iss4/art17/ (accessed 12 December 2013).

van Noordwijk, M., Woomer P., Cerri C., Bernoux, M. and Nugroho, K. (1997) Soil carbon in the humid tropical forest zone. *Geoderma* 79, 187–225.

van Noordwijk, M., Van Roode, M., McCallie, E.L. and Lusiana, B. (1998a) Erosion and sedimentation as multiscale, fractal processes: implications for models, experiments and the real world. In: Penning de Vries, F., Agus, F. and Kerr, J. (eds) *Soil Erosion at Multiple Scales, Principles and Methods for Assessing Causes and Impacts*. CAB International, Wallingford, UK, pp. 223–253.

van Noordwijk, M., Martikainen, P., Bottner, P., Cuevas, E., Rouland, C. and Dhillion, S.S. (1998b) Global change and root function. *Global Change Biology* 4, 759–772.

van Noordwijk, M., Suyamto, D.A., Lusiana, B., Ekadinata, A. and Hairiah, K. (2008) Facilitating agroforestation of landscapes for sustainable benefits: tradeoffs between carbon stocks and local development benefits in Indonesia according to the FALLOW model. *Agriculture Ecosystems and Environment* 126, 98–112.

van Noordwijk, M., Lusiana, B., Villamor, G., Purnomo, H. and Dewi, S. (2011) Feedback loops added to four conceptual models linking land change with driving forces and actors. *Ecology and Society* 16, r1. (http://www.ecologyandsociety.org/vol16/iss1/resp1/, accessed 12 December 2013).

van Noordwijk, M., Dewi, S., Lusiana, B., Harja, D., Agus, F., Rahayu, S., Hairiah, K., Maswar, M., Robiglio, V., Hyman, G., White, D. et al. (2012a) *Recommendations on the Design of National Monitoring Systems Relating the Costs of Monitoring to the Expected Benefits of Higher Quality of Data*. World Agroforestry Centre (ICRAF), Bogor, Indonesia, 48 pp.

van Noordwijk, M., Leimona, B., Jindal, R., Villamor, G.B., Vardhan, M., Namirembe, S., Catacutan, D., Kerr, J., Minang, P.A. and Tomich, T.P. (2012b) Payments for environmental services: evolution towards efficient and fair incentives for multifunctional landscapes. *Annual Review of Environment and Resources* 37, 389–420.

van Noordwijk, M., Agus, F., Dewi, S. and Purnomo, H. (2013) Reducing emissions from land use in Indonesia: motivation, policy instruments and expected funding streams. *Mitigation and Adaptation Strategies for Global Change* 19, 677–692.

van Noordwijk, M., Matthews, R.B., Agus, F., Farmer, J., Verchot, L., Hergoualc'h, K., Persch, S., Tata, H.L., Lusiana, B., Widayati, A. and Dewi, S. (2014) Mud, muddle and models in the knowledge value-chain to action on tropical peatland issues. *Mitigation and Adaptation Strategies for Global Change* 19, 887–905.

van Noordwijk, M., Minang, P.A. and Hairiah, K. (2015) Swidden transitions in an era of climate change debate. In: Cairns, M.F. (ed.) *Shifting Cultivation and Environmental Change: Indigenous People, Agriculture and Forest Conservation*. Routledge, Abingdon, UK (in press).

Van Oost, K., Verstraeten, G., Doetterl, S., Notebaert, B., Wiaux, F., Broothaerts, N. and Six, J. (2012) Legacy of human-induced C erosion and burial on soil–atmosphere C exchange. *Proceedings of the National Academy of Sciences USA* 109, 19492–19497.

van Wesemael, B., Paustian, K., Andrén, O., Cerri, C.E.P, Dodd, M., Etchevers, J., Goidts, E., Grace, P., Kätterer, T., McConkey, B.G. et al. (2011) How can soil monitoring networks be used to improve predictions of organic carbon pool dynamics and CO_2 fluxes in agricultural soils? *Plant and Soil* 338, 247–259.

Venter, O. and Koh, L.P. (2012) Reducing emissions from deforestation and forest degradation (REDD+): game changer or just another quick fix? *Annals of the New York Academy of Sciences* 1249, 137–150.

Verbist, B., Poesen, J., van Noordwijk, M., Widianto Suprayogo, D., Agus, F. and Deckers, J. (2010) Factors affecting soil loss at plot scale and sediment yield at catchment scale in a tropical volcanic agroforestry landscape. *Catena* 80, 34–46.

Wang, S., Wilkes, A., Zhang, Z., Chang, X., Lang, R., Wang, Y. and Niu, H. (2011) Management and land use change effects on soil carbon in northern China's grasslands: a synthesis. *Agriculture, Ecosystems and Environment* 142, 329–340.

Winiwarter, V. (2000) Soils in ancient Roman agriculture: analytical approaches to invisible properties. In: Nowotny, H. and Weiss, M. (eds) *Shifting Boundaries of the Real: Making the Invisible Visible*. vdf Hochschulverlag, Zurich, pp. 137–156.

Winjum, J.K., Brown, S. and Schlamadinger, B. (1998) Forest harvests and wood products: sources and sinks of atmospheric carbon dioxide. *Forest Science* 44, 272–284.

World Bank (2012) Carbon Sequestration in Agricultural Soils, Report No 67395. World Bank, Washington, DC.

Wunder, S. (2005) Payments for environmental services: some nuts and bolts. *CIFOR Occasional Paper 42*. Centre for International Forestry Research, Bogor, Indonesia.

Ziegler, A., Phelps, J., Yuen, J., Webb, E.L., Lawrence, D., Fox, J., Bruun, T., Leisz, S., Mertz, O., Dressler, W. *et al*. (2012) Carbon outcomes of major land-cover transitions in SE Asia: great uncertainties and REDD+ policy implications. *Global Change Biology* 18, 3087–3099.

Index

1D-ICZ model 114
4 × 40 challenge 2, 7

accounting rules 372–373
actors 54
    policy 72–77
Africa 307–309, 311
    biomass conversion 310–311
    management 311
    monitoring 310
    soil carbon status 309–310
        distribution 134–135
        functions 133
        management 135
    soil organic matter (SOM) 132–133, 136–138
agricultural
    land-use change (LUC) 28, 72, 128, 243–244
        annual crops and pasture 247–249
        southern grasslands 244, 245, 248–249
        sugarcane production 249–251
        tillage 248, 252–254
    practices 37–39, 229–230, 265–266, 287–288, 293, 307–309, 310, 354
        cover crops 281–282, 290–291
        erosion 227–229
        fallow frequency reduction 291
        fertilizers 100–101, 292–293, 316, 317, 328, 331, 334–336, 340, 355
        greenhouse gases (GHG) 120–121
        manure 105
        organic farming 124, 171–172
        residue management 291–292
        rice production 338–340
        soil carbon sequestration (SCS) 121–123, 174–175
        soil formation 92–93
        tillage 248, 252–254, 268, 280–281, 288–290, 316, 354, 355–356
        water buffalo grazing 299–301
    productivity 2, 5, 12–13, 20–21, 27, 61, 132–133, 138
        distribution 134–135
        functions 133
        management 135
        pest and disease control 270
        soil organic carbon (SOC) 12–13, 20–21, 133, 135–136
        soil organic matter (SOM) 132–133, 136–138
    transition curves 26–28, 42–43, 49–50
        carbon stocks 30–31
        critical thresholds 41–42
        ecosystem services 28
        global consequences 37
        good agricultural practices 37–41
        soil diversity 31
        stages 28–31, 33–36, 38–39
aluminium (Al) 104
apatite 87

biodegradation 155–159
biodiversity 61, 141–143, 149, 299–301, 302–303
    characterization of 143–145
    Convention on Biological Diversity (CBD) 121, 343, 345
    decline 2
    ecosystem services 147–149
    microbial 102–104
    organic matter 145–148
    soil organic carbon (SOC) 14–16, 21, 271

biofuels 41
biomass crops 14, 299, 301

C:N ratio 99
CANDY 204
carbon *see* soil organic carbon (SOC)
carbon dioxide (CO_2) 321–322
    emissions 7, 16, 120, 128, 301
    InfoCrop model 336–338
    soil carbon sequestration (SCS) 4–5, 39–40, 120–124, 128, 216–218, 235–236, 238–240, 269–270, 287–288, 334
        agricultural practices 121–123
        barriers to implementation 124–127
        carbon sink saturation 238
        China 356–358
        effectiveness 239–240
        land-use change (LUC) 128, 252–256
        leakage 239
        non-performance 239
        spatial dimension 127–128
        verification 239
        Verified Carbon Standard (VCS) 127
    trading 67, 69, 183–185
Century 136–137, 191, 202, 204, 205, 206, 207, 208, 334–335
challenges 1–2
    4 × 40 2, 7
    soil organic carbon (SOC) 18–20
chemical degradation 228
China
    carbon sequestration 356
        funding limitation 356–357
        knowledge gap 357
        policy challenge 357–358
    policy 357–358
        conservation tillage 355–356
        ecological construction 356
        fertile soil project 355
        soil testing and fertilizer recommendation (STFR) 355
    soil carbon management 353
        agricultural practice 354
        forestry 354
        land recovery 355
        land use 354
        pasture management 354
    soil organic matter (SOM) 314, 322–324
        benefits 316–322
        crop productivity 317
        ecosystem services 320–322
        historic depletion 315–316
        importance 314–315
        increase in 316
        nutrient retention 317–319

chronosequences 86–88
clean development mechanism (CDM) 124, 126
climate change 16, 21, 61
    Inter-governmental Panel for Climate Change (IPCC) 31, 119–120, 123, 189, 206–207, 208, 236, 237, 311, 331, 344–345, 372
    mitigation 2, 61, 119–121, 183
        biochar 119, 123
        chipped ramial wood (CRW) 119, 123–124
        clean development mechanism (CDM) 124, 126
        Convention on Biological Diversity (CBD) 121
        Inter-governmental Panel for Climate Change (IPCC) 31, 119–120, 123, 189, 206–207, 208, 236, 237, 311, 331, 344–345, 372
        Kyoto Protocol (KP) 124, 301, 368
        organic farming 124
        sequestration *see* soil carbon sequestration (SCS)
    soil carbon impacts 235–236, 240
        future responses 236–238
        global carbon cycle 236
        net primary productivity (NPP) 236, 237
    soil carbon sequestration (SCS) 4–5, 39–40, 120–124, 128, 216–218, 235–236, 238–240, 269–270, 287–288, 334
        agricultural practices 121–123
        barriers to implementation 124–127
        carbon sink saturation 238
        China 356–358
        effectiveness 239–240
        land-use change (LUC) 128, 252–256
        leakage 239
        non-performance 239
        spatial dimension 127–128
        verification 239
        Verified Carbon Standard (VCS) 127
COMET-VR 208
Common Agricultural Policy (CAP) 73–74, 303, 343, 347
common reed production 301–302
Convention on Biological Diversity (UNCBD) 121, 343, 345
Convention to Combat Desertification (UNCCD) 343, 345
copper (Cu) 105
cost–benefit analysis (CBA) 214, 218–220
cost-effectiveness analysis 214, 218–220
cover crops 281–282, 290–291
critical zone (CZ) 82–83
    observatories 93–94, 113–114, 116

DAISY 204
data sharing 71–72
degradation 2–3, 6, 7, 34–35, 134, 224, 227–229, 230–332, 268, 364
    agricultural practices 93
    climate 89–90
    control 269
    topography 88–89, 166
    wind erosion 161–163, 166, 228
        measurement of 163
        process physics 163–164
        soil organic carbon (SOC) stocks 163
        spatial variability 164–166
        temporal variability 164
dissolved organic carbon (DOC) 228, 229–230
DNDC 202, 204, 205, 354

ECCOSSE 204, 205, 206
economics
    carbon trading 67, 69, 183–185
    drivers 19
    markets 360–362, 375
        certified emission reduction units 367–368
        integrity of accounting rules 372–373
        international markets 369–370
        Kyoto Protocol 368
        performance-based incentive options 373–375
        price volatility 369
        reducing emissions from deforestation and forest degradation (REDD) 368–369, 372, 373–374
        transaction costs 370–372
        value chain 362–367
        voluntary markets 369
    total economic value (TEV) 181–182, 214
    valuation 179–181, 185–186, 220–221
        climate regulation 216–218
        cost–benefit analysis (CBA) 214, 218–220
        cost-effectiveness analysis 214, 218–220
        discounting 183
        ecosystem services 214–215, 218
        insurance value 181–183
        marginal abatement cost curve (MACC) 217–218
        market creation 183–185
        multi-criteria analysis (MCA) 214, 218–220
        production function approach 215–216, 218
        social cost of carbon (SCC) 217
        total economic value (TEV) 181–182, 214
        total output 181
        value chain 362–367

ecosystem services 2–6, 7, 26, 28, 49, 68, 109, 142, 265, 297–298, 302–303, 320–322, 356
    biodiversity 147–149, 271, 302–303
    incentives 271
    natural capital 266–267
    soil-based processes 267–268
    soil carbon management 271–272, 277–278, 277–278
        biodiversity 271, 299–301
        climate regulation 269–270
        cover crops 281–282
        disease control 270
        erosion control 269
        pest control 270
        pollutant attenuation 270, 320–321
        soil structure 268–269
        sugarcane production 278–279
        tillage 280–281
    valuation of 214–215, 218
energy supply 2, 5, 14, 21, 27
    biofuels 41
    biomass crops 14
environmental conditions 6
erosion 2–3, 6, 7, 34–35, 134, 224, 227–229, 231–323, 268, 364
    agricultural practices 93
    climate 89–90
    control 269
    topography 88–89, 166
    wind erosion 161–163, 166, 228
        measurement of 163
        process physics 163–164
        soil organic carbon (SOC) stocks 163
        spatial variability 164–166
        temporal variability 164
eucalyptus forests 251
Europe 348–349
    paludicultures 297–298, 303–304
        benefits 302–303
        land-use options 299–302
    policy 343–344, 346
        Common Agricultural Policy (CAP) 73–74, 303, 343, 347
        Roadmap to Resource-Efficient Europe (RRE) 343, 347–348
        soil thematic strategy 343, 346–347

fallow frequency reduction 291
fertilizers 100–101, 292–293, 316, 317, 355
    long-term fertilizer experiments (LTFEs) 328, 331, 334–336, 340
food supply 2, 5, 27, 61
    agricultural practices 37–39, 67–68
    food-deficient regions 12
    soil organic carbon (SOC) 12–13, 20–21, 133
    sustainable production 63–65, 317

formation of soil 82–85, 95, 110
   critical zone (CZ) 82–83
      observatories 93–94
   function development 91–92
   human influences 92–93
   parent material 83–84
   regolith 83, 88
   soil horizons 83
   soil organic carbon (SOC) 87–88
   soil strain (expansion) 86–87
   variables
      climate 89–91
      organisms 85–86
      time 86–88
      topography 88–89
Framework Convention on Climate Change (UNFCCC) 343, 344–345, 361, 362–363, 370
fuel supply 2, 5, 14, 21, 27
   biofuels 41
   biomass crops 14

global environment facilities (GEF) 70–71, 76–77, 197, 207
   GEFSOC modelling system 309–310, 331, 340
Global Soil Partnership (GSP) 65, 71, 73, 345–346
greenhouse gases (GHG) 2, 63, 74, 120–121, 189, 203, 208, 298, 311, 322, 344, 349, 367–368
   carbon dioxide (CO_2) 321–322
      emissions 7, 16, 120, 128, 301
      InfoCrop model 336–338
      sequestration *see* soil carbon sequestration (SCS)
      trading 67, 69, 183–185
   methane (CH_4) production 7, 120, 328, 338–340
   nitrous oxide (N_2O) production 7, 120, 128, 143, 147, 290, 292–293, 319

historical management 169–170, 329
   ecological agriculture 172–173
   humus 170–171
   mineralist period 171
   organic farming movement 171–172
hydrologic response units (HRU) 113
hydrology 108
   chemical weathering 109–112
   hydrologic response units (HRU) 113
   ion exchange 110–111
   modelling 112–116
      obstacles 115–116
   transformation 109–112
   water filtration 109–112
HYDRUS 114

India 328–329, 340
   black soil region (BSR) 328, 329, 331, 335, 340
   Indi-Gangetic plains (IGP) 328, 331, 335, 340
   InfoCrop model 336–338
   long-term fertilizer experiments (LTFEs) 328, 331, 334, 340
   methane (CH_4) production 338–340
   soil organic carbon (SOC) 331
      Century 334–335
      RothC 335–336
   soil organic matter
      dynamics 329–331
InfoCrop model 336–338
infrared reflective spectroscopy 196–197
innovation 54–57
Inter-governmental Panel for Climate Change (IPCC) 31, 119–120, 123, 189, 206–207, 208, 236, 237, 311, 331, 344–345, 372
ion exchange 110–111

Kenya 307–309, 311
   biomass conversion 310–311
   management 311
   monitoring 310
   soil carbon status 309–310
Kyoto Protocol (KP) 124, 301, 368

land degradation neutral world 64
land management 50–51
   global consequences 37
   land-use change (LUC) 28, 72, 128, 243–244, 256–257, 354, 368
      annual crops and pasture 247–249
      Atlantic forests 244, 246–247, 249–251
      Cerrado 244, 245, 247–248, 251–256
      forest plantations 251–252
      soil carbon sequestration (SCS) 252–256, 364–366
      southern grasslands 244, 245, 248–249
      sugarcane production 249–251
      tillage 248, 252–254
   soil fertility
      collapse 34–35
      decline 33–34
      recovery 35–36
   soil organic carbon (SOC) 18
   sustainable land management (SLM) 63–64
land-use change (LUC) 28, 72, 128, 243–244, 256–257, 354, 368
   annual crops and pasture 247–249
   Atlantic forests 244, 246–247, 249–251
   Cerrado 244, 245, 247–248, 251–256
   forest plantations 251–252
   soil carbon sequestration (SCS) 252–256, 364–366

southern grasslands 244, 245, 248–249
sugarcane production 249–251
tillage 248, 252–254

management 5–6, 47–48, 57, 265–266,
    287–288, 293
  agricultural practices 37–39, 229–230, 265–266,
      287–288, 293, 307–309, 310, 354
    cover crops 281–282, 290–291
    erosion 227–229
    fallow frequency reduction 291
    fertilizers 100–101, 292–293, 316, 317,
        328, 331, 334–336, 340, 355
    greenhouse gases (GHG) 120–121
    manure 105
    organic farming 124, 171–172
    residue management 291–292
    rice production 338–340
    soil carbon sequestration (SCS)
        121–123, 174–175
    soil formation 92–93
    tillage 248, 252–254, 268, 280–281,
        288–290, 316, 354, 355–356
    water buffalo grazing 299–301
  best practice 49–51, 69–70
  bottlenecks 51–55
  ecosystem services 266–268, 271–272,
      277–278, 297–298
    biodiversity 271
    climate regulation 269–270
    erosion control 269
    pollutant attenuation 270
    soil structure 268–269
  incentives 271
  innovation 54–57
  paludicultures 297–298, 303–304
    benefits 302–303
    land-use options 299–302
  scale 51–52, 55–57
  socio-economic factors 52–54
manure 105
marginal abatement cost curve (MACC) 217–218
markets 360–362, 375
  certified emission reduction units 367–368
  integrity of accounting rules 372–373
  international markets 369–370
  Kyoto Protocol 368
  performance-based incentive options
      373–375
  price volatility 369
  reducing emissions from deforestation and
      forest degradation (REDD)
      368–369, 372, 373–374
  transaction costs 370–372
  value chain 362–363
    erosion 364

  land use 363–364, 365–366
  offsets 366–367
  peatland soil 364
  voluntary markets 369
methane (CH_4) production 7, 120, 328, 338–340
microaggregates 101–102, 111–112
microbial biomass 102–104, 111, 143
mitigation 2, 61, 119–121, 183
  biochar 119, 123
  chipped ramial wood (CRW) 119, 123–124
  clean development mechanism (CDM)
      124, 126
  Convention on Biological Diversity
      (CBD) 121
  Inter-governmental Panel for Climate
      Change (IPCC) 31, 119–120, 123,
      189, 206–207, 208, 236, 237, 311,
      331, 344–345, 372
  Kyoto Protocol (KP) 124, 301, 368
  organic farming 124
  soil carbon sequestration (SCS) 4–5, 39–40,
      120–124, 128, 216–218, 235–236,
      238, 269–270, 287–288, 334
    agricultural practices 121–123
    barriers to implementation 124–127
    carbon sink saturation 238
    China 356–358
    effectiveness 239–240
    land-use change (LUC) 128, 252–256
    leakage 239
    non-performance 239
    spatial dimension 127–128
    verification 239
    Verified Carbon Standard (VCS) 127
Mitigation of Climate Change in Agriculture
    (MICCA) 69–70
modelling 197
  climate change 237
  hydrology 112–116
    1D-ICZ model 114
    HYDRUS 114
    obstacles 115–116
    Penn State Integrated Hydrologic
        Model (PIHM) 113–114, 115
    PROSUM 115
  soil organic matter (SOM) 136–137,
      202–203, 209–210
    application of 204–206
    CANDY 204
    Century 136–137, 191, 202, 204, 205,
        206, 207, 208, 334–335
    COMET-VR 208
    DAISY 204
    DNDC 202, 204, 205, 354
    ECOSSE 204, 205, 206
    GEFSOC modelling system 309–310,
        331, 340

modelling (*continued*)
    history of 203–204
    InfoCrop model 336–338
    NCSOIL 204
    PEATSTASH 204
    practical applications 208–209
    RothC 69, 70, 203–204, 205, 206, 207, 208, 335–336
    scale 206–207
    SUNDIAL 205
    uncertainties 207–208
monitoring 188–189, 198, 310
    analytical methods 195–196
        novel 196–197
    minimum detectable difference (MDD) 188, 191
    modelling 197
    soil monitoring networks (SMNs) 188, 189–190
        error 194
        sampling design 191–194
        sampling depth 194–195
        scale 190–191
multi-criteria analysis (MCA) 214, 218–220
multilateral environment agreements (MEA) 64
mycorrhizal fungi 85–86

NCSOIL 204
nitrogen (N) 99–101
    availability 87
    C:N ratio 99
    carbon interactions 105–106
    excess 105
    hydrology 108
        chemical weathering 109–112
        hydrologic response units (HRU) 113
        ion exchange 110–111
        modelling 112–116
        transformation 109–112
        water filtration 109–112
    retention 317–319
nitrous oxide (N_2O) production 7, 120, 128, 143, 147, 290, 292–293, 319
North America 287–288, 293
    cover crops 290–291
    fallow frequency reduction 291
    nitrous oxide (N_2O) emissions 292–293
    residue management 291–292
    tillage 288–289
nutrient
    aluminium (Al) 104
    C:N ratio 99, 136
    carbon *see* soil organic carbon (SOC)
    cation exchange capacity (CEC) 101
    copper (Cu) 105
    cycling 4, 12, 98–99
        complexation 104–105
        excess organic inputs 105
        nitrogen/carbon interactions 105–106
        organic carbon, role of 101
        physical properties of soil 101–102
        soil biota 102–104
        soil organic matter (SOM) 99–101, 102–104
    fertilizers 100–101, 292–293, 316, 317, 355
        long-term fertilizer experiments (LTFEs) 328, 331, 334–336, 340
    hydrology 108
        chemical weathering 109–112
        hydrologic response units (HRU) 113
        ion exchange 110–111
        modelling 112–116
        transformation 109–112
        water filtration 109–112
    manure 105
    nitrogen (N) 99–101
        availability 87
        C:N ratio 99
        carbon interactions 105–106
        excess 105
        hydrology 108–116
        retention 317–319
    phosphorus (P) 99
        apatite 87
        availability 87, 102
        retention 317–319
    potassium (K) 87
    sodium (Na) 87
    sulfur (S) 98, 99
    supply 67–68

organic farming 124, 171–172
organic matter 3–4, 40, 48–49, 224–225, 265–266, 314, 322–324
    agricultural productivity 132–133, 136–138
    benefits 316–322
    biodiversity 145–148
    buffering function 28
    crop productivity 317
    depletion 315–316
    ecosystem services 320–322
    historical management 169–170, 329
        ecological agriculture 172–173
        humus 170–171
        mineralist period 171
        organic farming movement 171–172
    importance 314–315
    increases in 316, 331
    microbial biomass 102–104
    nutrient reservoir 99–101, 317–319

organisms
    microbial biomass 102–104, 111, 143
    mycorrhizal fungi 85–86
    soil formation 85–86

paludicultures 297–298, 303–304
    benefits 302–303
    land-use options 299
        biofuels 301
        common reed production 301–302
        sphagnum biomass 302
        water buffalo grazing 299–301
    peatland utilization 298
    principles of 298–299
particulate organic carbon (POC) 228, 229–230
peatland management 348–349, 364
    paludicultures 297–298, 303–304
        benefits 302–303
        land-use options 299–302
PEATSTASH 204
Penn State Integrated Hydrologic Model (PIHM) 113–114, 115
performance-based incentive options 373–375
phosphorus (P) 99
    apatite 87
    availability 87, 102
    retention 317–319
physical degradation 228
policy 52, 60–61, 343–344, 346, 349
    actors 78
        institutions 72–74
        governance 74–77
    China 355–356, 357–358
    Common Agricultural Policy (CAP) 73–74, 303, 343, 347
    data sharing 71–72
    economics
        carbon trading 67, 69, 183–185
        drivers 19
        markets 360–375
        total economic value (TEV) 181–182, 214
        valuation 179–185, 185–186, 214–221, 362–367
    global environment facilities (GEF) 70–71, 76–77
    Global Soil Partnership (GSP) 65, 71, 73, 345–346
    imperative 61–63, 77
    Inter-governmental Panel for Climate Change (IPCC) 31, 119–120, 123, 189, 206–207, 208, 236, 237, 311, 331, 344–345, 372
    land degradation neutral world 64
    land-use change (LUC) 72, 343, 344–345
    Mitigation of Climate Change in Agriculture (MICCA) 69–70

    multilateral environment agreements (MEA) 64
    national action plans 64, 76
    profile 77
        international sustainable development 66–67
        regional patterns 66
        socio/cultural context 65–66
    rationale 67–69, 78
    recommendations 77–78
    reducing emissions from deforestation and forest degradation (REDD) 69, 77, 368–369
    Roadmap to Resource-Efficient Europe (RRE) 343, 347–348
    soil monitoring networks (SMNs) 70
    soil thematic strategy 343, 346–347
    support 69–72
    sustainable production 63–65
    United Nations
        Convention on Biological Diversity (UNCBD) 121, 343, 345
        Convention to Combat Desertification (UNCCD) 343, 345
        Framework Convention on Climate Change (UNFCCC) 343, 344–345, 361, 362–363, 370
    Verified Carbon Standard (VCS) 69
    World Overview of Conservation Approaches and Technologies (WOCAT) 69
pollutants 154, 159
    biodegradation 155–159, 270, 320–321
    filtration 109–112, 155
    purification 4–5, 13
    sorption 155–157
    xenobiotic compounds 157, 158–159
pollution 4–5, 320–321
population growth 1–2, 11, 26, 27, 231–232
potassium (K) 87
precipitation 90
production function approach 215–216, 218
productivity 2, 5, 12–13, 20–21, 27, 61, 132–133, 138
    distribution 134–135
    functions 133
    management 135
    pest and disease control 270
    soil organic carbon (SOC) 12–13, 20–21, 133, 135–136
    soil organic matter (SOM) 132–133, 136–138
PROSUM 115

reducing emissions from deforestation and forest degradation (REDD) 69, 77, 368–369, 372, 373–374
rice production 338–340

Roadmap to Resource-Efficient Europe (RRE) 343, 347–348
RothC 69, 70, 203–204, 205, 206, 207, 208, 335–336

sequestration (SCS) 4–5, 39–40, 120–124, 128, 216–218, 235–236, 238–240, 269–270, 287–288, 334
    agricultural practices 121–123
    barriers to implementation 124–127
    carbon sink saturation 238
    China 356–358
    effectiveness 239–240
    land-use change (LUC) 128, 252–256
    leakage 239
    non-performance 239
    spatial dimension 127–128
    verification 239
    Verified Carbon Standard (VCS) 127
social cost of carbon (SCC) 217
socio/cultural context 65–66, 169, 173–175
socio-economic factors 52–54, 56
sodium (Na) 87
soil
    carbon storage 26, 226–227
    degradation 2–3, 11
    diversity 31
    fertility
        collapse 34–35
        decline 33–34
        recovery 35–36
    formation 82–85, 95, 110
        critical zone (CZ) 82–83, 93–94
        function development 91–92
        human influences 92–93
        parent material 83–84
        regolith 83, 88
        soil horizons 83
        soil organic carbon (SOC) 87–88
        soil strain (expansion) 86–87
        variables 85–91
    function 3–4
        development 91–92
    hydrology 108
        chemical weathering 109–112
        hydrologic response units (HRU) 113
        ion exchange 110–111
        modelling 112–116
        transformation 109–112
        water filtration 109–112
    nutrient cycling 4, 12, 98–99
        complexation 104–105
        excess organic inputs 105
        nitrogen/carbon interactions 105–106
        organic carbon, role of 101
        physical properties of soil 101–102
        soil biota 102–104
        soil organic matter (SOM) 99–101, 102–104
    structure 3–4, 30, 268–269
    see also management, soil organic carbon (SOC)
soil carbon sequestration (SCS) 4–5, 39–40, 120–124, 128, 216–218, 235–236, 238–240, 269–270, 287–288, 334
    agricultural practices 121–123
    barriers to implementation 124–127
    carbon sink saturation 238
    China 356–358
    effectiveness 239–240
    land-use change (LUC) 128, 252–256
    leakage 239
    non-performance 239
    spatial dimension 127–128
    verification 239
    Verified Carbon Standard (VCS) 127
soil monitoring networks (SMNs) 70
soil organic carbon (SOC) 10–12, 21–22, 42–43, 48–49, 82, 224–225, 309, 360–362
    accumulation 6
    agricultural practices 37–39, 229–230, 265–266, 287–288, 293, 307–309, 310, 354
        cover crops 281–282, 290–291
        erosion 227–229
        fallow frequency reduction 291
        fertilizers 100–101, 292–293, 316, 317, 328, 331, 334–336, 340, 355
        greenhouse gases (GHG) 120–121
        manure 105
        organic farming 124, 171–172
        residue management 291–292
        rice production 338–340
        soil carbon sequestration (SCS) 121–123, 174–175
        soil formation 92–93
        tillage 248, 252–254, 268, 280–281, 288–290, 316, 354, 355–356
        water buffalo grazing 299–301
    agricultural productivity 2, 5, 12–13, 20–21, 27, 61, 132–133, 138
        distribution 134–135
        functions 133
        management 135
        pest and disease control 270
        soil organic carbon (SOC) 12–13, 20–21, 133, 135–136
        soil organic matter (SOM) 132–133, 136–138
    analysis of 125–126
    benefits 11–12, 28
    biodiversity 61, 141–143, 149, 299–301, 302–303

characterization of 143–145
Convention on Biological Diversity
  (CBD) 121, 343, 345
decline 2
ecosystem services 147–149
microbial 102–104
organic matter 145–148
soil organic carbon (SOC) 14–16, 21, 271
challenges 18–20, 230–232
climate 16, 21
critical thresholds 41–42
decline 33–34, 37, 229–231
distribution 134–135, 226–227
economics
  carbon trading 67, 69, 183–185
  drivers 19
  markets 360–375
  total economic value (TEV) 181–182, 214
  valuation 179–185, 185–186, 214–221, 362–367
energy supply 14, 21
erosion 2–3, 6, 7, 34–35, 134, 224, 227–229, 231–323, 268, 364
  agricultural practices 93
  climate 89–90
  control 269
  topography 88–89, 166
  wind erosion 161–166, 228
fluxes 229–230
historical management 169–170, 329
  ecological agriculture 172–173
  humus 170–171
  mineralist period 171
  organic farming movement 171–172
hydrology 108
  chemical weathering 109–112
  hydrologic response units (HRU) 113
  ion exchange 110–111
  modelling 112–116
  transformation 109–112
  water filtration 109–112
interactions 16–18
land-use change (LUC) 28, 72, 128, 243–244, 256–257, 354, 368
  annual crops and pasture 247–249
  Atlantic forests 244, 246–247, 249–251
  Cerrado 244, 245, 247–248, 251–256
  forest plantations 251–252
  soil carbon sequestration (SCS) 252–256, 364–366
  southern grasslands 244, 245, 248–249
  sugarcane production 249–251
  tillage 248, 252–254
management 5–6, 47–48, 57, 265–266, 287–288, 293
  best practice 49–51, 69–70
  bottlenecks 51–55

ecosystem services 266–272, 277–278, 297–298
incentives 271
innovation 54–57
paludicultures 297–304
scale 51–52, 55–57
socio-economic factors 52–54, 56
models see modelling
monitoring 188–189, 198, 310
  analytical methods 195–197
  minimum detectable difference (MDD) 188, 191
  modelling 197
  soil monitoring networks (SMNs) 188, 189–195
nutrient cycling 4, 12, 98–99
  complexation 104–105
  excess organic inputs 105
  nitrogen/carbon interactions 105–106
  organic carbon, role of 101
  physical properties of soil 101–102
  soil biota 102–104
  soil organic matter (SOM) 99–101, 102–104
policy see policy
priorities 20–21
  recovery 35–36
recovery 35–36
reduction 11
soil carbon sequestration (SCS) 4–5, 39–40, 120–124, 128, 216–218, 235–236, 238–240, 269–270, 287–288, 334
  agricultural practices 121–123
  barriers to implementation 124–127
  carbon sink saturation 238
  China 356–358
  effectiveness 239–240
  land-use change (LUC) 128, 252–256
  leakage 239
  non-performance 239
  spatial dimension 127–128
  verification 239
  Verified Carbon Standard (VCS) 127
soil formation 87–88
threats 6
transition curves 26–28, 42–43, 49–50
  carbon stocks 30–31
  critical thresholds 41–42
  ecosystem services 28
  global consequences 37
  good agricultural practices 37–41
  soil diversity 31
  stages 28–31, 33–36, 38–39
valuation 179–181, 185–186, 220–221
  climate regulation 216–218
  cost–benefit analysis (CBA) 214, 218–220

soil organic carbon (SOC) (*continued*)
    cost-effectiveness analysis 214, 218–220
    discounting 183
    ecosystem services 214–215, 218
    insurance value 181–183
    marginal abatement cost curve (MACC) 217–218
    market creation 183–185
    multi-criteria analysis (MCA) 214, 218–220
    production function approach 215–216, 218
    social cost of carbon (SCC) 217
    total economic value (TEV) 181–182, 214
    total output 181
    value chain 362–367
  water 13–14, 21
soil organic matter (SOM) 3–4, 40, 48–49, 224–225, 265–266, 314, 322–324
  agricultural productivity 132–133, 136–138
  benefits 316–322
  biodiversity 145–148
  buffering function 28
  crop productivity 317
  depletion 315–316
  ecosystem services 320–322
  historical management 169–170, 329
    ecological agriculture 172–173
    humus 170–171
    mineralist period 171
    organic farming movement 171–172
  importance 314–315
  increases in 316, 331
  microbial biomass 102–104
  nutrient reservoir 99–101, 317–319
soil thematic strategy 343, 346–347
sorption 155–157
South American biomes 243–244, 256–257, 277–278
  annual crops and pasture 247–249
  Atlantic forests 244, 246–247, 249–251
  Cerrado 244, 245, 247–248, 251–256
  cover crops 281–282
  forest plantations 251–252
  soil carbon sequestration (SCS) 252–256
  southern grasslands 244, 245, 248–249
  sugarcane production 249–251, 278–279
  tillage 248, 252–254, 280–281
sub-Saharan Africa 132, 138
  soil organic carbon (SOC)
    distribution 134–135
    functions 133
    management 135
  soil organic matter (SOM) 132–133, 136–138
sugarcane production 249–251, 278–279

sulfur (S) 98, 99
SUNDIAL 205

temperature 89–91
tillage 248, 252–254, 268, 280–281, 288–290, 316, 354, 355–356
total economic value (TEV) 181–182, 214
transition curves 26–28, 42–43, 49–50
  carbon stocks 30–31
  critical thresholds 41–42
  ecosystem services 28
  global consequences 37
  good agricultural practices 37–39
    carbon sequestration potential 39–40
    organic matter 40
    sustainable yield gap reduction 40–41
  soil diversity 31
  stages 28–31
    collapse 34–35, 38
    decline 33–34, 38
    recovery 35–36, 38–39

United Nations
  Convention on Biological Diversity (UNCBD) 121, 343, 345
  Convention to Combat Desertification (UNCCD) 343, 345
  Framework Convention on Climate Change (UNFCCC) 343, 344–345, 361, 362–363, 370

valuation 179–181, 185–186, 220–221
  climate regulation 216–218
  cost–benefit analysis (CBA) 214, 218–220
  cost-effectiveness analysis 214, 218–220
  discounting 183
  ecosystem services 214–215, 218
  insurance value 181–183
  marginal abatement cost curve (MACC) 217–218
  market creation 183–185
  multi-criteria analysis (MCA) 214, 218–220
  production function approach 215–216, 218
  social cost of carbon (SCC) 217
  total economic value (TEV) 181–182, 214
  total output 181
  value chain 362–367
Verified Carbon Standard (VCS) 69, 127

water 108
  catchment scale management 13
  chemical weathering 109–112
  erosion 228

hydrologic response units (HRU) 113
ion exchange 110–111
modelling 112–116
quality 154, 159
  biodegradation 155–159, 270, 320–321
  filtration 109–112, 155
  purification 4–5, 13
  sorption 155–157
soil formation 89–91
soil organic carbon (SOC) 13–14, 21
supply 2, 5, 154
transformation 109–112

wind erosion 161–163, 166, 228
  measurement of 163
  process physics 163–164
  soil organic carbon (SOC)
    stocks 163
  spatial variability 164–166
  temporal variability 164
World Overview of Conservation Approaches and
  Technologies (WOCAT) 69

xenobiotic compounds 157, 158–159